FUNDAMENTAL
INTERACTIONS

A Memorial Volume for
Wolfgang Kummer

FUNDAMENTAL
INTERACTIONS

A Memorial Volume for

Wolfgang Kummer

Editors

Daniel Grumiller
Anton Rebhan

Vienna University of Technology, Austria

Dimitri Vassilevich

Universidade Federal do ABC, Brazil &
St Petersburg State University, Russia

World Scientific

NEW JERSEY · LONDON · SINGAPORE · BEIJING · SHANGHAI · HONG KONG · TAIPEI · CHENNAI

Published by

World Scientific Publishing Co. Pte. Ltd.

5 Toh Tuck Link, Singapore 596224

USA office: 27 Warren Street, Suite 401-402, Hackensack, NJ 07601

UK office: 57 Shelton Street, Covent Garden, London WC2H 9HE

British Library Cataloguing-in-Publication Data
A catalogue record for this book is available from the British Library.

1006078255

ISBN-13 978-981-4273-07-7
ISBN-10 981-4273-07-4

Printed in Singapore by B & JO Enterprise

Contents

Foreword by the Editors   vii

I   Gauge Field Theory and Particle Physics   **1**

1.   Noncovariant Gauges at Zero and Nonzero Temperature   3
     P. V. Landshoff

2.   Non-Relativistic Bound States: The Long Way Back from
     the Bethe–Salpeter to the Schrödinger Equation   9
     A. Vairo

3.   Distended/Diminished Topologically Massive Electrodynamics 25
     S. Deser

4.   Dynamical Spin   33
     P. G. O. Freund

5.   Quantum Corrections to Solitons and BPS Saturation   41
     A. Rebhan, P. van Nieuwenhuizen and R. Wimmer

6.   Gauging Noncommutative Theories   75
     H. Grosse and M. Wohlgenannt

7.   Topological Phases and Contextuality Effects in Neutron
     Quantum Optics   91
     H. Rauch

8. First Class Constrained Systems and Twisting of Courant Algebroids by a Closed 4-Form 115

    M. Hansen and T. Strobl

9. Some Local and Global Aspects of the Gauge Fixing in Yang–Mills-Theories 145

    D. N. Blaschke, F. Delduc, F. Gieres, M. Schweda and S. P. Sorella

10. Frozen Ghosts in Thermal Gauge Field Theory 175

    P. V. Landshoff and A. Rebhan

II   Classical and Quantum Gravity **181**

11. Wolfgang Kummer and the Vienna School of Dilaton (Super-)Gravity 183

    L. Bergamin and R. Meyer

12. Order and Chaos in Two Dimensional Gravity 215

    R. B. Mann

13. 2-D Midisuperspace Models for Quantum Black Holes 231

    J. Gegenberg and G. Kunstatter

14. Global Solutions in Gravity. Euclidean Signature 249

    M. O. Katanaev

15. Thoughts on the Cosmological Principle 267

    D. J. Schwarz

16. When Time Emerges 277

    C. Faustmann, H. Neufeld and W. Thirring

17. Towards Noncommutative Gravity 293

   D. V. Vassilevich

18. Superembedding Approach to Superstring in $AdS_5 \times S^5$
   Superspace 303

   I. A. Bandos

19. Heterotic (0,2) Gepner Models and Related Geometries 335

   M. Kreuzer

20. Canonical Analysis of Cosmological Topologically
   Massive Gravity at the Chiral Point 363

   D. Grumiller, R. Jackiw and N. Johansson

III   Wolfgang Kummer and the Physics Community   375

21. Wolfgang Kummer at CERN 377

   H. Schopper

22. Wolfgang Kummer and the Little Lost Lane Boy 381

   K. Lane

23. Mitigation of Fossil Fuel Consumption and Global
   Warming by Thermal Solar Electric Power Production
   in the World's Deserts 387

   J. Steinberger

24. (My) Life with Wolfgang Kummer 399

   M. Schweda

25. Schubert in Stony Brook and Kinks in Vienna 403

   P. van Nieuwenhuizen

Author Index 407

Foreword by the Editors

The true masters of mathematical physics always knew how to isolate the physical content of complicated mathematical arguments, but unfortunately the majority of theoreticians in Europe are to this day sometimes over-fascinated by the mathematical aspects of the physical description of nature. Wolfgang Kummer[1]

Wolfgang Kummer had a rare gift: he recognized good people — while being one himself — and good ideas — while having them himself.

Wolfgang Kummer was born between the two World Wars on October 15, 1935, in Krems near Vienna, Austria. Until his graduation in 1958 at the Vienna University of Technology Wolfgang witnessed the changes from Austro-fascism (1934–1938) via Third Reich (1938–1945) and occupation by the Allied Forces (1945–1955) to a free and neutral Austria (since 1955). In 1960 he finished his PhD in Theoretical Physics under the supervision of Walter Glaser and Ludwig Flamm. In July 1960 he married Lore Kummer, who supported him until his death in 2007.

From 1961 on Wolfgang's career was closely connected with CERN, where he stayed as CERN fellow until 1964. Despite of bleak initial conditions after World War II — especially the brain drain and the latent anti-intellectual environment in Austria — Wolfgang Kummer became quickly one of the youngest full professors ever in 1968 at the age of 33, *after* he spent two years as director of the newly founded Austrian Institute for High Energy Physics of the Austrian Academy of Sciences. He managed to direct that institute simultaneously with his duties as full professor at the Vienna University of Technology until 1971, before leaving for an extended visit to the Pennsylvania University.

The visit to the United States was very inspiring for Wolfgang, and he regretted that he could not return after the turn of the millenium — but despite of several invitations he refused, partially because he was deeply disappointed by the moral climate in the country, where human rights were sacrificed for a misguided sense of security, and political correctness was gradually replacing the intellectual freedom.

After his inspiring period in the US he produced a work which to this day remains his best-cited paper: his promotion of axial gauge in non-abelian quantum field theories,[2] which led to the avoidance of ghosts.

In the 1980ies Wolfgang returned to the CERN Council as its Vice-President at the point in time when the new supercollider SPS was getting into shape. From 1985 to 1987, Wolfgang was president of the CERN Council, which was only briefly interrupted by the consequences of a terrorist attack at the Vienna airport on December 26 in 1985, where Wolfgang was one of the victims. More about Wolfgang's role at CERN, his amazingly quick recovery through skiing and his cultural activities as pianist and trained tenor are recounted in Part III of this Volume.

While Wolfgang had numerous academic and administrative positions, such as being first Secretary and then President of the High Energy Board of the European Physical Society from 1995–1999, he was especially responsible for building up a group of theoretical high energy physics at the Vienna University of Technology, covering a broad range of research in quantum field theory, string theory, and (mainly two-dimensional) quantum gravity. The stimulating atmosphere in the group which we all enjoyed was entirely due to Wolfgang's leadership. Wolfgang himself had contributed foundational work in quantum gauge field theory. For instance, besides his well-known work on axial gauge that we mentioned already he proposed a lepton number violating neutrino model, now known as "Zee model",[3] and he was very fond of toponium,[4] a bound-state of a top-anti-top-quark pair.

Since the early 1990's Wolfgang has mainly worked on two-dimensional gravity, to which he made pioneering contributions summarized in an extensive review article.[5]

Wolfgang has been unceasingly productive much beyond his official retirement as professor for theoretical physics in 2003, not sparing himself despite the fact that his health was deteriorating sharply. He also remained active as a member of the Austrian Academy of Sciences and chairman of the Advisory Board of the Institute for High Energy Physics (HEPHY), Vienna.

Wolfgang was not a person to idolize others, but if he had something like an idol it would probably be Victor Weisskopf for whom he wrote an obituary in the CERN courier.[1,6] In private discussions he once stated:

"Victor Weisskopf was among the people who I would call a 'good person'. He was honest, modest, friendly, forgiving, humorous and listened to others, but never subscribed to the postmodern 'anything-goes' mentality, was not afraid to make clear decisions and could be relentless in a positive way." We knew another person with these attributes. Wolfgang Kummer.

References

1. W. Kummer, Victor Weisskopf: looking back on a distinguished career. *CERN Cour.* **42N5** (2002) 28–31.
2. W. Kummer, Ghost Free Nonabelian Gauge Theory, *Acta Phys. Austriaca* **41** (1975) 315–334.
3. W. Konetschny and W. Kummer, Nonconservation of Total Lepton Number with Scalar Bosons, *Phys.Lett.* **B70** (1977) 433.
4. W. Mödritsch and W. Kummer, Relativistic and gauge independent off-shell corrections to the toponium decay width, *Nucl.Phys.* **B430** (1994) 3–12.
5. D. Grumiller, W. Kummer and D. Vassilevich, Dilaton gravity in two-dimensions, *Phys.Rept.* **369** (2002) 327–430.
6. W. Kummer, A rich inheritance, *CERN Cour.* **42** (2002) SUPPL29–32.

Brief overview of this volume

Part I of the volume contains ten contributions on gauge field theory and particle physics. We mention here some of them as a guide through the volume, in order of appearance. The first contribution "Noncovariant gauges at zero and nonzero temperature" by P. Landshoff touches one of Wolfgangs main topics, namely ghost-free gauges in non-abelian gauge theory. At finite temperature additional complications arise, which are presented concisely. The contribution by A. Vairo discusses in detail another of Wolfgang's favorite topics, non-relativistic bound states like positronium and quarkonia, and the use of the Bethe–Salpeter equation for describing them. The next contributions deal with topics that were of great interest to Wolfgang, even though he did not contribute to them with research papers: S. Deser introduces various Chern–Simons terms for extensions of QED in three dimensions and P.G.O. Freund discusses dynamical spin, i.e., the possibility to build all particles from spinless constituents, based upon his unpublished results from 1981. The remaining contributions in Part I deal with quantum corrections to solitons (A. Rebhan, P. van Nieuwenhuizen and R. Wimmer), the gauging of noncommutative theories (H. Grosse and M. Wohlgenannt), neutron quantum optics (H. Rauch), σ-models and Wess–Zumino–Courant algebroids (M. Hansen and T. Strobl), some aspects of gauge-fixing in Yang–Mills theories (D. Blaschke et al.) and, returning to the first contribution of this Memorial Volume, frozen ghosts in thermal gauge theory (P.V. Landshoff and A. Rebhan).

Part II of the volume contains ten contributions on classical and quantum gravity. Again we mention some of them, in order of appearance. The first contribution "Wolfgang Kummer and the Vienna School of Dilaton (Super-)Gravity" by L. Bergamin and R. Meyer provides an extensive review to Wolfgang's main research activities in his last two decades, dilaton gravity in two dimensions. This contribution is followed by related work on order and chaos in 2-dimensional gravity (R. Mann), 2-dimensional midisuperspace models for quantum black holes (J. Gegenberg and G. Kunstatter) and global solutions in Euclidean gravity (M. Katanaev). Then D.J. Schwarz, one of Wolfgang's most successful former students, provides mind-provoking "Thoughts on the Cosmological Principle". W. Thirring, together with C. Faustmann and H. Neufeld, discuss "When time emerges" by studying geodesics in spacetimes which allow for signature change. The simplicity of their 2-dimensional example surely would have appealed to Wolfgang. On a sidenote, this contribution is certainly the one with the biggest age gap between two of its authors: W. Thirring (82) and C. Faustmann (23). The remaining contributions in Part II deal with noncommutative gravity (D.V. Vassilevich), superstrings in $AdS_5 \times S^5$ superspace (I.A. Bandos), heterotic $(0,2)$ Gepner models (M. Kreuzer) and the constraint analysis of cosmological topologically massive gravity (D. Grumiller, R. Jackiw and N. Johansson).

Some reminiscences on Wolfgang's life within the physics community are scattered throughout the volume. In the final Part III we have collected five contributions that focus almost exclusively on Wolfgang's life and his interests. The first contribution by H. Schopper reports reminiscences of Wolfgang's activities as president of the CERN council in the 1980ies. K. Lane narrates a personal encounter and collaboration with Wolfgang in the 1970ies. J. Steinberger provides an exposition of a topic that was of great personal interest to Wolfgang and his wife Lore, namely the conservation of our environment — more specifically, he writes about fuel consumption, global warming and solar power production in deserts. M. Schweda delivered a speech in honor of Wolfgang on the occasion of his official retirement, which is printed in this volume for the first time. The volume concludes with a personal and perceptual account by P. van Nieuwenhuizen, "Schubert in Stony Brook and Kinks in Vienna".

Daniel Grumiller, Anton Rebhan and Dimitri Vassilevich   March 2009

PART I

Gauge Field Theory and Particle Physics

Chapter 1

Noncovariant Gauges at Zero and Nonzero Temperature

P. V. Landshoff

Department of Applied Mathematics and Theoretical Physics,
Cambridge CB3 0WA, UK
E-mail: pvl@damtp.cam.ac.uk

I review the formalism for gauge-field perturbation theory in noncovariant gauges, particularly the temporal axial gauge. I show that, even at zero temperature, there are complications and it is not known whether a formalism exists for handling these that is correct for all calculations. For thermal field theory in the imaginary time formalism, there are different difficulties, whose solution so far is known only up to lowest order in the coupling g.

1.1. Introduction

Noncovariant gauges in which the gauge field A_μ is defined to satisfy

$$n.A = n^\mu A_\mu = 0 \qquad (1.1)$$

where n^μ is some fixed 4-vector of unit length, have been pioneered by Wolfgang Kummer[1] and later widely used in nonabelian gauge theories.[2] This is partly because Faddeev-Popov ghosts are believed to decouple in such gauges,[3] and partly because a suitable choice of the direction of n^μ often seems to simplify calculations. In thermal field theory, in particular, it is natural to choose, in the rest frame of the ensemble under study,

$$n^\mu = (1, 0, 0, 0) \qquad (1.2)$$

The corresponding gauge is called the temporal gauge.

However, there are severe complications with calculating in noncovariant gauges. Indeed, it is not even clear that a consistent universal calculation scheme exists, even for ordinary perturbation theory at zero temperature. The basic problem is that a naive derivation of the gauge-field propagator gives

$$D_{\mu\nu}(k) = \left[-g_{\mu\nu} + \frac{k_\mu n_\nu + n_\mu k_\nu}{n.k} - n^2 \frac{k_\mu k_\nu}{(n.k)^2} \right] \frac{1}{k^2 + i\epsilon} \qquad (1.3)$$

and we do not know how to handle the double pole at $n.k = 0$. Traditionally, it was assumed[4] that it is correct to apply a principal-value prescription. This gives the right answer for simple calculations, but not when there are Feynman graphs where two gauge-field lines carry the same momentum k, so that $D_{\mu\nu}(k)$ has to be squared. This was shown[5,6] by comparing the calculation in Feynman gauge and in temporal gauge of a Wilson loop in next-to-leading order, where the sensitive graph is that shown in figure 1.

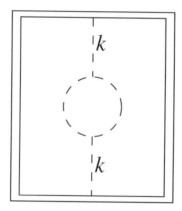

Figure 1.1   Graph in the calculation of a Wilson loop in which a propagator of momentum k is squared

Here I will first review progress, or the lack of it, with this difficulty. I then go on to discuss how things are even more complicated with thermal field theory in the temporal gauge.

1.2. Temporal gauge at finite temperature

The propagator (1.3) satisfies

$$n^\mu D_{\mu\nu}(k) = 0 = D_{\mu\nu}(k)n^\nu \qquad (1.4)$$

In particular, in the temporal gauge only the spacelike components are nonzero:

$$D_{ij}(k) = \frac{1}{k^2 + i\epsilon} \left(\delta_{ij} - \frac{k_i k_j}{k_0^2} \right) \qquad (1.5)$$

The apparent property that

$$D_{00} = D_{0i} = D_{i0} = 0 \qquad (1.6)$$

together with the absence of ghosts, is why the gauge is supposed to be useful: it makes calculations much simpler. However, it is not clear that (1.6) can be assumed to be true.

Certainly, non-leading-order calculations are usually delicate. In my calculation of the Wilson loop,[6] having established that the principal-value prescription

$$\frac{1}{k_0^2} \longrightarrow \frac{1}{2}\left(\frac{1}{(k_0 + i\eta)^2} + \frac{1}{(k_0 - i\eta)^2} \right) \qquad (1.7)$$

gave the wrong answer, I assumed a prescription that was fairly similar:

$$\frac{1}{k_0^2} \longrightarrow \frac{1}{k_0^2 + \eta^2} \qquad (1.8)$$

I found that individual graphs diverged as powers of $1/\eta^2$ when $\eta \to 0$. They must be calculated carefully. The limit must be taken only right at the end of the calculation, after Feynman's $\epsilon \to 0$. In particular, powers of η from the loop integration for the graph in figure 1.1 must be retained, as they multiply powers of $1/\eta$ coming from the k integration. In the end, all the divergences cancel and the same answer is obtained as in Feynman gauge.

The prescription (1.8) was chosen so as to give the correct answer for this calculation, but there is no derivation of it from first principles and therefore no guarantee that the same will be true for all other calculations.

Attempts to derive a prescription typically start with a gauge close to the one that is wanted. For example[7,8] one might start by imposing

$$A_0 = \eta \frac{1}{\partial_3} \frac{1}{\partial^2} \partial.\mathbf{A} = \eta \frac{1}{\partial_3} \frac{1}{\partial^2} \partial.\mathbf{A}^L \qquad (1.9)$$

with $\mathbf{A}^L$ the longitudinal field,

$$A_i^L = \partial_i \frac{1}{\partial^2} \partial.\mathbf{A}$$

and let $\eta \to 0$ at the end of any calculation. One can then eliminate A_0 from the Lagrangian, calculate the Hamiltonian and write down the corresponding equations of motion. Eliminating A_0 results in the loss of an equation of motion, which turns out to be Gauss's law. When the gauge-field coupling g is switched off, this is just $\partial.\dot{\mathbf{A}} = 0$. This cannot be imposed as an operator condition. Instead one requires that its matrix element vanishes:

$$\langle P'|\partial.\dot{\mathbf{A}}|P\rangle = 0 \qquad (1.10)$$

Here P and P' denote any pair of physical states, and for setting up perturbation theory it is sufficient to consider asymptotic states. The equations of motion ensure that, if this constraint is imposed at any time, say $t = 0$, it remains satisfied at other times. The constraint (1.10) enables one to pick a complete set of physical states, not uniquely, but most simply one might choose those states that contain no longitudinal gauge particles.

Canonical quantisation then leads to the prescription

$$\frac{1}{k_0^2} \longrightarrow \left(\frac{1}{k_0 + i\eta/k_3}\right)^2 \qquad (1.11)$$

This is the so-called Vienna prescription.[9] It again gives the correct Wilson loop to next-to-leading order,[10] but again one cannot be sure how generally it may be applied. The problem is that, until $\eta \to 0$, one cannot be sure that it is valid to neglect the elements D_{00}, D_{0i}, D_{i0} of the gauge-field propagator, nor indeed the ghosts.

The conclusion then is that, attractive as noncovariant gauges may seem to be, using them for calculations is, to say the least, problematical even at zero temperature.

1.3. Thermal field theory in temporal gauge

Thermal field theory is formulated by starting with the grand partition function

$$Z = \sum_i \langle i|e^{-(H-\mu N)/T}|i\rangle \qquad (1.12)$$

from which nearly all (though not all[11]) the interesting properties of the system under study may be calculated. Here T is the temperature, and we use units in which Boltzmann's constant $k_B = 1$. The system's Hamiltonian is H and N is some conserved quantum number, such as baryon number, with μ the corresponding chemical potential. The states $|i\rangle$ are a complete orthonormal set of physical states of the system. In scalar field theory all states are physical and so

$$Z = \text{tr } e^{-(H-\mu N)/T} \qquad (1.13)$$

which is invariant under changes in the choice of orthonormal basis of states. In the case of gauge theories there are unphysical states, for example longitudinally-polarised photons or gluons, which must be excluded from the summation in (1.12). So then

$$Z = \operatorname{tr} \mathbb{P} \, e^{-(H - \mu N)/T} \tag{1.14}$$

where $\mathbb{P}$ is a projection operator onto physical states. The presence of $\mathbb{P}$ can make things more complicated.

For scalar field theory in the so-called imaginary-time formalism of thermal field theory, the Feynman rules are just as at zero temperature, except that round each loop of a graph the usual loop-momentum integration undergoes the replacement

$$\int \frac{d^4 k}{(2\pi)^4} \to iT \sum_n \int \frac{d^3 k}{(2\pi)^3} \tag{1.15}$$

Here the summation is over discrete values

$$k_0 = n\pi T \qquad n = 0, \pm 2, \pm 4, \ldots \tag{1.16}$$

which has the consequence that the propagator is periodic in the time t:

$$D(t, \mathbf{k}) = D(t - i/T, \mathbf{k}) \tag{1.17}$$

Were this rule to apply also to a gauge-field theory in the temporal gauge, there would obviously be a difficulty, since the summation would include a contribution from $n = 0$, that is $k_0 = 0$, where the zero-temperature propagator has a double pole.

The transverse gauge-field propagator does behave similarly to a scalar field, with

$$[D_n^T(\mathbf{k})]_{ij} = - \left(\delta_{ij} - \frac{k_i k_j}{\mathbf{k}^2} \right) \frac{1}{\pi^2 n^2 T^2 + \mathbf{k}^2} \qquad n = 0, \pm 2, \pm 4, \ldots \tag{1.18}$$

and it has the periodicity (1.17). However,[12] this is not the case for the the longitudinal field: because of the presence of the projection operator in (1.14), canonical quantisation results in a longitudinal propagator $[D^L(t, \mathbf{k})]_{ij}$ that is not periodic in t. But there is a simplification to compensate for this complication: $[D_n^L(\mathbf{k})]_{ij}$ does not have a double pole at $n = 0$. Indeed, it is regular there, but n is not restricted to even values: $[D_n^L(\mathbf{k})]_{ij} = (k_i k_j / \mathbf{k}^2) D_n^L(k_3)$, with

$$D_n^L(k_3) = \begin{cases} 1/(4T^2) & n = 0 \\ -i\epsilon(k_3)/(2\pi n T^2) & n \text{ even} \\ -1/(\pi^2 n^2 T^2) + i\epsilon(k_3)/(2\pi n T^2) & n \text{ odd} \end{cases} \tag{1.19}$$

The non-vanishing of $[D_n^L(\mathbf{k})]_{ij}$ for odd n means that it is easier not to work with $[D_n(\mathbf{k})]_{ij}$ and perform summations over the various n associated with the different lines in a Feynman graph, but instead to work with propagators $[D^L(t,\mathbf{k})]_{ij}$ and integrate over the times associated with the various vertices.

Another complication that has to be taken into account is the constraint (1.10) on the physical states. At zero temperature, or when one is working in the real-time thermal formalism, one uses asymptotic states and this constraint is sufficient. But the imaginary-time formalism rather uses interaction-picture states and so the Gauss operator is now

$$G^a(t,\mathbf{x}) = \boldsymbol{\partial}.\dot{\mathbf{A}}^a(t,\mathbf{x}) - gf^{abc}\mathbf{A}^b(t,\mathbf{x}).\dot{\mathbf{A}}^c(t,\mathbf{x}) \qquad (1.20)$$

and it turns out[12] that one needs to impose a set of constraints

$$\langle P'|\prod_{i=1}^{N} G^{a_i}(0,\mathbf{x})|P\rangle = 0 \quad N = 1,2,\ldots \qquad (1.21)$$

As one increases the accuracy of one's calculation to higher powers of g, one needs to go up to higher and higher values of N. The physical states can no longer be taken as those with purely-transverse gauge particles. The solution of (1.21) is known only for low-order calculations for which it is sufficient to go up to $N = 2$.

References

1. W Kummer, Acta Physica Austriaca, 14 (1961) 149
2. G Leibbrandt, Reviews of Modern Physics 59 (1987) 1067
3. W Konetschny and W Kummer, Nuclear Physics B100 (1975) 106 and B124 (1977) 145
4. W Kummer, Acta Physica Austriaca 41 (1975) 315
5. S Carracciolo, G Curci and P Menotti, Physics Letters 113B (1982) 311
6. P V Landshoff, Physics Letters B169 (1986) 69
7. P V Landshoff, Physics Letters B227 (1989) 427
8. I Lazzizzera, Physics Letters B210 (1988) 188
9. P Gaigg and M Kreuzer, Phys Lett B205 (1988) 530
10. H Hüffel, P V Landshoff and J C Taylor, Physics Letters B217 (1989) 147
11. P V Landshoff and J C Taylor, Nuclear Physics B430 (1994) 683
12. K A James and P V Landshoff, Physics Letters B251 (1990) 167

Chapter 2

Non-Relativistic Bound States: The Long Way Back from the Bethe–Salpeter to the Schrödinger Equation

Antonio Vairo

Physik Department, Technische Universität München,
85748 Garching, Germany
E-mail: antonio.vairo@ph.tum.de

I review, in a personal perspective, the history of the theory of non-relativistic bound states in QED and QCD from the Bethe–Salpeter equation to the construction of effective field theories.

2.1. Introduction

The study of bound states and, in particular, of non-relativistic bound states has accompanied the quantum theory from its beginning through all its subsequent turning points up to what is now the Standard Model of particle physics. At the beginning it was the description of the hydrogen atom that led to the foundation of quantum mechanics, later the Lamb shift contributed to the development of relativistic field theories and renormalization, which eventually led to the foundation of Quantum Electrodynamics (QED); similarly, in the seventies, quarkonium played a special role in the foundation of Quantum Chromodynamics (QCD). The special role of non-relativistic bound states in particle physics is due to the striking experimental signatures that they provide and the fact that analytical (perturbative) methods are able to describe the relevant features of these signatures.

Despite this, it has proven very difficult to carry out theoretical analyses of a precision comparable with the data, in part due to the high quality of the data, but largely owing to the difficulties in performing bound-state calculations. These may be traced back to the presence of different energy scales that make it a challenge to maintain a consistent book-keeping in the calculations.

Let us consider a non-relativistic particle of mass m that propagates in a potential V (in the case of a Coulomb potential: $V = -\alpha/r$). If the momentum of the particle is non relativistic, then $p \sim mv$, $v \ll 1$ being the velocity of the particle. In the threshold region, the velocity is such that $mv^2 \sim V$. The balance between kinetic energy and potential creates the bound state: the particle propagator G cannot be computed order by order in V, but comes from resumming all potential insertions in the free propagator $G_0 = 1/(E - p^2/2m)$:

$$G = G_0 + G_0 V G. \qquad (2.1)$$

The function $G = 1/(E - p^2/2m - V)$ exhibits poles in correspondence of the bound-state energies $E_n \sim mv^2$ ($= -m\alpha^2/2n^2$ in the Coulombic case, which implies $v \sim \alpha$ and $1/r \sim m\alpha$), the residues at the poles, $\phi_n^* \phi_n$, satisfy the equation:

$$E_n \phi_n = \left(\frac{p^2}{2m} + V \right) \phi_n, \qquad (2.2)$$

which is the Schrödinger equation for a non-relativistic bound state whose wave function is ϕ_n.

Hence, the non-relativistic dynamics of a particle close to threshold is characterized by a hierarchy of energy scales: $m \gg mv \gg mv^2$. The scale of the mass is sometimes called "hard", the scale of the typical momentum transfer, or inverse size of the system, mv, is called "soft" and the scale mv^2 is called "ultrasoft".

At the level of non-relativistic quantum mechanics, m does not play any dynamical role, for the kinetic energy is $p^2/2m$ rather than $\sqrt{p^2 + m^2}$. The contributions from the other scales are accounted for by the Schrödinger equations (2.1) or (2.2). The solutions of the Schrödinger equation are non-relativistic bound states of typical energy of order mv^2 and typical momentum (or inverse size) of order mv.

One may expect that a more complicated picture will emerge in a relativistic field theory, although the leading dynamics should be still described by a Schrödinger equation. In a relativistic field theory description of the bound state, we will have, besides the bound state, other degrees of freedom, for instance photons (in QED) and gluons (in QCD) emitted and exchanged by the bound state; for each of them, modes associated to each of the energy scales, m, mv and mv^2 will appear. We shall discuss bound states in relativistic field theories in the next section.

2.2. The Bethe–Salpeter equation

Let us consider a particle and an antiparticle (e.g. an electron and a positron or a quark and an antiquark) that interact near threshold. In the centre-of-mass frame, their momenta p and energies E are small compared to their masses m: $p/m \sim v \ll 1$. We assume that we may express the interaction perturbatively in terms of Feynman diagrams. This is always the case in QED, but does not need to be so in QCD where, at the typical hadronic scale Λ_{QCD}, perturbation theory breaks down. Non-relativistic bound states in QCD are made by heavy quarks: this means that at least the quark mass is larger than Λ_{QCD}. A bound state of a heavy quark and a heavy antiquark is called quarkonium (examples are charmonium, a charm-anticharm bound state, and bottomonium, a bottom-antibottom bound state; a top-antitop bound state, which would be toponium, has no time to form due to the rapid top quark weak decay, however, near threshold, the bound-state enhancement should be visible in the top-antitop production cross section). A perturbative treatment of quarkonium, which requires $mv, mv^2 \gg \Lambda_{\mathrm{QCD}}$, is justified only for top-antitop pairs near threshold and possibly for the ground state of bottomonium.

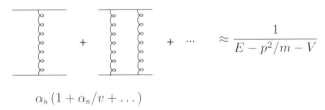

$$\alpha_{\mathrm{s}}\left(1 + \alpha_{\mathrm{s}}/v + \dots\right)$$

Figure 2.1   Resummed propagator near threshold.

How does the bound state emerge in a near threshold interaction? For certain sets of graphs, like those in Fig. 2.1, the perturbative expansion breaks down when $\alpha_{\mathrm{s}} \sim v$ (for definiteness, we will consider here and in the following figures the QCD case: continuous lines stand for quarks and antiquarks, and the curly lines for gluons; the strong coupling constant is α_{s}). The summation of all α_{s}/v contributions leads to the appearance of a bound-state pole of order $mv^2 \sim m\alpha_{\mathrm{s}}^2$ in the resummed propagator. Indeed, in the leading non-relativistic limit, when the quark/antiquark propagators can be approximated by $\dfrac{i}{\pm p^0 + E/2 - \mathbf{p}^2/2m + i\epsilon}\dfrac{1 \pm \gamma^0}{2}$ and the gluon exchange by $\dfrac{i}{\mathbf{q}^2}$ (close to threshold we may expand in $|q^0|/|\mathbf{q}| \ll 1$; γ^0 is a

Dirac matrix) the Green's function shown in Fig. 2.1 satisfies Eq. (2.1).

Beyond the leading non-relativistic limit, diagrams will be much more complicated to calculate and contributions from the different energy scales will get entangled. This happens for any diagram, but the annihilation diagram shown in Fig. 2.2 provides a rather immediate way to see it. Assuming that the incoming quarks are near threshold, the different gluons entering the diagram are characterized by different scales: the annihilation gluons have a typical energy of order m; binding gluons carry the momentum of the incoming quarks, which is of order mv, and ultrasoft gluons, sensitive to the intermediate bound state, have energies of the order of the binding energy, i.e. mv^2.

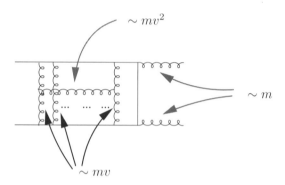

Figure 2.2   Annihilation diagram contributing to the quarkonium decay width.

The entanglement of the different energy modes makes it difficult to organize a full relativistic calculation of the bound state in QED or QCD. An equation suitable for bound states in field theory, formally similar to (2.1), was suggested almost sixty years ago by H. Bethe and E. Salpeter:[1]

$$G = G_0 + G_0 K G, \qquad (2.3)$$

where G is the two-particle Green's function, G_0 the product of the free propagators of the two particles and the kernel K is the sum of all amputated irreducible two-particle diagrams. Equation (2.3) does not represent an expansion, because, like (2.1), a bound state emerges only from the sum of all interactions, at least those shown in Fig. 2.1. However, unlike (2.1), the Bethe–Salpeter equation is not homogeneous in the momentum scale and an exact solution is unknown. To make the Bethe–Salpeter equation useful, the strategy has been to isolate from K a kernel K_c containing the leading contribution responsible for the formation of the bound state, i.e.

the Coulomb potential, and expand around it (see, for instance, Ref. 2). The most refined approach in this strategy can be found in Ref. 3 (see also Ref. 4): K_c is chosen in such a way that the corresponding Bethe–Salpeter equation, $G_c = G_0 + G_0 K_c G_c$, may be solved in an analytically closed form, and the full Green's function expanded around the exact solution:

$$G = G_c + G_c \delta K G, \qquad (2.4)$$

where $\delta K = K - K_c$. Since, in the non-relativistic limit, K_c becomes the Coulomb potential, G_c is nothing else than a relativistic modification of the solution of Eq. (2.1) for V equal to the Coulomb potential, which is known since long time.[5] The difference between Eq. (2.3) and Eq. (2.4) is that the latter is a perturbative expansion in the kernel while the former is not.

The Bethe–Salpeter equation was the only systematic tool to treat bound states in field theory until the end of the eighties. However, around that time, it became increasingly clear that perturbative calculations for QED bound states, to which the Bethe–Salpeter equation had been mostly applied, could not be push beyond the reached limit if not at the cost of a formidable amount of work. It shows the difficulty of the approach the fact that going from the calculation of the $m\alpha^5$ correction in the hyper-fine splitting of the positronium ground state[6] to the $m\alpha^6 \ln \alpha$ term[7,8] took twenty-five years! The main problem was the lack of an efficient way of disentangling the contributions coming from the different energy scales and organize them in a perturbative expansion (techniques for asymptotic expansions of Feynman integrals near threshold would be developed later[9]): each Feynman diagram would contribute to the observables with a series in the coupling constant. No obvious counting rules were available even for the leading term of the series. Also gauge invariance did not provide a useful organizational tool, since it was very cumbersome to isolate gauge-invariant subsets of diagrams in K.[10–13]

In the late seventies and eighties, systematic calculations of quarkonium observables started (for a recent review see Ref. 14). The complicated dynamics of QCD made it more apparent that a treatment based on the Bethe–Salpeter equation was inadequate to perform high-precision quarkonium calculations. First, not all of the quarkonium scales are in general perturbative, lower ones may not be, so that a separation of scales is necessary to achieve factorization. Second, even if a perturbative treatment would be possible (like for the bottomonium ground state and for $t\bar{t}$ threshold production), the number and topology of diagrams makes the calculation prohibitive. It was felt that somehow going back to the Schrödinger equation and identifying a quarkonium potential would lead to a more treatable problem. In Refs. 15–19, a quarkonium potential was derived from the quark-antiquark scattering amplitude. In the same years, focusing in particular on toponium and $t\bar{t}$ threshold physics, a similar program was carried out by W. Kummer and collaborators[20-24] (see also the Ph.D. thesis in Ref. 25). In this case, the starting point was the Bethe–Salpeter equation and the generalization to QCD of the solution of the Bethe–Salpeter equation for positronium found in Ref. 3. Still, the goal was not the solution of the Bethe–Salpeter equation itself, but the derivation of a potential, facing, in the process, some of the problems that, in a few years, would have led to (and found a solution with) the construction of effective field theories for non-relativistic bound states. Among the problems mentioned or addressed at that time were the infrared sensitivity of the potential, the inclusion of a finite decay width (in Refs. 22, 23, one can find addressed, for the first time in a formal way, how to include the top-quark instability beyond leading order), gauge invariance. The infrared sensitivity of the potential will be discussed in Sec. 2.5.

2.3. NRQED/NRQCD

In QED and QCD, one may take advantage of the hierarchy of scales that characterizes non-relativistic bound states by expanding Green's functions in the ratios of low energy scales over large energy scales. Working out these expansions, however, turns out to be cumbersome and does not lead to a straightforward and easy way to organize the calculation. If such an expansion is instead implemented at the Lagrangian level, it leads to the construction of an effective field theory (EFT). In the effective field theory the large scale is integrated out from the beginning and does not appear anymore in the Green's function. The terms in the EFT Lagrangian are organized as an expansion in powers of the inverse of the large scale that has been integrated out leading to a straightforward power counting.

The first EFT introduced for non-relativistic bound states in QED and QCD has been non-relativistic QED/QCD (NRQED/NRQCD).[26] The large scale that is integrated out in NRQED/NRQCD is the mass m of the bound-state constituents. The degrees of freedom of NRQED/NRQCD are non-relativistic fermions and antifermions, and photons/gluons of energy and momentum smaller than m; they build up the operators O_n of the Lagrangian. The Lagrangian is organized as an expansion in $1/m$:

$$\mathcal{L}_{\text{NRQED/NRQCD}} = \sum_n c_n(\alpha_s(m), \mu) \times \frac{O_n(\mu)}{m^n}. \qquad (2.5)$$

Since, once O_n has been run down to energies lower than m, the expectation value of O_n scales like mv or smaller scales, Eq. (2.5) provides, for any physical observable, a perturbative expansion in the ratio of the scale mv or smaller scales over m. The Wilson coefficients c_n are non analytical in the scale m and function of the factorization scale μ. They are calculated by equating, "matching", amplitudes in QED/QCD with amplitudes in NRQED/NRQCD order by order in $1/m$ and in the coupling constant since in both theories we have that α, $\alpha_s(m) \ll 1$. The matching may be performed on scattering amplitudes, hence in a manner completely independent of the bound state. This is not surprising: the formation of the bound state takes place at a scale, mv, which is much smaller than m.

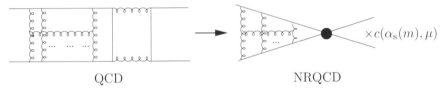

$$\times c(\alpha_s(m), \mu)$$

QCD   NRQCD

Figure 2.3   Matching to NRQCD.

The diagram in Fig. 2.2 corresponds, via the optical theorem, to the imaginary part of the diagram shown on the left in Fig. 2.3. The same process would be described in NRQCD by the diagram shown on the right in Fig. 2.3, i.e. by a diagram where the two hard gluons coming from the annihilation are replaced by a contact interaction. The difference between the two diagrams is compensated by the Wilson coefficient $c \sim \alpha_s(m)^2$.

As our example may suggest, NRQCD is particularly well suited to describe heavy quarkonium decay and production.[27,28] It is in the theory of quarkonium production that NRQCD has perhaps achieved its major success by explaining, in the nineties, the quarkonium production data at the Tevatron by a new mechanism allowed by the symmetries of NRQCD, the octet mechanism, but missed by previous approaches (see Ref. 14 and references therein).

Applications of NRQED have started in the nineties and with time have led to many new results (for some early works, see Refs. 29–31). However, the progress in high precision calculations in NRQED/NRQCD has been slowed down by two major shortcomings: first, the fact that soft and ultra-soft degrees of freedom still remain entangled in NRQED/NRQCD, second, the use in early NRQED/NRQCD calculations of a cut-off regularization scheme. The first difficulty led to a power counting that was non homo-geneous and to perturbative calculations that still involved two scales. To overcome this difficulty, lower energy EFTs were developed; we shall dis-cuss some of them in Sec. 2.5. The second difficulty, on one hand, pushed the development of lattice NRQCD[32] (see Ref. 33 for recent results on the bottomonium spectrum), on the other hand, addressed analytical studies towards a consistent formulation of NRQCD in dimensional regularization.

2.4. The bound state in dimensional regularization

Surprisingly, it was only few years after NRQCD had been introduced that an EFT for mesons made of a single heavy quark, the heavy quark effec-tive theory (HQET), was formulated.[34] In the two-fermion sector, the La-grangian of HQET contains the same operators as the NRQCD Lagrangian. However, HQET is a quite different theory from NRQCD: HQET contains only a single dynamical scale, Λ_{QCD}, which governs its power counting. As a consequence, the kinetic energy, which is of order Λ_{QCD}^2/m, is suppressed with respect to the binding energy, which is of order Λ_{QCD}, while, in a non-relativistic bound state, the two are of the same order.   .

It is precisely because, in HQET, propagators are expanded in the ki-netic energy that we may use dimensional regularization in loop calcula-tions. This has led to a rapid, vast and very successful use of the HQET in precision studies of D and B mesons.[35] Instead, keeping the kinetic energy in the denominators of the propagators, as the power counting of NRQCD seems to suggest, turns out to be disastrous and leads to the break down of the power counting. The reason is that, in dimensional regularization, integrals are not cut-off at high momenta and hard scale poles are going to contribute if present in the denominators. Once, this had been realized in Ref. 36, it became also clear that the way out was to compute the matching to NRQCD in the same way as the matching to the HQET, i.e. order by order in $1/m$. Since both in NRQCD and in the HQET the matching con-ditions are computed in the same way, the two Lagrangians are the same: not only the operators of the two theories coincide in the two-fermion sec-tor, but also their matching coefficients do. Obviously, in order to compute observables with the NRQCD Lagrangian, the usual non-relativistic power counting rules, different from the HQET ones, should be used.

Having understood how to treat the bound state in dimensional regularization, opened, finally, the doors to analytical high-precision calculations also for non-relativistic bound states in NRQED/NRQCD.

2.5. pNRQED/pNRQCD

The problem of disentangling the soft from the ultrasoft scale in NRQED/NRQCD was addressed immediately after dimensional regularization was established as an useful tool for non-relativistic bound state calculations also. The history and details of the developments that have ultimately led to the construction of EFTs for the ultrasoft degrees of freedom of NRQED/NRQCD have been recollected in Ref. 37 and we refer the interested reader to it. Here, we would like just to stress the importance that the process of $t\bar{t}$ production near threshold (see Ref. 14 and references therein) has played in these developments, providing the only near threshold, heavy quark-antiquark system in nature entirely accessible in perturbation theory. As it was mentioned before, this very special feature of the $t\bar{t}$ system near threshold had already been appreciated by the groups working on the subject at the beginning of the nineties and, in particular, by the Vienna group." .

In the following, in order to illustrate some general features, we will concentrate on the EFTs for ultrasoft degrees of freedom of NRQED/NRQCD known as potential NRQED[38,39] and potential NRQCD[40,41] (for an alternative formulation see Ref. 42 and the review in Ref. 43). The large scale that is integrated out in pNRQED/pNRQCD is the typical momentum transfer of the bound state, which, in coordinate space, is associated with the inverse of the typical distance r between the two heavy particles. The degrees of freedom of pNRQED/pNRQCD are non-relativistic fermions and antifermions, and photons/gluons of energy and momentum smaller than mv. They build up the operators $O_{k,n}$ of the Lagrangian; the operators may be also chosen to be explicitly gauge invariant. The Lagrangian is organized as an expansion in $1/m$, inherited from NRQED/NRQCD, and an expansion in r (multipole expansion), which is characteristic of the new EFT:

$$\mathcal{L}_{\text{pNRQED/pNRQCD}} = \sum_{k,n} \frac{1}{m^k} \times c_k(\alpha_s(m), \mu) \times V_n(r, \mu, \mu') \times r^n O_{k,n}(\mu').$$

(2.6)

Since, once $O_{k,n}$ has been run down to the lowest energy mv^2, the expectation value of $O_{k,n}$ scales like mv^2, Eq. (2.6) provides, for any physical observable, a perturbative expansion in the ratio of mv^2 over mv or m. The Wilson coefficients c_k are those inherited from NRQED/NRQCD, the Wilson coefficients V_n are the new ones of pNRQED/pNRQCD. They are non analytical in the scale r and function of the new factorization scale μ'. They are calculated by matching, order by order in r, Green's function in NRQED/NRQCD with Green's function in pNRQED/pNRQCD. In pN-RQED, the matching may be also done order by order in α. In pNRQCD, $\alpha_s(mv) \ll 1$ holds only for tightly bound states (short-range quarkonia, e.g. the bottomonium ground state or $t\bar{t}$ near threshold), while, in general, higher excited quarkonium states (long-range quarkonia) are not accessible by perturbation theory. This means that we can rely on an expansion in $\alpha_s(mv)$ only for the former states, while for the latter states the matching has to be done in a non-perturbative fashion.

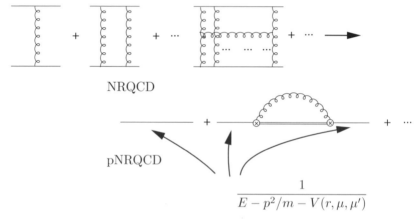

Figure 2.4   Matching to pNRQCD.

Let us consider the NRQCD diagram shown in Fig. 2.3 and the part of it where soft gluons are exchanged between the quark and antiquark. The sum of all soft-gluon exchanges would be described in pNRQCD by the diagram shown on the right in Fig. 2.4 where the single line stands for a quark-antiquark propagator in a color-singlet configuration, $1/(E - p^2/m - V)$, the double line for a quark-antiquark pair in a color-octet configuration, the curly line for ultrasoft gluons and the circle with a cross for a chromoelectric dipole interaction $\sim \phi^\dagger \mathbf{r} \cdot \mathbf{E}\, \phi$ that comes from multipole expanding the gluon fields in the NRQCD Lagrangian. The non-analytical behaviour in r of the NRQCD diagram is reproduced in pNRQCD by the Wilson coefficient V. Since V, together with p^2/m, makes up the pole of the quark-antiquark propagator, the interpretation of V is obvious: V is the potential describing the interaction in the heavy quark-antiquark pair. At leading order in the multipole expansion, when we neglect diagrams involving ultrasoft gluons, the equation of motion of a non-relativistic fermion-antifermion pair is nothing else than the Schrödinger equation (2.2).

The Schrödinger equation is the equation governing non-relativistic bound states in quantum mechanics. The full relativistic description provided by field theory, which is richer and much more complex, is given by the Bethe–Salpeter equation. This complexity arises from the entanglement of different energy scales. Once the contributions of all these scales have been separated/factorized, we are left with an EFT of the ultrasoft degrees of freedom. The Schrödinger equation naturally emerges as the equation of motion of these ultrasoft degrees of freedom. But, because the EFT contains all the richness and complexity of the field theory, although unfolded in a systematic and organized way, the Schrödinger equation, which we have gotten from the EFT, is much more than the Schrödinger equation of quantum mechanics we have started with. First, the EFT provides a proper, field theoretically founded, definition of the potential: the potential is the Wilson coefficient of the dimension six operator of the EFT, containing two fermion and two antifermion fields, that encodes all contributions coming from modes whose energies and momenta are larger than the binding energy. It undergoes renormalization, develops scale dependence and satisfies renormalization group equations, which, in perturbation theory, allow to resum potentially large logarithms. Moreover, the EFT accounts also for effects that cannot be cast in a Schrödinger equation and that are due to the coupling of the fermion-antifermion pair with the other ultrasoft degrees of freedom.

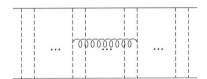

Figure 2.5   QCD diagrams responsible for the infrared sensitivity of the static potential.

In QCD, ultrasoft effects affect also the static potential. As first observed in Ref. 44, they come from the "non-Abelian Lamb shift"[24] diagrams displayed in Fig. 2.5. At fixed order in perturbation theory, the diagrams are infrared divergent; at order α_s^4, the leading logarithmic correction is[45]

$$\delta V(r, \mu') = -\frac{3}{r} \frac{\alpha_s(\mu')}{\pi} \alpha_s^3 (1/r) \ln(r\mu'). \qquad (2.7)$$

The result shows clearly the non-physical nature of the potential, which depends on the renormalization scale μ'. The potentially large logarithms, $\ln(r\mu')$, have been resummed by means of renormalization group equations in Ref. 46; subleading corrections have been calculated in Ref. 47. In physical observables, like the static energy or the quarkonium mass, the scale dependence of Eq. (2.7) cancels against ultrasoft contributions coming from the second diagram in the pNRQCD part of Fig. 2.4.

Higher-order terms in the relativistic expansion may be computed systematically in the EFT. Again, the full complexity and symmetries of the underlying field theory are not lost in the expansion. So, for instance, relativistic invariance imposes specific constraints on the Wilson coefficients/potentials of the EFT,[48,49] which can be tested on the lattice.[50,51]

Applications of pNRQCD and, more in general, of EFTs for the ultrasoft degrees of freedom of NRQCD have led to a plethora of new results in quarkonium physics (see Refs. 14, 37, 52–55 for some recent reviews) and, in particular, in $t\bar{t}$ threshold production (see Refs. 56, 57, 66, 67 for the present status of the art). Also QED calculations have remarkably benefited from the EFT approach and corrections of very high order in perturbation theory have been calculated in the last years for many observables after decades of very slow or no progress. As an example, we mention that for the hyperfine splitting of the positronium ground state the terms of order α^6, $\alpha^7 \ln^2 \alpha$ and $\alpha^7 \ln \alpha$ have been calculated (for recent reviews on positronium precision studies and further references we refer to Refs. 58, 59).

In Fig. 2.6, we summarize the hierarchy of EFTs for bound states in QED and, for heavy quarks, in QCD.

2.6. Outlook

The history of non-relativistic bound states in the quantum theory had in the last century a peculiar spiral behaviour. It started with the Schrödinger equation of the hydrogen atom and seemed to have written its ultimate chapter with the Bethe–Salpeter equation in the fifties. However, in face of the enormous difficulties in treating bound states in field theory by means of the Bethe–Salpeter equation, a long journey started in the seventies that took us back to the Schrödinger equation. This coming back, however, was not like closing a circle, it was more like building up a spiral. The

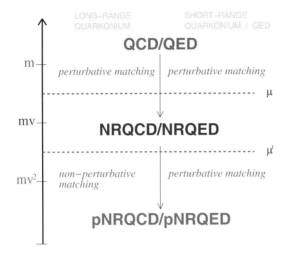

Figure 2.6   EFTs for bound states in QED and, for heavy quarks, in QCD.

Schrödinger equation we have come back to, encompasses all the complexity of the Bethe–Salpeter equation, all the richness of field theory, in the elegant and systematic setting of non-relativistic effective field theories. The counting rules and structure of the EFTs have allowed us to perform calculation with unprecedented precision, where higher-order perturbative calculations were possible, and to systematically factorize short from long range contributions where observables were sensitive to the non-perturbative, infrared dynamics of QCD.

Non-relativistic EFTs have become nowadays the standard tool to treat non-relativistic bound states. Besides QED bound states and quarkonium, these include hadronic atoms like pionium,[60] nucleon-nucleon systems,[61,62] non-relativistic bound states at finite temperature[63–65] and many others. The modern history of non-relativistic bound states is far from being finished and still needs to be told in its full extent.

Acknowledgments

I thank André Hoang for comments. I acknowledge financial support from the European Research Training Network FLAVIA*net* (FP6, Marie Curie Programs, Contract MRTN-CT-2006-035482) and from the DFG cluster of excellence "Origin and structure of the universe" (www.universe-cluster.de).

References

1. H. A. Bethe and E. Salpeter, *Phys. Rev.* **82**, 309, (1951).
2. G. P. Lepage, *Phys. Rev.* **A16**, 863, (1977). doi: 10.1103/PhysRevA.16.863.
3. R. Barbieri and E. Remiddi, *Nucl. Phys.* **B141**, 413, (1978). doi: 10.1016/0550-3213(78)90036-6.
4. A. Vairo, *Found. Phys.* **28**, 829, (1998). doi: 10.1023/A:1018858105862.
5. J. Schwinger, *J. Math. Phys.* **5**, 1606, (1964).
6. R. Karplus and A. Klein, *Phys. Rev.* **87**, 848, (1952). doi: 10.1103/PhysRev.87.848.
7. W. E. Caswell and G. P. Lepage, *Phys. Rev.* **A20**, 36, (1979). doi: 10.1103/PhysRevA.20.36.
8. G. T. Bodwin and D. R. Yennie, *Phys. Rept.* **43**, 267, (1978). doi: 10.1016/0370-1573(78)90151-5.
9. M. Beneke and V. A. Smirnov, *Nucl. Phys.* **B522**, 321, (1998). doi: 10.1016/S0550-3213(98)00138-2.
10. S. Love, *Ann. Phys.* **113**, 153, (1978). doi: 10.1016/0003-4916(78)90253-1.
11. G. Feldman, T. Fulton, and D. L. Heckathorn, *Nucl. Phys.* **B167**, 364, (1980). doi: 10.1016/0550-3213(80)90239-4.
12. G. Feldman, T. Fulton, and D. L. Heckathorn, *Nucl. Phys.* **B174**, 89, (1980). doi: 10.1016/0550-3213(80)90191-1.
13. A. Vairo, Gauge invariance on bound state energy levels, (1995). arXiv:hep-ph/9609264.
14. N. Brambilla et al., Heavy quarkonium physics, (CERN, 2005). CERN-2005-005.
15. S. N. Gupta and S. F. Radford, *Phys. Rev.* **D24**, 2309, (1981).
16. S. N. Gupta and S. F. Radford, *Phys. Rev.* **D25**, 3430, (1982).
17. W. Buchmüller, Y. J. Ng, and S. H. H. Tye, *Phys. Rev.* **D24**, 3003, (1981).
18. J. Pantaleone, S. H. H. Tye, and Y. J. Ng, *Phys. Rev.* **D33**, 777, (1986).
19. S. Titard and F. J. Yndurain, *Phys. Rev.* **D49**, 6007, (1994).
20. W. Kummer, *Nucl. Phys.* **B179**, 365, (1981). doi: 10.1016/0550-3213(81)90010-9.
21. W. Kummer and W. Mödritsch, *Z. Phys.* **C66**, 225, (1995). doi: 10.1007/BF01496596.
22. W. Mödritsch and W. Kummer, *Nucl. Phys.* **B430**, 3, (1994). doi: 10.1016/0550-3213(94)90647-5.
23. W. Kummer and W. Mödritsch, *Phys. Lett.* **B349**, 525, (1995). doi: 10.1016/0370-2693(95)00267-O.
24. W. Kummer, W. Mödritsch, and A. Vairo, *Z. Phys.* **C72**, 653, (1996).
25. W. Mödritsch, Quantum field theory near thresholds of ultra-heavy: top-antitop and beyond, *Ph.D. thesis*, (University of Vienna, 1995).
26. W. E. Caswell and G. P. Lepage, *Phys. Lett.* **B167**, 437, (1986).
27. G. T. Bodwin, E. Braaten, and G. P. Lepage, *Phys. Rev.* **D46**, 1914, (1992).
28. G. T. Bodwin, E. Braaten, and G. P. Lepage, *Phys. Rev.* **D51**, 1125, (1995).
29. P. Labelle, G. P. Lepage, and U. Magnea, *Phys. Rev. Lett.* **72**, 2006, (1994). doi: 10.1103/PhysRevLett.72.2006.

30. P. Labelle, Order α^2 nonrelativistic corrections to positronium hyperfine splitting and decay rate using nonrelativistic quantum electrodynamics, *Ph.D. thesis*, (Mc Gill University, 1994). UMI-94-16736.

31. T. Kinoshita and M. Nio, *Phys. Rev.* **D53**, 4909, (1996).

32. B. A. Thacker and G. P. Lepage, *Phys. Rev.* **D43**, 196, (1991).

33. A. Gray et al., *Phys. Rev.* **D72**, 094507, (2005). doi: 10.1103/PhysRevD.72. 094507.

34. N. Isgur and M. B. Wise, *Phys. Lett.* **B232**, 113, (1989). doi: 10.1016/ 0370-2693(89)90566-2.

35. M. Neubert, *Phys. Rept.* **245**, 259, (1994).

36. A. V. Manohar, *Phys. Rev.* **D56**, 230, (1997).

37. N. Brambilla, A. Pineda, J. Soto, and A. Vairo, *Rev. Mod. Phys.* **77**, 1423, (2005). doi: 10.1103/RevModPhys.77.1423.

38. A. Pineda and J. Soto, *Phys. Lett.* **B420**, 391, (1998).

39. A. Pineda and J. Soto, *Phys. Rev.* **D59**, 016005, (1999).

40. A. Pineda and J. Soto, *Nucl. Phys. Proc. Suppl.* **64**, 428, (1998).

41. N. Brambilla, A. Pineda, J. Soto, and A. Vairo, *Nucl. Phys.* **B566**, 275, (2000).

42. M. E. Luke, A. V. Manohar, and I. Z. Rothstein, *Phys. Rev.* **D61**, 074025, (2000). doi: 10.1103/PhysRevD.61.074025.

43. A. H. Hoang, Heavy quarkonium dynamics, (2002). arXiv:hep-ph/0204299.

44. T. Appelquist, M. Dine, and I. J. Muzinich, *Phys. Rev.* **D17**, 2074, (1978).

45. N. Brambilla, A. Pineda, J. Soto, and A. Vairo, *Phys. Rev.* **D60**, 091502, (1999).

46. A. Pineda and J. Soto, *Phys. Lett.* **B495**, 323, (2000).

47. N. Brambilla, X. Garcia i Tormo, J. Soto, and A. Vairo, *Phys. Lett.* **B647**, 185, (2007). doi: 10.1016/j.physletb.2007.02.015.

48. N. Brambilla, D. Gromes, and A. Vairo, *Phys. Rev.* **D64**, 076010, (2001).

49. N. Brambilla, D. Gromes, and A. Vairo, *Phys. Lett.* **B576**, 314, (2003).

50. Y. Koma and M. Koma, *Nucl. Phys.* **B769**, 79, (2007). doi: 10.1016/j. nuclphysb.2007.01.033.

51. Y. Koma, M. Koma, and H. Wittig, Relativistic corrections to the static potential at $O(1/m)$ and $O(1/m^2)$, *PoS.* **LAT2007**, 111, (2007).

52. A. Vairo, *Eur. Phys. J.* **A31**, 728, (2007). doi: 10.1140/epja/i2006-10200-0.

53. J. Soto, *Eur. Phys. J.* **A31**, 705, (2007). doi: 10.1140/epja/i2006-10255-9.

54. A. Vairo, *Int. J. Mod. Phys.* **A22**, 5481, (2007). doi: 10.1142/ S0217751X07038748.

55. N. Brambilla, NRQCD and quarkonia, (2007). arXiv:hep-ph/0702105.

56. A. H. Hoang, Top threshold physics, *PoS.* **TOP2006**, 032, (2006).

57. A. Pineda and A. Signer, *Nucl. Phys.* **B762**, 67, (2007). doi: 10.1016/j. nuclphysb.2006.09.025.

58. S. G. Karshenboim, *Int. J. Mod. Phys.* **A19**, 3879, (2004). doi: 10.1142/ S0217751X04020142.

59. A. A. Penin, *Int. J. Mod. Phys.* **A19**, 3897, (2004). doi: 10.1142/ S0217751X04020154.

60. J. Gasser, V. E. Lyubovitskij, and A. Rusetsky, *Phys. Rept.* **456**, 167, (2008). doi: 10.1016/j.physrep.2007.09.006.
61. S. R. Beane, P. F. Bedaque, W. C. Haxton, D. R. Phillips and M. J. Savage, "From hadrons to nuclei: Crossing the border," arXiv:nucl-th/0008064.
62. E. Epelbaum, H.-W. Hammer, and U.-G. Meissner, Modern theory of nuclear forces, *to appear in Rev. Mod. Phys.*, (2008).
63. N. Brambilla, J. Ghiglieri, A. Vairo, and P. Petreczky, *Phys. Rev.* **D78**, 014017, (2008). doi: 10.1103/PhysRevD.78.014017.
64. M. A. Escobedo and J. Soto, Non-relativistic bound states at finite temperature (I): the hydrogen atom, (2008). arXiv:0804.0691.
65. M. Laine, How to compute the thermal quarkonium spectral function from first principles?, (2008). arXiv:0810.1112.
66. A. H. Hoang, A. V. Manohar, I. W. Stewart and T. Teubner, *Phys. Rev.* **D65** (2002) 014014. arXiv:hep-ph/0107144.
67. A. H. Hoang, *Acta Phys. Polon.* **B34** (2003) 4491. arXiv:hep-ph/0310301.

Chapter 3

Distended/Diminished Topologically Massive Electrodynamics

Stanley Deser[1]

Physics Department, Brandeis University, Waltham, MA 02454 and Lauritsen Laboratory, California Institute of Technology, Pasadena, CA 91125, USA
E-mail: deser@brandeis.edu

We extend topologically massive electrodynamics, both by adding a higher derivative action to cast the entire three-term model in Chern-Simons (CS) form, and by embedding it in an AdS background. It can then be written as the sum of two CS terms, one of which vanishes at the "chiral" point, in analogy with its gravitational topologically massive counterpart. Separately, we shrink the model to pure CS electrodynamics, interacting with Einstein gravity and point massive charged sources.

3.1. Introduction

Theories involving Chern-Simons (CS) terms have remained popular ever since their introduction, in both gravitational and vector incarnations, over a quarter-century ago [1]. Most recently, there has been a great deal of work on extension of the original models from flat to AntideSitter (AdS) backgrounds; for an incomplete list, see [2,3]. Separately, standard topologically massive electrodynamics (TME) has been augmented by higher derivative, but CS-like, terms [4]. Here, we consider two "opposite" variants: First, we combine these two extensions of TME to express it as a "2-CS" chiral sum. Second, we drop the Maxwell part of the action, and consider the CS term alone, but now interacting with Einstein gravity, constituting a different sort of pure CS model, in presence of point massive charges. In both sectors, the field equations exhibit a field-current identity and are entirely

---

[1] Dedicated to the memory of Wolfgang Kummer, intrepid explorer of (a somewhat) lower dimension. This paper is also dedicated to the living: for Roman Jackiw on his 70th birthday.

decoupled, so the total system remains soluble. Accordingly, the resulting metric and vector potential have conical structure.

3.2. E(xtended)TME

Gravitational (tensor) and electrodynamical (vector) field models are often both similar and different; this is also true of their CS properties. The most relevant difference is that the gravitational CS actions is of third derivative order - higher than the Einstein action - whereas it is the opposite for vectors, whose CS term is first order, lower than the Maxwell action's. This is reflected in constructing their "pure CS" extensions, requiring respectively cosmological/higher derivative additions to the topologically massive two-term models. However, even for TME, as we shall see, adding a cosmological background simulates TMG in this context. For simplicity, we consider only the abelian vector case.

We begin with the vector CS action,

$$I_{CS}(B) = \int d^3x \, \epsilon^{\mu\nu\alpha} B_\mu \partial_\nu B_\alpha \,. \tag{3.1}$$

The resulting field equation,

$$F^\mu(B) \equiv \frac{1}{2}\epsilon^{\mu\nu\alpha} F_{\nu\alpha}(B) = 0 \tag{3.2}$$

states that field space is "flat," with a pure gauge vector potential. Next, generalize B_μ to be a combination

$$B_\mu^\pm(A) \equiv m^{-1/2} F_\mu(A) \pm m^{1/2} A_\mu \tag{3.3}$$

of the fundamental variable A_μ. The parameter m has dimensions of mass, needed to give A_μ its canonical dimension. We have also allowed for separate combinations $B_\pm$ which could be further generalized by allowing for two separate mass values, $m_\pm$. Our conventions are $(-++)$ signature, $\epsilon^{012} = 1$; the background is (initially) flat.

The action (3.1) with $B(A)$ as in (3.3) consists of three terms,

$$I_{CS}^\pm\left(B^\pm(A)\right) = m^{-1} \int d^3x \, \epsilon^{\mu\nu\alpha} \left\{ F_\mu(A)\partial_\nu F_\alpha(A) + m^2 A_\mu\partial_\nu A_\alpha \right\}$$
$$\pm \int d^3x \, F_{\mu\nu}^2(A)$$
$$\equiv \{m^{-1}I_{ECS} + mI_{CS}\} \mp 4I_{MAX} \,. \tag{3.4}$$

This result confirms for spin 1 that "everything is CS" in $D = 3$; even the Maxwell action is the difference of two CS terms. The two-mass generalization would permit even more flexibility in the relative coefficients.

The dynamics of the various three-term models in (3.4) was analyzed in [4], whose results we summarize for completeness. Pure I_{ECS} leads to a null-propagating field strength, $\Box F_{\mu\nu} = 0$, and hence does allow excitations, distinct from Maxwell's where $F^{\mu\nu}$ is of course also divergence-free. This model differs from pure I_{CS} in not being topological: for example, its action is metric - dependent and more fundamentally, it shows no interesting large gauge behavior, owing to the pure field-strength dependence, even in the nonabelian version. The combination $I_{ECS} + I_{MAX}$ also differs from the original TME: it contains a massive ghost excitation as well as the photon mode. Combining $I_{ECS} + I_{CS}$ does not add further excitations to that of I_{ECS} alone: instead, the field strength now propagates massively. Finally, the full three-term action depends on the two relative internal coefficients, and there are in general three masses, though there can be a degeneracy for suitable tuning. In all cases a ghost is unavoidable.

The above results are easily checked explicitly from the field equations, with the usual decompositions of the potentials into invariant, pure gauge and constraint components: since each term is separately gauge invariant, all excitations are as well, and depend only on the transverse vector potential, effectively the indexless scalar in $A_i^T = \epsilon^{ij}\partial_j S$.

3.3. C(osmological)ETME

We now introduce a nontrivial - AdS - gravitational background. The relevant aspect of this generalization is the appearance of a second dimensional parameter, the cosmological constant, $\Lambda \equiv -\ell^{-2}$.

Let us modify our previous discussion to follow the gravitational "2-CS" formulation of [2][a]. There, the variable corresponding to $B^{\pm}$ of (3.3) is a very similar combination of $B_\mu^{\pm}$, namely $\omega_\mu^{ab}(e) \pm \ell^{-1}\epsilon^{abc}e_{\mu c}$, where $\omega(e)$ is the spin connection constructed from the dreibein $e_{\mu c}$. Note the required inverse length, which we mostly set to unity. In this fashion, we get two gravitational CS combinations

$$I_{\pm}[\omega(e) \pm e] = I_{GCS}[\omega(e)] \mp I_{GR}[e], \qquad (3.5)$$

where $I_{GCS} \sim \frac{1}{2}\int d^3x(\epsilon\omega\partial\omega + \ldots)$, is the (third derivative) gravitational CS term, I_{GR} is the Einstein action including the cosmological term (pro-

---
[a]This was also found by D. Grumiller and R. Jackiw (unpublished.)

portional to Λ) but with the "wrong" sign required by TMG to ensure ghost freedom. To construct the cosmological topologically massive action, a mass parameter m—distinct from ℓ^{-1}—is introduced by hand to yield, from (3.5),

$$2I_{CTMG} = (1 + m^{-1})I^- + (1 - m^{-1})I^+$$
$$= -\int d^3x \sqrt{-g}(R - \Lambda) + m^{-1}I_{GCS} \qquad (3.6)$$

in Planck units. This CS doublet degenerates to a single term at either "chiral" value $m = \pm 1$.

Returning to our vector case, we define the extended variable to be the $m = \ell^{-1}$ value of (3.3),

$$B_\mu^\pm(A) = f_\mu(A) \pm A_\mu . \qquad (3.7)$$

[The other effect of the nontrivial background, say $g_{\mu\nu} = \phi^2 \eta_{\mu\nu}$, is that $f_\mu(A)$ is here a covariant vector, like A_μ so it acquires a factor ϕ^{-1} . Hence I_{ECS} is scaled by ϕ^{-2}, while $I_{MAX} \sim \phi^{-1}$ and of course I_{CS} is metric-independent. These extra factors are not directly relevant to our discussion.] The analog of (3.4) is simply obtained by replacing m by ℓ^{-1} there. Consequently, the CETME action is the combination

$$8I = (4m\ell - 1)I^+ + (4m\ell + 1)I^- \qquad (3.8)$$

where m is the mass parameter of TME, and we have restored ℓ explicitly. This parallels the gravitational form (3.6) except for the dimensionally dictated $m \to \frac{1}{m}$ there. This is the 3-term analog of TMG, and all three terms must be present. The $m \to 0$ limit is of course Maxwell, but ordinary 2-term TME is obtained only in the singular $\ell \to 0$ limit, while for gravity, it is the infinite mass limit that yields the (cosmological) Einstein action.

At the chiral points $4m\ell = \pm 1$, one of the actions vanishes, exactly as for chiral gravity. The physics of ordinary two-term CTME at the chiral point is laid out in [2], where it is shown to be in one-to-one correspondence with linearized CTMG at the latter's chiral point.

3.4. Sources

So far, we have studied our models in a source-free context. We now include sources, in the diminished pure CS plus gravity context.

It is instructive to first analyze the relevant similarities to - and differences from - the gravitational case. Recall that for spin two, the highest, third derivative, term is the gravitational CS action; its variation is the Cotton - conformal curvature - tensor, whose vanishing implies the metric is conformally flat. The Einstein action instead, effectively resembles that of pure vector CS: in both cases, their variations are the respective "curvature terms," whose vanishing implies field flatness. [The Maxwell term has no gravitational analog since it does describe a single physical excitation.] It is therefore really the Einstein and vector CS terms that corresponded most closely in the two systems. In each case, there is a field - current identity, respectively

$$G_{\mu\nu} = T_{\mu\nu} \tag{3.9}$$

$$F^\mu = j^\mu \tag{3.10}$$

where the Einstein tensor in (3.9) equivalent (being its double dual) to the full curvature, so that spacetime is flat away from sources, and there is no interaction among localized masses [5]. Similar considerations hold in presence of a Λ term, except that the exterior now has constant curvature [6]. The same holds for the field strength in (3.10), and non-interaction among charges. Note the counterintuitive property of (3.10) that charges create magnetic, while currents create electric, fields: F^0 is the magnetic field $\epsilon^{ij} F_{ij}$, while F^i is the electric field $\epsilon^{ij} E_j$. Point charges are represented by a current

$$j^\mu = \Sigma e_A u_A^\mu(t) \delta^2(\mathbf{r}_\mu - \mathbf{r}_A(t)) \,. \tag{3.11}$$

Note that j^μ is actually a metric-independent contravariant vector density just like F^μ, so the tensor and vector equations are totally independent. Current conservation alone requires the particle worldlines to be continuous (albeit not necessarily future timelike), while covariant conservation constrains any point-like stress tensor to be that of (a sum of) standard particles [7]. As in gravity, while there is no interaction, the large-scale "geometry" is affected by the configurations: in gravity these are the well-known metrics with conical singularities at the sources, together with their boosted generalizations, as discussed in [5], and similarly, as we now see, for the vector potentials.[b]

The simplest case is a single static charge at the origin,

$$j^0 = e\delta^2(\mathbf{r}) \tag{3.12}$$

---

[b]A more detailed perspective on CS electrodynamics with point charges may be found in [8].

which generates a pure magnetic field, $\epsilon^{ij}F_{ij} \sim e\delta^2(\mathbf{r})$ whose vector potential is

$$A_i = \frac{-e}{2\pi}\epsilon^{ij}\partial_j \ln r + \partial_i \alpha. \qquad (3.13)$$

Clearly, the potential is a superposition of such contributions if there are more static particles. Note that there is no self-force here, since the interaction term $\int A_\mu j^\mu$ vanishes identically. The configuration can be sampled through its Aharonov-Bohm phase, proportional to the sum of the charges. A moving particle will generate an electric field as well; for a single source,

$$F^i = \epsilon^{ij}E_j = \epsilon^{ij}\dot{A}_j = eu^i\delta^2(\mathbf{r} - \mathbf{r}(t)). \qquad (3.14)$$

This corresponds to a time-dependent vector potential $A_i(t)$ in $A_0 = 0$ gauge, with step-function behavior.

The Einstein + CS + particle system is now easy to solve; as noted, the vector CS term, being topological, is metric-independent, as is the particle's j^μ, so the combined field equations decouple,

$$G^{\mu\nu} = mu^\mu u^\nu \delta^2(\mathbf{r}), \qquad F^\mu = eu^\mu \delta^2(\mathbf{r}) \qquad (3.15)$$

and reduce, for the single static charge, with $u^\mu = \delta^{0\mu}$, to the usual conical space with deficit angle proportional to the source's mass, but independent of any charge properties, and a "conical" vector potential proportional to the total charge but independent of mass, as described by (3.13). The extension to superposition of several static particles is immediate, though there are interesting global geometric complications and limitations on the mass - and perhaps also (color) charge - parameters, and even more for moving particles, despite the absence of true dynamics. Irrespective of the details of generic, distributed, interior sources, the exterior fields are those of a single particle carrying the total mass and charge values.

3.5. Summary

We have discussed two separate problems: the primary one was to obtain a "pure CS" formulation of vector models in $D = 3$ to include the Maxwell action. This required addition of a third-derivative CS-like term. In an AdS background, the same procedure further allowed for a two-CS formulation using the freedom afforded by presence of two mass parameters (m, ℓ^{-1}). Here as in gravity, at either special "chiral" point, one of the two CS terms vanishes. This is also the common point for which TMG and TME equations can be put into one-one correspondence.

Our second topic was that of "pure CS" in the literal sense of keeping only the CS vector term along with its corresponding gravitational term, namely the Einstein action (with or without Λ). This two-field system was coupled to charged point masses. Because the two fields are entirely decoupled (CS being topological), the resulting configurations are separate conical metric and vector potential "spaces", with (known) interesting geometric complications in the gravitational sector. The nonabelian vector side should also prove of interest.

Acknowledgments

I thank my collaborators on [2], S. Carlip, A. Waldron, and D. Wise, whose insights there have also been useful here. This work was supported by NSF grant PHY 07-57190 and DOE grant DE-FG02-92ER40701.

References

1. S. Deser, R. Jackiw, and S.Templeton, *Ann.Phys.*, **140**, 372, 1982; *Phys.Rev.Lett.*, **48**, 975, 1982.
2. S. Carlip, S. Deser, A. Waldron, and D. Wise, *Class. Quant. Grav.* **26**, 078008 (2009). hep-th 0803.3998; 0807.0486, *Phys.Lett.*, **B666**, 272, 2008.
3. P. Kraus and F. Larsen, hep-th/0508218,*JHEP* **0601**, 22, 2006;
   S.N. Solodukhin, hep-th/0509148 . *Phys.Rev.*, **D74**, 024015, 2006; W. Li, W. Song, and A. Strominger, hep-th 0801.4566;
   A. Strominger, hep-th 0808.0506;
   D. Grumiller and N. Johansson, hep-th 0805.2610;
   W. Li, W. Song, and A. Strominger, hep-th 0805.3101;
   P. Baekler, E.W. Mielke, and F.W. Hehl,*Nuovo Cim.* **B107**, 91, 1992;
   S. Cacciatori, M. Caldarelli, A. Giacomini, D. Klemm, and D.S. Mansi, hep-th/0507200,*J. Geom. Phys.*, **56**, 2523, 2006;
   S. Carlip, hep-th 0807.4152;
   D. Grumiller, R. Jackiw, and N. Johansson, hep-th 0806.4185, this Volume;
   G. Giribet, M. Kleban, and M. Porrati, hep-th 0807.4703v2.
   M. Henneaux, C. Martinez, and R. Troncoso, hep-th/0901.2874.
4. S. Deser and R. Jackiw, hep-th/9901125, *Phys. Lett.* **B451**, 73, 1999.
5. S. Deser, R. Jackiw, and G. t'Hooft, *Ann. Phys.*, **152**, 220, 1984.
6. S. Deser and R. Jackiw, *Ann. Phys.*, **153**, 405, 1984.
7. W. Tulczyew, *Acta. Phys Polon.* **18**, 393, 1971;
   C. Aragone and S. Deser, *Nucl. Phys.*, **B92**, 327, 1975;
   M. Gurses and F. Gursey, *Phys. Rev.*, **D11**, 967, 1975.
8. R. Jackiw, *Ann. Phys.*, **201**, 83, 1990.

Chapter 4

Dynamical Spin

Peter G. O. Freund

Enrico Fermi Institute and Department of Physics, University of Chicago, Chicago, IL 60637, USA
E-mail: `freund@theory.uchicago.edu`

The possibility of building all particles form spinless constituents is explored. Composite fermions are formed from bosonic carriers of electric and magnetic charge of a composite abelian gauge field. Internal attributes are accounted for by dimensional reduction from a higher-dimensional space-time in which the abelian gauge field is replaced by a composite higher-rank antisymmetric tensor field. The problem of building magnetically neutral fermions is considered.

4.1. Dedication

It is with great sadness that I dedicate this paper to the memory of my friend Wolfgang Kummer. Wolfgang and I met during our student days in Vienna, and as fate would have it, we both went to Geneva for our first post-doctoral appointment. There we collaborated on the paper "The phases of the proton's electromagnetic form factors in the time-like region", Nuovo Cimento **24**, 1160 (1962).

During some work Wolfgang and I did in those early computer-era days, we were asked to use the computing facilities sparingly. For numerical integration we therefore availed ourselves of the services of that marvelous one-of-a-kind CERN employee, Mr. Klein. This Mr. Klein could perform complex arithmetic operations in his head. A Holocaust survivor, he had used this unusual ability to earn his living in a circus in the postwar years. There he was spotted and recruited by CERN. Mr. Klein calculated our integrals to such accuracy, that in the end, the computer's streamlined task was reduced to not much more than confirming his estimates.

I cannot resist mentioning here another bond between Wolfgang and me: since we both grew up in Viennas, he in the Austrian capital and I in the Romanian city of Timişoara, the Habsburg Empire's former "Little Vienna," we both developed keen musical interests, we both sang, we both were baritones, and we both gave recitals with mezzo-sopranos. Unlike me though, Wolfgang had as his singing partner none other than that wonderful mezzo-soprano Helga Dernesch, who after singing with him was destined to become a major star of the Wiener Staatsoper.

To this memorial volume I decided to contribute a paper I wrote in 1981, which however may still be of some interest today. It has been available on SPIRES (EFI-81/07-CHICAGO, Feb 1981) for over a quarter of a century, but due to all kinds of complications it has never been published before. I chose this paper, because I clearly recall a discussion with Wolfgang about the ideas contained in it. In fact, his interest in this paper prompted me to give him the preprint of the original 1981 version reproduced below without any changes.

4.2. Introduction

The search for a simple way of accounting for the observed particle spectrum and interactions has led to ever more remote constituent and subconstituent models.[1] In order to account for the observed fermions it is usually assumed that some or all of the constituents are themselves fermions, and thus carry half-odd-integer spin. Here we wish to explore the opposite case where none of the constituents carry spin so that all angular momentum is of dynamical origin. Spinless bosons can bind into bosonic states of integer angular momentum. If amongst these components there are gauge bosons, then we have the possibility of nontrivial topological objects that carry magnetic charge. Together with electrically charged objects we then have the ingredients to build spinorial fermions.[2,3] For such a picture to make even remote phenomenological sense, a considerable "attribute" (i.e., flavor, color, etc) proliferation at the level of the electrically and magnetically charged constituents seems to be required. An elegant way to avoid such a proliferation is provided by higher-dimensional Kaluza-type theories.[4] Yet the idea of building fermions from electric and magnetic charges relies heavily on a 4-dimensional space-time. To extend this idea to higher dimensions we propose to replace the abelian vector gauge field of 4 dimensions by gauge fields of higher (totally antisymmetric) tensorial rank.[5] The corresponding carriers of electric and magnetic charge are then not point particles but extended objects, as we shall see. Since abelian structures are natural in this context, the nonabelian gauge fields of electroweak and strong interactions are to be viewed as composites. Both Bose and Fermi composites being possible, dynamical supersymmetry may also arise.

4.3. Magnetic charges in higher dimensional spaces

Consider a Minkowski space M_d with one time- and $d-1$ space-dimensions. Define over M_d a rank-n antisymmetric tensor potential $A_{\mu_1....\mu_n}$ $(n \leq d-1)$ or, equivalently, the n-form $A = A_{\mu_1....\mu_n} dx^{\mu_1} \wedge dx^{\mu_2} \wedge ...dx^{\mu_n}$. The field strengths are the components of the n+1-form $F = dA$. It's dual $*F$ is a $d-n-1$ form. Introducing the "electric" current n-form J and the "magnetic" current $d-n-2$-form K, the field equations are

$$d * F = *J, \quad dF = *K.$$

The n-form J can be restricted to "live" on a $(d_e + 1)$--dimensional submanifold of M_d, provided $d_e + 1 \geq n$. We shall consider here the "minimal" case $d_e + 1 = n$, and specifically that one of the dimensions of the submanifold is time-like (a proper-time) and d_e are space-like. At any proper-time the support of the electric charge is then d_e-dimensional. Similarly, the support of magnetic charge has at least $d_m = d - n - 3$ dimensions. Notice that

$$d_e + d_m = d - 4, \tag{1}$$

so that both pointlike electric and magnetic charges are possible only in 4-dimensions. In general $d_e \neq d_m$, but in every even dimension there exists an electric-magnetic-dual case in which F and $*F$ are both $\frac{d}{2}$ forms, so that $n = \frac{d}{2} - 1$ and $d_e = d_m = \frac{d-4}{2}$. It is this electric-magnetic-dual case that interests us here.

At this point we want to make precise what we mean by an electric or a magnetic field configuration and to find the counterparts of the Coulomb-electric and Dirac-magnetic (monopole) potentials. To this effect we first consider a static configuration such that at all times the support of J is the $(\frac{d}{2} - 1)$-hyperplane (our results are obviously generalizable to other J-supports)

$$x^1 = x^2 = ... = x^{\frac{d}{2}+1} = 0. \tag{2}$$

Here it is worthwhile to streamline our notation. The last coordinate x^d is designated as time, the metric signature is thus $(-...-+)$. Indices that range from 1 to $\frac{d}{2} + 1$ (from $\frac{d}{2} + 2$ to d) will be designated by letters from the beginning (middle) of the latin alphabet $a, b, c, ...(m, n, p, ...)$. Thus, e.g., the hyperplane equation (2) becomes $x^a = 0$. A set of totally antisymmetrized indices of either type will be indicated in a generic way by a square bracket containing one of them. Specifically, $[a]$ means $a_1 a_2 ... a_{\frac{d}{2}-1}$ with all a's ranging from 1 to $\frac{d}{2}+1$, and $[m]$ means $m_1 m_2 ... m_{\frac{d}{2}-1}$ with all m's ranging from $\frac{d}{2}+2$ to d. Finally, the Levi-Cività symbol for the first $\frac{d}{2}+1$ (last $\frac{d}{2}-1$) indices will be written as $\epsilon_{[a]bc}$ ($\epsilon_{[m]}$).

With this notation the only nonvanishing components of J in our static situation (2) are given by

$$J_{[m]} = \epsilon_{[m]} \frac{e}{\Omega_{\frac{d}{2}}} \delta(x^1)...\delta(x^{\frac{d}{2}+1}) \tag{3a}$$

where $\Omega_{\frac{d}{2}}$ is the $\frac{d}{2}$-dimensional total solid angle (area of unit $\frac{d}{2}$-sphere: $\Omega_2 = 4\pi....$). The field equations then yield

$$A_{[m]} = \frac{e}{r^{\frac{d}{2}-1}} (\frac{-2}{d-2}), \tag{3b}$$

with

$$r^2 = (x^1)^2 + (x^1)^2 + ...(x^{\frac{d}{2}+1})^2 \tag{3c}$$

The nonvanishing field components are all "electric" and of the form

$$E_a = F_{a[m]} = \frac{e x^a \epsilon_{[m]}}{r^{\frac{d}{2}+1}} \tag{3d}$$

independent of time and of the last $\frac{d-4}{2}$ space coordinates, as expected. The equations (3) define a Coulomb-electric field configuration. A Dirac-magnetic configuration with support in the same hyperplane requires a structure of K of the same type as Eq. (3a) for J but with the "electric charge" e replaced by the "magnetic charge" g. For the magnetic field

$$H_a = \frac{1}{\frac{d}{2}!} \epsilon_{ab[c]} F_{b[c]} = \frac{1}{(\frac{d}{2}-1)!} \epsilon_{ab[c]} \partial_b A_{[c]} \tag{4}$$

we require it to be of the same form as the Coulomb field (3d) but with $e \to g$:

$$\frac{1}{(\frac{d}{2}-1)!} \epsilon_{ab[c]} \partial_b A_{[c]} = \frac{g x_a}{r^{\frac{d}{2}+1}}. \tag{5}$$

We now have to solve these equations for $A_{[a]}$. As in the familiar 4-dimensional Dirac case, the Bianchi identities force us to introduce a string of singularities starting in each point of the support of K. For convenience we point all these strings along, say, the 3-direction. The proper Ansatz for $A_{[a]}$ is then

$$A_{[a]} = \epsilon_{[a]3b}x^b f(r,\xi), \quad \xi = \frac{x^3}{r}. \tag{6a}$$

Inserting this Ansatz into Eq. (5) we find

$$f(r,\xi) = r^{-\frac{d}{2}} F(\xi) \tag{6b}$$

with $F(\xi)$ obeying the differential equation

$$F'(\xi) - \frac{d}{2}\frac{\xi}{1-\xi^2}F(\xi) + \frac{g}{1-\xi^2} = 0. \tag{7}$$

Since $|\xi| \equiv |\frac{x^3}{r}| \leq 1$, it is convenient to introduce the variable

$$\theta = Arccos\xi \tag{6c}$$

and the function

$$G(\theta) \equiv F(\xi) \tag{6d}$$

The solution to Eq. (7) is then

$$G(\theta) = g(sin\theta)^{-\frac{d}{2}}[\int^\theta (sin\psi)^{\frac{d-2}{2}}d\psi + \lambda] \tag{6e}$$

with λ an integration constant that goes with the indefinite integral. The equations (6) determine the Dirac potentials. As an example for $d = 4$ we obtain the familiar Dirac result with the string along the positive (negative) 3-axis for $\lambda = -1$ ($\lambda = +1$). From the familiar recursion formula for the indefinite integral in (6e), $G(\theta)$ is periodic in θ for d an integer multiple of 4. For $d = 2$ (*mod* 4) the indefinite integral in $G(\theta)$ contains also a linear term in θ which can be brought to the main determination $0 < \theta < \pi$ by readjusting the integration constant λ.

At this point we have to consider some global problems. As defined above, the support of both electric and magnetic charges for even $d > 4$ are infinite $\frac{d-4}{2}$-dimensional hyperplanes, which is undesirable. But if the higher dimensions are to be unobservable, then $d - 4$ space-like dimensions must have compact topology (e.g. a torus). But as we saw, $d_e + d_m = d - 4$, so that an electric-magnetic charge pair can always fit into the "extra" compact space-like dimensions.

From the electric and magnetic charges in d dimensions we can construct spinorial fermions. One way to see that, is to replicate the Tamm-Fierz [2] arguments for our spread-out charges. Heuristically, upon dimensional reduction (i.e., compactification of the $d - 4$ dimensions in which the charges are extended) the $\frac{d}{2} - 1$-tensor field contains ordinary 4-dimensional abelian gauge fields. Spinors can then be constructed from Bose electric and magnetic charges in the usual way [2,3]. But these 4-dimensional spinors must originate in d-dimensional spinors; they cannot come from d-dimensional tensors by dimensional reduction. Both e and g have dimension of (d-dimensional action)$^{\frac{1}{2}}$ so that the ensuing d-dimensional Dirac quantization is meaningful.

4.4. Composite picture

The way to use the arguments above to construct composite models is as follows. Suppose one starts with a d-dimensional space-time d = even integer larger than 4. In this space there exist a set of scalar fields for which one can build a composite ($\frac{d}{2} - 1$)-rank antisymmetric tensor field or, alternatively, this field can be "elementary". There can further appear electric and magnetic extended objects ϵ and μ and the corresponding anti-objects $\bar{\epsilon}$ and $\bar{\mu}$. From $\epsilon\mu$, $\epsilon\bar{\mu}$, $\bar{\epsilon}\mu$, $\bar{\epsilon}\bar{\mu}$ one can construct spinor composites, from $\epsilon\bar{\epsilon}$, $\mu\bar{\mu}$ tensor composites. With suitable dynamics these composites may exhibit a "dynamical" supersymmetry. If d is large this may involve higher rank tensors. A dimensional reduction is precipitated one way or another[4] and in four dimensions we have a proliferation of composites since each spinor and tensor from d dimensions branches into many counterparts in 4 dimensions (just as in extended supergravities). In 4 dimensions the spectrum is very rich, the simplicity is restored in d dimensions. This picture is, of course, very similar to extended supergravity except that the gauged supersymmetry in the original d dimensions is viewed as dynamical, thus allowing higher rank tensors and spin-tensors, or higher spins in 4 dimensions. As it stands, this picture has a serious flaw: all fermions $\epsilon\mu$, $\epsilon\bar{\mu}$, $\bar{\epsilon}\mu$, $\bar{\epsilon}\bar{\mu}$ contain one unit each of electric and magnetic charge. All fermions of the theory must contain an odd number of these basic fermions and, as such, must carry odd, and therefore non-vanishing, electric *and* magnetic charges. Even though we have not as yet specified the detailed nature of the abelian gauge field in 4 dimensions whose sources these charges are, this is a serious difficulty.

We want to sketch here one possible way out. Consider (in four-dimensional space-time) a spherical shell of uniformly distributed electric charge. Classically this tends to explode and the Casimir effect is known to have the wrong sign[6] and thus does not stabilize the configuration. It has been noted recently by Agostinho Ferreira, Zimerman and Ruggiero[7] that in a distribution of both electric and magnetic charge along a spherical shell the Casimir effect is stabilizing. Specifically, they consider a spherical shell that is a perfect magnetic conductor at its polar caps, and a perfect electric conductor on the "ring" between these caps: On the ring is uniformly distributed the electric charge while the two polar caps support uniform distributions of magnetic charge $\tilde{g}$ and $-\tilde{g}$ respectively, so that the whole system is magnetically neutral. Here the Casimir effect is stabilizing. We observe that for this system the angular momentum does not vanish as it would, were the magnetic charges at the two polar caps to have the same sign. By adjusting e and $\tilde{g}$, we can fix the total angular momentum at $\frac{\hbar}{2}$, as would befit a spinor (as a model for the electron such a semiclassical argument requires much too large a size). One may object that each polar cap contributes to the total angular momentum, which violates angular momentum quantization (or equivalently, the Dirac quantization). This can be circumvented by postulating that such "polar caps" can never be isolated, but must always come in like- or opposite-charged pairs, as if they were doublets of a confining $SU(2)$ gauge theory. This is similar to what would happen in discussing usual Dirac quantization were one guaranteed that all magnetically charged particles are composites made of an even number of very closely bound inseparable constituents of equal magnetic charge. Obviously then, the Dirac quantization for "monopoles" would translate into a quantization for the constituents.

The challenge is now to construct a detailed model that implements the ideas presented above.

Following the completion of this work I received a Trieste preprint IC/80/180 from J.C. Pati, A. Salam and J. Strathdee, in which similar ideas are explored in a rather different way.

Note (August 13, 2008):
The results reported in section 2 of this paper have also been obtained independently by R. Nepomechie[8] and by C. Teitelboim[9] (see also P. Orland[10]).

The paper by Pati, Salam and Strathdee mentioned in the last sentence of the text has since appeared,.[11]

The 1981 work reported here was supported in part by the NSF: Grant No. PH-4-78-23669. I wish to thank Beth An Nakatsuka for her help with the retyping of the 1981 preprint.

References

1. See A. Salam, in *Proceedings International Conference on HEP Geneva, 1979*, CERN, Geneva 1979 Vol 2, p. 853.
2. I.E. Tamm, *Zs. F. Physik* **71** (1931 141; M. Fierz, Helv. Phys. Acta **17** (1944) 27.
3. R. Jackiw and C. Rebbi, *Phys. Rev. Lett.* **36** (1976) 116; G. P. Hasenfratz and G. 't Hooft, *Phys. Rev. Lett.* **36** (1976) 1119; A.S. Goldhaber, Phys. Rev. Lett **36** (1976) 1122.
4. B. de Witt, *Dynamical theories of groups and fields* (Gordon and Breach, New York, 1965) p. 139; P. Kerner, *Ann. Inst. H. Poincare* **9** (1968) 143; A. Trautman, Rep. Math. Phys. **1** (1970) 39; Y.M. Cho and P.G.O. Freund, *Phys. Rev.* **D12** (1975) 1711.
5. M. Kalb and P. Ramond, *Phys. Rev.* **D9** (1974) 2273.
6. T. Boyer, *Phys. Rev.* **174** (1968) 1754.
7. L. Agostinho Ferreira, A. H. Zimerman, J.R. Ruggiero, Sao Paulo Preprint IFT-P-15/80.
8. R. Nepomechie, *Phys. Rev.***D31** (1985) 1921.
9. C. Teitelboim, *Phys. Lett.* **B167** (1986) 69.
10. P. Orland, *Nucl. Phys.* **205[FS5]** (1982) 107.
11. J.C. Pati, A. Salam and J.A. Strathdee, *Nucl. Phys.* **B185** (1981) 416.

Chapter 5

Quantum Corrections to Solitons and BPS Saturation

A. Rebhan[1], P. van Nieuwenhuizen[2] and R. Wimmer[3]

[1] *Institut für Theoretische Physik*
Technische Universität Wien, A-1040 Wien, Austria

[2] *C.N. Yang Institute for Theoretical Physics*
Stony Brook University, Stony Brook, NY 11794-3840, USA

[3] *Laboratoire de Physique, ENS Lyon,*
46 allée d'Italie, F-69364 Lyon CEDEX 07, France

We review our work of the past decade on one-loop quantum corrections to the mass M and central charge Z of solitons in supersymmetric field theories: the kink, the vortex, and the monopoles (focussing on the kink and the monopoles here). In each case a new feature was needed to obtain BPS saturation: a new anomaly-like contribution to Z for the kink and the $N = 2$ monopole, the effect of classical winding of the quantum vortex contributing to Z, surface terms contributing to M of the $N = 4$ monopole and to Z of the $N = 2$ and $N = 4$ monopoles, and composite operator renormalization for the currents of the "finite" $N = 4$ model. We use dimensional regularization, modified to preserve susy and be applicable to solitons, and suitable renormalization conditions. In the mode expansion of bosonic and fermionic quantum fields, zero modes appear then as massless nonzero modes.

5.1. Introduction

In the beginning of the 1970's particle physicists became interested in solitons. Since Dirac's work on the quantization of the electromagnetic field in the late 1920's, particles had been associated with the Fourier modes

---

[1] E-mail: rebhana@tph.tuwien.ac.at

[2] E-mail: vannieu@insti.physics.sunysb.edu

[3] E-mail: robert.wimmer@ens-lyon.fr

of the second-quantized fields, and perturbation theory had been used to compute scattering amplitudes. However, for the strong interactions this approach could not be used because the coupling constant is larger than unity, and nonlinearities are essential. Thus particle physicists turned to solitons as representations of particles in strongly interacting field theories. This changed the emphasis from properties of scattering amplitudes of two or more solitons to properties of single solitons.[1,2]

Also in the early 1970's, the renormalizability of nonabelian gauge theory was proven, and supersymmetry (susy) was discovered. A natural question that arose was: are nonabelian gauge theories (and abelian gauge theories) with solitons also renormalizable? In susy theories some divergences cancel, so it seemed interesting to extend the theories with solitons to susy theories with the same solitons, and to study whether cancellations of radiative corrections did occur. In particular, the mass of a soliton gets corrections from the sum over zero-point energies of bosons and fermions. A formal proof had been constructed that in susy theories the sum of all zero-point energies cancels,[3] and it was conjectured that also the corrections to the mass of a soliton vanish in susy theories.[4] We shall see that this is an oversimplification, and that the mass of solitons already receives corrections at the one-loop level.

In addition to susy, also topology became a major area of interest in soliton physics. In 1973 Nielsen and Olesen[5] used the vortex solution, which is a soliton in $2+1$ dimensions based on the abelian Higgs model with a complex scalar field, to construct topologically stable extended particles. Ginzberg and Landau[6] had used this model to describe superconductivity in 1950, and Abrikosov[7] had found the vortex solution in 1957. Nielsen and Olesen embedded this vortex solution into $3+1$ dimensions, and obtained in this way stringlike excitation of the dual resonance model of particle physics with a magnetic field confined inside the tubes. 't Hooft wondered if their construction could be extended to non-abelian Higgs models, and in 1974 he[8] and Polyakov[9] discovered that the nonabelian Higgs model in $3+1$ dimensions with gauge group SU(2) and a real triplet of Higgs scalars contains monopoles, which are solitonic solutions with a magnetic charge. They contain a topological number, the winding number, which prevents them from decaying to the trivial vacuum. Similarly, the vortex solution in $2+1$ dimensions has a winding number, and even the kink (a soliton in $1+1$ dimensions[10–12]) is topologically stable. There exist also nontopological solitons[13] but we shall not discuss them. In 1975 Julia and Zee constructed dyons,[14] solitons in the SU(2) nonabelian Higgs model with an electric and a magnetic charge, and soon afterwards Prasad and Sommerfeld[15] found exact expressions for these solitons in the limit of vanishing $\lambda \varphi^4$ coupling constant (the PS limit). In 1976 Bogomolnyi[16] showed that in all these cases of topological solitons one can write the energy density as a sum of squares plus total derivatives. Requiring these squares to vanish leads to first-order differential equations for solitons, the Bogomolnyi equations, which are much easier to solve than the second-order field equations. He also noted that the total energy has a bound $H \geq |Z|$ where Z is the contribution from the total derivatives. This bound is called the BPS bound because for monopoles in the nonabelian Higgs model it can only be saturated for vanishing coupling constant λ. For a classical soliton at rest, H is equal to its mass M, and $M = |Z|$. Finally in 1978 Olive and Witten[17] noted that the total derivative terms in the Bogomolnyi expression for the energy density are the central charges of the susy algebra of the corresponding susy theories. These charges are Heisenberg operators, containing all perturbative and nonperturbative quantum corrections. By using results of the representation theory of superalgebras in terms of physical states, they proved that for topological solitons the BPS bound $M \geq |Z|$ must remain saturated at the quantum level: $M = |Z|$.

We shall calculate the one-loop corrections to M and Z, and show that they are indeed equal, but nonvanishing for susy kinks[18-21] and the $N = 2$ monopole.[22] These calculation are not meant as a check of the proof of Olive and Witten, but rather they are a test of whether our understanding of quantum field theory in the presence of solitons has progressed enough to obtain saturation of the BPS bound. As we shall see, this is a nontrivial issue. The vacuum expectation values of the Higgs scalars acquire local corrections in the presence of solitons, boundary terms contribute to M and Z, a new anomaly-like contribution to Z yields a finite correction, and composite operators require infinite renormalization to obtain a finite answer in the "finite" $N = 4$ susy model. We shall use the background field formalism to formulate background-covariant R_ξ gauges, and we shall use the extended Atiyah-Singer-Patodi[23-26] index theorem for noncompact spaces to calculate the sum over zero-point energies in the presence of solitons. We shall also introduce an extension of dimensional regularization which preserves susy and can be used for solitons.[27,a]

As we have discussed, solitons were initially proposed for describing hadrons, but when duality between electric and magnetic fields, and extended dualities in supersymmetric field theories, were developed, another

---

[a]Dimensional regularization in the context of (bosonic) solitons was employed before in Refs. 28, 29.

point of view emerged. It was conjectured by Montonen and Olive,[30] and Witten[31] that there exist dual formulations of field theories in which particles become solitons, and solitons become particles. Modern work in string theory has confirmed and extended this hypothesis in an amazing way.

5.2. The simplest case: the susy kink and its "new anomaly"

In order to test one's understanding of a quantum field theory, static quantities should be among the first to consider. In the following we shall consider two static quantities in one of the simplest quantum field theories with a soliton: the mass and the central charge of the susy kink at the one-loop level. This exercise has proved to be a surprisingly subtle topic with all kinds of pitfalls. Even when the same renormalization conditions were employed, different regularization methods led to contradictory results,[32-38] and this confusing state of matters lasted until the end of the 1990's, when the question was reopened by a work by two of us,[39] in which it was shown that the methods used to produce the most widely accepted result of zero corrections in the susy case were inconsistent with the known integrability of the bosonic sine-Gordon model.[10] Subsequently, the pitfalls of the various methods were sorted out,[18-21,27,40-44] which involved the discovery of an anomalous contribution to the central charge guaranteeing BPS saturation. In the following, we shall show how all this works out using dimensional regularization adapted to susy solitons.

5.2.1. Mass

The mass of a soliton is obtained by taking the expectation value of the Hamiltonian with respect to the ground state in the soliton sector[b]. In addition one needs the contribution from counter terms which are needed to renormalize the model.

As Hamiltonian we take the gravitational Hamiltonian (obtained by varying the action with respect to an external gravitational field). We write all fields $\varphi(x,t)$ as a sum of (static) background fields $\varphi_b(x)$ and quantum fields $\eta(x,t)$, and only retain all terms quadratic in quantum fields. For real bosonic fields the Hamiltonian density of the quantum fields is of the

---

[b]This state is often called the soliton vacuum, but this is a misnomer because vacua have by definition vanishing energy while the soliton has a nonvanishing mass. The vacuum is the state with vanishing energy in the sector without winding, but to avoid misunderstanding, we shall consistently call it the trivial vacuum.

form

$$\mathcal{H} = \frac{1}{2}\dot{\eta}\dot{\eta} + \frac{1}{2}\partial_x\eta\partial_x\eta + \cdots \qquad (5.1)$$

and using partial integration yields $\frac{1}{2}\partial_x\eta\partial_x\eta + \cdots = \partial_x(\frac{1}{2}\eta\partial_x\eta) - \frac{1}{2}\eta(\partial_x^2\eta + \cdots)$. The terms $-\frac{1}{2}\eta(\partial_x^2\eta + \cdots)$ are then equal to $-\frac{1}{2}\eta\ddot{\eta}$ if one uses the linearized field equations for the fluctuations, and the expectation value $-\frac{1}{2}\langle soliton|\eta\ddot{\eta}|soliton\rangle$ is equal to $+\frac{1}{2}\langle soliton|\dot{\eta}\dot{\eta}|soliton\rangle$. For a real (Majorana) fermion there are no background fields, and the Hamiltonian density is of the form $\mathcal{H} = \frac{1}{2}\bar{\psi}\gamma^1\partial_x\psi + \dots$. Again using the linearized field equations for the fermion's fluctuations, one finds $\mathcal{H} = \frac{i}{2}\psi^\dagger\dot{\psi}$ since $\bar{\psi} = \psi^\dagger i\gamma^0$ and $(\gamma^0)^2 = -1$. Thus the one-loop quantum corrections to the mass of a soliton, $M = \langle soliton| \int \mathcal{H}dx|soliton\rangle$, are of the generic form

$$M^{(1)} = \int \, [\langle\dot{\eta}\dot{\eta}\rangle + \frac{i}{2}\langle\psi^\dagger\dot{\psi}\rangle]dx$$

$$+ \text{ boundary terms } \int \partial_x(\frac{1}{2}\langle\eta\partial_x\eta\rangle)$$

$$+ \text{ counter terms } \Delta M. \qquad (5.2)$$

To define the infinite and finite parts of the one-loop corrections, we need a regularization scheme that preserves susy and is easy to work with. This singles out dimensional regularization. Usually one needs dimensional regularization by dimensional reduction to preserve susy, but that option is not available to us because the soliton occupies all space dimensions. Going up in dimensions in general violates susy, but there is a way around these objections which combines the virtues of both approaches. In all cases we consider, the susy action in $D + 1$ dimensions can be rewritten as a susy action in $D + 2$ dimensions. Then going down in dimensions, we use in $(D + \epsilon) + 1$ dimensions standard dimensional regularization. This scheme clearly preserves susy, and it leaves enough space for the soliton.

Let us see how things work out for the susy kink. The susy action (after eliminating the susy auxiliary field) is given by

$$\mathcal{L} = \frac{1}{2}\dot{\varphi}^2 - \frac{1}{2}(\partial_x\varphi)^2 - \frac{1}{2}U^2 - \frac{1}{2}\bar{\psi}\partial\!\!\!/\psi - \frac{1}{2}U'\bar{\psi}\psi,$$

$$\frac{1}{2}U^2 = \frac{\lambda}{4}(\varphi^2 - \mu_0^2/\lambda)^2, \qquad (5.3)$$

where ψ is a 2-component Majorana spinor and φ a real scalar field. This model has $N = (1, 1)$ susy in 1+1 dimensions, but the same expression for $\mathcal{L}$ can also be viewed as an $N = 1$ model in 2+1 dimensions. The operator

$\not{\partial}$ in the Dirac action and in the transformation law $\delta\psi = (\not{\partial}\varphi - U)\epsilon$ is then given by $\gamma^0\partial_0\varphi + \gamma^1\partial_x\varphi + \gamma^2\partial_y\varphi$.

The energy density obtained from the gravitational stress tensor reads

$$\mathcal{H} = \frac{1}{2}\dot{\varphi}^2 + \frac{1}{2}(\partial_k\varphi)^2 + \frac{1}{2}U^2 + \frac{1}{2}\bar{\psi}\gamma^k\partial_k\psi + \frac{1}{2}U'\bar{\psi}\psi \,, \quad k = 1, 2. \quad (5.4)$$

For the classical soliton solution we set $\dot{\varphi} = \psi = 0$ and denote φ_b by φ_K. The classical mass of the kink follows from the Bogomolnyi way of writing the classical Hamiltonian as a sum of squares plus a boundary term

$$\mathcal{H} = \frac{1}{2}(\partial_x\varphi_K)^2 + \frac{1}{2}U_K^2$$
$$= \frac{1}{2}(\partial_x\varphi_K + U_K)^2 - (\partial_x\varphi_K)U_K, \quad (5.5)$$

where $U_K = U(\varphi_K)$. Thus the classical field equation for the soliton reads $\partial_x\varphi_K + U_K = 0$, and the classical mass is

$$M_{cl} = -\int_{-\infty}^{+\infty} dx\, \partial_x\Big[\int_0^{\varphi_K(x)} U(\varphi')d\varphi'\Big] = \frac{2\sqrt{2}\mu_0^3}{3\lambda}. \quad (5.6)$$

The kink solution is given by $\varphi_K(x) = \frac{\mu}{\sqrt{\lambda}}\tanh\frac{\mu x}{\sqrt{2}}$ (where μ is the normalized mass introduced below), but we shall not need this. Decomposing φ into $\varphi_K(x) + \eta(x, y, t)$, we find for the terms quadratic in quantum fluctuations

$$\mathcal{H}^{(2)} = \frac{1}{2}\dot{\eta}\dot{\eta} + \frac{1}{2}(\partial_k\eta)(\partial_k\eta) + \frac{1}{2}(\frac{1}{2}U_K^2)''\eta\eta + \frac{1}{2}\bar{\psi}\gamma^k\partial_k\psi + \frac{1}{2}U_K'\bar{\psi}\psi, \quad (5.7)$$

where $k = 1, 2$. Partial integration, use of the linearized field equations for quantum fields η and ψ, and substitution of $\bar{\psi}\gamma^0 = -i\psi^\dagger$ yields

$$\mathcal{H}^{(2)} = \frac{1}{2}\dot{\eta}\dot{\eta} + \frac{1}{2}\partial_k(\eta\partial_k\eta) - \frac{1}{2}\eta\ddot{\eta} + \frac{i}{2}\psi^\dagger\dot{\psi}$$
$$\langle\mathcal{H}^{(2)}\rangle = \langle\dot{\eta}\dot{\eta} + \frac{1}{2}\partial_k(\eta\partial_k\eta) + \frac{i}{2}\psi^\dagger\dot{\psi}\rangle. \quad (5.8)$$

(We shall later choose a real (Majorana) representation for the Dirac matrices, and then ψ is real, thus $\psi^\dagger = \psi^T$.)

To renormalize the field theory with quantum fields η and ψ, one considers the trivial vacuum, and chooses as background field $\varphi_b = \frac{\mu}{\sqrt{\lambda}}$. There are terms with 2, 3, and 4 quantum fields, and the terms with three η's, or one η and two ψ's, can give a tadpole loop which is divergent and needs renormalization. We therefore decompose the bare mass μ_0^2 into a renormalized part μ^2 and a counter term $\Delta\mu^2$, and require that $\Delta\mu^2$ cancels all (finite as well as infinite) contributions of the tadpoles. The bosonic loop yields $\Delta\mu^2 = 3\lambda\langle\eta^2\rangle$ while the fermionic loop yields $\Delta\mu^2 = -2\lambda\langle\eta^2\rangle$

$$\mu_0^2 = \mu^2 + \Delta\mu^2,$$

$$\Delta\mu^2 = \lambda \int \frac{d^{2+\epsilon}k}{(2\pi)^{2+\epsilon}} \int \frac{-i}{k^2 + m^2 - i\epsilon}$$

$$= \lambda \int \frac{d^{1+\epsilon}k}{(2\pi)^{1+\epsilon}} \frac{1}{2\sqrt{k^2 + m^2}}.$$

No further renormalizations are needed, so the Z factors for λ, η and ψ are all unity. This is a particular set of renormalization conditions.

Having fixed $\Delta\mu^2$ in the trivial sector, we now return to the kink sector and find for the mass counter term at the one-loop level

$$\Delta M = \left(\frac{2\sqrt{2}}{3\lambda}\right)(\mu^2 + \Delta\mu^2)^{\frac{3}{2}} - M_{cl} = \frac{m\Delta\mu^2}{\lambda} \; ; \; m = \sqrt{2}\mu. \qquad (5.9)$$

We must now evaluate the terms in (5.8). We do this by expanding η and ψ into modes, but as we shall see, the sum $\langle\dot{\eta}\dot{\eta} + i\psi^\dagger\dot{\psi}\rangle$ can also be extracted from an index theorem.

The field equation for $\eta(x, y, t) = \phi(x)e^{ily}e^{-i\omega t}$ reads

$$-\partial_x^2\phi + (\tfrac{1}{2}U_K^2)''\phi = (\omega^2 - l^2)\phi. \qquad (5.10)$$

Actually, the field operator $-\partial_x^2 + (\frac{1}{2}U_K^2)''$ factorizes into $(-\partial_x + m\tanh\frac{mx}{2})(\partial_x + m\tanh\frac{mx}{2}) \equiv L_1^\dagger L_2$, and this allows explicit expressions for the zero mode $\phi_0(x)$ (satisfying $[-\partial_x^2 + (\frac{1}{2}U_K^2)'']\phi_0 = 0$, with $\omega_0^2 = 0$), the bound state $\phi_B(x)$ (with $\omega_B^2 = -\frac{3}{4}m^2$), and the continuous spectrum $\phi(k, x)$ (with $\omega_k^2 = k^2 + m^2$). However, we do not need explicit expressions for these functions. The mode expansion in $1+\epsilon$ spatial dimensions reads

$$\eta(x, y, t) = \int_{-\infty}^{\infty} \frac{d^\epsilon l}{(2\pi)^{\frac{\epsilon}{2}}} \left\{ \int_{-\infty}^{\infty} \frac{dk}{\sqrt{2\pi}} \frac{1}{\sqrt{2\omega_{kl}}} (a_{kl}\phi(k, x)e^{ily}e^{-i\omega_{kl}t} \right.$$

$$+ a_{kl}^\dagger \phi^*(k, x)e^{-ily}e^{i\omega_{kl}t})$$

$$+ \frac{1}{\sqrt{2\omega_{Bl}}}(a_{Bl}\phi_B(x)e^{ily}e^{-i\omega_{Bl}t} + a_{Bl}^\dagger\phi_B(x)e^{-ily}e^{i\omega_{Bl}t})$$

$$\left. + \frac{1}{\sqrt{2\omega_{0l}}}(a_{0l}\phi_0(x)e^{ily}e^{-i|l|t} + a_{0l}^\dagger\phi_0(x)e^{-ily}e^{i|l|t}) \right\}, \qquad (5.11)$$

where $\omega_{kl}^2 = k^2 + l^2 + m^2$, $\omega_{Bl}^2 = -\frac{3}{4}m^2 + l^2$, and $\omega_{0l}^2 = l^2$. The annihilation and creation operators a and $a^\dagger$ satisfy the usual commutation relations, for example $[a_{0l}, a_{0l'}^{\dagger}] = \delta^{\epsilon}(l - l')$. The functions $\phi_B(x)$ and $\phi_0(x)$ are normalized to unity, while the distorted plane waves $\phi(k, x)$ are normalized such that they become plain waves $e^{i(kx + \frac{1}{2}\delta(k))}$ for $x \to +\infty$ and $e^{i(kx - \frac{1}{2}\delta(k))}$ for $x \to -\infty$, satisfying the completeness relation[c]

$$\int_{-\infty}^{\infty} (|\phi(k, x)|^2 - 1) \frac{dk}{2\pi} + \phi_B^2(x) + \phi_0^2(x) = 0. \tag{5.12}$$

For the fermion we use a real (Majorana) representation of the Dirac matrices γ^{μ} which diagonalizes the iterated field equations in 2+1 dimensions

$$\gamma^1 = \begin{pmatrix} 1 & 0 \\ 0 & -1 \end{pmatrix}, \ \gamma^0 = \begin{pmatrix} 0 & -1 \\ 1 & 0 \end{pmatrix}, \ \gamma^2 = \begin{pmatrix} 0 & 1 \\ 1 & 0 \end{pmatrix} \tag{5.13}$$

and also makes ψ real. The field equation

$$(\partial\!\!\!/ + U_K')\psi = 0 \ ; \ U_K' = m \tanh \frac{mx}{2} \tag{5.14}$$

reads then in component form

$$\left. \begin{array}{l} (\partial_x + U_K')\psi_+ = (\partial_0 - \partial_y)\psi_- \\ (\partial_x - U_K')\psi_- = (\partial_0 + \partial_y)\psi_+ \end{array} \right\} \ \psi = \begin{pmatrix} \psi_+ \\ \psi_- \end{pmatrix} \tag{5.15}$$

The iterated field equation of ψ_+ is the same as the η field equation, while for ψ_- we find the conjugate field operator

$$\left. \begin{array}{l} (L_2^{\dagger}L_2 - \partial_y^2 + \partial_0^2)(\eta \text{ or } \psi_+) = 0 \\ (L_2 L_2^{\dagger} - \partial_y^2 + \partial_0^2)\psi_- = 0 \end{array} \right\} \ \begin{array}{l} L_2 = \partial_x + U_K' \\ L_2^{\dagger} = -\partial_x + U_K' \end{array} \tag{5.16}$$

Setting $\psi_{\pm} = \psi_{\pm}(x)e^{ily - i\omega t}$, the Dirac equation yields

$$\psi_-(k, x) = \frac{i(\partial_x + U_K')}{\omega_{kl} + l}\psi_+(k, x) \ ; \ \omega_{kl}^2 = k^2 + l^2 + m^2 \tag{5.17}$$

---

[c]The completeness relation[42] reads $\int \phi(k, x)\phi^*(k, x')\frac{dk}{2\pi} + \phi_B(x)\phi_B(x') + \phi_0(x)\phi_0(x') = \delta(x - x')$, and has been rewritten in terms of $(\phi(k, x)\phi^*(k, x') - e^{ik(x-x')})$ by bringing the delta function to the left-hand side. It follows that $\phi(k, x)$, $\phi_B(x)$ and $\phi_0(x)$ are orthonormal, for example $\int \phi(k, x)\phi^*(k', x)dx = 2\pi\delta(k - k')$.

The mode expansion of ψ in $1+\epsilon$ spatial dimensions is then given by

$$
\psi = \begin{pmatrix} \psi_+ \\ \psi_- \end{pmatrix} = \int_{-\infty}^{\infty} \frac{d^\epsilon l}{(2\pi)^{\frac{\epsilon}{2}}} \int_{-\infty}^{\infty} \frac{dk}{\sqrt{2\pi}} \Bigg\{
$$

$$
\frac{1}{\sqrt{2\omega_{kl}}} \Bigg[b_{kl} \begin{pmatrix} \sqrt{\omega_{kl}+l}\,\phi(k,x) \\ \sqrt{\omega_{kl}-l}\,is(k,x) \end{pmatrix} e^{ily} e^{-i\omega_{kl}t}
$$

$$
+ b_{kl}^\dagger \begin{pmatrix} \sqrt{\omega_{kl}+l}\,\phi(k,x)^* \\ \sqrt{\omega_{kl}-l}(-i)s(k,x)^* \end{pmatrix} e^{-ily} e^{i\omega_{kl}t} \Bigg]
$$

$$
+ \frac{1}{\sqrt{2\omega_{Bl}}} \Bigg[b_{Bl} \begin{pmatrix} \sqrt{\omega_{Bl}+l}\,\phi_B(x) \\ \sqrt{\omega_{Bl}-l}\,is_B(x) \end{pmatrix} e^{ily} e^{-i\omega_{Bl}t}
$$

$$
+ b_{Bl}^\dagger \begin{pmatrix} \sqrt{\omega_{Bl}+l}\,\phi_B(x) \\ \sqrt{\omega_{Bl}-l}(-i)s_B(x) \end{pmatrix} e^{-ily} e^{i\omega_{Bl}t} \Bigg]
$$

$$
+ \frac{1}{\sqrt{2|l|}} \Bigg[b_{0l} \begin{pmatrix} \sqrt{|l|+l}\,\phi_0(x) \\ 0 \end{pmatrix} e^{ily} e^{-i|l|t}
$$

$$
+ b_{0l}^\dagger \begin{pmatrix} \sqrt{|l|+l}\,\phi_0(x) \\ 0 \end{pmatrix} e^{-ily} e^{i|l|t} \Bigg] \Bigg\}
$$

$$(5.18)$$

where

$$
s(k,x) = \frac{(\partial_x + U_K')\phi(k,x)}{\omega_k} \ , \quad \omega_k^2 = k^2 + m^2. \tag{5.19}
$$

Several remarks are to be made

- we have extracted the same factors $\frac{1}{\sqrt{2\omega}}$ as for the boson;
- the normalization factors $\sqrt{\omega+l}$ and $\sqrt{\omega-l}$ are needed to satisfy the equal-time canonical anticommutation relations, as we shall check,

$$
\{\psi_\pm(x,y,t), \psi_\pm(x',y',t)\} = \delta(x-x')\delta^\epsilon(y-y'),
$$
$$
\{\psi_+(x,y,t), \psi_-(x',y',t)\} = 0 \tag{5.20}
$$

- we treat zero modes and nonzero modes on equal footing. In fact, the zero modes have become massless nonzero modes at the regularized level with energy $|l|$;
- there are no zero modes for ψ_-, while the zero modes of ψ_+ have only positive momenta l in the extra dimensions, yielding massless chiral domain-wall fermions, which are right-moving on the domain wall;

- the zero mode sector can also be written as

$$\int_{-\infty}^{\infty} \frac{dl}{(2\pi)^{\frac{\epsilon}{2}}} b_{0l} \phi_0(x) e^{il(y-t)} \qquad (5.21)$$

where for positive l, b_{0l} is an annihilation operator, but for negative l a creation operator $(b_{0,-l} = b_{0,l}^{\dagger})$;

- the normalization factor $\sqrt{\omega - l}$ for s_k in ψ_- is obtained as follows: given that $\psi_+(k,x)$ is written in terms of $\sqrt{\omega + l}\phi$, multiply $\psi_-(k,x) = \frac{i(\partial_x + U_K')}{\omega + l} \psi_+(k,x)$ in the numerator and denominator by $\sqrt{\omega - l}$

$$\sqrt{\omega + l}\psi_-(k,x) = \sqrt{\omega + l} \frac{\sqrt{\omega - l}}{\sqrt{\omega - l}} \frac{i(\partial_x + U_K')}{\omega + l} \phi(k,x) \qquad (5.22)$$

$$= \sqrt{\omega - l} \frac{i(\partial_x + U_K')}{\sqrt{\omega^2 - l^2}} \phi(k,x) = \sqrt{\omega - l}\, is(k,x);$$

- the reality of ψ is manifest. One can also write the spinors in the terms with $e^{i\omega t}$ as $\sqrt{\omega + l}\phi$ and $-\sqrt{\omega - l}\, is$ since $\phi(k,x)^* = \phi(-k,x)$ and thus also $s(k,x)^* = s(-k,x)$, which corresponds to $\psi = -C\bar{\psi}^T$ where $C = i\gamma^0$ is the charge conjugation matrix. (The relation $\phi(k,x)^* = \phi(-k,x)$ follows from the reflection symmetry $x \to -x$ of the action, but one can also read it off from the explicit expression for $\phi(k,x)$.[42])

Let us check that this mode expansion for $\psi_\pm$ is correct by assuming that the annihilation and creation operators satisfy the usual anticommutators, and verifying that we obtain $\delta(x - x')\delta^\epsilon(y - y')$ and zero in (5.20). We begin with

$$\{\psi_+(x,y,t), \psi_+(x',y',t)\} = \int \frac{dk}{2\pi} \int \frac{d^\epsilon l}{(2\pi)^\epsilon} \qquad (5.23)$$

$$\left[\frac{\omega_{kl} + l}{2\omega_{kl}} \{\phi(k,x)\phi^*(k,x')e^{il(y-y')} + \phi^*(k,x)\phi(k,x')e^{-il(y-y')}\} \right.$$

$$\left. + \{\frac{\omega_{Bl} + l}{2\omega_{Bl}}\phi_B(x)\phi_B(x') + \frac{|l| + l}{2|l|}\phi_0(x)\phi_0(x')\}(e^{il(y-y')} + e^{-il(y-y')}) \right].$$

Using $\phi^*(k,x) = \phi(-k,x)$, and changing the integration variable for the terms with $\phi^*(k,x)$ from k to $-k$, we find that all terms factorize into terms with $\omega + l$ times $e^{il(y-y')} + e^{-il(y-y')}$. The factors l in $\omega + l$ cancel by symmetric integration, and then also the terms with ω cancel. All terms are now proportional to $e^{il(y-y')}$, and integration over l yields the required

$\delta^\epsilon(y - y')$. One is left with

$$\int \phi(k, x)\phi^*(k, x')\frac{dk}{2\pi} + \phi_B(x)\phi_B(x') + \phi_0(x)\phi_0(x') \qquad (5.24)$$

which is indeed equal to $\delta(x - x')$.

For the $\{\psi_-, \psi_-\}$ anticommutator there are two differences: instead of $\phi(k, x)$ one has $s(k, x)$, and there are no zero modes. One finds

$$\int s(k, x)s^*(k, x')\frac{dk}{2\pi} + s_B(x)s_B(x'). \qquad (5.25)$$

This is again equal to $\delta(x - x')$, as it is the completeness relation for $L_2 L_2^\dagger$. One can also directly check this[d]. For the $\{\psi_-, \psi_+\}$ anticommutator one finds along the same lines

$$\{\psi_+(x, y, t), \psi_-(x', y', t)\} = \int \frac{dk}{2\pi} \int \frac{d^\epsilon l}{(2\pi)^\epsilon} \qquad (5.26)$$

$$\left[\frac{\sqrt{\omega_{kl}^2 - l^2}}{2\omega_{kl}} \left\{ \phi(k, x)(-i)s^*(k, x')e^{il(y-y')} + is(k, x)\phi^*(k, x')e^{-il(y-y')} \right\} \right.$$

$$\left. + \frac{\sqrt{\omega_{Bl}^2 - l^2}}{2\omega_{Bl}} \left\{ \phi_B(x)(-i)s_B(x')e^{il(y-y')} + is_B(x)\phi_B(x')e^{il(y-y')} \right\} \right].$$

Because there are now no terms linear in l which multiply the exponents $e^{il(y-y')}$, we can change the integration variables k and l to $-k$ and $-l$ in half of the terms, and, using $\phi(k, x)^* = \phi(-k, x)$ and $s(k, x)^* = s(-k, x)$, all terms cancel.

The calculation of the one-loop mass of the susy kink is now simple. We must evaluate

$$M^{(1)} = \int dx \int d^\epsilon y \langle \dot\eta\dot\eta + \frac{i}{2}\psi^T\dot\psi \rangle + \int d^\epsilon y \frac{1}{2}\langle \eta\partial_x\eta \rangle \Big|_{x=-\infty}^{x=\infty} + \frac{m}{\lambda}\Delta\mu^2 \quad (5.27)$$

The first term gives the sum over zero-point energies

$$\int dx\, d^\epsilon y\, \langle \dot\eta\dot\eta + \frac{i}{2}\psi^T\dot\psi \rangle = V_y \int dx \int \frac{dk}{2\pi} \int \frac{d^\epsilon l}{(2\pi)^\epsilon} \qquad (5.28)$$

$$\times \frac{\omega_{kl}}{2}\left[\phi^*(k, x)\phi(k, x) - \frac{\omega_{kl} + l}{2\omega_{kl}}\phi^*(k, x)\phi(k, x) - \frac{\omega_{kl} - l}{2\omega_{kl}}s^*(k, x)s(k, x) \right]$$

$$= V_y \int dx \int \frac{dk}{2\pi} \int \frac{d^\epsilon l}{(2\pi)^\epsilon}\frac{\omega_{kl}}{4}(|\phi(k, x)|^2 - |s(k, x)|^2)$$

---

[d] Use Eqs. (9) and (10) of Ref. 42 together with Eq. (5.19) above.

where V_y is the volume $\int d^\epsilon y$ of the extra dimensions and where only contributions from the continuous spectrum have remained. There is no contribution from the bound state because $\int dx(\varphi_B^2(x) - s_B^2(x))$ vanishes (partially integrate as in (5.31), there is no boundary term because $\varphi_B(x)$ falls off exponentially fast). There is also no contribution from the zero mode because the corresponding integral $\int dk d^\epsilon l \, l^2/|l|$ is a scaleless integral, and scaleless integrals vanish in dimensional regularization. Note that the terms proportional to a single power of l (arising from the $\sqrt{\omega + l}$ and $\sqrt{\omega - l}$ in (5.18)) drop out because they are odd in the loop momentum l; in the calculation of Z these terms will give a crucial contribution. The total derivative $\int dx \frac{\partial}{\partial x} \int d^\epsilon y \langle \eta \partial_x \eta \rangle$ does not contribute because $\eta \partial_x \eta = \frac{1}{2} \partial_x(\eta\eta)$, and $\langle \eta\eta \rangle$ can only depend on x as $\frac{1}{x}$, in which case the derivative ∂_k yields $\frac{1}{x^2}$ which vanishes for large x.[e] (In $3+1$ dimensions one can get a contribution because there the measure is $4\pi r^2$).

The expression in (5.28) is what in early approaches was believed to be zero, but which is actually infinite. Combining it with the counter term contribution in (5.27), the total mass per volume V_y becomes then

$$M^{(1)} = \int_{-\infty}^{\infty} \frac{dk}{2\pi} \int \frac{d^\epsilon l}{(2\pi)^\epsilon} \frac{\omega_{kl}}{4} \Delta\rho(k^2) + \frac{m}{\lambda} \Delta\mu^2 \qquad (5.29)$$

where

$$\Delta\rho(k^2) = \int_{-\infty}^{\infty} dx(|\phi(k,x)|^2 - |s(k,x)|^2) \qquad (5.30)$$

is the difference of spectral densities of ψ_+ and ψ_-. One can use an index theorem[23–26,45] to compute $\Delta\rho(k^2)$, or one can directly calculate it, using partial integration,

$$\begin{aligned}
\int |s(k,x)|^2 dx &= \int \frac{[(\partial_x + U')\phi^*(k,x)][(\partial_x + U')\phi(k,x)]}{\omega_k^2} dx \\
&= [\frac{\phi^*(k,x)(\partial_x + U')\phi(k,x)}{\omega_k^2}]\Big|_{x=-\infty}^{x=\infty} \\
&\quad + \int_{-\infty}^{\infty} \frac{\phi^*(k,x)(-\partial_x + U')(\partial_x + U')\phi(k,x)}{\omega_k^2} dx \\
&= \frac{2m}{k^2 + m^2} + \int_{-\infty}^{\infty} dx|\phi(k,x)|^2 .
\end{aligned} \qquad (5.31)$$

We used that since $\phi^*(k,x)\partial_x\phi(k,x) = ik$ and $U' \to \pm m$ as $x \to \pm\infty$, the terms with $\phi^*(k,x)\partial_x\phi(k,x)$ cancel, while the terms with U' add. Note

---
[e] Actually, $\langle \eta\eta \rangle$ falls off even faster then $1/x$, namely exponentially fast.[20,42]

that $\Delta\rho(k^2)$ is nonvanishing, because $|\phi(k,x)|^2$ of the continuous spectrum in the second line of (5.31) does not vanish as $x \to \pm\infty$. With this result for the difference of spectral densities we obtain

$$
\begin{aligned}
M^{(1)} &= \frac{m}{2} \int \frac{dk\,d^\epsilon l}{(2\pi)^{1+\epsilon}} \left[-\frac{\sqrt{k^2+l^2+m^2}}{k^2+m^2} + \frac{1}{\sqrt{k^2+l^2+m^2}} \right] \\
&= \frac{m}{2} \int \frac{dk\,d^\epsilon l}{(2\pi)^{1+\epsilon}} \left[\frac{-l^2}{(k^2+m^2)\sqrt{k^2+l^2+m^2}} \right].
\end{aligned}
\tag{5.32}
$$

Note that the extra dimensions, needed to maintain susy at the regularized level, have produced a nonvanishing correction proportional to the square of the momentum in the extra dimensions! Using the standard formula for dimensional regularization

$$
\int \frac{d^n l}{(l^2+\mathcal{M}^2)^\alpha} = \pi^{n/2}(\mathcal{M}^2)^{\frac{n}{2}-\alpha} \frac{\Gamma(\alpha-\frac{n}{2})}{\Gamma(\alpha)},
\tag{5.33}
$$

we find for the l integral

$$
\begin{aligned}
\int \frac{d^\epsilon l\, l^2}{(l^2+\mathcal{M}^2)^{\frac{1}{2}}} &= \int \frac{d^\epsilon l}{(l^2+\mathcal{M}^2)^{-\frac{1}{2}}} - \mathcal{M}^2 \int \frac{d^\epsilon l}{(l^2+\mathcal{M}^2)^{\frac{1}{2}}} \\
&= \pi^{\frac{\epsilon}{2}}(\mathcal{M}^2)^{\frac{\epsilon}{2}+\frac{1}{2}} \left(\frac{-\epsilon}{\epsilon+1} \right) \frac{\Gamma(\frac{1}{2}-\frac{\epsilon}{2})}{\Gamma(\frac{1}{2})},
\end{aligned}
\tag{5.34}
$$

where $\mathcal{M}^2 = k^2+m^2$. We are left with the k integral

$$
\int dk(k^2+m^2)^{\frac{\epsilon}{2}-\frac{1}{2}} = -\frac{2}{\epsilon} \quad \text{for } \epsilon \to 0.
\tag{5.35}
$$

The factors ϵ and $\frac{1}{\epsilon}$ cancel, and the final result is

$$
M^{(1)} = -\frac{m}{2\pi}.
\tag{5.36}
$$

5.2.2. *Central Charge*

The central charge is one of the generators of the susy algebra. To construct the latter, we begin with the Noether current for rigid susy. If one integrates its time-component over space, one obtains the susy charge Q, but it is advantageous to postpone this integration and first evaluate the susy variation of the susy current, $\delta j^\mu = -i[j^\mu, \bar{Q}\epsilon]$.[20] Extracting ϵ, and integrating over space yields the $\{Q,Q\}$ anticommutators.

In order to regularize the quantum corrections, we first construct the $\{Q,Q\}$ anticommutators in $2+1$ dimensions, and then descend to $(1+\epsilon)+1$ dimensions. In $1+\epsilon$ dimensions, the translation generators P_y in the direction of the ϵ extra dimensions are still present, and they are added to the central charge Z_x which one naively finds in $1 + 1$ dimensions. As we shall show, $-P_y + Z_x$ is the regularized central charge. In loop calculations P_y will give a finite but nonvanishing contribution. For bosons in the loop, symmetric integration over l gives a vanishing result, but for fermions in the loop, a factor l coming from the derivative $\frac{\partial}{\partial y}$ in P_y combines with another factor l coming from the normalization factors $\sqrt{\omega + l}$ and $\sqrt{\omega - l}$ of the spinors ψ_+ and ψ_- to give a factor l^2. Integration over l yields then a nonvanishing contribution,

$$\langle P_y \rangle \sim \int \langle \psi_+ \partial_y \psi_+ + \psi_- \partial_y \psi_- \rangle \sim \int \frac{l^2}{\omega} \Delta \rho(k^2) \neq 0 \,. \qquad (5.37)$$

This result has the same form as one encounters in the calculation of the chiral triangle anomaly using dimensional regularization, namely a factor l^2 in the numerator which yields a factor n as $n \to 0$, and a divergent loop integral which gives a factor $\frac{1}{n}$. We therefore refer to the term in (5.37) as an anomaly-like contribution, or, less precisely, as an anomaly. (There is no anomaly in the conservation of the central charge current, just as there is no anomaly in the ordinary susy current, or the stress tensor, but there is an anomaly in the conservation of the *conformal* current[21]). The final result for the one-loop contributions to the regularized central charge is equal to the one-loop mass correction, and thus BPS saturation continues to hold at the quantum level.

In earlier work, not enough attention was paid to careful regularization, and extra terms, such as the occurrence of P_y, were missed. Any other regularization scheme should also lead to BPS saturation if one is careful enough. Of course, one should specify the same renormalization conditions in the calculation of $M^{(1)}$ and $Z^{(1)}$; in our case this means that we again remove tadpoles by decomposing μ_0^2 into $\mu^2 + \Delta\mu^2$. Let us now show the details for the kink.

The Noether current (in $2 + 1$ dimensions) is given by $j^\mu = -\partial\!\!\!/\varphi\gamma^\mu\psi - U\gamma^\mu\psi$, and with the representation in (5.13) we find for the two spinor components

$$j_+^0 = (\dot{\varphi} - \partial_y\varphi)\psi_+ + (\partial_x + U)\psi_-$$
$$j_-^0 = (\dot{\varphi} + \partial_y\varphi)\psi_- + (\partial_x - U)\psi_+ \qquad (5.38)$$

We can evaluate the variation of $j_\pm^0$ either by transforming the fields in $j_\pm^0$ under rigid susy transformations, or by evaluating the anticommutators with $Q_\pm = \int j_\pm^0(x', y', t)dx'dy'$. We follow the latter approach. Using the equal-time canonical (anti)commutation relations

$$[\dot\varphi(x',y',t),\varphi(x,y,t)] = \frac{1}{i}\delta(x'-x)\delta(y'-y)$$

$$\{\psi_\pm(x',y',t),\psi_\pm(x,y,t)\} = \delta(x'-x)\delta(y'-y)$$

$$\{\psi_+(x',y',t),\psi_-(x,y,t)\} = 0 \tag{5.39}$$

one finds straightforwardly, after partial integration of $\frac{\partial}{\partial x'}$ and $\frac{\partial}{\partial y'}$ derivatives,

$$\{Q_+,j_+\} = \dot\varphi^2 - 2\dot\varphi\partial_y\varphi + (\partial_y\varphi)^2 + (\partial_x\varphi)^2 + 2U\partial_x\varphi + U^2$$
$$-2i\psi_+\partial_y\psi_+ - 2iU\psi_+\psi_- + i\psi_+\partial_x\psi_- + i\psi_-\partial_x\psi_+ \tag{5.40}$$

The right-hand side can be written in terms of the densities of the Hamiltonian, translation generator P_y, and naive central charge Z_x as follows

$$\{Q_+,j_+\} = 2\mathcal{H} - 2\mathcal{P}_y + 2\mathcal{Z}_x$$
$$2\mathcal{H} = \dot\varphi^2 + (\partial_y\varphi)^2 + (\partial_x\varphi)^2 + U^2 - 2iU\psi_+\psi_-$$
$$+i\psi_+\partial_x\psi_- + i\psi_-\partial_x\psi_+ - i\psi_+\partial_y\psi_+ + i\psi_-\partial_y\psi_-$$
$$-2\mathcal{P}_y = -2\dot\varphi\partial_y\varphi - i\psi_+\partial_y\psi_+ - i\psi_-\partial_y\psi_-$$
$$2\mathcal{Z}_x = 2U\partial_x\varphi \tag{5.41}$$

The sum of the last term of $2\mathcal{H}$ and $-2\mathcal{P}_y$ cancels, but we have added these terms to obtain the complete expressions for $\mathcal{H}$ and $\mathcal{P}_y$. One can check that $\mathcal{H}$ and $\mathcal{P}_y$ generate the correct time- and space- translations of φ, $\dot\varphi$, ψ_+, and ψ_-. The other susy anticommutators are given by

$$\{Q_-,j_-\} = 2\mathcal{H} + 2\mathcal{P}_y - 2\mathcal{Z}_x \tag{5.42}$$
$$\{Q_+,j_-\} = 2\mathcal{P}_x + 2\mathcal{Z}_y \tag{5.43}$$

where $2\mathcal{Z}_y = 2U\partial_y\varphi$.

Integrating over x and y, and using two-component spinors we obtain

$$\frac{1}{2}\{Q,Q\} = -(\gamma^\mu\gamma^0)P_\mu + (\gamma^2\gamma^0)Z_x - (\gamma^1\gamma^0)Z_y \tag{5.44}$$

where $P_0 = H$, and this clearly demonstrates that $Z_x - P_y$ and $P_x + Z_y$ are the regulated versions of Z_x and P_x, respectively.

The naive central charge Z_x receives no quantum corrections. This was observed by several authors. To demonstrate this, we expand $\varphi = \varphi_K + \eta$

and $\mu_0^2 = \mu^2 + \Delta\mu^2$, and find to second order in η

$$\mathcal{Z}_x = U\partial_x\varphi = \partial_x\Big(\int^\varphi U(\varphi')d\varphi'\Big)$$

$$= U_K\partial_x\varphi_K + \partial_x(U_K\eta) + \frac{1}{2}\partial_x(U_K'\eta^2) - \frac{\Delta\mu^2}{\sqrt{2\lambda}}\partial_x\varphi_K \quad (5.45)$$

The first term yields classical BPS saturation, since it is just minus the total derivative in (5.5). Taking the expectation value in the kink ground state, the term linear in η vanishes, and the last two terms give, after integration over x and y,

$$\langle Z_x^{(1)}\rangle = \Big[m\langle\eta^2(x\to\infty)\rangle - 2\Delta\mu^2\frac{\mu}{\sqrt{2\lambda}}\Big]V_y. \quad (5.46)$$

We used that $U_K' \to \pm m$ and $\varphi_K \to \pm\mu/\sqrt{\lambda}$ as $x \to \pm\infty$. Recalling that $\mu = m/\sqrt{2}$, and $\Delta\mu^2 = \lambda\langle\eta^2\rangle$ in the trivial vacuum, we see that $\langle Z_x^{(1)}\rangle$ vanishes. The tadpole renormalization in the trivial vacuum, and thus also far away from the kink, cancels the contribution from the naive central charge.

However, we get a nonvanishing correction from P_y. The bosonic fluctuation do not contribute

$$\langle P_y^{bos}\rangle = \int\langle\dot\eta\partial_y\eta\rangle dx d^\epsilon y \sim \int\frac{\omega l}{2\omega}|\phi(k,x)|^2 dk d^\epsilon l = 0 \quad (5.47)$$

due to symmetric integration over l. But from the fermions we get a nonvanishing contribution

$$\langle P_y^{ferm}\rangle = \int\frac{i}{2}\langle\psi_+\partial_y\psi_+ + \psi_-\partial_y\psi_-\rangle dx d^\epsilon y$$

$$= \frac{1}{2}\int\frac{dk}{2\pi}\frac{d^\epsilon l}{(2\pi)^\epsilon}\Big(\frac{l(\omega+l)|\varphi(k,x)|^2}{2\omega} + \frac{l(\omega-l)|s(k,x)|^2}{2\omega}\Big)dx d^\epsilon y$$

$$= \frac{1}{2}\int\frac{dk}{2\pi}\frac{d^\epsilon l}{(2\pi)^\epsilon}\frac{l^2}{2\omega}(|\varphi(k,x)|^2 - |s(k,x)|^2)dx d^\epsilon y \quad (5.48)$$

This is the same expression as we found for $M^{(1)}$, hence BPS saturation holds.

Repeating the same calculation for $N = 2$ susy φ^4 kinks, one finds[18,27] that BPS saturation holds without anomalous contributions from $\langle P_y\rangle$, because in these models the extra fields lead to a complete cancellation of $\Delta\rho(k^2)$. However, in the 1+1-dimensional $N = 2$ CP^1 model with so-called twisted mass term,[46] $\Delta\rho(k^2)$ is instead twice the amount found in the minimally susy kink models.[f] The appearance of an anomalous contribution in the $N = 2$ twisted-mass CP^1 model[47,48] has to do with the fact that the $N = 2$ CP^1 model provides an effective field theory for confined monopoles[49-52] of 3+1-dimensional $N = 2$ SU(2)×U(1) gauge theories, which in turn are related[53] by holomorphicity to 't Hooft-Polyakov $N = 2$ monopoles, and for the latter we shall indeed find anomalous contributions to the central charge in what follows.

5.3. Boundary terms and composite operator renormalization for supersymmetric monopoles

We now discuss susy monopoles in 3+1 dimensions, and study how BPS saturation is realized when one-loop quantum corrections are included. From what has been learned from the kink, one might expect that if one defines proper renormalization conditions and takes again into account an anomaly-like contribution to the central charge, BPS saturation will follow. This turns out to be the case for the $N = 2$ monopole and leads us to correcting once again previous results in the literature[g], but, surprisingly enough, for the $N = 4$ monopole in the "finite" $N = 4$ super Yang-Mills theory, there are divergences left in boundary contributions, and these can only be canceled, it seems, by introducing a new concept in the study of solitons, which was not necessary before: infinite composite operator renormalization of the stress tensor and the central charge current. For the $N = 2$ model, all surface contributions, which are individually divergent, cancel nicely.

Composite operator renormalization of the stress tensor and the central charge is no contradiction to the lore that "conserved currents don't renormalize", because that applies only to internal currents, not to spacetime ones. The stress tensor appearing in the susy algebra can be written as the sum of an improved stress tensor, which is traceless, and "improvement terms" corresponding to $R\varphi^2$ terms in the action in curved space. While the improved stress tensor turns out to be finite, the non-traceless part renormalizes multiplicatively in the $N = 4$ model, and just happens to be finite as well in the $N = 2$ case. Thus this new feature of composite opera-

---

[f]Another special feature of the $N = 2$ twisted-mass CP^1 model is that the nonrenormalization of $\langle Z_x \rangle$ involves fermionic boundary terms.[47]

[g]In contrast to the susy kink, the few explicit calculations of one-loop corrections to the $N = 2$ monopole were all agreeing on a null result in a minimal renormalization scheme.[54,55]

tor renormalization in the $N = 4$ model does not upset the BPS saturation
of the $N = 2$ model that was obtained previously without it. One could of
course start with the improved currents at the classical level, but this would
change the traditional value of the classical mass of the 't Hooft-Polyakov
monopole.

5.3.1. *The $N = 2$ monopole*

The action of the $N = 2$ super Yang-Mills model in 3+1 dimensions can be
obtained in a simple way by applying dimensional reduction to the action
of minimal super Yang-Mills theory in 5+1 dimensions

$$\mathcal{L} = -\frac{1}{4}F_{MN}^2 - \bar{\lambda}\Gamma^M D_M \lambda; \qquad \lambda = \begin{pmatrix} \psi \\ 0 \end{pmatrix}$$

$$= -\frac{1}{4}F_{\mu\nu}^2 - \frac{1}{2}(D_\mu P)^2 - \frac{1}{2}(D_\mu S)^2 - \frac{1}{2}g^2(S \times P)^2$$
$$-\bar{\psi}\gamma^\mu D_\mu \psi - g\bar{\psi}\gamma_5(P \times \psi) - ig\bar{\psi}(S \times \psi). \tag{5.49}$$

with ψ a 4-component complex spinor and $(S \times P)^a = \epsilon^{abc}S^b P^c$. We
decomposed A_M^a into (A_μ^a, P^a, S^a) and used a particular representation of
the Dirac matrices in $5+1$ dimensions.[22] In the topologically trivial sector
we take S^3 as the Higgs field with vev v (and S^1, S^2 the would-be Goldstone
fields). In the soliton sector, the energy density for a static configuration
with nonvanishing A_j^a and S^a can be written as

$$\mathcal{H} = \frac{1}{4}\left(F_{ij}^a + \epsilon_{ijk}D_k S^a\right)^2 - \frac{1}{2}\partial_k\left(\epsilon_{ijk}F_{ij}^a S^a\right). \tag{5.50}$$

Thus the Bogomolnyi equation for a monopole residing in A_j and S reads

$$F_{ij}^a + \epsilon_{ijk}D_k S^a = 0. \tag{5.51}$$

The asymptotic behavior of A_j and S is given by

$$A_j^a = \epsilon_{aij}\frac{x^j}{gr^2} + \dots, \qquad F_{ij}^a = -\epsilon_{ijk}\frac{x^a x^k}{gr^4} + \dots,$$

$$S^a = \frac{x^a v}{r} - \frac{x^a}{gr^2} + \dots, \qquad D_k S^a = \frac{x^a x^k}{gr^4} + \dots, \tag{5.52}$$

where the suppressed subleading terms are exponentially decreasing for
large radius r, and the classical mass of the monopole reads

$$M_{cl} = \frac{4\pi v}{g} = \frac{4\pi m}{g^2} \tag{5.53}$$

with $m = gv$.

The susy algebra can be obtained as before by varying the time component of the Noether current and afterwards integrating over space. One obtains[h]

$$\frac{1}{2}\{Q^\alpha, Q^\dagger_\beta\} = \delta^\alpha_\beta P_0 - (\gamma^k \gamma^0)^\alpha{}_\beta P_k - (\gamma_5 \gamma^0)^\alpha{}_\beta U + i(\gamma^0)^\alpha{}_\beta V \qquad (5.54)$$

where

$$U = \int d^3x\, \partial_k \left[\frac{1}{2}\epsilon^{ijk} F_{ij} \cdot S + F_{k0} \cdot P\right],$$

$$V = \int d^3x\, \partial_k \left[\frac{1}{2}\epsilon^{ijk} F_{ij} \cdot P - F_{k0} \cdot S\right], \qquad (5.55)$$

and $P_\mu = \int T_{\mu 0}\, d^3x$ so that $P_0 = H > 0$. To make contact with the usual form of the susy algebra for N-extended susy,

$$\frac{1}{2}\{Q^{Ai}, Q^{Bj}\} = \epsilon^{AB} Z^{ij}, \quad Z^{ij} = -Z^{ij} \text{ complex}$$

$$\frac{1}{2}\{Q^{Ai}, \bar{Q}_{\dot{B}j}\} = \delta^i_j (\sigma^\mu)^A{}_{\dot{B}} P_\mu, \qquad (5.56)$$

note that our complex Q^α can be written in terms of Majorana $Q^{j\alpha}$ as $Q^\alpha = (Q^1 + iQ^2)^\alpha$, and $Q^{\alpha j} = (Q^{Aj}, \bar{Q}_{\dot{A}j})$ in terms of two-component spinors. Then $Z^{12} = -Z^{21} = -U + iV$ for the $N = 2$ model, whereas we already see that for the $N = 4$ model to be discussed below there will be 6 complex (12 real) central charges. Classically only U is nonvanishing, and BPS saturation holds for the above monopole solution.[i]

For calculating quantum corrections, we use an "R_ξ" gauge-fixing term

$$\mathcal{L}_{\text{fix}} = -\frac{1}{2\xi}\left(D_M(A)a^M\right)^2 = -\frac{1}{2\xi}\left(D_\mu(A)a^\mu + gP \times p + gS \times s\right)^2. \quad (5.57)$$

We have written "R_ξ" in quotation marks because a genuine R_ξ gauge-fixing term would have a factor ξ in front of $gP \times p$ and $gS \times s$. The above form is advantageous to keep the SO(5,1) symmetry of the theory prior to dimensional reduction. We shall set $\xi = 1$ in which case the kinetic terms in the fluctuation equations become diagonal (in a genuine R_ξ gauge, this is also true for $\xi \neq 1$).

---

[h]For $\{Q^\alpha, Q^\beta\}$ one finds the integral of a total derivative of a bilinear in fermions, $\int \partial_j (\psi^T C \gamma^0 \psi) d^3x (\gamma^j C^{-1})^{\alpha\beta}$. Since $\langle \psi\psi \rangle$ vanishes, we shall omit this term from the algebra.

[i]Using a suitable representation of the Dirac matrices, the right-hand side of (5.54) takes on the form $\begin{pmatrix} P_0 + \sigma^k P_k & iU + V \\ -iU + V & P_0 - \sigma^k P_k \end{pmatrix}$. For vanishing P_k one obtains $P_0^2 \geq U^2 + V^2$, hence in general, $M^2 \geq U^2 + V^2$

The field equations for the fluctuations $a_m = \{a_i, s\}$, $i = 1, 2, 3$, read

$$\left((\partial_0^2 - \partial_5^2 - D_{\underline{\ell}}^2)\delta_{mn} - 2gF_{\underline{mn}} \times \right) a_{\underline{n}} = 0, \tag{5.58}$$

where $D_{\underline{\ell}}^{ab} = (D_i^{ab}, igS^{ab})$ with $D_i^{ab} = \partial_i \delta^{ab} + \epsilon^{acb}A_\mu^c$ and $S^{ab} = \epsilon^{acb}S^c$. They can be written in spinor notation as

$$(\bar{\slashed{D}}\slashed{D} + \partial_5^2 - \partial_0^2)\bar{\slashed{a}} = 0; \quad \bar{\slashed{D}}\slashed{D} = D_{\underline{m}}^2 + \frac{1}{2}\bar{\sigma}^{\underline{mn}}gF_{\underline{mn}}, \tag{5.59}$$

where $\bar{\slashed{a}} = \bar{\sigma}^{\underline{m}}a_{\underline{m}}$, $\bar{\sigma}^{\underline{mn}} = \frac{1}{2}(\bar{\sigma}^{\underline{m}}\sigma^{\underline{n}} - \bar{\sigma}^{\underline{n}}\sigma^{\underline{m}})$ with $\bar{\sigma}^{\underline{m}} = (\vec{\sigma}, -i\mathbf{1})$ and $\sigma^{\underline{m}} = (\vec{\sigma}, i\mathbf{1})$ in the 4-dimensional Euclidean space labeled by the index $\underline{m}$. Furthermore,

$$(\slashed{D}\bar{\slashed{D}} + \partial_5^2 - \partial_0^2)\slashed{q} = 0; \quad \slashed{D}\bar{\slashed{D}} = D_{\underline{m}}^2 \tag{5.60}$$

for the remaining quartet of bosonic fields $q_m = (a_0, p, b, c)$, where b, c are Faddeev-Popov ghost fields.

For the spinors we find

$$\slashed{D}\psi_+ = (\partial_0 - \partial_5)\psi_- \quad , \quad \bar{\slashed{D}}\psi_- = (\partial_0 + \partial_5)\psi_+ \ , \tag{5.61}$$

where

$$\slashed{D}^{ab} = \sigma^{\underline{m}}D_{\underline{m}}^{ab} = \sigma^k D_k^{ab} + igS^{ab}, \quad \bar{\slashed{D}}^{ab} = \bar{\sigma}^{\underline{m}}D_{\underline{m}}^{ab} = \sigma^k D_k^{ab} - igS^{ab}. \tag{5.62}$$

Iterating (5.61) we have

$$\bar{\slashed{D}}\slashed{D}\,\psi_+ = (\partial_0^2 - \partial_5^2)\psi_+ \ , \quad \slashed{D}\bar{\slashed{D}}\,\psi_- = (\partial_0^2 - \partial_5^2)\psi_- \ , \tag{5.63}$$

so the two columns of $\bar{\slashed{a}}$ have the same field equations as ψ_+, and the two columns of $\slashed{q}$ have the same field equations as ψ_-, the analogous situation as we found for the susy kink.

One can now construct the gravitational stress tensor $T_{\mu\nu}$ and consider the terms in the Hamiltonian density T_{00} which are quadratic in quantum fields. For the bosons, there are terms of the form $\partial a \partial a$ and terms of the form $a\partial^2 a$. Partially integrating the former, we can use the field equations for the fluctuations to obtain the following result[45]

$$M^{1-\mathrm{loop}} = \int d^3x \langle a_0 \partial_0^2 a_0 - a_j \partial_0^2 a_j - p\partial_0^2 p - s\partial_0^2 s - b\partial_0^2 c - (\partial_0^2 b)c$$

$$+ \frac{i}{2}\psi^\dagger \overleftrightarrow{\partial}_0 \psi \rangle$$

$$+ \lim_{r\to\infty} \frac{1}{4}4\pi r^2 \frac{\partial}{\partial r}\langle a_0^2 + a_j^2 + p^2 + s^2 + 2bc - \frac{2}{3}a_j^2 \rangle, \tag{5.64}$$

where we used that $\partial_j \partial_k \langle a_j a_k \rangle = \frac{1}{3}\partial_k^2 \langle a_j^2 \rangle$ and $\langle a_j a_0 \rangle = 0$ for large r.

The bulk contributions give the sum over zero-point energies of all quantum fields. Fermions have the mode expansion

$$\psi = \begin{pmatrix} \psi_+ \\ i\psi_- \end{pmatrix} = \int \frac{d^\epsilon \ell}{(2\pi)^{\epsilon/2}} \int \frac{d^3 k}{(2\pi)^{3/2}} \frac{1}{\sqrt{2\omega}} \sum \left\{ b_{kl} e^{-i(\omega t - \ell y)} \begin{pmatrix} \sqrt{\omega + \ell} \, \chi_k^+ \\ -\sqrt{\omega - \ell} \, \chi_k^- \end{pmatrix} \right.$$

$$\left. + d_{kl}^\dagger e^{i(\omega t - \ell y)} \begin{pmatrix} \sqrt{\omega + \ell} \, \chi_k^+ \\ \sqrt{\omega - \ell} \, \chi_k^- \end{pmatrix} \right\} + \text{bound states} + \text{zero modes} , \quad (5.65)$$

where the sum refers to the two possible polarizations of the χ_k's. On the other hand, the mode expansion of the bosonic fields a_j and s (which we combined into ϕ) only involves χ_k^+ and the one of the quartet q only χ_k^-. This leads to

$$M^{(1)\text{bulk}} = V_y \int d^3 x \int \frac{d^3 k \, d^\epsilon \ell}{(2\pi)^{3+\epsilon}} \frac{\omega}{2} (\mathcal{N}_+ |\chi_k^+|^2(x) + \mathcal{N}_- |\chi_k^-|^2(x)) \quad (5.66)$$

where $\mathcal{N}_+ = 4 - 2$ from a_m and ψ_+, and $\mathcal{N}_- = 1 + 1 - 1 - 1 - 2$ from $q_m = (a_0, p, b, c)$ and ψ_-. The result thus involves only a difference of spectral densities which can be evaluated by an index theorem[25,26,45,54,55]

$$\Delta\rho(k^2) = \int d^3 x (|\chi_k^+|^2(x) - |\chi_k^-|^2(x)) = \frac{-4\pi m}{k^2(k^2 + m^2)}. \quad (5.67)$$

On the other hand, all surface contributions (in the present $N = 2$ case) cancel,[45]

$$M^{(1)\text{surface}} = \lim_{r\to\infty} \frac{1}{4} 4\pi r^2 \frac{\partial}{\partial r} \langle a_0^2 + a_j^2 + p^2 + s^2 + 2bc - \frac{2}{3} a_j^2 \rangle$$

$$= (-1 + 3 + 1 + 1 - 2 - 2) \lim_{r\to\infty} \pi r^2 \frac{\partial}{\partial r} \langle s^2 \rangle = 0, \quad (5.68)$$

where we have used that the propagators of all bosonic fields become the same for large r, since only terms of order $1/r$ can contribute, whereas $F_{\mu\nu}^a$ falls off as $1/r^2$. (The contribution of $\langle a_0^2 \rangle$ is minus $\langle s^2 \rangle$ because of the metric $\eta^{\mu\nu}$ in the canonical commutation relations of the creating and annihilation operators.) Hence, $M^{(1)\text{bulk}}$ is the complete, but still unrenormalized, one-loop result.

The momentum integral that we are left with to evaluate upon insertion of (5.67) into (5.66) is UV divergent, and the required counter term ΔM comes from the renormalization of $M_{\text{cl.}} = 4\pi v_0/g_0$. We clearly need Z_g and Z_v. In the background field formalism which we have been using, one has the well-known relation $Z_g = Z_A^{-1/2}$, so we could first determine Z_A by requiring that all loops with two external background fields A_μ^3 are cancelled at zero external momentum (the quantum fields in these loops are all massive, so there are no IR problems). We did this in Ref. 45 even at arbitrary ξ, but one can get Z_g also from the known one-loop formula of the β-function

$$Z_g = 1 - g^2 \{ \frac{11}{3} - \frac{2}{3} n_{\text{Maj.ferm.}} - \frac{1}{6} n_{\text{real scalars}} \} C_2(\text{SU}(2)) \frac{I}{2}$$
$$= 1 - 2g^2 I \quad \text{for } N = 2; \quad \text{but } Z_g = 1 \text{ for } N = 4, \qquad (5.69)$$

where

$$I \equiv \int \frac{d^{4+\epsilon}k}{(2\pi)^{4+\epsilon}} \frac{-i}{(k^2 + m^2)^2} = \int \frac{d^{4+\epsilon}k_E}{(2\pi)^{4+\epsilon}} \frac{1}{(k_E^2 + m^2)^2} = -\frac{1}{8\pi^2} \frac{1}{\epsilon} + O(\epsilon^0). \qquad (5.70)$$

The value of Z_v is equal to Z_S, because constant v's are a special case of arbitrary background fields S^3 (more precisely, in the trivial sector only the combination $v + S^3$ occurs). At $\xi = 1$, the value of Z_S is equal to Z_A because all relevant background-field vertices are contained in $a_M D_m^2 a^M$ which is SO(5,1) invariant. (At $\xi \neq 1$, Z_S becomes ξ-dependent, while Z_A is ξ-independent, because it is given by the β-function.) Since $Z_g Z_S^{1/2} = 1$, the mass $m = gv$ does not renormalize (at $\xi = 1$), and thus

$$\Delta M = \frac{4\pi m}{Z_g^2 g^2} - \frac{4\pi m}{g^2} = \frac{4\pi m}{g^2} 4g^2 I = 16\pi m I. \qquad (5.71)$$

The mass correction to the $N = 2$ monopole is finally given by

$$M^{(1)} = 2 \int d^3 x \int \frac{d^3 k\, d^\epsilon \ell}{(2\pi)^{3+\epsilon}} \frac{\sqrt{k^2 + \ell^2 + m^2}}{2} \Delta\rho(k) + \Delta M$$
$$= -2 \frac{m}{\pi} \frac{\Gamma(-\frac{1}{2} - \frac{\epsilon}{2})}{(2\pi^{\frac{1}{2}})^\epsilon \Gamma(-\frac{1}{2})} \int_0^\infty dk (k^2 + m^2)^{-\frac{1}{2} + \frac{\epsilon}{2}} + 16\pi m I$$
$$= \left(-\frac{1}{1+\epsilon} + 1 \right) 16\pi m I = -\frac{2m}{\pi} + O(\epsilon). \qquad (5.72)$$

The one-loop corrections to the original expression in (5.55) for the central charge U of the $N = 2$ monopole cancel completely[j] against the counterterms due to ordinary renormalization, but the translation operator P_y in the extra ϵ dimensions gives again a nonvanishing "anomalous" contribution which exactly matches $M^{(1)}$, in complete analogy to the case

---

[j]The first graph above (5.88) yields a divergence $-4g^2 IU$, but wave function renormalization of S and A_μ in U yields a counterterm $4g^2 IU$. This cancellation was worked out first in Ref. 55, but it only works for $N = 2$, while $N = 4$ involves new issues[45,56] that we shall discuss below.

of the susy kink (see Eq. 5.48)[k],

$$U^{(1)} = P_y = \int \frac{d^3k \, d^\epsilon \ell}{(2\pi)^{3+\epsilon}} \frac{\ell^2}{2\sqrt{k^2 + \ell^2 + m^2}} \Delta\rho(k^2) = -\frac{2m}{\pi} + O(\epsilon). \quad (5.73)$$

Clearly, BPS saturation holds for the $N = 2$ monopole at the one-loop level. However, the finite nonvanishing correction to both the mass and the central charge had been missed in all the literature preceding Ref. 22, although closer inspection reveals that the commonly accepted null result was in conflict with the low-energy effective action of $N = 2$ super Yang-Mills theory obtained some time ago by Seiberg and Witten.[57,58]

5.3.2. *The $N = 4$ monopole*

We now turn to the monopole in $N = 4$ super Yang-Mills theory in 3+1 dimensions, where the naive expectation of vanishing one-loop corrections to mass and central charge in the end turns out to be correct. However, how this comes about is highly nontrivial, and in several ways the properties of the $N = 4$ case are opposite to the $N = 2$ case, with dramatic consequences.

We begin by following the same steps as in the $N = 2$ case. The action of $N = 4$ super Yang-Mills theory in 3+1 dimensions is most easily obtained by applying dimensional reduction to the $N = 1$ super Yang-Mills theory in 9+1 dimensions, yielding

$$\mathcal{L} = -\frac{1}{4}F_{MN}^2 - \frac{1}{2}\bar{\lambda}\Gamma^M D_M \lambda \qquad (5.74)$$

$$= -\frac{1}{4}F_{\mu\nu}^2 - \frac{1}{2}(D_\mu S_{\mathsf{j}})^2 - \frac{1}{2}(D_\mu P_{\mathsf{j}})^2 - \frac{1}{2}\bar{\lambda}^I \not{D}\lambda^I + \text{interactions}.$$

where we decomposed A_M^a into $(A_\mu^a, S_{\mathsf{j}}^a, P_{\mathsf{j}}^a)$, with 3 adjoint scalars S_{j} and 3 pseudoscalars P_{j}, $\mathsf{j} = 1, 2, 3$, instead of only one of each in the $N = 2$ case. The 16-component adjoint Majorana-Weyl spinor λ^a has been decomposed into four 4-component Majorana spinors λ^{aI} with $I = 1, \ldots, 4$, with a factor $\frac{1}{2}$ in front of their action because of the Majorana property. The

---

[k]However, in contrast to the case of the susy kink, if one combines the integral with $\Delta\rho$ in the mass correction (5.72) with the integral representation of the counter term ΔM, one does not obtain a factor ℓ^2 in the numerator as in (5.73).

susy algebra reads

$$\frac{1}{2}\{Q^{\alpha I}, Q^{\beta J}\} = \delta^{IJ}(\gamma^\mu C^{-1})^{\alpha\beta} P_\mu$$

$$+i(\gamma_5 C^{-1})^{\alpha\beta}(\alpha_\mathbf{j})^{IJ}\int d^3x\, U_\mathbf{j} - (C^{-1})^{\alpha\beta}(\beta_\mathbf{j})^{IJ}\int d^3x\, V_\mathbf{j}$$

$$+(C^{-1})^{\alpha\beta}(\alpha_\mathbf{j})^{IJ}\int d^3x\, \tilde{V}_\mathbf{j} + i(\gamma_5 C^{-1})^{\alpha\beta}(\beta_\mathbf{j})^{IJ}\int d^3x\, \tilde{U}_\mathbf{j}$$

$$-\frac{1}{8}\int d^3x\,(\bar{\lambda}\Gamma^0\Gamma_{PQ}D_R\lambda)(\Gamma^{PQR}C^{-1})^{\alpha\beta}, \qquad (5.75)$$

where the last term is on-shell a total derivative[1] of the form $\partial_\rho(\bar{\lambda}\Gamma^{0RS}\lambda)$. Since the expectation value of this term contains the spinor trace $\mathrm{tr}(\Gamma^{0RS}\not{k})$, which vanishes, we drop this term from now on. In (5.75) we have used a particular[22] representation of the 32×32 Dirac matrices Γ^M in terms of the 4×4 Dirac matrices, involving the matrices $\alpha_\mathbf{j}$ and $\beta_\mathbf{j}$ which are proportional to the matrices $\eta_\mathbf{j}^{IJ}$ and $\bar{\eta}_\mathbf{j}^{IJ}$ which 't Hooft introduced for the construction of instantons. The $\alpha_\mathbf{j}$ and $\beta_\mathbf{j}$ represent the 6 generators of SO(4): totally antisymmetric 4×4 matrices, purely imaginary, and either self-dual (α) or anti-self-dual (β). The indices I and J are raised and lowered with the Euclidean metric δ^{IJ} and δ_{IJ}, and finally $\alpha, \beta = 1, \dots, 4$ are the spinor indices in 3+1 dimensions. Clearly, we have 12 real central charges

$$U_\mathbf{j} = \partial_i(S_\mathbf{j}^a \frac{1}{2}\epsilon^{ijk}F_{jk}^a), \qquad \tilde{U}_\mathbf{j} = \partial_i(P_\mathbf{j}^a F_{0i}^a)$$

$$V_\mathbf{j} = \partial_i(P_\mathbf{j}^a \frac{1}{2}\epsilon^{ijk}F_{jk}^a), \qquad \tilde{V}_\mathbf{j} = \partial_i(S_\mathbf{j}^a F_{0i}^a). \qquad (5.76)$$

In the $N = 2$ case, Eqs. (5.55), only the sums $U_3 + \tilde{U}_3$ and $V_3 + \tilde{V}_3$ appeared, but here they split into parts with different tensor structures, half of them with α matrices, the other half with β's.

We set $S_3^a = v$ for adjoint color index $a = 3$ in the topologically trivial sector, and locate the monopole inside the fields A_j^a and S_3^a. For quantum calculations we use again the background field formalism as in the $N = 2$ case above, which now gives $Z_v = Z_S = Z_A = Z_g = 1$, since the β-function for the $N = 4$ model vanishes.

The gravitational stress tensor yields the Hamiltonian density, which we write again after use of the linearized field equations for fluctuations as time

---

[1]Use $\Gamma^{RS}\Gamma^N D_N\lambda = 0 = \Gamma^{RSN}D_N\lambda + \Gamma^R D^S\lambda - \Gamma^S D^R\lambda$ to write all terms as $\bar{\lambda}\Gamma^{RST}\lambda$. Then use $\bar{\lambda}\Gamma^{RST}D_N\lambda = \frac{1}{2}\partial_N(\bar{\lambda}\Gamma^{RST}\lambda)$.

$$M^{1-\text{loop}} = \int d^3x \langle a_0 \partial_0^2 a_0 - a_j \partial_0^2 a_j - p_{\mathbf{j}} \partial_0^2 p_{\mathbf{j}} - s_{\mathbf{j}} \partial_0^2 s_{\mathbf{j}} - b \partial_0^2 c - (\partial_0^2 b)c$$

$$+ \frac{i}{2} (\lambda^I)^T \partial_0 \lambda^I \rangle$$

$$+ \lim_{r \to \infty} \frac{1}{4} 4\pi r^2 \frac{\partial}{\partial r} \langle a_0^2 + a_j^2 + p_{\mathbf{j}}^2 + s_{\mathbf{j}}^2 + 2bc - \frac{2}{3} a_j^2 \rangle, \qquad (5.77)$$

where the only differences with the $N = 2$ model are that there are now three times as many scalar and pseudoscalar fields and we have four real 4-component spinors instead one one complex 4-component spinor. However, the consequences could not have been more severe. The sum of the zero-point energies (the bulk contribution in (5.77)) vanishes for the $N = 4$ case: the fields associated with the field operator $\bar{\not{D}}\not{D}$ are a_j, s_3, λ_+^I and yield in eq. (5.66) $\mathcal{N}^+ = 3 + 1 - 4 = 0$, instead of $3 + 1 - 2 = 2$, while the fields associated with $\not{D}\bar{\not{D}}$ are $a_0, s_1, s_2, p_{\mathbf{j}}, b, c$, and λ_-^I yielding $\mathcal{N}^- = 1 + 1 + 1 + 3 - 1 - 1 - 4 = 0$, too, instead of $1 + 1 - 1 - 1 - 2 = -2$. On the other hand, for $N = 4$ the boundary terms no longer vanish, since instead of $-1 + 3 + 1 + 1 - 2 - 2 = 0$ we now have $-1 + 3 + 3 + 3 - 2 - 2 = 4$, yielding

$$M^{(1)\text{surface}} = \lim_{r \to \infty} \pi r^2 \frac{\partial}{\partial r} \langle 4(s_1^a)^2 \rangle. \qquad (5.78)$$

For large r, the difference between the operators $\not{D}\bar{\not{D}}$ and $\bar{\not{D}}\not{D}$ is due to F_{mn}, see (5.59) and (5.60), which falls off like $1/r^2$, so the bosonic propagators in the background covariant $\xi = 1$ gauge have a common asymptotic form,

$$\langle a_M^a(x) a_N^b(y) \rangle \simeq \eta_{MN} G^{ab}(x, y), \qquad \langle b^a(x) c^b(y) \rangle \simeq -G^{ab}(x, y), \qquad (5.79)$$

with

$$G^{ab}(x, y) = \langle x | \frac{-i}{-\Box + m^2 - \frac{2m}{r}} (\delta^{ab} - \hat{x}^a \hat{x}^b) + \frac{i}{\Box} \hat{x}^a \hat{x}^b | y \rangle. \qquad (5.80)$$

Inserting complete sets of momentum eigenstates, the same procedure as used to calculate anomalies from index theorems[45] yields for the r-dependent part of $G^{aa}(x, x)$

$$\langle s_1^a(x) s_1^a(x) \rangle \simeq 2 \int \frac{d^{4+\epsilon}k}{(2\pi)^{4+\epsilon}} \frac{-i}{(k^2 + m^2) + 2ik^\mu \partial_\mu - \partial_\mu^2 - \frac{2m}{r}} \to 2 \frac{2m}{r} I$$

$$(5.81)$$

with I given in (5.70). (The overall factor 2 is due to tracing $\delta^{ab} - \hat{x}^a \hat{x}^b$.)

Hence, we have arrived at a divergent result for the mass of the $N = 4$ monopole,

$$M^{1-\text{loop}} = M^{(1)\text{surface}} = -16\pi m I. \tag{5.82}$$

Ordinary renormalization, namely renormalization of the parameters in the action, is of no help, since, as we have seen, all Z factors which helped to make the $N = 2$ result finite, are unity in the $N = 4$ case.

The solution is extra-ordinary renormalization, namely renormalization of the stress tensor density, and also of the central charge density (and, in fact, all currents of the corresponding susy multiplet) as composite operators. In the literature it has been shown that the improved stress tensor[59] does not renormalize,[59-63] which we extend to the statement that none of the improved currents in the susy multiplet renormalize.[56,64] However, the currents in the susy algebra displayed above are nonimproved currents, and to construct improved currents one must add improvement terms to the unimproved currents. In order to find those for both the stress tensor and the central charges, we go back one step and begin with our unimproved Noether susy currents j^μ, to which we add the improvement terms $\Delta j_{\text{imp}}^\mu$,

$$j_{\text{imp}}^\mu = j^\mu + \Delta j_{\text{imp}}^\mu = \frac{1}{2}\Gamma^{RS}F_{RS}^a\Gamma^\mu\lambda^a - \frac{2}{3}\Gamma^{\mu\nu}\partial_\nu(A_{\mathcal{J}}^a\Gamma^{\mathcal{J}}\lambda^a) \tag{5.83}$$

where we use a 10-dimensional notation in which $A_{\mathcal{J}}$, $\mathcal{J} = 5, \ldots 10$, comprises all scalars and pseudoscalars in the model. The Γ-matrices are 32×32-matrices but we in fact deal with the dimensionally reduced theory so that the sum over ν runs only from 0 to 3. Both j^μ and j_{imp}^μ are on-shell conserved. (Use the Bianchi identity $\Gamma^{RS\mathcal{J}}A_{\mathcal{J}} \times F_{RS} = 0$). In addition, the improved (ordinary, not conformal) susy currents are on-shell gamma-traceless, $\Gamma_\mu j_{\text{imp}}^\mu = 0$. This can be verified by using $\Gamma_\mu\Gamma^{\rho\sigma}\Gamma^\mu = 0$ and $F_{\rho\mathcal{J}} = D_\rho A_{\mathcal{J}}$. One finds $\Gamma_\mu j_{\text{imp}}^\mu = 2\Gamma^{\rho\mathcal{J}}(D_\rho A_{\mathcal{J}}) \cdot \lambda + 2\Gamma^{\mathcal{J}\mathcal{K}}(A_{\mathcal{J}} \times A_{\mathcal{K}}) \cdot \lambda - 2\Gamma^\nu\partial_\nu(A_{\mathcal{J}}\Gamma^{\mathcal{J}} \cdot \lambda)$ which indeed vanishes on-shell, where $\Gamma^\rho D_\rho\lambda = -\Gamma^{\mathcal{K}}A_{\mathcal{K}} \times \lambda$. From the susy variation of $\Delta j_{\text{imp}}^\mu$ we find the improvement terms in $T_{\mu\nu}$ and the central charges. Switching to 3+1-dimensional notation, we find[45]

$$\Delta T_{00}^{\text{impr}} = -\frac{1}{6}\partial_j^2(A_{\mathcal{J}}A^{\mathcal{J}}) \tag{5.84}$$

$$\Delta U_3^{\text{impr}} = -\frac{1}{3}\left[U_3 + \int \frac{i}{8}\partial_i(\epsilon^{ijk}\bar{\lambda}\alpha^1\gamma_{jk}\lambda)d^3x\right]. \tag{5.85}$$

It is important to note that we do not start with improved Noether currents at the classical level. Indeed, the standard result for the classical value of the mass of a monopole is only obtained when the unimproved stress tensor is used, and also the classical value of the improved and unimproved central charge differ.[m] However, even when starting with unimproved currents, we have to expect improvement terms as counterterms, since we can write the unimproved currents as $j_\mu = j_\mu^{\text{impr}} - \Delta j_\mu^{\text{impr}}$ and we expect the improved part to be finite, and the improvement terms $\Delta j_\mu^{\text{impr}}$ to renormalize multiplicatively. Denoting the common Z factor for all improvement terms in the susy current, the stress tensor and the central charges by Z_{impr}, the composite operator counterterms to mass and central charge will be given by

$$\Delta T_{00} = -(Z_{\text{impr}} - 1)\Delta T_{00}^{\text{impr}}, \quad \Delta U = -(Z_{\text{impr}} - 1)\Delta U^{\text{impr}}, \quad (5.86)$$

where the overall minus sign is due to having written $j_\mu = j_\mu^{\text{impr}} - \Delta j_\mu^{\text{impr}}$. We shall now show by a detailed calculation[56] that a single factor Z_{impr} removes the divergences in mass and central charge and thus ensures BPS saturation, but it would be interesting to check by an explicit (though laborious) calculation that there are no more composite operator counterterms in the renormalization of the full susy algebra, and to find the superspace formulation from which this follows.

To determine Z_{impr} we decompose $U = U_3 = \int \partial_i (S_3^a \frac{1}{2} \epsilon^{ijk} F_{jk}^a) d^3 x$ as $U^{\text{impr}} - \Delta U^{\text{impr}}$ where

$$U^{\text{impr}} = \frac{2}{3}\left[U - \int \frac{i}{16} \partial_i (\epsilon^{ijk} \bar{\lambda} \alpha^1 \gamma_{jk} \lambda) d^3 x \right] \qquad (5.87)$$

$$\Delta U^{\text{impr}} = -\frac{1}{3}\left[U + \int \frac{i}{8} \partial_i (\epsilon^{ijk} \bar{\lambda} \alpha^1 \gamma_{jk} \lambda) d^3 x \right] \equiv -\frac{1}{3}(U + F).$$

For the one-loop composite operator renormalization of U we thus have to consider four classes of proper diagrams: graphs with the bosonic U or the fermionic F inserted in proper one-loop diagrams with external bosonic background fields or external fermionic fields. The number of potentially divergent graphs with bosonic-bosonic structure is 31, with fermionic-fermionic structure it is 2, while there is one graph with U insertion and external fermions, and 3 graphs with F insertion and external bosonic fields. Of the set of 31 graphs, some vanish by themselves, some vanish in the background covariant $R_{\xi=1}$ gauge, some subsets of diagrams are finite, and if we only consider graphs with one external field S_3^3 and one external gauge field A_μ^3, only one graph survives!

---

[m]This is clear from the fact that ΔU^{impr} involves the bosonic term U.

The set of one-loop graphs to be evaluated and their divergent contributions (obtained in the topologically trivial sector) is as follows (denoting gauge fields A_μ by wavy lines, the scalar field S_3 by a dashed line, and fermions λ by straight lines),

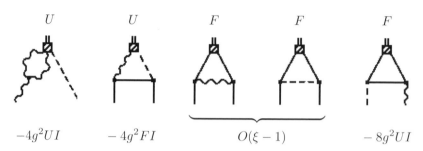

$$-4g^2UI \qquad -4g^2FI \qquad\qquad O(\xi-1) \qquad\qquad -8g^2UI$$

to which one needs to add a term

$$Z_\lambda F = -8g^2 FI + O(\xi - 1), \tag{5.88}$$

since in $N = 4$ theory there is nontrivial ordinary wave function renormalization Z_λ for fermions, whereas bosonic fields do not obtain ordinary wave function renormalization in the gauge $\xi = 1$.

This leads to the (ξ-independent) composite operator counterterms,

$$\Delta U_{c.o.r.} = 4g^2(U + F)I, \quad \Delta F_{c.o.r.} = 8g^2(U + F)I. \tag{5.89}$$

From this it follows that $Z_{\text{impr}} = 1 + 12g^2 I$, while the improved part of U, $U^{\text{impr}} = \frac{2}{3}U - \frac{1}{3}F$, is indeed finite.

In the monopole background, the classical values of $-U$ and M are equal and given by $4\pi m/g^2$ without renormalization of m or g. Since $U = U^{\text{impr}} - \Delta U^{\text{impr}}$, composite operator renormalization produces the contribution $12g^2 I \times (+\frac{1}{3}) \times 4\pi m/g^2 = 16\pi m I$ to $-U$ as well as M, which indeed cancels the divergence obtained above in (5.82). Thus we have $M^{(1)} = -U^{(1)}(= 0)$ and BPS saturation is verified.

Having found that composite operator renormalization is needed for the $N = 4$ case, we should of course go back to the $N = 2$ calculation (and also all other one-loop calculations for solitons performed so far), and make sure that in these cases there is no new contribution that could upset the BPS saturation obtained previously. For the kink in 1+1 dimensions, it is easy to check that there are no divergent one-loop diagrams for composite operator renormalization of the stress tensor (but it turns out that there is in fact a need for composite operator renormalization in the local energy density of 3+1 dimensional kink domain walls, which does however not contribute to the integrated total energy[65]). For the vortex in 2+1 dimensions,[66-68] all currents are finite in dimensional regularization. But for the $N = 2$ monopole, no composite operator renormalization is needed for two reasons: (1) the improvement terms of the central charge $\Delta \mathcal{U}_{N=2,\text{impr}}$ are proportional to the central charge $\mathcal{U}_{N=2}$ itself and (2) the central charge $\mathcal{U}_{N=2}$ is a finite operator due to ordinary renormalization. To prove these statements consider the central charge density for the $N = 2$ model

$$\mathcal{U}_{N=2} = \frac{1}{2}\epsilon_{ijk}\partial_i(S^a F_{jk}^a) + \partial_i(P^a F_{i0}^a) = \mathcal{U} - \tilde{\mathcal{U}}. \tag{5.90}$$

The improvement terms in the susy current are given by

$$\Delta j_{N=2,\text{impr}}^{\mu} = -\frac{2}{3}\Gamma^{\mu\nu}\partial_\nu[(P\Gamma^5 + S\Gamma^6)\lambda]$$

$$= -\frac{2}{3}\gamma^{\mu\nu}\partial_\nu[(P\gamma^5 + iS)\lambda]. \tag{5.91}$$

The susy variation of $\Delta j_{N=2,\text{impr}}^{\mu}$ yields $\Delta U_{N=2,\text{impr}}$. Using $\delta P = \delta A_5 = \bar{\lambda}\Gamma^5\epsilon = \bar{\lambda}\gamma_5\epsilon$, $\delta S = \delta A_6 = \bar{\lambda}\Gamma_6\epsilon = i\bar{\lambda}\epsilon$ and $\delta\lambda = \frac{1}{2}\Gamma^{PQ}F_{PQ}\epsilon$ we find

$$\Delta j_{N=2,\text{impr}}^0 = \gamma^{0j}\partial_j[\gamma_5\lambda(\bar{\lambda}\gamma_5\epsilon) - \lambda(\bar{\lambda}\epsilon) + (P\gamma_5 + iS)\frac{1}{2}\gamma^{\rho\sigma}F_{\rho\sigma}\epsilon]$$

$$= \gamma^{0j}\partial_j[-\frac{1}{4}(\bar{\lambda}\mathcal{O}^I\lambda)(\gamma_5\mathcal{O}_I\gamma_5 - \mathcal{O}_I)$$

$$+ P\gamma_5\gamma^{0k}F_{0k} + iS\frac{1}{2}\gamma^{kl}F_{kl}]\epsilon$$

$$\sim \gamma_5\partial_j(PF_{0j} - \frac{1}{2}\epsilon^{jkl}SF_{kl})\epsilon, \tag{5.92}$$

because the terms with $\mathcal{O}_I \sim \gamma_{kl}$ cancel in the first line. Thus $\Delta\mathcal{U}_{N=2,\text{impr}}$ is proportional to $\mathcal{U}_{N=2}$, so there are no fermionic terms in $\Delta\mathcal{U}_{N=2,\text{impr}}$. Both $\mathcal{U}$ and $\tilde{\mathcal{U}}$ produce divergence proportional to F, but their sum cancels. Since ordinary renormalization already gave counterterms which make $U_{N=2}$ finite, we do not need composite operator counterterms, so we can set $Z_{\text{impr}} = 1$, leaving the results for the $N = 2$ monopole unchanged.

5.4. Conclusions

The one-loop corrections to the mass and central charge of kinks, vortices (not discussed here, but treated in Refs. 66–68), and monopoles in $N = 2$ and $N = 4$ super Yang-Mills theory satisfy the BPS bound. To obtain this result, we needed to carefully regularize the susy field theories, which in our choice of regularization scheme meant that we needed to take extra dimensions into account. In these extra dimensions the modes of bosonic and fermionic quantum fields had extra momenta, and the square of these extra momenta gave an extra contribution to the 1-loop central charge. In addition we found that boundary terms contributed to the mass of the $N = 4$ monopole. These boundary terms were divergent, and we needed multiplicative composite operator renormalization of the improvement terms in the stress tensor to obtain a finite quantum mass. The same composite operator renormalization was needed for the central charge. For the $N = 2$ monopole, all boundary terms canceled, and there was no composite operator renormalization, but the sum over zero point energies in the bulk was divergent, and standard renormalization counter terms cancelled these divergences. For the susy kink, boundary terms could not even appear because the classical kink solution falls off exponentially fast.

We found that the 1-loop corrections to the susy kink and $N = 2$ monopole are nonzero. In the literature it was assumed, or proofs were proposed, that these corrections vanish. Our results for the $N = 2$ monopole agree with results based on holomorphicity by Seiberg-Witten,[57] which also require a nonvanishing correction to the mass and central charge (although this was noticed only subsequently[22]). This raises the question whether our results are consistent with Zumino's general proof that the sum over zero-point energies must vanish in any susy theory.[3] This proof is based on path integrals and does not take into account regularization. Hence, it is not clear that there is a disagreement. There is a way of understanding our nonvanishing result. If one encloses the kink in a large box, and imposes susy boundary conditions, one finds a spurious boundary energy which one must subtract to obtain the true mass of the susy soliton.[20] Dimensional regularization by itself subtracts this spurious boundary energy.

A superspace treatment of solitons would be useful, but we have found problems in gauge theories with a superspace R_ξ gauge if solitons are present.[69] A superspace treatment of the anomalies in the superconformal currents of the kink has been given in collaboration with Fujikawa,[70,71] see also Shizuya.[72–75]

Our methods could perhaps be applied to D-branes,[76] at least the D-branes that are solitons. Also extension to finite temperature is interesting; in fact, we have found new surprises for kink domain walls at finite temperature.[65]

Acknowledgments

We thank Yu-tin Huang for assistance in writing up this review and acknowledge financial support from the Austrian Science Foundation FWF, project nos. J2660-N16 and P19958.

References

1. R. Rajaraman, *Solitons and Instantons* (North-Holland, Amsterdam, 1982).
2. C. Rebbi and G. Soliani (eds.), *Solitons and particles* (World Scientific, Singapore, 1984).
3. B. Zumino, "Supersymmetry and the Vacuum", *Nucl. Phys.* **B89**, 535 (1975).
4. A. D'Adda, R. Horsley and P. Di Vecchia, "Supersymmetric magnetic monopoles and dyons", *Phys. Lett.* **B76**, 298 (1978).
5. H. B. Nielsen and P. Olesen, "Vortex-line models for dual strings", *Nucl. Phys.* **B61**, 45 (1973).
6. V. L. Ginzburg and L. D. Landau, "On the Theory of superconductivity", *Zh. Eksp. Teor. Fiz.* **20**, 1064 (1950).
7. A. A. Abrikosov, "On the Magnetic properties of superconductors of the second group", *Sov. Phys. JETP* **5**, 1174 (1957).
8. G. 't Hooft, "Magnetic monopoles in unified gauge theories", *Nucl. Phys.* **B79**, 276 (1974).
9. A. M. Polyakov, "Particle spectrum in quantum field theory", *JETP Lett.* **20**, 194 (1974).
10. R. F. Dashen, B. Hasslacher and A. Neveu, "The particle spectrum in model field theories from semiclassical functional integral techniques", *Phys. Rev.* **D11**, 3424 (1975).
11. L. D. Faddeev and V. E. Korepin, "Quantum theory of solitons", *Phys. Rept.* **42**, 1 (1978).
12. J. L. Gervais and B. Sakita, "Extended particles in quantum field theories", *Phys. Rev.* **D11**, 2943 (1975).
13. T. D. Lee and Y. Pang, "Nontopological solitons", *Phys. Rept.* **221**, 251 (1992).
14. B. Julia and A. Zee, "Poles with Both Magnetic and Electric Charges in Nonabelian Gauge Theory", *Phys. Rev.* **D11**, 2227 (1975).
15. M. K. Prasad and C. M. Sommerfield, "An exact classical solution for the 't Hooft monopole and the Julia-Zee dyon", *Phys. Rev. Lett.* **35**, 760 (1975).
16. E. B. Bogomolnyi, "Stability of classical solutions", *Sov. J. Nucl. Phys.* **24**, 449 (1976).
17. E. Witten and D. Olive, "Supersymmetry algebras that include topological charges", *Phys. Lett.* **B78**, 97 (1978).
18. H. Nastase, M. Stephanov, P. van Nieuwenhuizen and A. Rebhan, "Topological boundary conditions, the BPS bound, and elimination of ambiguities in the quantum mass of solitons", *Nucl. Phys.* **B542**, 471 (1999).

19. N. Graham and R. L. Jaffe, "Energy, central charge, and the BPS bound for 1+1 dimensional supersymmetric solitons", *Nucl. Phys.* **B544**, 432 (1999).
20. M. A. Shifman, A. I. Vainshtein and M. B. Voloshin, "Anomaly and quantum corrections to solitons in two-dimensional theories with minimal supersymmetry", *Phys. Rev.* **D59**, 045016 (1999).
21. A. Rebhan, P. van Nieuwenhuizen and R. Wimmer, "The anomaly in the central charge of the supersymmetric kink from dimensional regularization and reduction", *Nucl. Phys.* **B648**, 174 (2003).
22. A. Rebhan, P. van Nieuwenhuizen and R. Wimmer, "A new anomalous contribution to the central charge of the $N = 2$ monopole", *Phys. Lett.* **B594**, 234 (2004).
23. M. F. Atiyah, V. K. Patodi and I. M. Singer, "Spectral asymmetry and Riemannian Geometry 1", *Math. Proc. Cambridge Phil. Soc.* **77**, 43 (1975).
24. C. Callias, "Index theorems on open spaces", *Commun. Math. Phys.* **62**, 213 (1978).
25. E. J. Weinberg, "Parameter counting for multimonopole solutions", *Phys. Rev.* **D20**, 936 (1979).
26. E. J. Weinberg, "Index calculations for the fermion–vortex system", *Phys. Rev.* **D24**, 2669 (1981).
27. A. Rebhan, P. van Nieuwenhuizen and R. Wimmer, "One-loop surface tensions of (supersymmetric) kink domain walls from dimensional regularization", *New J. Phys.* **4**, 31 (2002).
28. M. Lüscher, "Dimensional regularization in the presence of large background fields", *Ann. Phys.* **142**, 359 (1982).
29. A. Parnachev and L. G. Yaffe, "One-loop quantum energy densities of domain wall field configurations", *Phys. Rev.* **D62**, 105034 (2000).
30. C. Montonen and D. I. Olive, "Magnetic Monopoles as Gauge Particles?", *Phys. Lett.* **B72**, 117 (1977).
31. E. Witten, "Dyons of charge $e\theta/2\pi$", *Phys. Lett.* **B86**, 283 (1979).
32. J. F. Schonfeld, "Soliton masses in supersymmetric theories", *Nucl. Phys.* **B161**, 125 (1979).
33. R. K. Kaul and R. Rajaraman, "Soliton energies in supersymmetric theories", *Phys. Lett.* **B131**, 357 (1983).
34. H. Yamagishi, "Soliton mass distributions in (1+1)-dimensional supersymmetric theories", *Phys. Lett.* **B147**, 425 (1984).
35. C. Imbimbo and S. Mukhi, "Index theorems and supersymmetry in the soliton sector", *Nucl. Phys.* **B247**, 471 (1984).
36. A. Uchiyama, "Nonconservation of supercharges and extra mass correction for supersymmetric solitons in (1+1) dimensions", *Prog. Theor. Phys.* **75**, 1214 (1986).
37. J. Casahorrán, "Nonzero quantum contribution to the soliton mass in the SUSY sine-Gordon model", *J. Phys.* **A22**, L413 (1989).
38. L. J. Boya and J. Casahorrán, "Kinks and solitons in SUSY models", *J. Phys.* **A23**, 1645 (1990).
39. A. Rebhan and P. van Nieuwenhuizen, "No saturation of the quantum Bogomolnyi bound by two-dimensional $N = 1$ supersymmetric solitons", *Nucl.*

Quantum Corrections to Solitons and BPS Saturation 73

Phys. **B508**, 449 (1997).

40. A. Litvintsev and P. van Nieuwenhuizen, "Once more on the BPS bound for the susy kink", hep-th/0010051.

41. A. S. Goldhaber, A. Litvintsev and P. van Nieuwenhuizen, "Mode regularization of the susy sphaleron and kink: Zero modes and discrete gauge symmetry", *Phys. Rev.* **D64**, 045013 (2001).

42. A. S. Goldhaber, A. Litvintsev and P. van Nieuwenhuizen, "Local Casimir energy for solitons", *Phys. Rev.* **D67**, 105021 (2003).

43. A. S. Goldhaber, A. Rebhan, P. van Nieuwenhuizen and R. Wimmer, "Clash of discrete symmetries for the supersymmetric kink on a circle", *Phys. Rev.* **D66**, 085010 (2002).

44. M. Bordag, A. S. Goldhaber, P. van Nieuwenhuizen and D. Vassilevich, "Heat kernels and zeta-function regularization for the mass of the SUSY kink", *Phys. Rev.* **D66**, 125014 (2002).

45. A. Rebhan, P. van Nieuwenhuizen and R. Wimmer, "Quantum mass and central charge of supersymmetric monopoles: Anomalies, current renormalization, and surface terms", *JHEP* **0606**, 056 (2006).

46. N. Dorey, "The BPS spectra of two-dimensional supersymmetric gauge theories with twisted mass terms", *JHEP* **9811**, 005 (1998).

47. C. Mayrhofer, A. Rebhan, P. van Nieuwenhuizen and R. Wimmer, "Perturbative Quantum Corrections to the Supersymmetric CP^1 Kink with Twisted Mass", *JHEP* **0709**, 069 (2007).

48. M. Shifman, A. Vainshtein and R. Zwicky, "Central charge anomalies in 2D sigma models with twisted mass", *J. Phys.* **A39**, 13005 (2006).

49. A. Hanany and D. Tong, "Vortices, instantons and branes", *JHEP* **07**, 037 (2003).

50. R. Auzzi *et al.*, "Nonabelian superconductors: Vortices and confinement in N = 2 SQCD", *Nucl. Phys.* **B673**, 187 (2003).

51. R. Auzzi, S. Bolognesi, J. Evslin and K. Konishi, "Nonabelian monopoles and the vortices that confine them", *Nucl. Phys.* **B686**, 119 (2004).

52. M. Shifman and A. Yung, "Localization of non-abelian gauge fields on domain walls at weak coupling (D-brane prototypes II)", *Phys. Rev.* **D70**, 025013 (2004).

53. M. Shifman and A. Yung, "Non-abelian string junctions as confined monopoles", *Phys. Rev.* **D70**, 045004 (2004).

54. R. K. Kaul, "Monopole mass in supersymmetric gauge theories", *Phys. Lett.* **B143**, 427 (1984).

55. C. Imbimbo and S. Mukhi, "Index theorems and supersymmetry in the soliton sector. 2. Magnetic monopoles in (3+1)-dimensions", *Nucl. Phys.* **B249**, 143 (1985).

56. A. Rebhan, R. Schöfbeck, P. van Nieuwenhuizen and R. Wimmer, "BPS saturation of the N = 4 monopole by infinite composite-operator renormalization", *Phys. Lett.* **B632**, 145 (2006).

57. N. Seiberg and E. Witten, "Electric-magnetic duality, monopole condensation, and confinement in N=2 supersymmetric Yang-Mills theory", *Nucl. Phys.* **B426**, 19 (1994).

74   *A. Rebhan, P. van Nieuwenhuizen and R. Wimmer*

58. N. Seiberg and E. Witten, "Monopoles, duality and chiral symmetry breaking in N=2 supersymmetric QCD", *Nucl. Phys.* **B431**, 484 (1994).
59. C. G. Callan, Jr., S. Coleman and R. Jackiw, "A new improved energy-momentum tensor", *Ann. Phys.* **59**, 42 (1970).
60. D. Z. Freedman, I. J. Muzinich and E. J. Weinberg, "On the energy-momentum tensor in gauge field theories", *Ann. Phys.* **87**, 95 (1974).
61. D. Z. Freedman and E. J. Weinberg, "The energy-momentum tensor in scalar and gauge field theories", *Ann. Phys.* **87**, 354 (1974).
62. J. C. Collins, "The energy-momentum tensor revisited", *Phys. Rev.* **D14**, 1965 (1976).
63. L. S. Brown, "Dimensional regularization of composite operators in scalar field theory", *Ann. Phys.* **126**, 135 (1980).
64. T. Hagiwara, S.-Y. Pi and H.-S. Tsao, "Regularizations and superconformal anomalies", *Ann. Phys.* **130**, 282 (1980).
65. A. Rebhan, A. Schmitt and P. van Nieuwenhuizen, YITP-SB-09-03, to appear.
66. D. V. Vassilevich, "Quantum corrections to the mass of the supersymmetric vortex", *Phys. Rev.* **D68**, 045005 (2003).
67. A. Rebhan, P. van Nieuwenhuizen and R. Wimmer, "Nonvanishing quantum corrections to the mass and central charge of the $N = 2$ vortex and BPS saturation", *Nucl. Phys.* **B679**, 382 (2004).
68. S. Ölmez and M. Shifman, "Revisiting Critical Vortices in Three-Dimensional SQED", *Phys. Rev.* **D78**, 125021 (2008).
69. A. S. Goldhaber, A. Rebhan, P. van Nieuwenhuizen and R. Wimmer, "Quantum corrections to mass and central charge of supersymmetric solitons", *Phys. Rept.* **398**, 179 (2004).
70. K. Fujikawa and P. van Nieuwenhuizen, "Topological anomalies from the path integral measure in superspace", *Ann. Phys.* **308**, 78 (2003).
71. K. Fujikawa, A. Rebhan and P. van Nieuwenhuizen, "On the nature of the anomalies in the supersymmetric kink", *Int. J. Mod. Phys.* **A18**, 5637 (2003).
72. K. Shizuya, "Superfield formulation of central charge anomalies in two-dimensional supersymmetric theories with solitons", *Phys. Rev.* **D69**, 065021 (2004).
73. K. Shizuya, "Topological-charge anomalies in supersymmetric theories with domain walls", *Phys. Rev.* **D70**, 065003 (2004).
74. K. Shizuya, "Central charge and renormalization in supersymmetric theories with vortices", *Phys. Rev.* **D71**, 065006 (2005).
75. K. Shizuya, "Effect of quantum fluctuations on topological excitations and central charge in supersymmetric theories", *Phys. Rev.* **D74**, 025013 (2006).
76. J. Polchinski, "Dirichlet-Branes and Ramond-Ramond Charges", *Phys. Rev. Lett.* **75**, 4724 (1995).

Chapter 6

Gauging Noncommutative Theories

Harald Grosse[1] and Michael Wohlgenannt[2]

[1] *University of Vienna, Faculty of Physics,*
Boltzmanngasse 5, A-1090 Vienna, Austria

[2] *Dipartimento di Scienze e Tecnologie Avanzate,*
Università del Piemonte Orientale
and
I.N.F.N. - Gruppo Collegato di Alessandria
Via V. Bellini 25/G, I-15100 Alessandria, Italy

We review some recent developments concerning the renormalizability of noncommutative quantum field theories. And present a consequential formulation of noncommutative gauge theories.

6.1. Introduction

One of us (Harald Grosse) remembers very well the early first interactions with Prof. Wolfgang Kummer. Being at the time a very young student I entered the Rochester congress in 1968, which was held in Vienna and a young Professor from the Technical University approached me and asked me to help with the organization.

Later on I learnt several times from contributions of Wolfgang to solve the ultraviolet problem of weak interactions and from his work on the special axial gauge. After Wolfgang turned to two-dimensional gravity, we had more intensive contacts which led to common publications.

Wolfgang tried in his work to attack the unsolved problems of quantum field theory. My recent work on noncommutative models goes exactly in that direction. Therefore, I am sure he would have enjoyed seeing that the Landau ghost problem can be solved this way and noncommutative

---

[1] E-mail: harald.grosse@univie.ac.at
[2] E-mail: miw@mfn.unipmn.it

gauge models can be formulated and lead to promising new directions. In this contribution, we want to present a brief and biased summary of some of these recent developments which lead to an ongoing project on gauge models.

As already indicated, four dimensional quantum field theory still suffers from severe problems: From infrared and ultraviolet divergences as well as from the divergence of the renormalized perturbation expansion. Despite the impressive agreement between theory and experiments and despite many attempts, these problems are not settled and remain a big challenge for theoretical physics. Furthermore, attempts to formulate a quantum theory of gravity have not yet been fully successful. It is astonishing that the two pillars of modern physics, quantum field theory and general relativity, seem to be incompatible. This convinced physicists to look for more general descriptions: After the formulation of supersymmetry and supergravity, string theory was developed, and anomaly cancellation forced the introduction of six additional dimensions. On the other hand, loop gravity was formulated, and led to spin networks and space-time foams. Both approaches are not fully satisfactory. A third impulse came from noncommutative geometry developed by Alain Connes, providing a natural interpretation of the Higgs effect at the classical level. This finally led to noncommutative quantum field theory, which is the subject of this contribution. It allows to incorporate fluctuations of space into quantum field theory. There are of course relations among these three developments. In particular, the field theory limit of string theory leads to certain noncommutative field theory models (NCFT), and some models defined over fuzzy spaces are related to spin networks.

The argument that space-time should be modified at very short distances can be traced back to Schrödinger and Heisenberg and even to Riemann's habilitation. Noncommutative coordinates appeared already in the work of Peierls for the magnetic field problem, and are obtained after projecting onto a particular Landau level. Pauli communicated this to Oppenheimer, whose student Snyder[1] wrote down the first deformed space-time algebra preserving Lorentz symmetry. After the development of noncommutative geometry by Connes,[2] it was first applied in physics to the integer quantum Hall effect. Gauge models on the two-dimensional noncommutative tori were formulated, and the relevant projective modules over this space were classified. Filk[3] developed Feynman rules for canonically deformed four dimensional scalar field theory, and Doplicher, Fredenhagen and Roberts[4] published their work on deformed spaces. The subject experienced a major boost after one realized that string theory leads to noncommutative field theory under certain conditions,[5,6] and the subject developed very rapidly. However, some unexpected features such as IR/UV mixing arise upon quantization. In 2000 Minwalla, van Raamsdonk and Seiberg realized[7] that perturbation theory for field theories defined on the Moyal plane faces a serious problem. The planar regular contributions $(B = 1)$ show the standard singularities which can be handled by a renormalization procedure; B is the number of boundary components. The planar nonregular $(B > 2)$ one loop contributions are finite for generic momenta, however they become singular at exceptional momenta. The usual UV divergences are then reflected in new singularities in the infrared, which is called IR/UV mixing. This spoils the usual renormalization procedure: Inserting many such loops to a higher order diagram generates singularities of any inverse power. Without imposing a special structure such as supersymmetry, the renormalizability seems lost.

However, progress was made when one of us (H.G.) and R. Wulkenhaar were able to give a solution of this problem for the special case of a scalar four dimensional theory defined on the Moyal-deformed space $\mathbb{R}^4_\theta$,[8]

$$[x_\mu \overset{\star}{,} x_\nu] = i\theta_{\mu\nu} , \tag{6.1}$$

where $\theta_{\mu\nu} = -\theta_{\nu\mu} \in \mathbb{R}$ and $\star$ denotes the Moyal star product

$$(a \star b)(x) := \int d^4y \, \frac{d^4k}{(2\pi)^4} a(x + \tfrac{1}{2}\theta \cdot k) b(x + y) \, e^{iky} . \tag{6.2}$$

In this contribution, we will concentrate on Moyal-deformed Euclidean spaces only. The IR/UV mixing is taken into account through a modification of the free Lagrangian by adding an oscillator term with parameter Ω. The proof follows ideas of Polchinski. There are indications that a constructive procedure might be possible and give a nontrivial ϕ^4 model.[9]

Nonperturbative aspects of NCFT have also been studied in recent years. The most significant and surprising result is that the IR/UV mixing can lead to a new phase denoted as "striped phase",[10] where translational symmetry is spontaneously broken. The existence of such a phase has indeed been confirmed in numerical studies.[11,12] To understand better the properties of this phase and the phase transitions, further work and better analytical techniques are required, combining results from perturbative renormalization with nonperturbative techniques. Here a particular feature of scalar NCFT is very suggestive: The field can be described as a hermitian matrix, and the quantization is defined nonperturbatively by integrating over all such matrices. This provides a natural starting point for

nonperturbative studies. In particular, it suggests and allows to apply ideas
and techniques from random matrix theory.[13,14]

Then, we will discuss a formulation of gauge theories related to the
approach to NCFT presented here. We start with noncommutative ϕ^4
theory with additional oscillator potential. We couple an external gauge
field to the scalar field using covariant coordinates. As in the classical case,
we extract the dynamics of the gauge field from the divergent contributions
to the 1-loop effective action. The effective action is calculated using a heat
kernel expansion.[15] The technical details are presented in Refs. 16, 17. The
ongoing project on the BRST quantization of a related model[18] will also
be discussed. This is a joint project with the group of M. Schweda at the
Vienna University of Technology.

6.2. Renormalization of ϕ^4-theory

We briefly sketch the methods used in Ref. 8 proving the renormalizability
for scalar field theory on $\mathbb{R}^4_\theta$. The IR/UV mixing was taken into account
through a modification of the free Lagrangian, by adding an oscillator term
which modifies the spectrum of the free Hamiltonian:

$$S = \int d^4x \left(\frac{1}{2} \partial_\mu \phi \star \partial^\mu \phi + 2\Omega^2(\tilde{x}_\mu \phi) \star (\tilde{x}^\mu \phi) + \frac{\mu^2}{2} \phi \star \phi + \frac{\lambda}{4!} \phi \star \phi \star \phi \star \phi \right)(x) , \tag{6.3}$$

where $\tilde{x}_\mu = (\theta^{-1})_{\mu\nu} x^\nu$. The model is covariant under the Langmann-
Szabo[19] duality relating short and long distance behavior. At $\Omega = 1$, the
model becomes self-dual and connected to integrable models.

The renormalization proof proceeds by using a matrix base, which leads
to a dynamical matrix model of the type:

$$S[\phi] = (2\pi\theta)^2 \sum_{m,n,k,l \in \mathbb{N}^2} \left(\frac{1}{2} \phi_{mn} \Delta_{mn;kl} \phi_{kl} + \frac{\lambda}{4!} \phi_{mn} \phi_{nk} \phi_{kl} \phi_{lm} \right) , \tag{6.4}$$

where

$$\Delta_{\substack{m^1 \ n^1 ; \ k^1 \ l^1 \\ m^2 \ n^2 ; \ k^2 \ l^2}} = \left(\mu^2 + \frac{2+2\Omega^2}{\theta}(m^1+n^1+m^2+n^2+2) \right) \delta_{n^1 k^1} \delta_{m^1 l^1} \delta_{n^2 k^2} \delta_{m^2 l^2} \tag{6.5}$$

$$- \frac{2-2\Omega^2}{\theta} \left(\sqrt{k^1 l^1} \, \delta_{n^1+1,k^1} \delta_{m^1+1,l^1} + \sqrt{m^1 n^1} \, \delta_{n^1-1,k^1} \delta_{m^1-1,l^1} \right) \delta_{n^2 k^2} \delta_{m^2 l^2}$$

$$- \frac{2-2\Omega^2}{\theta} \left(\sqrt{k^2 l^2} \, \delta_{n^2+1,k^2} \delta_{m^2+1,l^2} + \sqrt{m^2 n^2} \, \delta_{n^2-1,k^2} \delta_{m^2-1,l^2} \right) \delta_{n^1 k^1} \delta_{m^1 l^1} .$$

The interaction part becomes a trace of product of matrices, and no oscillations occur in this basis. The propagator obtained from the free part is

quite complicated; in 4 dimensions it is:

$$
G_{\substack{m^1\,n^1\,;\,k^1\,l^1\\m^2\,n^2\,;\,k^2\,l^2}} = \frac{\theta}{2(1+\Omega)^2} \sum_{v^1=\frac{|m^1-l^1|}{2}}^{\frac{m^1+l^1}{2}} \sum_{v^2=\frac{|m^2-l^2|}{2}}^{\frac{m^2+l^2}{2}}
\tag{6.6}
$$

$$
+ B\left(1+\tfrac{\mu^2\theta}{8\Omega} + \tfrac{1}{2}(m^1+k^1+m^2+k^2) - v^1-v^2,\, 1+2v^1+2v^2\right)
$$

$$
\times\, {}_2F_1\left(\begin{array}{c} 1+2v^1+2v^2\,,\ \tfrac{\mu^2\theta}{8\Omega}-\tfrac{1}{2}(m^1+k^1+m^2+k^2)+v^1+v^2 \\ 2+\tfrac{\mu^2\theta}{8\Omega}+\tfrac{1}{2}(m^1+k^1+m^2+k^2)+v^1+v^2 \end{array}\middle|\, \frac{(1-\Omega)^2}{(1+\Omega)^2}\right)
$$

$$
\times \left(\frac{1-\Omega}{1+\Omega}\right)^{2v^1+2v^2}
$$

$$
\times \prod_{i=1}^{2} \delta_{m^i+k^i,n^i+l^i}\sqrt{\binom{n^i}{v^i+\frac{n^i-k^i}{2}}\binom{k^i}{v^i+\frac{k^i-n^i}{2}}\binom{m^i}{v^i+\frac{m^i-l^i}{2}}\binom{l^i}{v^i+\frac{l^i-m^i}{2}}}\ .
$$

The propagator shows asymmetric decay properties, cf. Fig. (6.1). They

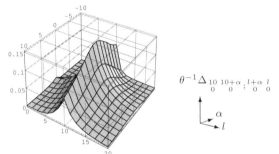

$$\theta^{-1}\Delta_{\substack{10\ 10+\alpha\\0\ \ 0}}{}_{;}{}_{\substack{l+\alpha\ l\\0\ \ 0}}$$

Figure 6.1   Decay properties, $\Omega = 0.1$, $\mu = 0$

decay exponentially in particular directions (in l-direction in Fig. (6.1)), but have power law decay in others (in α-direction). These decay properties are crucial for the perturbative renormalizability of the models.

The proof follows the ideas of Polchinski.[20] The quantum field theory corresponding to the action (6.4) is defined by the partition function

$$
Z[J] = \int \left(\prod_{m,n} d\phi_{mn}\right) \exp\left(-S[\phi] - \sum_{m,n}\phi_{mn}J_{nm}\right).
\tag{6.7}
$$

The strategy due to Wilson consists in integrating in the first step only those field modes ϕ_{mn} which have a matrix index bigger than some scale $\theta\Lambda^2$. The result is an effective action for the remaining field modes which depends on Λ. One can now adopt a smooth transition between integrated and not integrated field modes so that the Λ-dependence of the effective action is given by a certain differential equation, the Polchinski equation.

Renormalization amounts to prove that the Polchinski equation admits a regular solution for the effective action which depends on only a finite number of initial data. This requirement is hard to satisfy because the space of effective actions is infinite dimensional and as such develops an infinite dimensional space of singularities when starting from generic initial data.

The Polchinski equation can be solved iteratively in perturbation theory. Graphically, it is depicted in Fig. (6.2). The graphs are graded by the num-

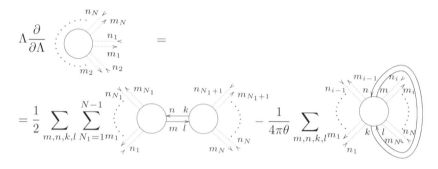

Figure 6.2   Polchinski equation

ber of vertices and the number of external legs. Only graphs with a smaller number of vertices and a bigger number of legs contribute to the Λ-variation of a graph on the lhs of Fig. (6.2). A general graph is thus obtained by iteratively adding a propagator to smaller building blocks, starting with the initial ϕ^4-vertex, and integrating over Λ. The propagators are differentiated cut-off propagators $Q_{mn;kl}(\Lambda)$ which vanish (for an appropriate choice of the cut-off function) unless the maximal index is in the interval $[\theta\Lambda^2, 2\theta\Lambda^2]$. Since fields carry two matrix indices and propagators four, the graphs are ribbon graphs familiar from matrix models.

It can then be shown that the cut-off propagator $Q(\Lambda)$ is bounded by $\frac{C}{\theta\Lambda^2}$. This was achieved numerically in Ref. 8 and later confirmed analytically in Ref. 21. A nonvanishing frequency parameter Ω is required for such a decay behavior. As the volume of each two-component index $m \in \mathbb{N}^2$ is bounded by $C'\theta^2\Lambda^4$ in graphs of the above type, the power counting degree of divergence is (at first sight) $\omega = 4S - 2I$, where I is the number of propagators and S the number of summation indices.

It is important to note that given three indices of a propagator $Q_{mn;kl}(\Lambda)$ the fourth one is determined by $m+k = n+l$. For simple planar graphs one finds that $\omega = 4-N$, where N is the number of external legs. At a closer look, however, one encounters a difficulty concerning completely inner vertices, which require additional index summations. The graph shown in Fig. (6.3) entails four independent summation indices p_1, p_2, p_3 and q,

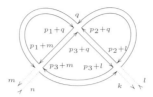

Figure 6.3 Pretzel

whereas for the powercounting degree $2 = 4 - N = 4S - 5 \cdot 2$ we should only have $S = 3$ of them. But due to the quasi-locality of the propagator (the exponential decay in l-direction in Fig. (6.1)), the sum over q for fixed m can be estimated without the need of the volume factor. Remarkably, the quasi-locality of the propagator not only ensures the correct powercounting degree for planar graphs, it also renders all nonplanar graphs superficially convergent. E.g., in the nonplanar graphs in Fig. (6.4) the summation over

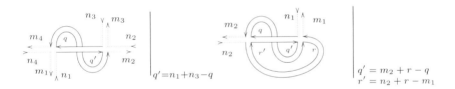

Figure 6.4 Nonplanar graphs

q and q, r, resp., is of the same type as over q in Fig. (6.3) so that the graphs in Fig. (6.4) can be estimated without any volume factor.

After all, we have obtained the powercounting degree of divergence

$$\omega = 4 - N - 4(2g + B - 1) \tag{6.8}$$

for a general ribbon graph, where g is the genus and B the number of holes of the Riemann surface on which the graph is drawn . Both are directly determined by the graph. It should be stressed that although the number (6.8) follows from counting the required volume factors, its proof in our scheme is not so obvious:[22] The procedure consists of adding a new cut-off propagator to a given graph. The topology (B, g) has many possibilities to arise from the topologies of the smaller parts for which one has estimates by induction. Moreover, the boundary conditions for the integration have to be correctly chosen to confirm (6.8).

The powercounting behavior is good news because it implies that all planar nonregular graphs are superficially convergent. Which is not true for $\Omega = 0$. However, this does not mean that all problems are solved: Power counting (6.8) suggests that the remaining planar two- and four-leg graphs which are divergent independent of their matrix indices. An infinite number of adjusted initial data would be necessary in order to remove these divergences. Fortunately, a more careful analysis shows that the powercounting behavior is improved by the index jump along the trajectories of the graph. E.g., the index jump for the graph (6.3) is defined as $J = \|k - n\|_1 + \|q - l\|_1 + \|m - q\|_1$. Then, the amplitude is suppressed by a factor of order $\left(\dfrac{\max(m, n \ldots)}{\theta \Lambda^2} \right)^{\frac{J}{2}}$ compared with the naive estimate. Thus, only planar four-leg graphs with $J = 0$ and planar two-leg graphs with $J = 0$ or $J = 2$ are divergent (the number of jumps is even). For these cases, a discrete Taylor expansion about the graphs with vanishing indices is employed. Only the leading terms of the expansion, i.e. the reference graphs with vanishing indices, are divergent whereas the difference between original graph and reference one is convergent. Thus, only the reference graphs must be integrated in a way that involves initial conditions. For example, if the contribution to the rhs of the Polchinski equation, Fig. (6.2) is given by:

$$\Lambda \frac{\partial}{\partial \Lambda} A^{(2)\text{planar},1\text{PI}}_{mn;nk;kl;lm}[\Lambda] = \sum_{p \in \mathbb{N}^2} \left(\begin{array}{c} \text{graph} \end{array} \right) (\Lambda), \qquad (6.9)$$

the Λ-integration is performed as follows:

$$A^{(2)\text{planar},1\text{PI}}_{mn;nk;kl;lm}[\Lambda] = \qquad (6.10)$$

$$= -\int_{\Lambda}^{\infty} \frac{d\Lambda'}{\Lambda'} \sum_{p \in \mathbb{N}^2} \left(\begin{array}{c} \text{graph} \end{array} - \begin{array}{c} \text{graph} \end{array} \right) [\Lambda']$$

$$+ \begin{array}{c} \text{graph} \end{array} \left[\int_{\Lambda_R}^{\Lambda} \frac{d\Lambda'}{\Lambda'} \sum_{p \in \mathbb{N}^2} \left(\begin{array}{c} \text{graph} \end{array} \right) [\Lambda'] + A^{(2,1,0)1\text{PI}}_{00;00;00;00}[\Lambda_R] \right].$$

Only one initial condition, $A^{(2,1,0)1\text{PI}}_{00;00;00;00}[\Lambda_R]$, is required for an infinite number of planar four-leg graphs (distinguished by the matrix indices). We need one further initial condition for the two-leg graphs with $J = 2$ and two more initial conditions for the two-leg graphs with $J = 0$ (for the leading quadratic and the subleading logarithmic divergence). This is one condition more than in a commutative ϕ^4-theory, and this additional condition justifies a posteriori our starting point of adding one new term to the action (6.3), the oscillator term Ω.

6.2.1. *The Landau ghost*

Knowing the relevant/marginal couplings, we can compute Feynman graphs with sharp matrix cut-off $\mathcal{N}$. The most important question concerns the β-function appearing in the renormalization group equation which describes the cut-off dependence of the expansion coefficients $\Gamma_{m_1 n_1;...;m_N n_N}$ of the effective action when imposing normalisation conditions for the relevant and marginal couplings. We have[23]

$$\lim_{\mathcal{N}\to\infty} \left(\mathcal{N}\frac{\partial}{\partial\mathcal{N}} + N\gamma + \mu_0^2\beta_{\mu_0}\frac{\partial}{\partial\mu_0^2} + \beta_\lambda\frac{\partial}{\partial\lambda} + \beta_\Omega\frac{\partial}{\partial\Omega} \right)$$
$$\times \Gamma_{m_1 n_1;...;m_N n_N}[\mu_0,\lambda,\Omega,\mathcal{N}] = 0 , \tag{6.11}$$

where

$$\beta_\lambda = \mathcal{N}\frac{\partial\lambda}{\partial\mathcal{N}} , \beta_\Omega = \mathcal{N}\frac{\partial\Omega}{\partial\mathcal{N}} ,$$
$$\beta_{\mu_0} = \frac{\mathcal{N}}{\mu_0^2}\frac{\partial\mu_0^2}{\partial\mathcal{N}} , \gamma = \mathcal{N}\frac{\partial\ln\mathcal{Z}}{\partial\mathcal{N}} .$$

The couplings depend on the physical parameters $\mu_{\text{phys}},\lambda_{\text{phys}},\Omega_{\text{phys}}$ and the cut-off $\mathcal{N}$. The wavefunction renormalization is denoted by $\mathcal{Z}$. To one-loop order one finds[23]

$$\beta_\lambda = \frac{\lambda_{\text{phys}}^2}{48\pi^2}\frac{(1-\Omega_{\text{phys}}^2)}{(1+\Omega_{\text{phys}}^2)^3} , \beta_\Omega = \frac{\lambda_{\text{phys}}\Omega_{\text{phys}}}{96\pi^2}\frac{(1-\Omega_{\text{phys}}^2)}{(1+\Omega_{\text{phys}}^2)^3} , \tag{6.12}$$

$$\beta_\mu = -\frac{\lambda_{\text{phys}}\left(4\mathcal{N}\ln(2) + \frac{(8+\theta\mu_{\text{phys}}^2)\Omega_{\text{phys}}^2}{(1+\Omega_{\text{phys}}^2)^2}\right)}{48\pi^2\theta\mu_{\text{phys}}^2(1+\Omega_{\text{phys}}^2)} , \gamma = \frac{\lambda_{\text{phys}}}{96\pi^2}\frac{\Omega_{\text{phys}}^2}{(1+\Omega_{\text{phys}}^2)^3} .$$

Hence, the ratio of the coupling constants $\frac{\lambda}{\Omega^2}$ remains bounded along the renormalization group flow up to first order. Starting from given small values for Ω_R,λ_R at $\mathcal{N}_R$, the frequency grows in a small region around $\ln\frac{\mathcal{N}}{\mathcal{N}_R} = \frac{48\pi^2}{\lambda_R}$ to $\Omega \approx 1$. The coupling constant approaches $\lambda_\infty = \frac{\lambda_R}{\Omega_R^2}$, which can be made small for sufficiently small λ_R. This leaves the chance of a nonperturbative construction[9] of the model. In particular, the β-function vanishes at the self-dual point $\Omega = 1$,[24] indicating special and interesting properties of the model.

6.3. Induced gauge theory

Since elementary particles are most successfully described by gauge theories it is a big challenge to formulate consistent gauge theories on noncommutative spaces. Let u be a unitary element of the algebra such that the scalar fields ϕ transform covariantly:

$$\phi \mapsto u^* \star \phi \star u, \ u \in \mathcal{G}. \tag{6.13}$$

The approach employed here makes use of two basic ideas. First, it is well known that the $\star$-multiplication of a coordinate - and also of a function, of course - with a field is not a covariant process. The product $x^\mu \star \phi$ will not transform covariantly. The introduction of covariant coordinates $\tilde{X}_\nu = \tilde{x}_\nu + A_\nu$ finds a remedy to this situation. The gauge field A_μ and hence the covariant coordinates transform in the following way:

$$A_\mu \mapsto iu^* \star \partial_\mu u + u^* \star A_\mu \star u, \quad \tilde{X}_\mu \mapsto u^* \star \tilde{X}_\mu \star u. \tag{6.14}$$

Using covariant coordinates we can construct an action invariant under gauge transformations from (6.3) - note that $\partial_\mu f = -i[\tilde{x}_\mu, f]_\star$:

$$S = \int d^4x \left(\frac{1}{2}\phi \star [\tilde{X}_\nu, [\tilde{X}^\nu, \phi]_\star]_\star + \frac{\Omega^2}{2}\phi \star \{\tilde{X}^\nu, \{\tilde{X}_\nu, \phi\}_\star\}_\star \right.$$
$$\left. + \frac{\mu^2}{2}\phi \star \phi + \frac{\lambda}{4!}\phi \star \phi \star \phi \star \phi \right)(x). \tag{6.15}$$

Secondly, we apply the heat kernel formalism. The gauge field A_μ is an external, classical gauge field coupled to ϕ. In the classical case, the divergent terms determine the dynamics of the gauge field.[25,26] There have already been attempts to generalise this approach to the noncommutative realm.[27-29] However, the results there are not applicable, since we have modified the free action and expand around $-\nabla^2 + \Omega^2 \tilde{x}^2$ rather than $-\nabla^2$.

The regularised one loop effective action for the model defined by the classical action (6.15) is given by

$$\Gamma^\epsilon_{1l}[\phi] = -\frac{1}{2} \int_\epsilon^\infty \frac{dt}{t} \text{Tr} \left(e^{-tH} - e^{-tH^0} \right). \tag{6.16}$$

For the effective potential H we have the expression

$$\frac{\theta}{2} \frac{\delta^2 S}{\delta \phi^2} \equiv H = H^0 + \frac{\theta}{2} V. \tag{6.17}$$

The effective action is calculated as a power series in the potential V up to fourth order using the Duhamel formula:

$$
\Gamma_{1l}^\epsilon = \frac{\theta}{4} \int_\epsilon^\infty dt\ \mathrm{Tr}\ V e^{-tH^0} - \frac{\theta^2}{8} \int_\epsilon^\infty \frac{dt}{t} \int_0^t dt'\ t'\ \mathrm{Tr}\ V e^{-t'H^0} V e^{-(t-t')H^0}
$$
$$
+ \frac{\theta^3}{16} \int_\epsilon^\infty \frac{dt}{t} \int_0^t dt' \int_0^{t'} dt''\ t''\ \mathrm{Tr}\ V e^{-t''H^0} V e^{-(t'-t'')H^0} V e^{-(t-t')H^0} \quad (6.18)
$$
$$
- \frac{\theta^4}{32} \int_\epsilon^\infty \frac{dt}{t} \int_0^t dt' \int_0^{t'} dt'' \int_0^{t''} dt'''\ t'''\ \mathrm{Tr}\ V e^{-t'''H^0} V e^{-(t''-t''')H^0}
$$
$$
\times V e^{-(t'-t'')H^0} V e^{-(t-t')H^0}.
$$

Higher order terms are already finite. As before, the calculations are performed in the matrix basis, where the star product is just a matrix product. After a suitable rescaling, all the operators depend, beside on θ, only on the following three parameters:

$$
\rho = \frac{1 - \Omega^2}{1 + \Omega^2}, \ \tilde{\epsilon} = \epsilon(1 + \Omega^2), \ \tilde{\mu}^2 = \frac{\mu^2 \theta}{1 + \Omega^2}. \quad (6.19)
$$

We concentrate on terms involving only the gauge field and assume $\lambda = 0$. The explicit calculation is very tedious and is given in detail in Refs. 16, 17. Although the method is not manifestly gauge invariant, various terms from different orders add up to a gauge invariant final expression:

$$
\Gamma_{1l}^\epsilon = \frac{1}{192\pi^2} \int d^4x \left\{ \frac{24}{\tilde{\epsilon}\,\theta} (1 - \rho^2)(\tilde{X}_\nu \star \tilde{X}^\nu - \tilde{x}^2) \right. \quad (6.20)
$$
$$
+ \ln \epsilon \left(\frac{12}{\theta} (1 - \rho^2)(\tilde{\mu}^2 - \rho^2)(\tilde{X}_\nu \star \tilde{X}^\nu - \tilde{x}^2) \right.
$$
$$
\left. \left. + 6(1 - \rho^2)^2 \big((\tilde{X}_\mu \star \tilde{X}^\mu)^{\star 2} - (\tilde{x}^2)^2\big) + \rho^4 F_{\mu\nu} F^{\mu\nu} \right) \right\},
$$

where the field strength is given by $F_{\mu\nu} = [\tilde{x}_\mu, A_\nu]_\star - [\tilde{x}_\nu, A_\mu]_\star + [A_\mu, A_\nu]_\star$. Our main result is summarised in Eq. (6.20): The logarithmically divergent part is an interesting candidate for a renormalisable gauge interaction. As far as we know, this action did not appear before in string theory. The sign of the term quadratic in the covariant coordinates may change depending on whether $\tilde{\mu}^2 \lessgtr \rho^2$. This reflects a phase structure. The matrix model in the limit $\Omega = 1$ ($\rho = 0$) is of particular interest. In the limit $\Omega \to 0$, we obtain just the standard deformed Yang-Mills action. One of the problems of quantising action (6.20) is connected to the tadpole contribution, which is non-vanishing and hard to eliminate. The Orsay group also considered the 1-loop effective action in the case $\Omega \neq 0$, for a complex model. They calculated the divergent contributions in x-space by evaluating relevant Feynman diagrams and arrived at the same result.[30]

An appropriate rescaling $\tilde{X}_\alpha \to \frac{\sqrt{2\sqrt{3}}}{\sqrt{\theta}} \tilde{X}_\alpha$ and $\tau \equiv -\sqrt{3}\frac{1-\rho^2}{\rho^2}$ leads to the equations of motion

$$D_\nu F^{\sigma\nu} = \tau \tilde{X}^\sigma + \tau^2 \{\tilde{X}^\sigma, \tilde{X}_\nu \star \tilde{X}^\nu\}_\star, \qquad (6.21)$$

where we have assumed for simplicity $\tilde{\mu} = 0$ and used $D_\nu F^{\sigma\nu} = -[\tilde{X}_\nu, [\tilde{X}^\sigma, \tilde{X}^\nu]_\star]_\star$. In Ref. 31, the matter fields have been included in order to find some solutions. However, the gauge part (6.21) alone also exhibits a number of solutions currently under investigation, such as $su(2)$.

A similar model has been discussed in Ref. 18. This model includes an oscillator potential for the gauge fields, $\tilde{x}^2 A^2$, and for the ghosts. Other terms occuring here are missing. Hence, the considered action is not gauge invariant, but a BRST invariance could be established. These "missing" terms may nevertheless come into the game through one loop corrections. A brief sketch of this model follows.

6.3.1. *BRST quantization*

A BRST invariant model for noncommutative $U(1)$ gauge theory has been introduced in Ref. 18. Its action is given by

$$\Gamma^{(0)} = \int d^4x \left(\frac{1}{4}F_{\mu\nu} \star F_{\mu\nu} + s(\bar{c} \star \partial_\mu A_\mu) - \frac{1}{2}B^2 + \frac{\Omega^2}{8}s(\tilde{c}_\mu \star C_\mu) \right), \quad (6.22)$$

where $C_\mu = \{\{\tilde{x}_\mu, A_\nu\}_\star, A_\nu\}_\star + [\{\tilde{x}_\mu, \bar{c}\}_\star, c]_\star + [\bar{c}, \{\tilde{x}_\mu, c\}_\star]_\star$. These terms modify the free theory, the propagators of the involved fields. The action is - by constructed - invariant under the following noncommutative BRST transformations

$$sA_\mu = \partial_\mu c - ig[A_\mu, c]_\star, \quad s\bar{c} = B,$$
$$sc = igc \star c, \quad sB = 0, \quad s\tilde{c}_\mu = \tilde{x}_\mu. \qquad (6.23)$$

Hence, we have $s^2 = 0$. The multiplier field B implements the gauge fixing. In the limit $\tilde{c}_\mu \to 0$, this reduces to the usual Feynman gauge, $\partial_\mu A^\mu - B = 0$. The field $\tilde{c}_\mu$ with mass dimension 1 and ghost number -1 is yet another Lagrange multiplier. It has been introduced in order to rescue the BRST invariance of the x^2 terms for ghosts and gauge field contained in C_μ. These terms also modify the free theory. All propagators are given by the Mehler kernel (6.24), where the momentum non-conservation becomes transparent.

The index structure for the gauge field propagator is trivial and given by the Kronecker delta $\delta^{\alpha\beta}$. In momentum space, the Mehler Kernel reads

$$\tilde{K}_M(p,q) = \frac{\omega^3}{8\pi^2}\int_\epsilon^\infty \frac{d\alpha}{\sinh^2\alpha}\exp\left\{-\frac{\omega}{4}u^2\coth\frac{\alpha}{2} - \frac{\omega}{4}v^2\tanh\frac{\alpha}{2}\right\}, \quad (6.24)$$

where $\omega = \frac{\theta}{\Omega}$, and $u = p-q$, $v = p+q$. We use a UV regulator $\epsilon = 1/\Lambda^2$ in order to cut the integration over the auxiliary Schwinger parameter. The action provides the weights for a three-photon ($\tilde{V}_{\mu\nu\sigma}^{3A}$) and a four-photon vertex ($\tilde{V}_{\mu\nu\sigma\tau}^{4A}$), and also for a ghost – two photon vertex ($\tilde{V}_\mu^c$). The explicit expressions will be presented in a forthcoming publication.

Preliminary results. Let us discuss some preliminary results to one-loop. Remarkably, the tadpole contribution (one external photon) is divergent. After Taylor expansion of the external field, the divergent contributions are given by - summation over all indices is implied:

$$T = \int d^4p\tilde{A}_\rho(p)\int d^4k\int d^4k'\tilde{K}_M(k,k')\delta_{\sigma\tau}(\tilde{V}_{\rho\sigma\tau}^{3A}(p,-k',k) + \tilde{V}_\rho^c(p,-k',k))$$

$$= -\frac{c}{\epsilon}\int d^4x\,\tilde{x}_\mu A^\mu - \ln\epsilon\frac{c\Omega\theta}{4+\Omega^2}\int d^4x\,\tilde{x}^2\tilde{x}_\mu A^\mu + \mathcal{O}(\epsilon^0). \quad (6.25)$$

This suggests to introduce the counterterms $\tilde{x}_\mu A^\mu$ and $\tilde{x}^2\tilde{x}_\mu A^\mu$.

The contribution for the gauge loop with two external photon legs and one photon vertex is given by

$$D = \int d^4p\int d^4q\tilde{A}_\rho(p)\tilde{A}_\epsilon(-q)\int d^4k_1\int d^4k_2\tilde{K}_M(k_1,k_2)\delta_{\sigma\tau}$$

$$\times \tilde{V}_{\rho\sigma\tau\epsilon}^{4A}(p,-k_1,k_2,-q). \quad (6.26)$$

To extract the divergent contributions, we perform a Taylor expansion of the integrand depending on q, except for the Mehler kernel, around $p = q$. The divergent contributions are given by

$$D_{(1)} = \frac{c'\,\omega}{\epsilon}\int d^4p\,\tilde{A}_\mu(p)\tilde{A}_\mu(-p), \quad (6.27)$$

$$D_{(2)} = \ln\epsilon\frac{c'}{2\bar{\omega}}\int d^4p\,\tilde{A}_\mu(p)\tilde{A}_\mu(-p)\,\tilde{p}^2, \quad (6.28)$$

$$D_{(3)} = -\ln\epsilon\frac{c'}{2}\left(4 - \frac{\theta^2}{\omega\bar{\omega}}\right)\int d^4p\,\tilde{A}_\rho(p)(-\partial^{q^2}\tilde{A}_\rho)(-p), \quad (6.29)$$

with $\bar{\omega} = \omega(1 + \frac{\theta^2}{4\omega^2})$ suggesting further counterterms. Remarkably, all the counterterms are contained in the induced action (6.20). There are more one-loop diagrams which have to be studied. Then, the exact coefficients of the calculated counterterms can be compared with the induced action. Of course, it is important to study the behavior at higher loops. Especially the IR behaviour is of interest in order to see whether IR/UV mixing is present in this model. For this aim, we look at a diagram with two gauge loops of type (6.26). We proceed similar as in the scalar case: To first approximation, momentum is conserved at the vertex, and we obtain

$$D \sim \int d^4 p \tilde{A}_\mu(p) \tilde{A}_\mu(-p) \int d^4 u \int d^4 v \tilde{K}_M(u, v) \left(\cos \frac{p\theta u}{2} - \cos \frac{p\theta v}{2} \right). \quad (6.30)$$

The first term corresponds to the planar regular part and the second one to the nonregular one. The latter one coincides - except for the sign - with the corresponding contribution in the scalar case (with oscillator term). Therefore, the same estimates, which showed that IR/UV mixing does not occur there, seem to be applicable here. This point is studied at the moment and gives rise to the hope that the discussed gauge model is free of IR/UV divergences.

References

1. H. S. Snyder, Quantized Space-Time, *Phys. Rev.* **71**, 38–41, (1947).
2. A. Connes, Noncommutative differential geometry, *Inst. Hautes Etudes Sci. Publ. Math.* **62**, 257, (1986).
3. T. Filk, Divergencies in a field theory on quantum space, *Phys. Lett.* **B376**, 53–58, (1996).
4. S. Doplicher, K. Fredenhagen, and J. E. Roberts, The quantum structure of space-time at the Planck scale and quantum fields, *Commun. Math. Phys.* **172**, 187–220, (1995).
5. V. Schomerus, D-branes and deformation quantization, *JHEP.* **06**, 030, (1999).
6. N. Seiberg and E. Witten, String theory and noncommutative geometry, *JHEP.* **09**, 032, (1999).
7. S. Minwalla, M. Van Raamsdonk, and N. Seiberg, Noncommutative perturbative dynamics, *JHEP.* **02**, 020, (2000).
8. H. Grosse and R. Wulkenhaar, Renormalisation of ϕ^4 theory on noncommutative $\mathbb{R}^4$ in the matrix base, *Commun. Math. Phys.* **256**, 305–374, (2005).
9. J. Magnen and V. Rivasseau, Constructive ϕ^4 field theory without tears, *Annales Henri Poincare.* **9**, 403–424, (2008). doi: 10.1007/s00023-008-0360-1.
10. S. S. Gubser and S. L. Sondhi, Phase structure of non-commutative scalar field theories, *Nucl. Phys.* **B605**, 395–424, (2001).
11. W. Bietenholz, F. Hofheinz, and J. Nishimura, Phase diagram and dispersion relation of the non- commutative lambda ϕ^4 model in $d = 3$, *JHEP.* **06**, 042, (2004).

12. X. Martin, A matrix phase for the ϕ^4 scalar field on the fuzzy sphere, *JHEP.* **04**, 077, (2004).

13. J. Ambjorn, Y. M. Makeenko, J. Nishimura, and R. J. Szabo, Nonperturbative dynamics of noncommutative gauge theory, *Phys. Lett.* **B480**, 399–408, (2000).

14. H. Steinacker, Quantized gauge theory on the fuzzy sphere as random matrix model, *Nucl. Phys.* **B679**, 66–98, (2004).

15. P. B. Gilkey, Invariance theory, the heat equation and the Atiyah-Singer index theorem. (1995).

16. H. Grosse and M. Wohlgenannt, Induced gauge theory on a noncommutative space, *Eur. Phys. J.* **C52**, 435–450, (2007).

17. M. Wohlgenannt, Induced Gauge Theory on a Noncommutative Space, *J. Phys. Conf. Ser.* **103**, 012008, (2008).

18. D. N. Blaschke, H. Grosse, and M. Schweda, Non-commutative U(1) gauge theory on $\mathbb{R}^4_\theta$ with oscillator term and BRST symmetry, *Europhys. Lett.* **79**, 61002, (2007). doi: 10.1209/0295-5075/79/61002.

19. E. Langmann and R. J. Szabo, Duality in scalar field theory on noncommutative phase spaces, *Phys. Lett.* **B533**, 168–177, (2002).

20. J. Polchinski, Renormalization and effective Lagrangians, *Nucl. Phys.* **B231**, 269–295, (1984).

21. V. Rivasseau, F. Vignes-Tourneret, and R. Wulkenhaar, Renormalization of noncommutative ϕ^4-theory by multi- scale analysis, *Commun. Math. Phys.* **262**, 565–594, (2006).

22. H. Grosse and R. Wulkenhaar, Power-counting theorem for non-local matrix models and renormalisation, *Commun. Math. Phys.* **254**, 91–127, (2005).

23. H. Grosse and R. Wulkenhaar, The beta-function in duality-covariant non-commutative ϕ^4 theory, *Eur. Phys. J.* **C35**, 277–282, (2004).

24. M. Disertori, R. Gurau, J. Magnen, and V. Rivasseau, Vanishing of beta function of non commutative $\phi(4)^4$ theory to all orders, *Phys. Lett.* **B649**, 95–102, (2007). doi: 10.1016/j.physletb.2007.04.007.

25. A. H. Chamseddine and A. Connes, The spectral action principle, *Commun. Math. Phys.* **186**, 731–750, (1997).

26. E. Langmann, Generalized Yang-Mills actions from Dirac operator determinants, *J. Math. Phys.* **42**, 5238–5256, (2001).

27. D. V. Vassilevich, Non-commutative heat kernel, *Lett. Math. Phys.* **67**, 185–194, (2004).

28. V. Gayral and B. Iochum, The spectral action for Moyal planes, *J. Math. Phys.* **46**, 043503, (2005).

29. D. V. Vassilevich, Heat kernel, effective action and anomalies in noncommutative theories, *JHEP.* **08**, 085, (2005).

30. A. de Goursac, J.-C. Wallet, and R. Wulkenhaar, Noncommutative induced gauge theory, *Eur. Phys. J.* **C51**, 977–987, (2007).

31. H. Grosse and R. Wulkenhaar, 8D-spectral triple on 4D-Moyal space and the vacuum of noncommutative gauge theory. (2007).

Chapter 7

Topological Phases and Contextuality Effects in Neutron Quantum Optics

Helmut Rauch

Vienna University of Technology,
Atomic Institute of the Austrian Universities
Stadionallee 2, 1020 Wien, Austria
E-mail: rauch@ati.ac.at

Neutron matter-wave optics provides the basis for new quantum experiments and a step towards applications of quantum phenomena. Most experiments have been performed with a perfect crystal neutron interferometer where widely separated coherent beams can be manipulated individually. Various geometric phases have been measured and their robustness against fluctuation effects has been proven, which may become a useful property for advanced quantum communication. Quantum contextuality for single particle systems shows that quantum correlations are to some extent more demanding than classical ones. In this case entanglement between external and internal degrees of freedom offers new insights into basic laws of quantum physics. Non-contextuality hidden theories can be rejected by arguments based on the Kochen-Specker theorem.

7.1. Introduction

Quantum physics describes nature by means of the Schrödinger equation

$$ i\hbar\frac{\partial|\psi\rangle}{\partial t} = \mathcal{H}|\psi\rangle \,, \tag{7.1} $$

where $\mathcal{H}$ denotes the interaction Hamiltonian between physical entities, e.g. particle-like neutrons, with the nuclei of a target. For elastic scattering processes a solution of this equation can be given by the partial wave method, where the ψ-function can be written as an incident plane wave with a spherical diffracted wave[1]

$$ \psi = e^{ikz} + f_k(\vartheta)\frac{e^{ikr}}{r} \,, \tag{7.2} $$

with

$$f_k = \frac{1}{k} \sum_{l=0}^{\infty} (2l + 1)e^{i\delta_l} \sin \delta_l P_l(\cos \Theta) \,, \qquad (7.3)$$

where P_l denotes the Legendre polynomial of order l and δ_l the phase-shift caused by the interaction.

In this article we deal with the interaction of slow neutrons with matter where only the $l = 0$ (s-wave) contribution has to be considered. This restriction is justified because the wavelength of the neutrons is much larger than the range of strong interaction and because all experimental facts confirm isotropic scattering in the center of mass system. Therefore, the wave function can be written in the form

$$\psi = e^{ikz} + b\frac{e^{ikr}}{r} \,, \qquad (7.4)$$

where the scattering length b is a constant related to the phase shift between the incident plane and reflected spherical wave, $b = \lim_{k \to 0} -(\sin \delta_0/k) \simeq -\delta_0/k$ (see Fig. 7.1). Since there is yet no complete theory of strong interaction the phase shifts have to be taken as parameters describing the individual interaction. Isotropic scattering can be described within the first Born approximation by scattering from a point-like potential in form

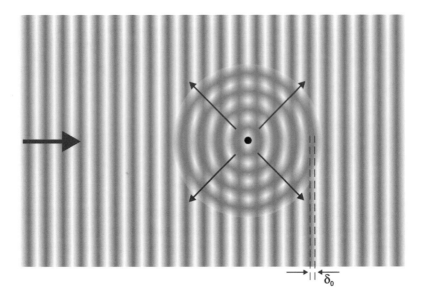

Figure 7.1   Phase shift due to interaction of slow neutrons with a nucleus.

of a Fermi δ-potential

$$V(\vec{r}) = \frac{2\pi\hbar^2}{m} b\delta(\vec{r}) \,. \tag{7.5}$$

In a more general sense the incident wave has to be described by a wave packet which is the general solution of Eq. (7.5) for the interaction free region

$$\psi(\vec{r}, t) = (2\pi)^{-3/2} \int a(\vec{k}, \omega) e^{i(\vec{k}\vec{r} - \omega t)} d^3k d\omega \,. \tag{7.6}$$

It represents the coherent superposition of plane waves and, in terms of quantum optics, a multimode coherent state that can be seen as a quasi-classical state. $|a(\vec{k}, \omega)|^2 = g(\vec{k}, \omega)$ denotes the density of states.

The first-order, two-point, two-time auto-correlation function relating the physical situation at $(\vec{r}, t)$ and $(\vec{r}', t)$ is given as[2]

$$G^{(1)}(\vec{r}, t, \vec{r}', t') = T_r\{\hat{\rho}\psi^*(\vec{r}, t)\psi(\vec{r}', t')\} \,, \tag{7.7}$$

where the density operator $\hat{\rho}$ can be written as

$$\hat{\rho} = \int p(\vec{r}, t)\psi^*(\vec{r}, t)\psi(\vec{r}, t)d^3r dt \,. \tag{7.8}$$

$p(\vec{r}, t)$ denotes the classical probability of finding a quantum system at $\vec{r}$ and t, e.g. the beam profile or the time structure of the beam.

The auto-correlation function of the wave function gives ($\Delta = \vec{r} - \vec{r}', \tau = t - t'$)

$$\Gamma^{(1)}(\Delta, \tau) = \langle\psi^*(\vec{r}, t)\psi(\vec{r}', t')\rangle = \int g(k, \omega) e^{i(\vec{k}\vec{\Delta} - \omega_k \tau)} d^3k d\omega \,. \tag{7.9}$$

For a stationary beam this relation simplifies to (Cittert-Zernike theorem)

$$\Gamma^{(1)}(\Delta) = \int g(k) e^{i\vec{k}\vec{\Delta}} d^3k \,. \tag{7.10}$$

The characteristic dimension of this function defines the coherence length

$$\Delta_c^2 = \frac{\int \Delta^2 |\Gamma^{(1)}(\Delta)| d\Delta}{\int |\Gamma^{(1)}(\Delta)| d\Delta} \,. \tag{7.11}$$

Since the wavelength, λ, and especially the coherence length of the wave packets are considerably larger than the interatomic distances in condensed matter a simultaneous interaction with many nuclei takes place Fig. 7.2. This justifies the definition of a coherent b_c and an incoherent b_i scattering

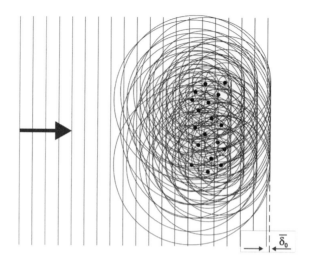

Figure 7.2   Mean phase shift of a neutron wave interacting with many nuclei.

length which are related to a mean phase shift $\langle \delta_0 \rangle$ and to its variance $\langle \delta_0^2 \rangle - \langle \delta_0 \rangle^2$.

$$b_c \simeq -\frac{\langle \delta_0 \rangle}{k} \ , \tag{7.12}$$

$$b_i^2 \simeq -\frac{\langle \delta_0^2 \rangle - \langle \delta_0 \rangle^2}{k^2} \ . \tag{7.13}$$

In the case of spin incoherence this reads as

$$b_c = \frac{I+1}{2I+1}b_+ + \frac{I}{2I+1}b_- \ , \tag{7.14}$$

$$b_i = \frac{\sqrt{I(I+1)}}{2I+1}(b_+ - b_-) \ . \tag{7.15}$$

where b_+ and b_- denote the scattering lengths for interaction in the $I + 1/2$ and $I - 1/2$ channel, respectively (I ... nuclear spin). In the forward direction the scattered waves overlap coherently as far as $\langle \delta_0 \rangle$ is concerned and in diffracted directions they also overlap coherently when the scattering centers are regularly arranged and the momentum transfer Q relates to the reciprocal lattice constant τ ($\vec{Q} = 2\pi\vec{\tau}$ ... Bragg equation). Since the neutron wave interacts with many nuclei a mean interaction potential can be defined (optical potential, i.e. Ref. 3)

$$\bar{V} = \int \frac{2\pi\hbar^2}{m} b_c \delta(\vec{r}) d^3 r = \frac{2\pi\hbar^2}{m} b_c N \ , \tag{7.16}$$

where N denotes the number of target atoms within the unit volume. Thus, for a stationary situation the time-independent Schrödinger equation has to be solved for a step potential

$$-\frac{\hbar^2}{2m}\nabla^2\psi(r) + \bar{V}(r)\psi(r) = E\psi(r) , \qquad (7.17)$$

where E denotes the energy-eigenvalue of the motion which has a continuous spectrum when a wave packet is considered $E = \hbar^2 k^2/2m$. Inside the material the wave number becomes $K^2 = 2m[E - \bar{V}(r)]/\hbar^2$ and, therefore, an index of refraction can be defined

$$n = \frac{K}{k} = \sqrt{1 - \frac{\bar{V}}{E}} \simeq 1 - \frac{\bar{V}}{2E} = 1 - \lambda^2\frac{Nb_c}{2\pi} , \qquad (7.18)$$

where the second part of this equation follows from $\bar{V} \ll E$. From this a phase shift, χ, between a wave transmitted through a material of thickness D and a free travelling wave can be derived

$$\chi = \vec{\Delta} \cdot \vec{k} = (k - K)D = k(1 - n)D = Nb_c\lambda D . \qquad (7.19)$$

In a rather general sense the phase shift can be calculated by a path integral along a loop[4]

$$\chi = \oint \vec{k}\, d\vec{s} \qquad (7.20)$$

where k denotes the canonical momentum of the neutron.

In a similar way, a phase shift due to the static magnetic interaction $\bar{V}_{mag} = -\vec{\mu}_n\vec{B}(\vec{r})$ can be defined for a homogeneous field B as, e.g. Ref. 5

$$\chi_{mag} = \pm 2\pi m\mu_n BD/\hbar^2 , \qquad (7.21)$$

and which is half of the Larmor precession angle indicating its connection to spinor properties (see Sec. 7.3.1).

Gravitational and Coriolis interaction also cause a phase shift due to the related interactions; $\bar{V}_{grav} = m\vec{g}\vec{r}$, where $\vec{g}$ denotes the graviational acceleration and $\bar{V}_{cor} = -\hbar\Omega(\vec{r}\times\vec{k})$, where Ω is the Earth rotation frequency. The related phase shifts between the two beams in an interferometer read as: (e.g. Ref. 6)

$$\chi_{grav} = -2\pi\lambda m^2 g A_0 \sin\alpha/\hbar^2 + 4\pi m\Omega A_0 \cos\Theta_L \cos\alpha/\hbar . \qquad (7.22)$$

A_0 denotes the area encircled by the beams, α is the angle of this area against horizontality, and Θ_L is the colatitude angle where the experiment is carried out.

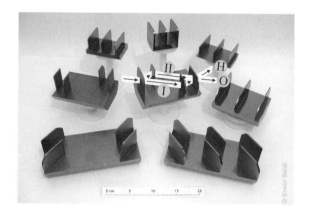

Figure 7.3   Various types of silicon perfect crystal neutron interferometers.

7.2. Neutron Interferometry

Perfect crystal neutron interferometry[5,7] allows application of different interactions to the widely separated coherent beams and, therefore, the realization of many basic quantum experiments where quantum phases and quantum topology can be studied in detail. Figure 7.3 shows different perfect crystal neutron interferometers used in the course of our experiments. The perfect arrangement of the atoms within a perfect silicon crystal allows the calculation of wave functions behind the individual plates and behind the whole interferometer. This calculation is based on the dynamical diffraction theory first developed for X-rays and electrons and later on adapted to neutrons.[6,8] The related wave functions have quite a difficult structure due to the mutual interferences of internal wave fields, but for the use of such interferometers for quantum measurements the following relations can explain the action of such a device. The wave function behind the interferometer in the forward (0) direction is composed of wave functions arising from beam paths I and II

$$I \propto |\psi_I + \psi_{II}|^2 \, , \tag{7.23}$$

where beam path I is transmitted-reflected-reflected and beam path II is reflected-reflected-transmitted, respectively. Due to symmetry relations both wave functions have to be equal concerning amplitude and phase, thus $\psi_I = \psi_{II}$. When a phase shift, χ, is applied to one beam one gets

$$I \propto |\psi_I|^2 (1 + \cos \chi) \, . \tag{7.24}$$

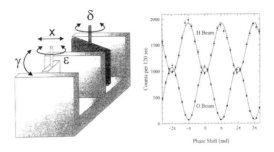

Figure 7.4   Arrangement of phase shifters within the interferometer and observed interference pattern.

Small deviations from the perfectness of such a system are unavoidable; small variations of the lattice constant, small rotations of the lattice planes, small temperature gradients, and small vibrations. Therefore, the observed interference pattern has to be written as

$$I = A + B \cos(\chi + \chi_0) \,, \qquad (7.25)$$

where A, B and χ_0 are internal parameters of the individual set-up. Figure 7.4 shows that a nearly ideal situation can be achieved when the adjustment procedure is done properly. Differently shaped crystals permit interferometers with rather long beam paths (20cm), large beam separations (8cm) and double loop interferometer set-ups (Figure 7.3).

Most of the experiments have been performed with thermal neutrons having an energy of about 0.025 eV, a wavelength of about 1.8Å, and an energy spread of $\delta E/E \simeq 0.01$. Fission neutrons are slowed down within a moderator of a reactor and a thermal beam is extracted and monochromatized afterwards. All the experiments are done in the single particle regime since the phase space density of any neutron sources is rather small (10^{-15}), which means that only one neutron at a time is within the apparatus and the next one is still in the uranium nucleus of the reactor fuel. Even so, the results indicate that each neutron behind the interferometer has information about the physical situation in both widely separated beam paths. The arrangement clearly demonstrates wave-particle duality since the neutron after fission and during moderation behaves as particle, inside the interferometer as a wave and behind it, in the detector, as a particle again.

7.3. Experimental Results from the 20th Century

7.3.1. *Spinor Symmetry: Neutrons in a Magnetic Field*

The interaction Hamiltonian can be written as

$$\mathcal{H}_{mag} = -\vec{\mu}_n \vec{B}(\vec{r}, t) , \tag{7.26}$$

which for a static and homogeneous field simplifies to

$$\mathcal{H}_{mag} = -\vec{\mu}_n \vec{B} \sigma_z , \tag{7.27}$$

and the neutron wave function evolves as

$$|\psi(t)\rangle = e^{-i\mathcal{H}t/\hbar}|\psi(0)\rangle = e^{-i\mu_n B \sigma_z/\hbar}|\psi(0)\rangle \tag{7.28}$$
$$= e^{-i\sigma_z \alpha/2/\hbar}|\psi(0)\rangle = |\psi(\alpha)\rangle ,$$

where α denotes the Larmor precession angle

$$\alpha = \frac{2\mu_n B t}{\hbar} = \frac{2\mu_n B L}{\hbar \nu} , \tag{7.29}$$

where L is the length of the interaction region and ν is the velocity of the neutrons. From relation Eq. (7.29) follows

$$|\psi(2\pi)\rangle = -|\psi(0)\rangle ,$$
$$|\psi(4\pi)\rangle = |\psi(0)\rangle , \tag{7.30}$$

showing the 4π-symmetry of spinor wave functions Fig. 7.5.

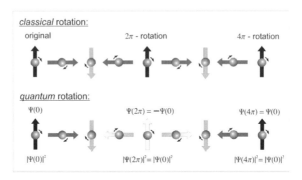

Figure 7.5   Schematic view of the 4π-symmetry.

This feature becomes measurable with the neutron interferometer as predicted by Aharonov and Susskind[9] and by Bernstein[10]

$$I_0 \propto \left| |\psi_0(0)\rangle + |\psi_0(\alpha)\rangle \right|^2 \propto 1 + \cos\frac{\alpha}{2} . \tag{7.31}$$

The related experiments have been done nearly simultaneously by our group[11] and by a U.S. one[12] and demonstrated this basic feature Fig. 7.6. It shows the experimental set-up and the first results giving $\alpha = (715.9 \pm 3.8)°$. It has been measured with unpolarized and with polarized neutrons which demonstrates the intrinsic feature of this phenomenon. Depending on the axis of quantization chosen the effect can be described by the index of refraction or a rotation around the magnetic field. Since the effect depends on the strength of the field the related phase shift has to be seen as a dynamical one in comparison with a geometric phase as discussed in Sec. 7.4.1.

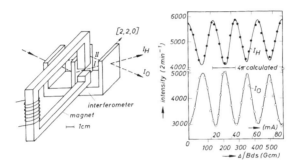

Figure 7.6 Experimental set-up and results of the first verification experiment of 4π-symmetry of spinor wave functions.[11]

7.3.2. *Spin Superposition*

In a famous article on the theory of measurement Wigner[13] brought attention to the curious situation of a superposition of a spin-up and a spin-down state since quantum mechanics predicts a state perpendicular to both initial states with an angle depending on the relative phase of those states, whereas classical physics predicts a mixture (Figure 7.7). When one uses

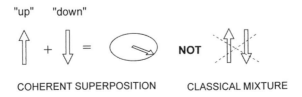

Figure 7.7 Sketch of quantum and classical spin-superposition.

polarized incident neutrons and rotates the direction of polarization in one
beam path by π the related wave function reads

$$|\psi'(\xi,\pi)\rangle = e^{i\chi}e^{-i\sigma_y\pi/2}|z\rangle = e^{i\chi}|-z\rangle , \qquad (7.32)$$

and the final polarization after superposition becomes $(\psi = \psi' + \psi'')$

$$\vec{P} = \frac{\langle\psi|\vec{\sigma}|\psi\rangle}{\langle\psi|\psi\rangle} = \begin{pmatrix} \cos\chi \\ \sin\chi \\ 0 \end{pmatrix} . \qquad (7.33)$$

It shows that an initially pure state in the $|z\rangle$-direction is transferred to
a pure state in the x,y-plane indicating again that information from both
beam parts are needed to explain these phenomena. The scheme and the
results of such an experiment are shown in Fig. 7.8.[14] The spin rotation in
this case achieved with by Larmor precession coils.[15]

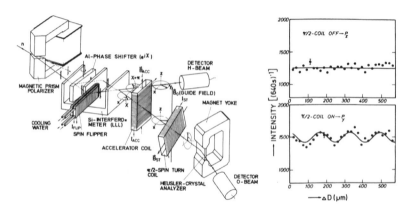

Figure 7.8   Experimental arrangement and results of the static spin-superposition experiment.[14]

A similar experiment has been done with a resonance Rabi-flipper.[16] In
this case, the spin reversal is caused by a time-dependent interaction and
is associated with an energy change of $\hbar\omega_r = 2|\mu|B_0$.[17,18] Therefore, the
wave function changes as:

$$\psi''(\chi,\omega_r) = e^{i\chi}e^{-i(\omega-\omega_r)t}|-z\rangle , \qquad (7.34)$$

and the final polarization rotates within the x,y-plane as

$$\vec{P} = \begin{pmatrix} \cos(\chi-\omega_r t) \\ \sin(\chi-\omega_r t) \\ 0 \end{pmatrix} , \qquad (7.35)$$

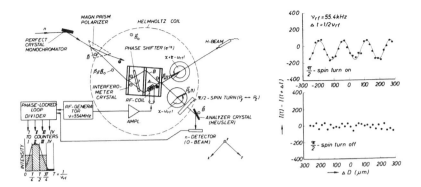

Figure 7.9 Experimental arrangement and results of the dynamic spin-superposition experiment indicating the stroboscopic measuring technique.[19]

which has been verified experimentally (Fig. 7.9[19]). Since one photon is exchanged between the neutron and the resonator, the question arises whether this can be used for beam path detection in the sense of Feynman's discussion of the double slit experiment.[20] It can be concluded "no" because phase information, ϕ, of the classical resonance field is needed to perform the stroboscopic registration of neutrons and then it follows from the particle number-phase uncertainty relation in its simplest form (e.g. Ref. 21) that a single exchange photon cannot be registered. For a more profound discussion see Refs. 22, 23.

7.3.3. *Magnetic Josephson Effect*

When two resonance flippers are inserted, one in each beam, and operated at a slightly different frequence the wave function becomes

$$\psi = \psi_0^I + \psi_0^{II} = e^{-i(\omega - \omega_{r1})t}|-z\rangle + e^{i\chi}e^{-i(\omega - \omega_{r2})t}|-z\rangle \,, \qquad (7.36)$$

which yields an intensity modulation as

$$I_0 \propto 1 + \cos[\chi + (\omega_{r1} - \omega_{r2})t] \,. \qquad (7.37)$$

Thus the intensity oscillates between forward (0) and diffracted beam (H) being driven by an extremely small frequency difference which relates in the case of our measurements to a small energy difference, $\Delta E = 8.6 \times 10^{-17}$ eV (Fig. 7.10).[24] A similar experiment has also been done by means of a neutron polarimeter set-up by a Japanese group.[25]

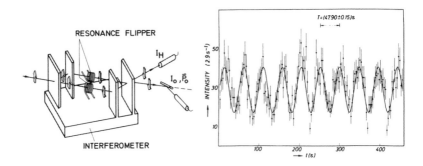

Figure 7.10   Double coil experiment verifying the magnetic Josephson effect.[24]

We can formulate the different energy transfer as a time-dependent phase shift

$$\Delta = \Delta_1 - \Delta_2 = (\omega_{r1} - \omega_{r2})t = \frac{2\mu_n \Delta B_0}{\hbar} t \,, \qquad (7.38)$$

which compares to the analog equation for the electric Josephson effect

$$\Delta\phi = \phi_1 - \phi_2 = \frac{2eV}{\hbar}t \,, \qquad (7.39)$$

where the rapidly oscillating tunneling current ($I_s = I_{Max} \sin\phi$) is driven by the electric potential between the tunnel junction.[26]

When an off-resonance field is applied multi-photon exchange processes occur which are related to dressed neutron phenomena. This behavior was tested in a dedicated experiment.[27]

7.4. Experiments from the 21st Century

7.4.1. Geometric Phases

The solution of the Schrödinger equation (Eq. (7.1)) for an adiabatic evolution can be written in the form

$$|\psi(t)\rangle = e^{i\phi(t)}|n(\vec{R}(t))\rangle \,, \qquad (7.40)$$

where $|n(\vec{R}(t))\rangle$ denotes the eigenstate of the instantaneous Hamiltonian $\mathcal{H}(\vec{R}(\sqcup))|\backslash(\vec{R}(\sqcup))\rangle = \mathcal{E}_\backslash|\backslash(\vec{R}(\sqcup))\rangle$ and $\phi(t)$ a generalized phase. Inserting this equation into Eq.(1) and integrating over a closed path C in parameter space $|\psi(\vec{R}(T))\rangle = |\psi(\vec{R}(0))\rangle$ one gets a separation into a dynamical phase (ϕ_d) accumulating the energy (momentum) change along the loop and a geometric phase (ϕ_g) which is independent from energy.[28] For a constant magnetic field only a dynamical phase exists, as shown in the previous sections.

$$\phi(T) = \arg\langle\psi(T)|\psi(0)\rangle = \frac{1}{\hbar}\int_0^T E_n(\vec{R}(t))dt$$

$$+i\oint d\vec{R}\langle n(\vec{R})|\vec{\nabla}_R|n(\vec{R})\rangle = \phi_d + \phi_g \; . \tag{7.41}$$

In the case of a slow change of the Hamiltonian (magnetic field) which corresponds to an adiabatic evolution the neutron spin will be pinned to the direction of the magnetic field

$$\vec{B}(t) = B\vec{n}(t) = B\begin{pmatrix} \cos\phi(t)\sin\Theta(t) \\ \cos\phi(t)\sin\Theta(t) \\ \cos\Theta(t) \end{pmatrix} \; , \tag{7.42}$$

with the eigenvectors

$$|\psi_\uparrow(\Theta,\phi)\rangle = \begin{pmatrix} \cos\frac{\Theta(t)}{2} \\ e^{i\phi(t)}\sin\frac{\Theta(t)}{2} \end{pmatrix} \; , \tag{7.43}$$

$$|\psi_\downarrow(\Theta,\phi)\rangle = \begin{pmatrix} \sin\frac{\Theta(t)}{2} \\ e^{i\phi(t)}\cos\frac{\Theta(t)}{2} \end{pmatrix} \; , \tag{7.44}$$

which yields

$$\left\langle\psi_\uparrow\left|\frac{\partial}{\partial\phi}\right|\psi_\uparrow\right\rangle = \frac{i}{2}(1 - \cos\Theta(t)) \; , \tag{7.45}$$

and when we move along a circle of latitude (Θ=constant) we obtain the well-known Berry phase

$$\phi_g = i\int_0^{2\pi} \frac{i}{2}(1 - \cos\Theta(t))d\phi = -\pi(1 - \cos\Theta) = -\Omega/2 \; , \tag{7.46}$$

i.e., the geometric phase is just half of the solid angle Ω enclosed by the path. A related neutron interferometric experiment has been performed by Wagh et al.[29] with Larmor precession coils rotated in an opposite sense in both coherent beams. Complete arrangement between theory and experiment has been achieved.

Aharonov and Anandan[30] generalized this approach to any cyclic evolution of a quantum system. They found that any excursion curve in Hilbert

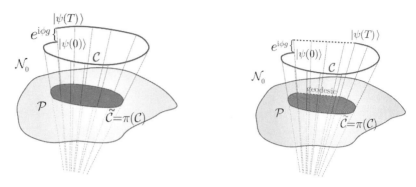

Figure 7.11   Sketch of a non-adiabatic (left) and a non-adiabatic and non-cyclic evolution (right) of a quantum system and indication of the related geometric phases.

space (C) having the same projections onto $\mathcal{P}$ have the same geometric phase modular 2π (Fig. 7.11)

$$\phi_g = 2\pi n + i \int_0^T \langle \psi(t)|\frac{d}{dt}|\psi(t)\rangle dt .\qquad(7.47)$$

Later on Samuel and Bhandari[31] generalized this formalism to non-cyclic and non-adiabatic evolutions. This formalism is strongly based on the centennial work of Pancharatnam.[32] In this case, a geodesic connection between and has to be drawn (Fig. 7.11). This means that the shortest possible path has to be chosen to define the enclosed area. When the initial and final states are orthogonal to each other a special treatment is necessary.[33,35]

A related experiment has been performed with a double loop interferometer where two phase shifters (PS) and an absorber (A) permit quite peculiar state excursions (Fig. 7.12).[34] The upper beam $|\psi_t^0\rangle$ of the first loop is used as a reference beam with adjustable phase η and as the incident wave for loop 2 where the transmitted beam, $|p\rangle$, becomes attenuated $(T = \exp(-\sigma_t N D))$ and phase-shifted (χ_2) and the orthogonal beam, $|p^\perp\rangle$ which becomes phase-shifted by χ_1. This gives an overlap of the reference beam $|\psi_{ref}\rangle$ and the loop 2 beam $|\psi_2\rangle$,

$$|\psi_{ref}\rangle \propto (|p\rangle + |p^\perp\rangle) = |q\rangle ,\qquad(7.48)$$

$$|\psi_2\rangle \propto (e^{i\chi_1} + \sqrt{T}e^{i\chi_2})|q\rangle ,\qquad(7.49)$$

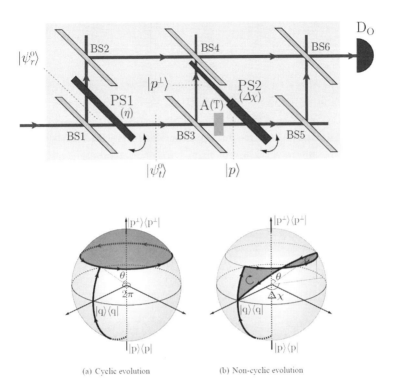

(a) Cyclic evolution   (b) Non-cyclic evolution

Figure 7.12   Sketch of a two-loop interferometer (above) and two examples of quantum evolutions to measure the geometric phase (left) and a non-cyclic phase (right).[34]

$$\phi_d = \arg\langle\psi_{ref}|\psi\rangle = \frac{\chi_1 + \chi_2}{2} \arctan\left[\frac{\chi_1 - \chi_2}{2}\left(\frac{1 - \sqrt{T}}{1 + \sqrt{T}}\right)\right] . \quad (7.50)$$

Since the dynamical phase can be written as

$$\phi_d = \frac{\chi_1 + T\chi_2}{1 + T} , \quad (7.51)$$

which becomes a constant when

$$\chi_1 + T\chi_2 = \text{const} . \quad (7.52)$$

A proper manipulation of phase shifters and the absorbers permit cyclic and non-cyclic evolutions on the Bloch sphere where the north pole and the south pole correspond to well defined paths along the upper, $|p^\perp\rangle$, and the lower beam paths $|p\rangle$ within the second interferometer loop. The absorber determines the latitude where the evolution driven by the phase shifter PS2 takes place. This allows us to write the absorption in the form

$$T = \tan^2 \frac{\Theta}{2} \, . \tag{7.53}$$

The geometric phase can be measured when closed cycles at different latitudes are chosen. Non-cyclic evolutions occur when such a rotation is stopped before a cycle is complete and then this endpoint has to be connected by a geodesis line to the equator. Figure 7.13 shows the results of such measurements.[34] They clearly define the geometric phase and the non-cyclic phase for situations shown in Fig. 7.12 (right).

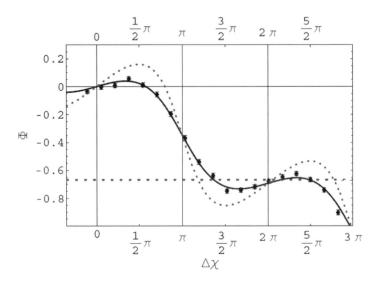

Figure 7.13   Results of the non-cyclic phase experiment according to non-cyclic excursion in Figure 7.12 in comparison with calculated values.[34]

The results indicate that geometric phases are well-defined and well-measurable quantities which may become even more important in future since they seem to be less sensitive to any fluctuation of external parameters.[36] A related experiment with bottled ultra-cold neutrons has been performed recently[37] (Fig. 7.14). Polarized ultra-cold neutrons rotate, guided by a magnetic field, around the axis of a Bloch sphere on the northern hemisphere, then they are flipped by π and rotate in the opposite direction on the southern hemisphere. This eliminates the dynamical phase and provides direct access to the geometric phase. This is measured without and

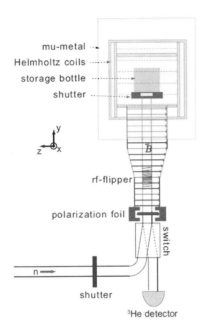

Figure 7.14   Experimental set-up to measure geometric phases with ultra-cold neutrons by means of a spin-echo method to balance the dynamical phase.[37]

with additional noisy fields which change slightly the direction but not the strength. This gave results as shown in Fig. 7.15.[37] This clearly indicates that the geometric phase becomes better defined when the neutron spends longer time within the noisy field, an effect opposite to the behavior of the dynamical phase.

7.4.2. *Contextuality and Kochen-Specker Phenomenon*

Entanglement of pairs of photons or material particles is a well-known phenomenon,[38–41] but entanglement means in a more general sense entanglement of different degrees of freedom. Therefore, entanglement can also exist between different degrees of freedom of a single particle system which yields to quantum contextuality.[42,43] Contextuality implies that the outcome of a measurement depends on the experimental context, i.e. the outcome of a previous or simultaneous experiment of another observable.[44] In this respect it is a more stringent demand than non-locality. In a related neutron experiment[45,46] the commuting observables of the spin path (s) and the

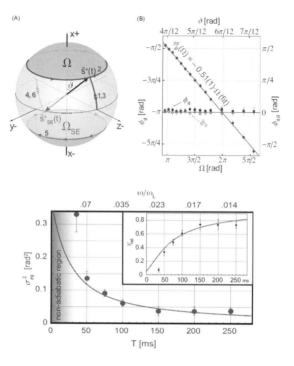

Figure 7.15   Experimental situation (left), results (middle) for the measurement of the geometric phase, and its stability against magnetic fluctuations (right).[37]

beam path (p) through the interferometer act as two independent degrees of freedom. Both represent a two-level system and can be described by Pauli spin matrices with the commutation relations

$$[\sigma_j^s, \sigma_k^p] = 0 \qquad \text{for} \qquad \{j, k\} = \{x, y\} , \tag{7.54}$$

$$[\sigma_x^s, \sigma_y^p, \sigma_y^s, \sigma_x^p] = 0 . \tag{7.55}$$

and, applied to a Bell-like state (Eq. (7.57)), one obtains the eigenvalue equations

$$\sigma_i^s \sigma_i^p |\psi\rangle = -|\psi\rangle \qquad\qquad i = x, y , \tag{7.56}$$

$$(\sigma_x^s \sigma_y^p)(\sigma_y^s \sigma_x^p)|\psi\rangle = -|\psi\rangle . \tag{7.57}$$

The related Bell-state can be produced within the interferometer when a polarized incident beam is split coherently into two beam paths (I and II) and the spin in one beam path is rotated by Larmor precession in the $-y$ and in the other beam path to the $+y$ direction (Fig. 7.16).

Then entangled Bell-states read as

$$|\psi\rangle = | \rightarrow \rangle \otimes |I\rangle \pm | \leftarrow \rangle \otimes |II\rangle \,, \qquad (7.58)$$

the other three (separable and non-separable ones) can be formulated similarly. In a full quantum tomographic analysis these Bell-like states have been measured as shown in Fig. 7.17.[47] This shows spin-path entanglement in spin-path joint measurements. In all these cases Bell-like inequalities can be formulated to demarcate a quantum world from a classical one.

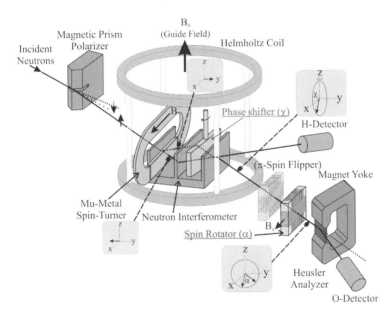

Figure 7.16   Experimental arrangement to produce and to analyze spin-path entangled neutron states.[45]

The phase shift χ between the beams and the spin rotation angle α are used as path and spin parameters[45]

$$-2 \leq S \leq 2 \,, \qquad (7.59)$$

$$-2\sqrt{2} \leq S \leq 2\sqrt{2} \,, \qquad (7.60)$$

$$S = E(\alpha_1, \chi_1) + E(\alpha_1, \chi_2) - E(\alpha_2, \chi_1) + E(\alpha_2, \chi_2) \,, \qquad (7.61)$$

$$E(\alpha, \chi) == \frac{N(\alpha, \chi) + N(\alpha + \pi, \chi + \pi) - N(\alpha, \chi + \pi) + N(\alpha + \pi, \chi)}{N(\alpha, \chi) + N(\alpha + \pi, \chi + \pi) + N(\alpha, \chi + \pi) + N(\alpha + \pi, \chi)} \qquad (7.62)$$

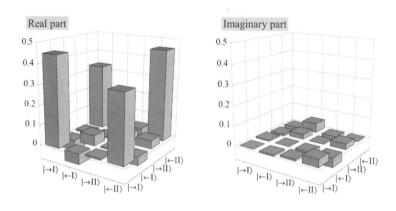

Figure 7.17   Spin-state reconstruction of an entangled neutron state.[47]

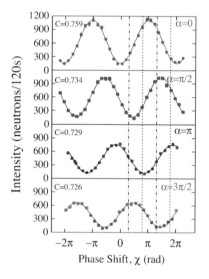

Figure 7.18   Experimental results of the contextuality experiment with spin-path entangled neutron states.[45]

The maximal violation towards the quantum mechanical description happens for the following parameters: $\alpha_1 = 0$, $\alpha_2 = \pi/2$, $\chi_1 = \pi/4$, and $\chi_2 = -\pi/4$. Typical results are shown in Fig. 7.18. Careful data analysis gave a value of $S = 2.051 \pm 0.019$, i.e. beyond the classical prediction. The reason why this value is considerably below $2\sqrt{2}$ lies in imperfections of the set-up. The contrast of the interference was about 74% mainly caused by stray fields of the spin rotator, and the degree of polarization was 95%. Nevertheless quantum contextuality has been demonstrated indicating an intrinsic correlation between the spin and the momentum (path) variables.

A more recent experiment46 dealt with the Kochen-Specker theorem[42] and the Mermin inequalities,[43] where even stronger violations of classical hidden variable theories can be verified. A related test of the Kochen-Specker theorem was formulated by Simon et al.[48] and realized for photons by Huang et al.[49] For neutron matter-waves a related proposal came from Basu et al.[50] A related experiment has been performed using a set-up similar to that shown in Fig. 7.15 but with the additional feature that the beam paths could be closed alternatively by means of an absorber sheet.[46] The measurement of the product observable $(\sigma_x^s \sigma_y^p) \cdot (\sigma_y^s \sigma_x^p)$ was done by measuring $(\sigma_z^s \sigma_z^p)$ and using a priori the non-contextuality relation. The measurable quantity is defined by a sum of product observables

$$C = \hat{I} - \sigma_x^s \sigma_x^p \sigma_y^s \sigma_y^p - (\sigma_x^s \sigma_y^p) \cdot (\sigma_y^s \sigma_x^p) . \tag{7.63}$$

In any experiment expectation values only can be measured. For non-contextual models the last term can be separated:

$$\langle (\sigma_x^s \sigma_y^p) \rangle \langle (\sigma_y^s \sigma_x^p) \rangle = \langle \sigma_x^s \rangle \langle \sigma_y^p \rangle \langle \sigma_y^s \rangle \langle \sigma_x^p \rangle , \tag{7.64}$$

which gives

$$C_{nc} = \pm 2 , \tag{7.65}$$

whereas quantum mechanics predicts

$$C_{qm} = 4 . \tag{7.66}$$

The measured value was

$$C_{exp} = 3.138 \pm 0.0115 , \tag{7.67}$$

which is well above the non-contextuality (classical) limit of 2 and provides an all-versus-nothing-type contradiction. It provides a Peres-Mermin proof of quantum-mechanics against non-contextual hidden variable theories.

A debate in literature[48,52] criticized the a priori use of the non-contextuality relation and in this connection the use of an absorber to measure this quantity. In a follow-up proposal[51] and subsequent experiment[53] the previous result (Eq. (7.67)) has been verified. In this case a quantum erasure has been used instead of an absorber and, therefore, all quantities required for Eq. (7.63) could be measured within the same context.

Years ago it has been shown that the Zeeman energy ($\hbar\omega = 2\mu B_0$) can be exchanged between the neutron and a resonance coil coherently.[18,54] This provides the basis for triple entanglement experiments using spin-path-energy as independent degrees of freedom.[54] These GHZ-states are a new tool for basic neutron quantum optics experiments and may be new components of quantum computing elements like NOTNOT gates. The energy states have geometric nature and may be rather robust under dissipative effects as shown in Sec. 7.4.1.

7.5. Discussion

Neutron interferometry is a powerful tool to measure quantum phenomena under new conditions. Single particle interference exhibits all basic features of this fundamental theory and multi-particle systems only complete these insights. Here, a review is given of measurements of various topological phases and their robustness against fluctuations and interactions with the environment. Indeed it appears that geometric phases are much more robust than the dynamical phase and this may have consequences for future quantum information systems. It has also been shown that entanglement between external and internal degrees of freedom in single particle systems can be achieved which leads to quantum contextuality, another important feature of quantum physics. Bell-like inequalities can be formulated and tested with distinct consequences for the understanding of multi-particle entanglement. Non-contextuality hidden variable theories can be turned down by precise measurements of Peres-Mermin inequalities which are based on the Kochen-Specker theorem.

Acknowledgments

Most of the experiments described in this review have been financially supported by the Austrian Science Fund (SFB 1513). Useful cooperation with my colleagues Yuji Hasegawa, Stefan Filipp, Juergen Klepp and my former teacher Wolfgang Kummer are gratefully acknowledged, and Ewald Balcar assisted in the preparation of the manuscript.

References

1. J. M. Blatt and V. F. Weisskopf, *Theoretical Nuclear Physics* (J. Wiley, New York, 1932).

2. R. J. Glauber, *Phys. Rev.* **130**, 2766 (1963).
3. V. f. Sears, V. F. Sears, *Can. J. Phys.* **56**, 1261 (1978).
4. R. P. Feynman, *Rev. Mod. Phys.* **20**, 367 (1948).
5. H. Rauch and S.A. Werner, *Neutron Interferometry* (Clarendon Press, Oxford, 2000).
6. J. L. Staudenmann, S. A. Werner, R. Colella and A. W. Overhauser, *Phys. Rev.* **A**21, 1419 (1980).
7. H. Rauch, W. Treimer and U. Bonse, *Phys. Lett.* **A**47, 369 (1974).
8. H. Rauch and D. Petrascheck, in *Neutron Diffraction*, Ed. H. Dachs (Springer Verlag, Heidelberg, 1978), *Top. Curr. Phys.* **6**, 303 (1978).
9. Y. Aharonov and L. Susskind, *Phys. Rev.* **158**, 1237 (1967).
10. H. J. Bernstein, *Phys. Rev. Lett.* **18**, 1102 (1967).
11. H. Rauch, A. Zeilinger, G. Badurek, A. Wilfing, W. Bauspiess and U. Bonse, *Phys. Lett.* 54**A**, 425 (1975).
12. S. A. Werner, R. Colella, A. W. Overhauser and C. F. Eagen, *Phys. Rev. Lett.* **35**, 1053 (1975).
13. E. P. Wigner, *Am. J. Phys.* **31**, 6 (1963).
14. J. Summhammer, G. Badurek, H. Rauch and U. Kischko, *Phys. Rev.* **A**27, 2532 (1983).
15. F. Mezei, *Z. Physik* **255**, 146 (1972).
16. I. I. Rabi, *Phys. Rev.* **51**, 652 (1937).
17. G. M. Drabkin, R. A. Zhitnikov, *Sov. Phys. JETP* **11**, 729 (1960).
18. B. Alefeld, G. Badurek and H. Rauch, *Z. Physik* **B**41, 231 (1981).
19. G. Badurek, H. Rauch and J. Summhammer, *Phys. Rev. Lett.* **51**, 1015 (1983).
20. R. P. Feynman, R. B. Leighton and K. Sands, *The Feynman Lectures on Physics*, Vol. III (Addison-Wesley, Reading, 1965).
21. P. Carruthers and M. N. Nieto, *Phys. Rev. Lett.* **53**, 411 (1968).
22. M. Namiki, S. Pascazio and H. Nakazato, *Decoherence and Quantum Measurement* (World Scientific, Singapore, 1987).
23. M. O. Scully, J. Schwinger and H. Walther, *Nature* **351**, 111 (1991).
24. G. Badurek, H. Rauch and D. Tuppinger, *Phys. Rev.* **A**34, 2600 (1986).
25. D. Yamazaki, T. Ebisawa, T. Kawai, S. Tasaki, M. Hino, T. Akayoshi and H. Achiwa, *Physica* **B**141-143, 186 (1998).
26. B. D. Josephson, *Rev. Mod. Phys.* **46**, 251 (1974).
27. J. Summhammer, K. A. Hamacher, H. Kaiser, H. Weinfurter, D. L. Jacobson and S. A. Werner, *Phys. Rev. Lett.* **75**, 3206 (1995).
28. M. V. Berry, *Proc. Royal Soc. London* **A**392, 45 (1984).
29. A. G. Wagh, V. C. Rakhecha, J. Summhammer, G. Badurek, H. Weinfurter, B. E. Allman, H. Kaiser, K. Hamacher, D. L. Jacobson and S. A. Werner, *Phys. Rev. Lett.* **78**, 755 (1997).
30. Y. Aharonov and J. S. Anandan, *Phys. Rev. Lett.* **58**, 1593 (1987).
31. J. Samuel and R. Bhandari, *Phys. Rev. Lett.* **60**, 2339 (1988).
32. S. Pancharatnam, *Proc. Ind. Acad. Sci.* **44**, 247 (1956).
33. R. Bhandari, *Phys. Rep.* **281**, 1 (1997).

34. S. Filipp, Y. Hasegawa, R. Loidl and H. Rauch, *Phys. Rev.* **A**72, 021602 (2005).
35. A. Shapere and F. Wilczek (Eds.), *Geometric Phases in Physics*, World Scientific, Singapure 1989.
36. G. De Chiara and G. M. Palma, *Phys. Rev. Lett.* **91**, 090404 (2003).
37. S. Filipp, J. Klepp, Y. Hasegawa, Ch. Plonka-Spekr, U. Schmidt, P. Geltenbort and H. Rauch, *Phys. Rev. Lett.* **102**, 030404 (2009).
38. A. Einstein, B. Podolski and N. Rosen, *Phys. Rev.* **47**, 777 (1935).
39. J. S. Bell, *Physics* **1**, 195 (1964).
40. J. F. Clauser and A. Shimony, *Rep. Prog. Phys.* **41**, 1881 (1978).
41. R. A. Bertlmann and A. Zeilinger, Eds., *Quantum [Un]speakables* (Springer, Berlin, 2002).
42. S. Kochen and F. P. Specker, *J. Math. Mech.* **17**, 59 (1967).
43. D. Mermin, *Phys. Rev. Lett.* **65**, 1838 (1990).
44. S. M. Roy and V. Singh, *Phys. Rev.* **A**48, 3379 (1993).
45. Y. Hasegawa, R. Loidl, G. Badurek, M. Baron and H. Rauch, *Nature* **425**, 45 (2003).
46. Y. Hasegawa, R. Loidl, G. Badurek, M. Baron and H. Rauch, *Phys. Rev. Lett.* **97**, 23041 (2006).
47. Y. Hasegawa, R. Loidl, G. Badurek, S. Filipp, J. Klepp and H. Rauch, *Phys. Rev.* **A**76, 052108 (2007).
48. C. Simon, M. Zukovski, H. Weinfurter and A. Zeilinger, *Phys. Rev. Lett.* **85**, 1783 (2000).
49. Y.-F. Huang, C.-F. Li, Y.-S. Zhang, J.-W. Pan and G.-C. Guo, *Phys. Rev. Lett.* **90**, 250401 (2003).
50. S. Basu, S. Bandhyopadhyay, G. Kar and D. Home, *Phys. Lett.* **A**279, 281 (2001).
51. A. Cabello, S. Filipp, H. Rauch and Y. Hasegawa, *Phys. Rev. Lett.* **100**, 130404 (2008).
52. C. Cinelli, M. Barbieri, R. Perris, P. Mataloni and F. De Martini, *Phys. Rev. Lett.* **95**, 240405 (2005).
53. H. Bartosik, Y. Hasegawa et al., (in preparation).
54. S. Sponar, J. Klepp, R. Loidl, S. Filipp, G. Badurek, Y. Hasegawa and H. Rauch, *Phys. Rev.* **A**78, 061604 (2008).

Chapter 8

First Class Constrained Systems and Twisting of Courant Algebroids by a Closed 4-Form

Markus Hansen[1] and Thomas Strobl[2]

[1] *Friedrich-Schiller-Universität Jena, Mathematisches Institut, Ernst-Abbe-Platz 2, 07743 Jena, Germany*

[2] *Université de Lyon, Université Lyon 1, CNRS UMR 5208, Institut Camille Jordan, 43 Boulevard du 11 Novembre 1918, F-69622 Villeurbanne Cedex, France*

In memoriam of Prof. Wolfgang Kummer

We show that in analogy to the introduction of Poisson structures twisted by a closed 3-form by Park and Klimcik-Strobl, the study of three dimensional sigma models with Wess-Zumino term leads in a likewise way to twisting of Courant structures by closed 4-forms H.

The presentation is kept pedagogical and accessible to physicists as well as to mathematicians, explaining in detail in particular the interplay of field transformations in a sigma model with the type of geometrical structures induced on a target. In fact, as we also show, even if one does not know the mathematical concept of a Courant algebroid, the study of a rather general class of 3-dimensional sigma models *leads* one to that notion by itself.

Courant algebroids became of relevance for mathematical physics lately from several perspectives — like for example by means of using generalized complex structures in String Theory. One may expect that their twisting by the curvature H of some 3-form Ramond-Ramond gauge field will become of relevance as well.

Preamble by Thomas Strobl

It was one of the meritorious goals of Prof. W. Kummer to promote promising students as soon as possible in the course of their studies. An

---

[1] E-mail: markus.hansen1@gmx.net
[2] E-mail: strobl@math.univ-lyon1.fr

important tool in this context were the three "Vorbereitungspraktika" (experimental, but optionally also theoretical internships of about six weeks length each), which one performs before tackling the diploma thesis at the Technical University of Vienna. I made two such internships with him, one on two-dimensional gravity models, upgraded into a diploma thesis, and another one on constrained systems. Pursuing further two-dimensional gravity models within my PhD with Kummer, a hidden target Poisson structure in such models became transparent in the work together with P. Schaller. While this gave rise to the Poisson sigma model (PSM) and its later twisting by a closed 3-form with C. Klimcik, we will consider a topological pendant of the PSM in three spacetime dimensions, the (as we will define it: also twisted) Courant sigma model (CSM) that generalizes the Chern Simons theory and can be also related to gravity models in a spacetime dimension by one more than two. We think that W. Kummer would have enjoyed such a development and extension of previous, in part joint activities (despite his probably slightly less mathematical or structural interests). In the present contribution we focus on rather concrete calculations within the Hamiltonian framework, which Kummer enjoyed very much in the last decades of his scientific work. We complement this, however, also with a coordinate-free reinterpretation, which turns out to be less obvious as one may think at first. Let me mention on this occasion also that I was always fascinated by the joy Kummer had in calculational projects, an enthusiasm, that, in the end, was very stimulating for the whole group surrounding him.

The present account is a report about an internship that I appointed to a promising student at the FSU Jena some years ago, who is now in his PhD and is coauthoring this article. Since on the one hand this enterprise relates in several ways to Kummer's activities and interests, in particular those tying me with him (cf. also above), and on the other hand its result provides the twisting of Courant algebroids (a lately much discussed mathematical notion, of relevance in several branches of modern physics and geometry) by closed 4-forms including the corresponding topological sigma model, we thought it well adapted as a contribution to a memorial volume for Wolfgang Kummer. A structurally further-going related analysis about the appearing current algebra, which would generalize a joint paper with A. Alekseev on two-dimensional current algebras to arbitrary dimensions, may be provided elsewhere.

8.1. Introduction

The Poisson sigma model (PSM)[1,2]

$$S_{PSM}[X^i, A_i] = \int_\Sigma A_i \wedge \mathrm{d}X^i + \frac{1}{2}\mathcal{P}^{ij} A_i \wedge A_j \qquad (8.1)$$

has become an important tool within mathematical physics. In the above, Σ is an oriented 2-manifold, X^i and A_i are a collection of 0-forms and 1-forms on it, respectively, and $\mathcal{P}^{ij}$ is a matrix depending on the X-fields in such a way that it satisfies a target space Jacobi identity (the brackets denote antisymmetrization)

$$\mathcal{P}^{l[i}\mathcal{P}^{jk]}{}_{,l} = 0. \qquad (8.2)$$

The PSM not only comprises a big class of two-dimensional gravity Yang-Mills gauge theory models,[3–5] it also served Kontsevich to find his famous formula[6] for the deformation quantization of Poisson manifolds by means of a perturbative expansion of its path integral, cf. Ref. 7. Finally, it is a prototype of a nonlinear gauge theory, which has lead e.g. to Lie algebroid extensions of ordinary Yang-Mills theories (i.e. non-topological and in any spacetime dimension, cf., e.g., Refs. 8, 9).

For a Hamiltonian formulation one needs to choose Σ to contain a factor $\mathbb{R}$, corresponding to "time" in some sense, and for simplicity we stick to $\Sigma = S^1 \times \mathbb{R}$, with a "periodic" coordinate σ and an evolution parameter τ.[a] Plugging $A_i = \lambda_i \mathrm{d}\tau + p_i \mathrm{d}\sigma$ into (8.1), we see that the spatial components $p_i(\sigma)$ of the A-fields are momenta canonically conjugate to the "string" fields $X^i(\sigma)$, while its τ-components λ_i serve as Lagrange multipliers for the following constraints

$$G^i(\sigma) = \partial X^i + \mathcal{P}^{ij}(X)p_j, \qquad (8.3)$$

where ∂ denotes a derivative w.r.t. σ and, on the r.h.s. the dependence on σ is understood. By means of the canonical Poisson brackets and using (8.2) one now easily verifies that the constraints are "first class",[11] meaning that they close w.r.t. the Poisson brackets of the field theory (with structural functions as coefficients), or, in more mathematical terms, their zero level surface defines a coisotropic submanifold in the original unconstrained symplectic phase space. Indeed, one finds

$$\{G^i(\sigma), G^j(\sigma')\} = -\delta(\sigma - \sigma')\,\mathcal{P}^{ij}{}_{,k}G^k(\sigma), \qquad (8.4)$$

[a]The open string Hamiltonian formulation is, albeit slightly more involved, still very similar. It carries more interesting mathematical structures, cf. Ref. 10, but these particular ones are not the focus of the present consideration.

with the structural functions being determined by the X^k-derivative of the $\mathcal{P}$-tensor, a feature typical in particular for gravity theories; only in the particular case when $\mathcal{P}^{ij}(X)$ is linear in X, $\mathcal{P}^{ij} = C^{ij}{}_k X^k$, one reobtains a Yang-Mills type gauge theory with structure constants given by $C^{ij}{}_k$ (S_{PSM} reduces to a topological BF-theory in this case).

In fact, the consideration can be even reversed: The condition (8.2), which turns $\mathcal{P}^{ij}$ into a Poisson bivector, is not only sufficient, it is also necessary for the constraints $G^i(\sigma) \approx 0$ to be of the first class. Indeed, it was this consideration that has lead Schaller-Strobl to find the general Poisson sigma model after noting corresponding similarities of some particular two-dimensional gravity or Yang-Mills models — which then simultaneously turned out to all carry this hidden Poisson target space geometry. In principle, Poisson geometry could have been invented by looking at the functional S_{PSM} and requiring it to define a first class constrained system.

In fact, this strategy was reapplied to disclose a new type of geometry,[12,b] namely what was later called twisted Poisson geometry.[14] Adding a Wess-Zumino term coming from a closed 3-form H to S_{PSM},

$$ S_{HPSM} = S_{PSM} + \int_\Sigma \mathrm{d}^{-1} H \,, \tag{8.5} $$

which can be interpreted as saying that the symplectic form from before is changed only by adding a transgression contribution from H to it,

$$ \omega = \oint_{S^1} \delta X^i(\sigma) \wedge \delta p_i(\sigma) \, \mathrm{d}\sigma + \frac{1}{2} \oint_{S^1} H_{ijk}(X(\sigma)) \, \partial X^i(\sigma) \, \delta X^j(\sigma) \wedge \delta X^k(\sigma) \, \mathrm{d}\sigma \,, \tag{8.6} $$

the constraints of the modified Lagrangian, which still have the form (8.3), are first class, *iff* the following generalization of (8.2) is satisfied:

$$ \mathcal{P}^{il} \partial_l \mathcal{P}^{jk} + \mathrm{cycl}(ijk) = \mathcal{P}^{ii'} \mathcal{P}^{jj'} \mathcal{P}^{kk'} H_{i'j'k'} \,. \tag{8.7} $$

In Ref. 14 it was shown that a couple of a bivector and a closed 3-form satisfying the above condition is in one-to-one correspondence with T^*M-projectable so called Dirac structures in split exact Courant algebroids[15–17] (cf. also Ref. 18 for details).

Courant algebroids became quite fashionable lately within some modern developments in geometry, like generalized complex structures and pure spinors (cf., e.g., Refs. 19, 20), but also branches within theoretical physics, such as in String Theory and supersymmetric sigma models (cf., e.g., Ref. 21). In fact, in Ref. 22 it was shown that the Courant bracket

---

[b]Cf. also Ref. 13 for another related, but historically independent consideration.

appears naturally within a certain type of current algebra on a phase space governed by the symplectic form (8.6). It was moreover found that maximal systems of first class constraints within this setting are then in bijection to Dirac structures, which explained also why the consideration in Ref. 12 yielded the Dirac structures as described in Ref. 14.

Courant algebroids or Courant structures are the first higher analogue of Poisson structures: while the latter ones correspond to so-called NPQ-manifolds of degree one, the former ones are equivalent to NPQ-manifolds of degree two, cf. Ref. 23 for details on this. Now, NPQ-manifolds are ideally suited for the construction of topological field theories following the so-called AKSZ-procedure.[24] While for the degree one case one obtains in this way the PSM (8.1), cf., e.g., Ref. 25, in the degree two case one obtains[26] the Courant sigma model (CSM):[27,28]

$$S_{CSM}[X^i, A^a, B_i] = \int_{\Sigma_3} B_i \wedge (\mathrm{d}X^i - \rho_a^i A^a) + \frac{1}{2}\eta_{ab} A^a \wedge \mathrm{d}A^b + \frac{1}{6}C_{abc} A^a \wedge A^b \wedge A^c.$$
(8.8)

Here Σ_3 is a 3-manifold, X^i, A^a, and B_i are collections of 0-forms, 1-forms, and 2-forms on it, respectively — where the number of scalar and 2-form fields is the same and possibly different to the number of 1-form gauge fields — and ρ_a^i, η_{ab}, and C_{abc} are structural functions of the Courant algebroid, its dependence being on the scalar X-fields in (8.8). These structural functions are to satisfy a sequel of coupled partial differential equations so as to give rise to the structure of a general (not necessarily exact) Courant algebroid.

We are not displaying and explaining those equations, the higher analogue of the equation (8.2) above, at this point. Rather, as we will show in detail, they can be found by applying the same strategy as the one leading from (8.1) to (8.2), which we briefly recalled above, but now applied in the context of the more elaborate action (8.8). So, without knowing yet what is a Courant algebroid, its defining conditions can be derived from requiring the 3-dimensional sigma model above, with a priori unrestricted structural functions ρ_a^i, η_{ab}, and C_{abc}, to have first class constraints. Moreover, now twisting the sigma model by a closed 4-form H,

$$S_{HCSM} = S_{CSM} + \int_{\Sigma_3} \mathrm{d}^{-1}H,$$
(8.9)

we will be lead to a higher analogue of the twisting of a Poisson structure as in (8.7), namely the twisting of the structure of a Courant algebroid by such a 4-form H.[c]

In the subsequent section we will first reconsider some general prototype of (potentially topological) sigma models in two and three dimensions of Σ, leading, under relatively mild assumptions, to (8.1) and (8.8), respectively. As a byproduct we will be able to determine the tensorial character — or the precise deviation thereof — of the coefficient objects in these two actions, which will turn out to be particularly essential in the three-dimensional context. The coefficient function η in (8.8), for example, will be seen to correspond to a fiber metric on a vector bundle E that serves as (part of) the target of the 3d sigma model. The C-coefficients, on the other hand, are found to have a highly non-tensorial transformation behaviour (cf. Eqs. (8.13) and (8.15) below).

In section 8.3 we perform the explicit Hamiltonian analysis of the sigma model S_{HCSM} (with yet unspecified structural functions ρ, η, C and H) and determine the necessary and sufficient conditions on these functions so as to render the constrained system first class — thus making the sigma model in particular also topological. These calculations will be performed for constant η (achievable by field redefinitions and corresponding e.g. to orthonormal frames in the above mentioned vector bundle E), since this simplifies the basic Poisson brackets and thus the ensuing calculations considerably. The drawback of this step is that the structural identities obtained are then known only in orthonormal frames.

This sounds less restrictive than it in fact *is*: The structural identities turn out to also contain derivative terms of the fiber metric η and *cannot* be reconstructed from knowing the structural equations in orthonormal frames only (where these extra terms vanish identically). It is here where the considerations of section 8.2 become essential. However, another related complication in this context is that the transformation property of the C-coefficients does not correspond to *any* product of sections of E. We will still be able to construct a (non-C^∞-linear) product on $\Gamma(E)$, the structure functions of which will agree with the Cs in orthonormal frames.

These questions will be dealt with in the final section to this contribution, putting together the facts from the two sections 8.2 and 8.3 before and providing a coordinate/frame independent or mathematical formulation of what one may call an H_4-twisted (or a Wess-Zumino-) Courant algebroid.

---

[c]We denote this 4-form again by H. It is not to be confused with the closed 3-form appearing in very particular Courant algebroids, namely split exact ones. Here we will find a *generalization* of a Courant structure, in complete analogy to the generalization of Poisson structures given by (8.7) and induced by the sigma model (8.5).

It is given by Definition 1 in section 8.4. In this context we also will take care of providing a *minimal* set of defining axioms, other structural identities being shown to follow from them. We conclude with a concrete example of an H_4-twisted Courant algebroid where H_4 is exact.

8.2. Field redefinitions and their geometric significance

8.2.1. *Two dimensional sigma models without background data on Σ*

Let us start with the simpler situation in two dimensions. We first want to address what kind of action functionals one can construct without any further structure than orientability of the base manifold Σ; we do want orientability for defining the integral. In particular, there will be no metric given on Σ, used in most known cases of action functionals already in the kinetic term of the non-interacting, "free" theory — but also likewisely in the standard type of sigma models, where one uses metrics on the base or source manifold Σ as well as on the target manifold M. We will consider functionals for 0-forms, 1-forms, and 2-forms in two dimensions. (In principle one could also consider local functionals defined for fields of other tensor type on Σ, even without using a metric, but we will not do this here). We will restrict ourselves to 0- and 1-forms, $(X^i)_{i=1}^n$ and $(A^\alpha)_{\alpha=1}^r$, respectively; in two dimensions this restriction is very mild, however, and we will comment on the small modifications when considering also 2-form fields at the end of the subsection.

Under these circumstances we are lead to consider functionals of the following type:

$$S[X^i, A_\alpha] = \int_\Sigma e_i^\alpha(X)\, A_\alpha \wedge \mathrm{d}X^i + \frac{1}{2}\mathcal{P}^{\alpha\beta}(X)\, A_\alpha \wedge A_\beta + \frac{1}{2}B_{ij}(X)\,\mathrm{d}X^i \wedge \mathrm{d}X^j ,$$
$$(8.10)$$

where the matrices e, $\mathcal{P}$ and B may at this stage depend arbitrarily on the scalar fields, the latter two being antisymmetric, certainly. This is the most general ansatz in the above mentioned context.

We now come to the first type of field transformations, namely transformations mapping 0-forms into 0-forms only. Being invertible (and sufficiently smooth) so as to constitute a permitted field redefinition, clearly this can be interpreted as a coordinate transformation on the target spanned by the n scalar fields. The target would be the range of possible values of X^i, which, a priori, would be an $\mathbb{R}^n$. Using transformations of the just mentioned type for an eventual gluing, and considering (8.10) as an appropriately understood locally valid expression only, we can generalize this to considering X as a map from Σ to a general n-dimensional (target) manifold M.

As a consequence from this consideration, the last term in (8.10) receives the interpretation of the pullback to Σ by X of a 2-form B on M.[d] To also give a geometric meaning to the other quantities in the above action, we consider transformations of the form $A_\alpha \mapsto M_\alpha^\beta(X) A_\beta$ (for invertible, smooth matrices M). Since there are no derivatives acting on the A-fields, they just imply a tensorial transformation property of the α-indices in e and $\mathcal{P}$. In particular, we may conclude that A_α, besides being 1-forms on Σ, corresponds to components (indexed by α) of sections in some rank r vector bundle E living over M (M_α^β corresponding to local frame changes in this bundle, moreover). This implies then that $\mathcal{P} \in \Gamma(\Lambda^2 E^*)$ and that $e \in \Gamma(E^* \otimes T^*M)$, where E^* denotes the bundle dual to E; e can equivalently be viewed as a map from E to T^*M.

Let us now, as the main restriction in this context, assume that this map e provides an isomorphism, $e \colon E \xrightarrow{\sim} T^*M$, which in particular implies that the number n of 0-form fields and the number r of 1-form fields need to be equal and that the then $n \times n$ matrices e are everywhere invertible. In fact, under this condition, e is seen to be nothing but a *vielbein* on M, and A_α then turns out to be the components of a 1-form in M in a potentially non-holonomic basis, while $\mathcal{P}$ becomes a bivector field: $E \cong T^*M$ implies $E^* \cong TM$, i.e. in a holonomic basis ∂_i of TM one has $\mathcal{P} = \frac{1}{2}\mathcal{P}^{ij}\partial_i \wedge \partial_j \in \Gamma(\Lambda^2 TM)$. A field redefinition of A_a of the form $A_i := e_i^\alpha(X)A_\alpha$, which induces a redefinition of the coefficient matrix in the quadratic A-term $e_\alpha^i e_\beta^j \mathcal{P}^{\alpha\beta} =: \mathcal{P}^{ij}$, where e_α^i is the inverse vielbein, is now seen to just correspond to a change from a general frame to a holonomic basis of TM. The action (8.10) is now seen to be identical to (8.5) with $H = \mathrm{d}B$ after these change of variables.

Diffeomorphisms of the target, $X^i \mapsto \widetilde{X}^i(X)$, can now be compensated directly with a corresponding redefinition of the A-fields in the holonomic frame, $A_i \mapsto \widetilde{A}_i = \frac{\partial X^j}{\partial \widetilde{X}_i} A_j$. Certainly this is in general *not* a symmetry of the action functional, since the explicit form of the matrices $\mathcal{P}^{ij}$ and B_{ij}

---

[d]Note that here certainly B is not a field but a fixed 2-form on M, which only encodes part of the kinetic and interaction terms for the scalar fields. This is as in String theory, where one denotes such a term by precisely the same symbols conventionally, and where then B becomes a 2-form field only on the *target* by means of a dynamics induced implicitly by string fluctuations.

as functions of X will change — except if the generating vector field of the diffeomorphism Lie annihilates the bivector field and the 2-form, in which case one has a rigid symmetry giving rise to Noether charges.[e]
There are further field redefinitions of less immediate geometric significance. One of these corresponds to a shift of the A-fields by terms proportional to dX. Such transformations are easily seen to change the B-contribution to (8.10) and they can be shown to even permit to get rid of this contribution altogether; this is by far less immediate and was in fact proven rigorously only for small enough B in Ref. 18 (but cf. also Ref. 29). Assuming this to hold true also for general B, it implies that only the deRham cohomology class of H, entering as a Wess-Zumino term in Ref. 12, has a physical significance, at least if no additional meaning is attributed to distinguished fields or target coordinates in the action (like it might happen in some particular gravitational applications, for example). The geometrical significance of these changes of H by an exact term dB is less immediate as well: Note, for example, that this permits to change a Poisson tensor $\mathcal{P}$ into one that is only dB-twisted Poisson. Still, there is some geometrical notion behind this, which, interestingly, relates in a different way again to Courant algebroids, the main subject of this article, in their interplay with sigma models: (possibly twisted) Poisson structures are particular so-called Dirac structures, a particular type of subbundles in exact Courant algebroids. As a bundle an exact Courant algebroid is isomorphic to $T^*M \oplus TM$, where the isomorphism corresponds to a splitting in an exact sequence and changes of this splitting correspond precisely to some $B \in \Omega^2(M)$ as above.[f] We refer to Ref. 18 for further details.
We conclude this subsection with some remarks on possible generalizations. The main assumption leading to an identification of (8.10) with (8.5) (for exact H) resulted from requiring $e\colon E \to T^*M$ to be an isomorphism. Even if the number n of scalar fields and the rank r of E are equal, e might still have a kernel, for example. In fact, if one permits such a kernel, one is lead to a somewhat more general sigma model than one of the form of the twisted PSM (8.5), namely one that is of the form of a so-called Dirac sigma model[18] — more precisely, to the part of it that was called topological there for not depending on additional background data like a metric on Σ

---

[e]The analogue of this in String theory with a Minkowski target are momentum and angular momentum, as the Noether charges of the Poincare isometry group.
[f]The above $E = T^*M$ enters this picture in so far as the Dirac structure corresponding to a bivector field or to a twisted Poisson structure provides by itself an isomorphism of T^*M into an appropriate subbundle of $T^*M \oplus TM$, so that, after the choice of a splitting, E can be identified with this subbundle of the exact Courant algebroid.

(cf., e.g., Eqs. (17)–(20) and Eq. (24) in Ref. 18). The restriction to $r = n$, on the other hand, seems less restrictive than one might believe at first sight. If $r < n$, it corresponds to $r = n$ with e having a kernel of dimension $n - r$ and correspondingly many A-fields not entering the action at all. If, on the other hand, $r > n$, one should be able to eliminate excess A-fields (at least up to potential global issues): namely those components in the kernel of e enter the action at most quadratically and only algebraically and then can be correspondingly eliminated with their own field equations. Suppose, for example, that A_1 and A_2 are not present in the $A \wedge \mathrm{d}X$ part of the action and that they enter (8.10) only via $\int_\Sigma A_1 \wedge A_2$. Thus, variation w.r.t. these two fields require them to vanish. Correspondingly, this term, and thus any A_1- and A_2-dependence in this example can be dropped without changing the physical content of the functional at all.

Finally we briefly comment on not considering also 2-form fields in the present context. In fact, in the spirit of this section, any 2-form field can enter an action as (8.10) only linearly, then being multiplied with some function $f(X)$. Variation w.r.t. this field yields a constraint $f(X) = 0$ which, in the smooth case, singles out a submanifold of M. The sigma model with such a 2-form field, or several of them, then just reduces effectively to one without those fields but defined on a smaller target, namely the one of the original M where the respective functions vanish. The situation can become more interesting, certainly, if the subspaces singled out by the vanishing of functions are singular and not just submanifolds. The explicit conditions on a PSM-type functional in the presence of such 2-form additions to be topological were studied in Ref. 30.

An action functional of the type (8.10) to have a maximal number of possible gauge symmetries and to not carry any propagating degrees of freedom poses certain conditions on the tensors on the target of M, which are most efficiently found in the Hamiltonian framework. In the case of (8.5) this lead to (8.7), for example. Although the absence of propagating degrees of freedom together with the absence of any background structure used for the definition of such a functional is sufficient to get topological sigma models, it is not always necessary. An example of a topological sigma model which uses a background metric on Σ (as well as a metric on M) is the G/G WZW model (cf., e.g., Ref. 31), or, more generally, the (full) Dirac sigma model.[18] The presence of such auxiliary structures can be used also as another argument for restriction to 0- and 1-forms in two dimensions: any 2-form is Hodge dual to a 0-form. This argument can be used, however, only in this *extended* context for a convincing exclusion of

2-form fields, where, on the other hand, there then are also uncountably more possibilities for the construction of an action functional out of 0- and 1-form fields than those parametrized in (8.10).

8.2.2. *Three dimensional sigma models*

We now turn to sigma models that can be defined without any background structures on an orientable three dimensional base manifold Σ. In analogy to before we consider functionals for 0-form fields $(X^i)_{i=1}^n$, 1-form fields $(A^a)_{a=1}^r$, and now also 2-form fields $(B_\alpha)_{\alpha=1}^s$ — the omission of top degree-form fields again poses essentially no restriction. A most general ansatz in this context takes the following form

$$S[X^i, A^a, B_\alpha] = \int_\Sigma e_i^\alpha \, B_\alpha \wedge \mathrm{d}X^i - \rho_a^\alpha \, B_\alpha \wedge A^a$$

$$+ \frac{1}{2}\eta_{ab} \, A^a \wedge \mathrm{d}A^b + \frac{1}{6}C_{abc} \, A^a \wedge A^b \wedge A^c$$

$$+ \frac{1}{2}\Lambda_{aij} \, A^a \wedge \mathrm{d}X^i \wedge \mathrm{d}X^j + \frac{1}{2}\Delta_{abi}A^a \wedge A^b \wedge \mathrm{d}X^i$$

$$+ \frac{1}{6}F_{ijk} \, \mathrm{d}X^i \wedge \mathrm{d}X^j \wedge \mathrm{d}X^k \qquad (8.11)$$

where e_i^α, ρ_a^α, η_{ab}, C_{abc}, Λ_{aij}, and F_{ijk} are functions of X, parametrizing the action functional. They have the obvious symmetry properties like e.g. C_{abc} being completely antisymmetric or η_{ab} being symmetric in the exchange of indices.

In analogy to before we restrict ourselves to the case of e_i^α being invertible. In addition, here we also require the likewise coefficient matrix η_{ab} nondegenerate as well. One may expect that relaxing one or the other of these conditions can lead to interesting generalizations — for example, in the two-dimensional setting this step permits the more general also topological Dirac sigma model — , but we will not pursue this here further. Instead, we will now make use of the non-degeneracy of e to again simplify the above action by means of appropriate field redefinitions.

First we introduce $B_i := e_i^\alpha \, B_\alpha$. This one can always do, certainly, but only in the invertible case we can use B_i as new fields, by introducing $\rho_a^i :=$ $\rho_a^\alpha e_\alpha^i$, where, as before, e_α^i can be regarded as inverse vielbein. In addition to replacing e by a unit matrix when it is invertible, field redefinitions also permit to put η_{ab} into constant normal form *and* to get rid of the terms with coefficient Λ and Δ altogether. Clearly, redefining B_i by $B_i - \frac{1}{2}\Lambda_{aji} \, A^a \wedge$ $\mathrm{d}X^j$, we eliminate the Λ-term while simultaneously we only have to change the coefficient Δ_{abi} to $\Delta_{abi}^{\text{new}} := \Delta_{abi} + \rho_a^j \Lambda_{bij}$. Similarly we can now get rid of the Δ-term by a subsequent shift $B_i \mapsto B_i - \frac{1}{2}\Delta_{abi}^{\text{new}} A^a \wedge A^b$, which now only changes the coefficient of the cubic A-term to $C_{abc}^{\text{new}} = C_{abc} + 3\rho_{[a}{}^i \Delta_{bc]i}$, where the brackets $[\ldots]$ denote antisymmetrization of the indices enclosed. In this manner we brought the above action already into the form (8.9) with $H = \mathrm{d}F$. We are thus left with the analysis of this action further on.[g]

We are now left with analysing the field transformations of more immediate geometrical significance. First of all there are again the diffeomorphisms of the target of the sigma model, certainly, which determine also the tensorial character of the index i in ρ_a^i as well as that $F \in \Omega^3(M)$, as anticipated already in the identification $H = \mathrm{d}F$ mentioned above. The diffeomorphisms induce a likewise transformation of B_i, while not effecting the A-fields. The latter 1-form fields take again values in some rank r vector bundle $E \to M$ (more precisely, $A \in \Omega^1(M, X^*E)$). We are left with analysing changes of quantities induced by transformations $A^a \mapsto \widetilde{A}^a$,

$$A^a = M_b^a(X)\widetilde{A}^b\,, \tag{8.12}$$

corresponding to changes of local frames in E. Obviously the Wess-Zumino term in (8.9), stemming from some $H \in \Omega_{\text{closed}}^4(M)$, is not effected by such transformations and we can focus on (8.8) for this purpose.

Note that at this point we have not yet put η into some normal form, since this would restrict the permitted local frames in E to orthogonal ones w.r.t. η viewed as a fiber metric on E. This will be an important issue since, as we will see in the end of the analysis within the present article, on the one hand many *different* geometrical quantities within the present setting will be seen to coincide in orthonormal frames (and in their index representation differ decisively from one another by derivatives of η_{ab} only), while, on the other hand, the Hamiltonian analysis simplifies drastically in orthonormal frames so that, at least for that purpose, we *do* want to restrain η to a constant normal form. But to be able to retrieve other involved objects in general frames again from there, we need to know their transformation properties w.r.t. a general transformation as in (8.12). We now first observe that a transformation of the form (8.12) induces also a nontrivial Δ-contribution to the action, namely one with $\Delta_{abi} = -\frac{1}{2}\widetilde{\eta}_{ab,i}$,[h]

---

[g]Depending on the context, we will consider this action — or likewisely (8.8) — as the one of the (twisted) Courant sigma model or just as a sigma model of this *form* with structural functions not yet fulfilling the identities needed to correspond to a (twisted) Courant algebroid.

[h]Note that this contribution is no more present if η_{ab} was already put to constants and if one restrains M_b^a to respect that, i.e. to correspond to orthogonal transformations. —

where

$$\widetilde{\eta}_{ab} = M_a^c M_b^d \eta_{cd} \tag{8.13}$$

denotes the components of η in the new frame. (It is also this equation, together with the required non-degeneracy, that justifies to regard η as a fiber metric on E). To get rid of the unwanted contribution in the action, we learnt above that we can do this by accompanying (8.12) by a B-field transformation, $B_i \to \widetilde{B}_i$ where

$$B_i = \widetilde{B}_i + \frac{1}{4}\widetilde{\eta}_{ab,i}\widetilde{A}^a \wedge \widetilde{A}^b . \tag{8.14}$$

This, on the other hand, by itself leads to a new additive contribution to the coefficient of the cubic A-term, thus rendering C_{abc} to have a *non-tensorial* transformation property; besides the obvious $\widetilde{\rho}_b^i = \rho_a^i M_b^a$ one finds

$$\widetilde{C}_{abc} = M_a^d M_b^e M_c^f C_{def} - 3 M_{[a}^d M_b^e M_{c],d}^f \eta_{ef} , \tag{8.15}$$

where from now on we use the further on useful abbreviated notation

$$f_{,a} \equiv \rho_a^i f_{,i} \tag{8.16}$$

for derivatives along letters of the beginning of the alphabet. So, while thus η is seen to correspond to a fiber metric on E, which, by an appropriate choice of M_a^b we can always put to some constant normal form and ρ is found to be an element of $\Gamma(E^* \otimes TM)$, or, equivalently, a vector bundle map

$$\rho \colon E \to TM , \tag{8.17}$$

the differential geometric meaning of C is much more intricate. We will clarify its meaning after having derived the equations the structural functions have to satisfy in an orthonormal basis in the subsequent section so as to render (8.9) topological.

Finally, we remark that also in the three dimensional context it is only the non-exact WZ-term that gives something qualitatively new. Here in three dimensions this is even relatively easy to see explicitly in the sigma model: A transformation $B_i \mapsto B_i + \frac{1}{6}F_{jki}\mathrm{d}X^j \wedge \mathrm{d}X^k$ adds a term of the form of the last one in (8.11). We also produce a nontrivial Λ-term in this manner, but we already know how to remove it by further B-field transformations. The upshot is that such a combined transformation only changes H to $H + \mathrm{d}F$ in (8.9). Indeed, the situation is very analogous to

---

In this article, we use the convention that $f_{,i}$ denotes the partial derivative of a function f w.r.t. X^i.

the geometry one finds from the two-dimensional sigma model, it is a higher analogue of it in several ways and we will display here only parts of the full story.

8.3. Hamiltonian analysis

8.3.1. Hamiltonian formulation

In this section we perform a Hamiltonian analysis of the action (8.9). For this purpose we choose $\Sigma_3 = \Sigma \times \mathbb{R}$ with Σ an oriented, compact 2-surface without boundary, and in a first step we only regard the usual local part of the action given by (8.8). Since the ensuing Hamiltonian formulation is much easier when η_{ab} is constant, we will assume this to be the case within this section. Also, since η is nondegenerate, we can use it also freely to raise and lower letters from the beginning of the alphabet, so, e.g.,

$$\rho^{ai}(X) = \eta^{ab}\rho_b^i(X), \qquad (8.18)$$

and, since η is constant in the given frame, this can be done also with quantities that are hit by derivatives. In the above, η with upper indices denotes, as usual, the inverse to η with lower indices (agreeing, at the same time, with one of the two having changed both index positions by the respective other one in the indicated way — a feature where the symmetry of η is essential for). As mentioned repeatedly already, except for appropriate smoothness conditions, at this point we do require nothing more of the coefficient functions of the X-fields in (8.8).

In fact, the action (8.8) is already in a Hamiltonian form. To see this we decompose the forms appropriately: $A^a = \mathcal{A}^a + \Lambda^a d\tau$, where $\mathcal{A}^a$ are 1-forms on Σ at a fixed value of the evolution parameter τ, and Λ^a likewise 0-forms. Analogously, we have $B_i = p_i + d\tau \wedge \lambda_i$, with 2-forms and 1-forms p_i and λ_i, respectively. Plugging this decomposition into (8.8), using $d = d_\Sigma + d\tau \wedge \partial_\tau$ and denoting the τ-derivative of a quantity by an overdot, $\partial_\tau \phi \equiv \dot{\phi}$, we find

$$S_{CSM} = \int_{\mathbb{R}} \left[\int_{\Sigma} p_i \dot{X}^i - \frac{1}{2}\eta_{ab}\mathcal{A}^a \wedge \dot{\mathcal{A}}^b + \lambda_i \wedge G^i + \Lambda_a H^a \right] \wedge d\tau, \quad (8.19)$$

with $G^i \equiv d_\Sigma X^i - \rho_a^i \mathcal{A}^a$ and $H^a \equiv d_\Sigma \mathcal{A}^a + \frac{1}{2}C_{bc}^a \mathcal{A}^b \wedge \mathcal{A}^c - \rho^{ai}p_i$. The first two terms are a symplectic potential; such a potential gives rise to a symplectic form by replacing τ-derivatives by differentials of the respective field (in field space, we will denote the corresponding exterior derivative by δ for clarity, as we did already in (8.6)) and taking the negative exterior derivative of the result.[i] Denoting the field $\mathcal{A}^a$ again by simply A^a, this evidently yields

$$\omega_{CSM} = \oint_\Sigma \delta X^i \wedge \delta p_i + \frac{1}{2} \oint_\Sigma \eta_{ab} \delta A^a \wedge \delta A^b \,. \tag{8.20}$$

The remaining two terms in (8.19) give rise to constraints only, with the λ_i and Λ_a being their Lagrange multiplier fields. Thus, in the simplified notation, the following currents have to vanish:

$$G^i(\sigma) = \mathrm{d}X^i - \rho_a^i(X)A^a \tag{8.21}$$

$$J^a(\sigma) = \mathrm{d}A^a + \frac{1}{2}C_{bc}^a(X)A^b \wedge A^c - \rho^{ai}(X)p_i \,. \tag{8.22}$$

Here X^i, A^a, and p_i are now functions, 1-forms, and 2-forms on the 2-surface Σ, respectively, and, correspondingly, also the suffix Σ has been dropped on the deRham differential. For some purposes it is useful to introduce test objects so as to obtain true functions on the field theoretic phase space. Let μ_i and φ_a be such a collection of test 1-forms and 0-forms on Σ, respectively, and set:[j]

$$G[\mu] := \oint_\Sigma \mu_i \wedge G^i \tag{8.23}$$

$$J[\varphi] := \oint_\Sigma \varphi_a J^a \,. \tag{8.24}$$

These functions on phase space $\mathcal{M}$ have to vanish for all choices of test objects (which can be considered as generalized labels for the constraints), which defines the constraint surface $\mathcal{C} \subset \mathcal{M}$; this is a consequence following from the action functional S_{CSM}.

We will in the following require that in addition also mutual Poisson brackets of the constraints vanish on $\mathcal{C}$, which is a restriction on the structural functions in the action. In the nomenclature of Dirac this is denoted as

$$G[\mu] \overset{!}{\approx} 0 \overset{!}{\approx} J[\varphi] \qquad \forall \mu_i, \varphi_a \,. \tag{8.25}$$

It means that also the Hamiltonian vector fields of the constraints, restricted to the constraint surface $\mathcal{C}$, are required to be tangent to it. In a more

---

[i]The pioneering work about a Hamiltonian formulation of gauge theories goes back to Dirac.[11] For a somewhat simplified version, applicable also in the present context, cf., e.g., Ref. 32.

[j]For simplicity, we consider a fixed frame and do not permit any of the test objects to depend on the X-field (or any other field) in what follows. — The integration symbol $\oint$ has been chosen, here and already before, so as to stress that there are no boundary contributions to the integral due to the choice of Σ.

mathematical language the first class property is tantamount to saying that the (here infinite dimensional) submanifold $\mathcal{C}$ of the original (here weakly symplectic) phase space manifold $\mathcal{M}$ is coisotropic.[k]

Twisting the sigma model by a closed 4-form as in (8.9), gives a contribution to the symplectic form only. In fact, the action (8.9) is uniquely valued only when $H = \mathrm{d}h$ is exact, in which case it amounts to adding the pullback of $h \in \Omega^3(M)$ by the map $X : \Sigma_3 \to M$ to the action (8.8). In this case, (8.9) with $\Sigma_3 = \mathbb{R} \times \Sigma$ is understood to be

$$S_{HCSM} = S_{CSM} + \int_{\mathbb{R}} \left[\int_{\Sigma} \frac{1}{2} h_{ijk} d_{\Sigma} X^i \wedge d_{\Sigma} X^j \dot{X}^k \right] \wedge d\tau, \qquad (8.26)$$

so that the new term clearly gives a contribution to the symplectic potential only. The corresponding contribution to the symplectic form depends on $\mathrm{d}h$ only,

$$\omega_{HCSM} = \omega_{CSM} + \frac{1}{2} \oint_{\Sigma} (h_{jil,k} + h_{kjl,i}) \delta X^k \wedge \delta X^l \wedge d_{\Sigma} X^i \wedge d_{\Sigma} X^j, \quad (8.27)$$

and can be defined for arbitrary closed H:

$$\omega_{HCSM} = \oint_{\Sigma} \delta X^i \wedge \delta p_i + \frac{1}{4} \oint_{\Sigma} H_{ijkl} \delta X^i \wedge \delta X^j \wedge \mathrm{d}X^k \wedge \mathrm{d}X^l + \frac{1}{2} \oint_{\Sigma} \eta_{\mathrm{ab}} \delta A^{\mathrm{a}} \wedge \delta A^{\mathrm{b}}. \tag{8.28}$$

This form remains (weakly) nondegenerate for any choice of H; closedness of H (on the target M) becomes necessary for the closedness of the symplectic form ω_{HCSM} on the field theoretic phase space $\mathcal{M}$. In the Hamiltonian formulation a Wess-Zumino term can be added without any integrability condition on the closed $d + 1$-form; this would arise upon geometric prequantization, for example.

8.3.2. Constraint algebra

To calculate Poisson brackets among the constraints, we first display the elementary Poisson brackets as they follow from the symplectic form (8.28). By standard methods one obtains, written in components,

$$\left\{ X^i(\sigma), \tilde{p}_j(\tilde{\sigma}) \right\} = \delta^i{}_j \, \delta(\sigma - \tilde{\sigma}), \qquad (8.29)$$

$$\left\{ \tilde{p}_i(\sigma), \tilde{p}_j(\tilde{\sigma}) \right\} = \frac{1}{2} H_{ijkl} X^k{}_{,\mu} X^l{}_{,\nu} \, \epsilon(\mu\nu) \, \delta(\sigma - \tilde{\sigma}), \qquad (8.30)$$

$$\left\{ A^a_\mu(\sigma), A^b_\nu(\tilde{\sigma}) \right\} = 2\eta^{ab} \, \epsilon(\mu\nu) \, \delta(\sigma - \tilde{\sigma}), \qquad (8.31)$$

---

[k]Cf., e.g., Ref. 33 for several equivalent characterizations of this notion within the finite dimensional setting.

with the other brackets vanishing. Here $A^a = A^a_\mu d\sigma^\mu$, $p_i = \tilde{p}_i \, d\sigma^1 \wedge d\sigma^2$, $\delta(\sigma - \tilde{\sigma})$ is the delta function w.r.t. the measure $d\sigma^1 \wedge d\sigma^2$, $\epsilon(\mu\nu)$ denotes the ϵ-symbol normalized according to $\epsilon(12) = 1$, and quantities on the r.h.s. are understood to depend on either σ or $\tilde{\sigma}$. Using again test objects,

$$\hat{X}[\alpha] := \int_\Sigma \alpha_i X^i, \qquad \hat{A}[\mu] := \int_\Sigma \mu_a \wedge A^a, \qquad \hat{P}[\varphi] := \int_\Sigma \varphi^i p_i, \qquad (8.32)$$

where φ^i, μ_a, and α_i are 0-, 1-, and 2-forms on Σ respectively, this can be rewritten as

$$\left\{ \hat{X}[\alpha], \hat{P}[\varphi] \right\} = \int_\Sigma \varphi^i \alpha_i, \qquad (8.33)$$

$$\left\{ \hat{P}[\varphi], \hat{P}[\tilde{\varphi}] \right\} = \frac{1}{2} \int_\Sigma \varphi^i \tilde{\varphi}^j H_{ijkl} dX^k \wedge dX^l, \qquad (8.34)$$

$$\left\{ \hat{A}[\mu], \hat{A}[\tilde{\mu}] \right\} = 2 \int_\Sigma \mu_a \wedge \tilde{\mu}^a, \qquad (8.35)$$

all other brackets between the elements (8.32) vanishing.

Now we are ready for the real calculation. Using the above elementary brackets, one computes those between the constraints (8.23), (8.24). We display here only the result of the somewhat lengthy calculation. One obtains:

$$\left\{ G[\mu], G[\tilde{\mu}] \right\} = \int_\Sigma \rho^{ai} \rho^j_a \, \mu_i \wedge \tilde{\mu}_j \qquad (8.36)$$

$$\left\{ G[\mu], J[\varphi] \right\} = \int_\Sigma G^j \wedge \mu_i \varphi^a \, \rho^i_{a,j} + A^b \wedge \varphi^a \mu_i \left(\rho^{ci} C_{abc} + 2\rho^i_{[a,b]} \right) \qquad (8.37)$$

$$\begin{aligned}
\left\{ J[\varphi], J[\tilde{\varphi}] \right\} = &- \int_\Sigma \left(J^f C^d_{ef} + G^i \wedge A^f C^d_{ef,i} \right) \varphi^e \tilde{\varphi}_d \\
&+ \int_\Sigma \frac{1}{2} (G^k \wedge G^l + 2G^{[k} \rho^{l]}_c \wedge A^c) \varphi_a \tilde{\varphi}_d \rho^{ai} \rho^{dj} H_{ijkl} \\
&- \int_\Sigma \varphi^b \tilde{\varphi}^a p_i \left(\rho^{ci} C_{abc} + 2\rho^i_{[a,b]} \right) \\
&- \int_\Sigma \varphi^a d\tilde{\varphi}^b \wedge A^c \left(C_{abc} + C_{bac} \right) \\
&+ \int_\Sigma \varphi^a \tilde{\varphi}^b A^c \wedge A^d \Big(\frac{1}{2} C_{bae} C^e_{cd} - C_{bad,c} - C_{a]cd,[b} \\
&\qquad\qquad + C_{aec} C_b{}^e{}_d + \frac{1}{2} \rho^i_a \rho^j_b \rho^k_c \rho^l_d H_{ijkl} \Big).
\end{aligned} \qquad (8.38)$$

We used the convention $\eta_{ab}C^b_{cd} = C_{acd}$ here (cf. also Eqs. (8.16) and (8.18), so that e.g. $C_{abc,d} \equiv C_{abc,i}\rho^i_d$). Now we can determine the necessary and sufficient conditions for the constraints to be first class.

Note in this context that the test objects can be chosen arbitrarily. In particular then the vanishing of (8.36) implies by a standard argument (the test objects being arbitrary) that

$$\rho^i_a\rho^j_b\eta^{ab} = 0. \qquad (8.39)$$

We remark in parenthesis that this certainly has to hold for any point in M since any such a point can be image of the map $X: \Sigma \to M$.

Next we regard (8.37), to vanish on (8.25), which in particular implies that the first term on the right hand side of Eq. (8.37) is zero on this surface. There is now one qualitatively more complicated step than the one in the 1+1 dimensional context of the Poisson sigma model. There the constraints were 1-forms and on the spatial slice S^1 there are no integrability conditions. Here, there are no integrability conditions for the 2-form constraints $J = 0$, Σ being two-dimensional, whereas applying the deRham differential δ to the 1-form constraints $G = 0$, leads, upon usage of these two equations (8.25), to

$$\frac{1}{2}\left(\rho^{ci}C_{cab} + 2\rho^i_{[a,b]}\right)A^a \wedge A^b + \rho^{aj}\rho^i_a p_j = 0. \qquad (8.40)$$

The second term was found to necessarily vanish in Eq. (8.39) above. We want to conclude from (8.37) that $\rho^{ci}C_{abc} + 2\rho^i_{[a,b]} = 0$, which, using that C_{abc} is completely antisymmetric, can be rewritten also as

$$\rho^j_a\rho^i_{b,j} - \rho^j_b\rho^i_{a,j} = C^c_{ab}\rho^i_c. \qquad (8.41)$$

It is, however, precisely this equation that also enters the integrability condition (8.40) and we want to make sure to avoid circular reasoning. We need to choose A^a at a given point on Σ sufficiently general to conclude (8.41) from the restriction of (8.37) to (8.25). The main difficulty at this point is that even at a given point p on Σ the 1-forms A^a cannot be chosen arbitrarily at this stage since they need to satisfy (8.40). However, what we can do is to choose them still sufficiently general: Let them be of the form $A^a := \lambda^a\alpha$ where α is some arbitrary 1-form on Σ at p; then clearly $A^a \wedge A^b \equiv 0$ (at p) and the given data at p can be extended into some neighborhood of p satisfying (8.25). On the other hand, with λ^a to be free at our disposal, we can now indeed conclude (8.41) from (8.37).

Also note that at *this* point the integrability conditions are always satisfied, which in particular implies that at a given point in Σ the 1-forms A^a

and the 2-forms p_i can now be chosen arbitrarily — still permitting choices for extensions of the fields into a neighborhood of that point such that (8.25) holds true (cf. Eqs. (8.21) and (8.22)). In particular, this implies that each line in (8.38) has to vanish separately on the constraint surface. In fact, the first two lines vanish by themselves already, and the third one reproduces just (8.41) — at least if we use that C_{abc} is completely antisymmetric in its three indices, which in fact is reinforced in the fourth line of (8.38).

Here some remark is in order: In the action that we used to derive the Hamiltonian system the coefficients C_{abc} entered already completely antisymmetrically. Still, the constraints (8.21) and (8.22) make sense also when C^a_{bc} is antisymmetric in the last two indices only. We performed the ensuing calculation in this relaxed setting. Then we find that the first class property enforces the antisymmetry in the first two indices as well, cf. the fourth line of (8.38), i.e. thus in all three indices. This is analogous to the situation in the Poisson sigma model: The constraints (8.3) are meaningful already in the more general setting of a general contravariant 2-tensor $\mathcal{P}^{ij}$. Also there the first class property enforces both, the antisymmetry of $\mathcal{P}^{ij}$ as well as the Jacobi identity. Both conditions there have a meaning in terms of Dirac structures: the first being the condition of isotropy, the other one an integrability condition (cf. also Ref. 22 for further details on this relation). In the three dimensional setting, there are two algebraic conditions of this kind now, Eq. (8.39) as well as the antisymmetry condition,

$$C_{abc} = -C_{bac}, \tag{8.42}$$

as well as two integrability conditions, Eq. (8.41) and

$$C^e_{ab}C^d_{ce} + C^d_{ab,c} + \text{cycl}(abc) = C_{cab,e}\eta^{ed} + \rho^{di}\rho^j_a\rho^k_b\rho^l_c H_{ijkl}, \tag{8.43}$$

enforced by the vanishing of the last line in (8.38).

8.4. Axioms of H_4-twisted Courant algebroids

In this section we want to extract the coordinate independent information contained in the structural identities obtained above. In section 8.2.2 we already discovered that the differential geometric setting is a vector bundle E over a base manifold M, equipped with a nondegenerate bilinear pairing η, a bundle map ρ, cf. Eq. (8.17), which we will call the anchor of E, and a closed 4-form H on M. The main task of this section is to give a meaning to the structural functions C^a_{bc} and the interplay of all the structural functions as dictated by the identities found above.

Let us be guided by the special well-known case of the Chern Simons theory. This is obtained from M being a point, H and ρ correspondingly zero, and (E, η) thus being just a vector space equipped with a non-degenerate bilinear form. In this case, C^a_{bc} correspond to structure constants of a Lie algebra — in accordance with this, Eq. (8.43) reduces to the Jacobi identity — and η is invariant w.r.t. the adjoint transformations of this Lie algebra, as expressed by Eq. (8.42).

The most near-at-hand generalization of the above scenario over a point would be that C^a_{bc} defines a product on the space of sections $\Gamma(E)$ of the bundle $E \to M$. However, this is in conflict with the transformation properties found in (8.15)! Let D^a_{bc} be structural functions of a product of sections, i.e. if e_a is a basis of sections in E and we denote the product by a bracket, one has

$$[e_a, e_b] = D^c_{ab} e_c . \tag{8.44}$$

With this definition it is clear that under a local change of basis

$$\widetilde{e}_a = M^b_a e_b \tag{8.45}$$

the first index of $D_{abc} = \eta_{ad} D^d_{bc}$ transforms in a C^∞-linear fashion, i.e. that $\widetilde{D}_{abc}$ will be simply proportional to M^d_a, a matrix M with a lower a-index (while the other indices can produce also derivatives of M-matrices etc. — cf. Eq. (8.50) below as a possible realization of this requirement). This is however *not* the case for C_{abc}, as we learn from Eq. (8.15).

In order to cure this deficiency of C to define structural functions of a product of sections, we want to make an ansatz using the other structural quantities at hand:

$$D_{abc} = C_{abc} + \alpha \eta_{ab,c} + \beta \eta_{bc,a} + \gamma \eta_{ca,b} . \tag{8.46}$$

We observe that a-derivatives of η transform in the following way[1]

$$\widetilde{\eta}_{ab,c} = M^d_a M^e_b M^f_c \eta_{de,f} + M^f_{a,d} M^e_b M^d_c \eta_{ef} + M^f_{b,d} M^e_a M^d_c \eta_{ef} . \tag{8.47}$$

Writing out the six terms coming from the antisymmetrization of the second term in (8.15), it is now easy to see, that the required C^∞-linearity implies $\gamma = -\alpha = \frac{1}{2}$, leaving β arbitrary at this point. With such a choice of constants, D thus defines a product by means of (8.44).

To fix the remaining constant, we regard the generalization of the ad-invariance condition for η. For this purpose we first express $\eta([e_a, e_b], e_c) +$

---

[1] By definition of a-derivatives, cf. Eq. (8.16), one has $(fg)_{,a} = f_{,a}g + fg_{,a}$. Note, however, that such type of derivatives do not commute. Instead, as a consequence of (8.41), one finds $f_{,ab} \equiv (f_{,a})_{,b} = f_{,ba} + C^c_{ba} f_{,c}$.

$\eta(e_b, [e_a, e_c])$ in terms of the structural functions D; using (8.44), this becomes identical to $D_{cab} + D_{bac}$. So it is the symmetrization over the first and the *third* index of D (at this point it is not clear that D will define an antisymmetric product — and in fact it will not — in which case one would be able to trade this into a symmetrization of the first and second index, as one is used to from Lie algebras, cf. also Eq. (8.42)). Using, on the other hand, (8.46) with the above choice for α and γ, we find

$$D_{abc} + D_{cba} = \eta_{ac,b} + (\beta - \frac{1}{2}) \left(\eta_{ab,c} + \eta_{cb,a}\right), \qquad (8.48)$$

since C_{abc} is completely antisymmetric as entering the action (8.8).[m] While the first term on the r.h.s. of (8.48) fits an ad-invariance condition very well, the other terms are disturbing in this context. It is thus comforting to see that they can be made to vanish by a unique choice of the still free constant β in our ansatz. Thus we are lead to[n]

$$D_{abc} = C_{abc} + \frac{1}{2} \left(\eta_{bc,a} + \eta_{ac,b} - \eta_{ab,c}\right). \qquad (8.49)$$

Under arbitrary changes (8.45) of frames, these coefficients transform according to

$$\tilde{D}_{abc} = M_a^d M_b^e M_c^f D_{def} + M_a^d \left(\eta_{de} M_{c,f}^e M_b^f - \eta_{de} M_{b,f}^e M_c^f + \eta_{ef} M_{b,d}^e M_c^f\right). \qquad (8.50)$$

We collect what we obtained up to now — it is already quite a lot, and all this is coming from the action functional and its transformation properties only: We have a vector bundle E over M together with an anchor map $\rho: E \to TM$. E is equipped with a fiber metric η and a product $[\cdot, \cdot]$ on its sections. This product is *not* antisymmetric. Rather, according to (8.49), we see that

$$\eta(\psi_1, [\psi_2, \psi_3] + [\psi_3, \psi_2]) = \rho(\psi_1)\eta(\psi_2, \psi_3), \qquad (8.51)$$

where ψ_i are arbitrary sections of E and $\rho(\psi_1)$ is the vector field $\psi_1^a \rho_a^i \partial_i$. (This follows from Eq. (8.49) as follows: In the case that all three sections

---

[m]There exists a more involved argument using only (8.42) in orthonormal frames and the transformation properties of the coefficients to arbitrary frames to arrive at this conclusion from milder assumptions on C_{ab}^c (as described at the end of the previous section). However, up to this point within this section all the argumentation can be done already at the level of the action. The Hamiltonian perspective will then be used only to extract the Jakobi condition.

[n]The formal analogy of the expression for the difference between C and D with the standard formula for a torsion-free, metrical connection in a holonomic frame is somewhat striking at this point.

are linearly independent, we can use them as part of a basis e_a. By construction, (8.49) holds in arbitrary frames. Symmetrization over the last two indices in (8.49) indeed yields $D_{abc} + D_{acb} = \eta_{bc,a} \equiv \rho_a^i \eta_{bc,i}$, which gives (8.51) for this case. Validity of that equation in degenerate cases of linear dependence now follows for example by continuity.) Since η is non-degenerate, this equation determines the symmetric part of the bracket uniquely. In a completely analogous manner we conclude from (8.49) (cf. Eq. (8.48) for $\beta = \frac{1}{2}$) the ad-invariance condition of the fiber metric w.r.t. the bracket on sections,

$$\eta([\psi_1, \psi_2], \psi_3) + \eta(\psi_2, [\psi_1, \psi_3]) = \rho(\psi_1)\eta(\psi_2, \psi_3) \,. \tag{8.52}$$

Note that the r.h.s. of the last two equations is identical. Thus, using a standard polarization argument (η being symmetric), we can rewrite these two equations according to

$$\eta([\psi', \psi], \psi) = \frac{1}{2}\rho(\psi')\eta(\psi, \psi) = \eta(\psi', [\psi, \psi]) \,, \tag{8.53}$$

valid for arbitrary two sections ψ, ψ' of E.

There is still one further important property of the bracket that one can conclude from the above definitions and transformation properties. It concerns the relation of $[\psi_1, f\psi_2]$ to $f[\psi_1, \psi_2]$, where f is an arbitrary function on M. Let us for this purpose choose ψ_1 and ψ_2 as the first two basis elements of a local frame e_a (we assume them to be linearly independent and again conclude on the case of proportional sections by continuity) and consider a change of frame (8.45) with $M_a^b = \delta_a^b$ if $(a, b) \neq (2, 2)$ and $M_2^2 = f$ (we assume f to be nonzero, at least in a neighborhood of our interest — otherwise $[\psi_1, f\psi_2]$ vanishes already by bilinearity of the bracket). Then (8.50) yields (for $a \neq 2$)

$$\eta(e_a, [\psi_1, f\psi_2]) = \widetilde{D}_{a12} = fD_{a12} + \eta(e_a, \psi_2)\rho(\psi_1)f \,, \tag{8.54}$$

since $M_1^e = \delta_1^e$ is constant and its derivative gives no contribution. Thus we find the following Leibniz property of the bracket:

$$[\psi_1, f\psi_2] = f[\psi_1, \psi_2] + (\rho(\psi_1)f)\,\psi_2 \,. \tag{8.55}$$

For later use we finally mention that (8.51) (or, equivalently, the second equality of (8.53)), can be also rewritten according to

$$[\psi, \psi] = \frac{1}{2}\rho^* d\eta(\psi, \psi) \,, \tag{8.56}$$

where, by definition, ρ^* of some 1-form $\alpha = \alpha_i dx^i$ is just $\alpha_i \rho^{ai} e_a$ (it is the fiberwise transpose of ρ with a subsequent use of η to identify E^* with E).

From this and (8.55) one may also conclude for example about the behavior of the bracket under multiplication of the first section w.r.t. a function:

$$[f\psi_1, \psi_2] = f[\psi_1, \psi_2] - (\rho(\psi_2)f)\,\psi_1 + \eta(\psi_1, \psi_2)\,\rho^*(\mathrm{d}f)\,. \tag{8.57}$$

This also puts us in the position to express the general product or bracket of two sections by means of the structural functions:

$$[\psi_1, \psi_2] = \left(\psi_1^b \psi_2^c D_{bc}^a + \rho(\psi_1)\psi_2^a - \rho(\psi_2)\psi_1^a + \rho^{ai}(\psi_1^b)_{,i}(\psi_2)_b\right) e_a\,, \tag{8.58}$$

where, certainly, $\rho(\psi_1)\psi_2^a = \rho_b^i \psi_1^b (\psi_2^a)_{,i}$ and $\rho^{ai} = \eta^{ab}\rho_b^i$.

We now turn to the structural identities that we obtained in the previous section. Here we need to emphasize that they were obtained in an *orthonormal* frame (or at least a frame where η_{ab} is constant). Clearly, terms of importance in a general frame may be absent in such a frame. One example is the Ad-invariance condition of the metric tensor η: The condition (8.52) becomes (cf. Eq. (8.48) for $\beta = \frac{1}{2}$)

$$D_{abc} + D_{cba} = \eta_{ac,b} \tag{8.59}$$

in an arbitrary local frame. Clearly the r.h.s. of this equation vanishes in an orthonormal frame and it is the question how one can recover it from knowing the condition in orthonormal frames only. On the other hand, we took great effort to derive transformation properties of all structural functions with respect to *general* changes of a frame bundle basis (8.45). Thus, we may proceed as follows in principle: We note that within an *orthonormal* frame $D_{abc} = C_{abc}$. Thus we can replace in all of the identities obtained in the previous section the structural functions C by D everywhere. Then we can apply the transformation formulas such as (8.50) to all these identities, transforming them to a general frame.

Let us illustrate this at the example of (8.42): Let us assume that the frame e_a is orthonormal, thus η_{ab} in particular constant, and $\widetilde{e}_a$ an arbitrary frame, so that $\widetilde{\eta}_{ab}$ as given by Eq. (8.13) is in general non-constant. Using the transformation property (8.50) we now compute

$$\widetilde{D}_{abc} + \widetilde{D}_{cba} = M_a^d M_b^e M_c^f (D_{def} + D_{fed}) + M_a^d \eta_{de} M_{c,f}^e M_b^f + M_c^d \eta_{de} M_{a,f}^e M_b^f\,, \tag{8.60}$$

where we made use of the fact that the last two terms in (8.50) give no contribution when symmetrized over indices a and c. In the orthonormal frame e_a we have $D_{def} + D_{fed} = C_{def} + C_{fed} = -(C_{dfe} + C_{fde})$, which *vanishes* due to (8.42). On the other hand, the remaining two terms on the r.h.s. of (8.60) combine into $(\widetilde{\eta}_{ac})_{,f} M_b^f$, which is nothing but $\rho(\widetilde{e}_b)\widetilde{\eta}_{ac}$.

Thus indeed from (8.42) and the transformation property (8.50) we find

$$\widetilde{D}_{abc} + \widetilde{D}_{cba} = \rho(\widetilde{e}_b)\widetilde{\eta}_{ac} , \qquad (8.61)$$

i.e. Eq. (8.59) as it is to hold in an arbitrary frame.

Thus we now could apply the same strategy on the other equations obtained in the previous sections, such as for example to (8.43). Using (8.50) we would find, after quite a lengthy calculation and on use of the other structural identities, that, miraculously, (8.43) would take the *same* form in an arbitrary frame (this certainly is partially due to the fact, how we presented that formula — it certainly could be rewritten in several inequivalent ways for constant metric coefficients η_{ab} such that this property holds no more true). On the other hand, if we use e.g. the *transformation property* that one *obtains* upon choosing $\beta = 0$ in (8.46), i.e. for

$$E_{abc} = C_{abc} + \frac{1}{2} \left(\eta_{ac,b} - \eta_{ab,c} \right) , \qquad (8.62)$$

which one might use as coefficients of another product as we found above, one would find (8.43) to become more complicated in an arbitrary frame. (Again one would have $E_{abc} = C_{abc}$ in an orthonormal frame, could thus replace all Cs by Es in (8.43), but now the Es would transform in a different way than the Ds, Eq. (8.50), which now would produce extra terms similarly to what happened in the transition from (8.42) to (8.61) above. Using $D = C$ in orthonormal frames and the transformation (8.50) on the other hand, will leave the equation form-invariant, in fact upon usage of the other identities obtained in the previous section). This observation may be used as another argument besides (8.48) for the choice $\beta = \frac{1}{2}$ in the definition of the bracket.

There is, however, a more direct route to arrive at the missing axioms as induced from the previous section. Before turning to it, but also in preparation for it, let us briefly reconsider the relation of the three different quantities C_{abc}, D_{abc}, and E_{abc} from a slightly more abstract perspective. First of all, we observe that according to its definition in (8.62), $E_{abc} = D_{a[bc]}$, so Es are nothing but the structural functions of the *antisymmetrization* of the product (8.44). So, if we denote by $[[\cdot, \cdot]]$ the bracket defined via E_{abc}, i.e.

$$[[e_a, e_b]] = E_{ab}^c e_c \qquad (8.63)$$

one has

$$[[\psi_1, \psi_2]] = \frac{1}{2} \left([\psi_1, \psi_2] - [\psi_2, \psi_1] \right) . \qquad (8.64)$$

This bracket is, by construction, antisymmetric, but, as mentioned already, its other properties are slightly more involved than those for the bracket $[\cdot,\cdot]$ — like e.g. instead of (8.52) one finds

$$\eta([[\psi_1,\psi_2]],\psi_3) + \eta(\psi_2,[[\psi_1,\psi_3]])$$
$$= \rho(\psi_1)\eta(\psi_2,\psi_3) - \frac{1}{2}\rho(\psi_2)\eta(\psi_3,\psi_1) - \frac{1}{2}\rho(\psi_3)\eta(\psi_1,\psi_2)\,, \quad (8.65)$$

for which reason we prefer to work with the previously introduced non-antisymmetric bracket. Finally, according to its definition, what is the relation of the coefficients C_{abc} with the bracket? As mentioned, C_{abc} are *not* the structure functions of any product of sections. However, as we see from the very definition of D_{abc} in (8.44), one has $C_{abc} = D_{[abc]}$. This implies that if one defines

$$C(\psi_1,\psi_2,\psi_3) := \frac{1}{6}\sum_{\sigma \in S_3}(-1)^{|\sigma|}\eta(\psi_{\sigma_1},[\psi_{\sigma_2},\psi_{\sigma_3}])$$
$$= \frac{1}{3}\eta(\psi_1,[[\psi_2,\psi_3]]) + \text{cycl}(123)\,, \quad (8.66)$$

where S_3 denotes the permutation group of three elements and $|\sigma|$ the parity of the permutation element σ, we have $C(e_a,e_b,e_c) \equiv C_{abc}$. So, Eq. (8.66) relates C in an arbitrary frame or as an abstract object to the other two brackets and the scalar product. Again, we remark that there is no way to induce a product from C, in contrast to D or E. (For example, $E(\psi_1,\psi_2,\psi_3) = \eta(\psi_1,[[\psi_2,\psi_3]])$ is, in contrast to (8.66), C^∞-linear in ψ_1, which thus permits to define the product $[[\cdot,\cdot]]$ on sections of the vector bundle E from it).

We now come to the frame independent, abstract formulation of the information contained in the three conditions (8.39), (8.41), (8.43). Clearly, in more abstract terms, (8.39) just states that

$$\rho \circ \rho^* = 0\,, \quad (8.67)$$

where $\rho\colon E \to TM$ was the anchor map and $\rho^*\colon T^*M \to E$ essentially its transpose, as introduced above. Here we used that ρ_a^i and η_{ab} have a tensorial transformation property, so that (8.39) in orthonormal frames applies the likewise formula in arbitrary frames. Next we turn to (8.41). Also this equation is not difficult to reinterpret. Let us for this purpose apply the map ρ to Eq. (8.58):

$$\rho([\psi_1,\psi_2]) = \left(\psi_1^b\psi_2^c D_{bc}^a + \rho(\psi_1)\psi_2^a - \rho(\psi_2)\psi_1^a\right)\rho_a^i\partial_i\,, \quad (8.68)$$

where we have already made use of (8.67) to get rid of the last term in (8.58). This equation holds true in *any* frame. Thus also in an orthonormal frame, where we can replace D by C and then make use of Eq. (8.41), yielding — in this orthonormal frame — :

$$\rho([\psi_1, \psi_2]) = \psi_1^b \psi_2^c \left(\rho_b^j \rho_{c,j}^i - \rho_c^j \rho_{b,j}^i \right) \partial_i + \rho(\psi_1)\psi_2^a \partial_i - \rho(\psi_2)\psi_1^a \rho_a^i \partial_i \,. \quad (8.69)$$

The r.h.s. is, however, nothing but the commutator of the vector fields $\rho(\psi_1)$ with $\rho(\psi_2)$. Thus we obtain, for an *arbitrary* choice of ψ_1, ψ_2 in $\Gamma(E)$,

$$\rho([\psi_1, \psi_2]) = [\rho(\psi_1), \rho(\psi_2)] \,. \quad (8.70)$$

Note that here we only had to use an orthonormal frame as an intermediary step. The resulting equation does no more show any dependence on the frame; it is obviously sufficient and necessary to guarantee (8.41) in view of our definition of the bracket — taking (8.67) for granted! In fact, we can even deduce (8.67) from (8.70): Setting $\psi_1 = \psi_2$ and using (8.56), we find (8.67) upon noting that ψ_1 can be chosen such that $d\eta(\psi_1, \psi_1)$ takes any possible value at a given point.

We will encounter a likewise fact in what follows next: the equation that we will extract from (8.43) will entail both, Eqs. (8.70) and (8.67). Certainly, such facts are true only upon usage of the Leibniz rules (8.55) and (8.57), which we derived from the general transformation and symmetry properties of D_{abc} above. We now turn to the final, most complicated condition, equation (8.43). One may remark also that it is the *only* place where the 4-form H enters finally.

To interpret Eq. (8.43) within our present setting, we may again remember to what it reduces for M being a point, when it becomes just the Jakobi identity for the Lie bracket. This may motivate to consider the following expression:

$$J(\psi_1, \psi_2, \psi_3) := [\psi_1, [\psi_2, \psi_3]] - [[\psi_1, \psi_2], \psi_3] - [\psi_2, [\psi_1, \psi_3]] \,. \quad (8.71)$$

Note that certainly with the bracket $[\cdot, \cdot]$ not being antisymmetric, there are several inequivalent ways of writing the Jakobiator. The above definition of J corresponds to the choice which measures the deviation of the bracket to satisfy a Leibniz property with respect to itself, i.e. that the adjoint transformation $\mathrm{ad}_\psi := [\psi, \cdot]$ is a derivation of the bracket.

To relate (8.43) to J, we compute $J_{abc} := J(e_a, e_b, e_c)$ with e_a being an *orthonormal* basis. In these frames we have $D_{abc} = C_{abc}$; using (8.44), (8.55), and (8.57), one then easily establishes the equivalence of (8.43) with

$$J_{abc} = \rho^{di} \rho_a^j \rho_b^k \rho_c^l H_{ijkl} e_d \,. \quad (8.72)$$

To relate this expression to one in a general basis, we first make use of Eqs. (8.55)–(8.57) to obtain

$$J(\psi_1, \psi_2, f\psi_3) = fJ(\psi_1, \psi_2, \psi_3) + ([\rho(\psi_1), \rho(\psi_2)] - \rho([\psi_1, \psi_2])) \, f \, \psi_3 \qquad (8.73)$$

$$J(\psi_1, f\psi_2, \psi_3) = fJ(\psi_1, \psi_2, \psi_3) + ([\rho(\psi_1), \rho(\psi_3)] - \rho([\psi_1, \psi_3])) \, f \, \psi_2$$
$$+ [\rho(\psi_1)\eta(\psi_2, \psi_3) - \eta([\psi_1, \psi_2], \psi_3) - \eta(\psi_2, [\psi_1, \psi_3])] \, \rho^*(\mathrm{d}f)$$
$$- \eta(\psi_2, \psi_3)[\rho^*(\mathrm{d}f), \psi_1] \qquad (8.74)$$

$$J(f\psi_1, \psi_2, \psi_3) = fJ(\psi_1, \psi_2, \psi_3) + ([\rho(\psi_2), \rho(\psi_3)] - \rho([\psi_2, \psi_3])) \, f \, \psi_1$$
$$- \rho^*(\mathrm{d}f) \, [\rho(\psi_2)\eta(\psi_1, \psi_3) - \eta([\psi_2, \psi_1], \psi_3) - \eta(\psi_1, [\psi_2, \psi_3])]$$
$$+ \eta(\psi_1, \psi_3)[\rho^*(\mathrm{d}f), \psi_2] - \eta(\psi_1, \psi_2)[\rho^*(\mathrm{d}f), \psi_3] \,. \qquad (8.75)$$

Since we already have the identities (8.70) and (8.52) at our disposal, we see that it is sufficient to show that

$$[\rho^*\mathrm{d}f, \psi] = 0 \,, \qquad (8.76)$$

for any $f \in C^\infty(M)$ and $\psi \in \Gamma(E)$, to obtain that J is $C^\infty(M)$-linear in each of its entries, i.e. that it is a tensorial object, $J \in \Gamma((E^*)^{\otimes 3} \otimes E)$. Since, on the other hand, the r.h.s. of (8.72) is constructed by means of purely tensorial objects, this equation then can immediately be considered as one valid for arbitrary frames, or, likewisely can be rewritten as

$$J(\psi_1, \psi_2, \psi_3) = \rho^* \left[H(\cdot, \rho(\psi_1), \rho(\psi_2), \rho(\psi_3)) \right] \,, \qquad (8.77)$$

valid for arbitrary sections $\psi_i \in \Gamma(E)$.

It thus remains to prove (8.76). We distinguish tow cases, η having indefinite or definite signature. Thus we first assume, that η has a indefinite signature. Let $f \in C^\infty(M)$ and $\psi \in \Gamma(E)$ be arbitrary, but with $\eta(\psi, \psi) \neq 0$. (The particular case with $\eta(\psi, \psi) = 0$ in (8.76) then follows by continuity from those cases). We put $\psi_1 = \eta(\psi, \psi)^{-1}\psi$. Because of Eqs. (8.55) and (8.67) it is sufficient to show (8.76) for ψ_1 instead of ψ. Now we choose $\psi_2, \psi_3 \in \Gamma(E)$ orthogonal to ψ_1 such that $\eta(\psi_2, \psi_2) = \eta(\psi_3, \psi_3) = 0$ and $\eta(\psi_2, \psi_3) = 1$. (This is always possible if the rank of E is not too small). Finally we complete ψ_1, ψ_2, ψ_3 by orthonormal sections to get a basis $e_a = \psi_a$ in $\Gamma(E)$. In this frame all components η_{ab} are constant (at least locally). The same holds true, if we replace ψ_2 and ψ_3 by $\widetilde{\psi}_2 = f\psi_2$ and $\widetilde{\psi}_3 = f^{-1}\psi_3$ (the case $f \equiv 0$ is trivial), yielding a new basis $\widetilde{\psi}_a$, where in fact then one obviously has $\widetilde{\eta}_{ab} = \eta_{ab}$. Hence we find by (8.73) and (8.74)

$$J(\psi_1, \widetilde{\psi}_2, \widetilde{\psi}_3) = ff^{-1}J(\psi_1, \psi_2, \psi_3) + \eta(\psi_2, \widetilde{\psi}_3)[\rho^*(\mathrm{d}f), \psi_1]$$
$$= J(\psi_1, \psi_2, \psi_3) + f^{-1}[\rho^*(\mathrm{d}f), \psi_1] \,,$$

and on the other hand due to (8.72) (note that $\eta_{ab} = \tilde{\eta}_{ab}$ is constant so that this formula can be applied in *both* frames)

$$J(\psi_1, \tilde{\psi}_2, \tilde{\psi}_3) = \rho^* \left[H(\cdot, \rho(\psi_1), \rho(f\psi_2), \rho(f^{-1}\psi_3)) \right] = J(\psi_1, \psi_2, \psi_3).$$

Together this proves (8.76).

The case of definite signature now either follows by a complexification argument (we can consider the sigma model with imaginary fields) or by completing ψ_1 to an arbitrary orthonormal frame and considering now $\tilde{\psi}_2 = \psi_2 \cos f + \psi_3 \sin f$ and $\tilde{\psi}_3 = -\psi_2 \sin f + \psi_3 \cos f$. The details of this second approach as well as the remaining cases — i.e. if there are no three linearly independent sections — are left as an exercise to the reader.

Now, finally, we are in the position to give a concise, abstract definition of an H_4-twisted Courant algebroid:

Definition 8.1. A Courant algebroid twisted by a closed 4-form H is a vector bundle $E \to M$ with fiber metric η, a bundle map $\rho: E \to TM$, and a bilinear product $[\cdot, \cdot]$ on $\Gamma(E)$ such that

$$[\psi_1, f\psi_2] = f[\psi_1, \psi_2] + (\rho(\psi_1)f)\,\psi_2 \tag{8.78}$$

$$\rho(\psi_1)\eta(\psi_2, \psi_3) = \eta(\psi_1, [\psi_2, \psi_3]) + \eta(\psi_1, [\psi_3, \psi_2]) \tag{8.79}$$

$$[\psi_1, [\psi_2, \psi_3]] = [[\psi_1, \psi_2], \psi_3] + [\psi_2, [\psi_1, \psi_3]]$$
$$+\rho^* \left[H(\cdot, \rho(\psi_1), \rho(\psi_2), \rho(\psi_3)) \right]. \tag{8.80}$$

Here we took care to provide a possible minimal set of axioms. Even in the known case of an ordinary (i.e. nontwisted) Courant algebroid, this has not always been the case in the mathematical literature. It is, however, a fact that reversing our considerations from before and starting with the above definition, *all* the equations of this section can be recovered. For example (8.56) is equivalent to (8.79), hence (8.57) can be deduced as before, and thus also Eqs. (8.73)–(8.75). Now (8.70) is a consequence of (8.80) and (8.73) due to the tensorial behaviour of $J(\psi_1, \psi_2, \psi_3)$, and (8.67) follows from (8.79) with $\psi_1 = \rho^*(\alpha)$ and (8.70). Furthermore, (8.76) can be obtained from (8.80) and (8.56) (note that $\eta(\psi_1, \psi_2)$ can be an arbitrary function, even when ψ_2 is fixed), and finally (8.52) follows from (8.74), (8.70), and (8.76).

At the end we shall discuss a concrete realization of such a twisted Courant algebroid. Let M be an arbitrary manifold, and consider $E = TM \oplus T^*M$. We define $\rho((u, \alpha)) = u$,

$$\eta((u, \alpha), (v, \beta)) = \alpha(v) + \beta(u),$$

and

$$[(u, \alpha), (v, \beta)] = \big([u, v]_{\mathrm{Lie}}, L_u \beta - L_v \alpha + \mathrm{d}(\alpha(v)) + h(u, v, \cdot)\big)$$

for some arbitrary 3-form h. This implies $\rho^*(\alpha) = (0, \alpha)$, and by a calculation recommended to the reader as an exercise one arrives at

$$J\left((u, \alpha), (v, \beta), (w, \gamma)\right) = (0, (\mathrm{d}h)(u, v, w, \cdot)) \ .$$

So for the case that the 3-form h is closed, one has an example of an ordinary Courant algebroid. In fact, this is just the split exact Courant algebroid mentioned in the Introduction (h being the closed 3-form mentioned in footnote c in particular and its deRham cohomology class is the Severa class which uniquely characterizes an exact Courant algebroid[17]).

If, on the other hand, we consider the above data for an arbitrary 3-form h, we find an example of an H-twisted Courant algebroid where the 4-form is simply $H = \mathrm{d}h$. (It is easy to verify that all the axioms in Definition 1 hold true in this case). This however implies that this example is one with an exact 4-form H only.

Such as twisted Poisson structures are best understood in terms of appropriate substructures in (split) exact Courant algebroids, H-twisted Courant structures can be understood as substructures of the next higher analogue of these kind of nested structures (particular degree three NPQ-manifolds in the corresponding language mentioned briefly in the Introduction). Moreover, in the twisted Poisson case it is only the cohomology class of the closed 3-form that plays an inherent role from that perspective, exact 3-forms can be "gauged away" by a change of the splitting (cf., e.g., Ref. 18). We expect a likewise feature for the 4-form H above within the one step higher analogue, so that it may be worthwhile to search for examples of H-twisted Courant structures with nonexact 4-forms also.

Acknowledgments

We are grateful to M. Grützmann for discussions and to the Erwin Schrödinger Institute in Vienna for hospitality in the context of the program "Poisson sigma models, deformations, Lie algebroids, and higher analogues" in 2007.

References

1. P. Schaller and T. Strobl, *Mod. Phys. Lett.*, A9:3129–3136, 1994.

2. N. Ikeda, *Ann. Phys.*, 235:435–464, 1994.
3. T. Klösch and T. Strobl, *Class. Quant. Grav.*, 13:965–984, 1996. Erratum ibid. 14 (1997) 825.
4. D. Grumiller, W. Kummer and D. V. Vassilevich, *Phys. Rept.*, 369:327–430, 2002. hep-th/0204253.
5. T. Strobl, *Gravity in Two Spacetime Dimensions*, Habilitationsschrift, Rheinisch-Westfälische Technische Hochschule Aachen, 1999.
6. M. Kontsevich, q-alg/9709040.
7. A. S. Cattaneo and G. Felder, *Commun. Math. Phys.*, 212:591, 2000.
8. T. Strobl, *Phys. Rev. Lett.*, 93:211601, 2004. hep-th/0406215.
9. C. Mayer and T. Strobl, Lie Algebroid Yang Mills with Matter Fields, submitted to Geometry and Physics.
10. A. S. Cattaneo and G. Felder, math.SG/0003023, 2000.
11. P. A. M. Dirac, *Lectures on Quantum Mechanics*, Yeshiva University, New York: Academic Press, 1967.
12. Ct. Klimcik and T. Strobl, *J. Geom. Phys.*, 43:341–344, 2002.
13. J.-S. Park, hep-th/0012141.
14. P. Severa and A. Weinstein, *Prog. Theor. Phys. Suppl.*, 144:145–154, 2001.
15. T. J. Courant, *Trans. Amer. Math. Soc.*, 319:631–661, 1990.
16. Z.-J. Liu, A. Weinstein and P. Xu, *J. Diff. Geom.*, 45:547–574, 1997.
17. P. Severa, *Letters to Weinstein*, http://sophia.dtp.fmph.uniba.sk/~severa/ letters/.
18. A. Kotov, P. Schaller and T. Strobl, *Commun. Math. Phys.*, 260:455–480, 2005. hep-th/0411112.
19. N. Hitchin, *Q.J. Math.*, 54 (2003) no. 3, 281-308.
20. M. Gualtieri, Ph.D. thesis, University of Oxford, 2003. math.DG/0401221.
21. U. Lindstrom, R. Minasian, A. Tomasiello and M. Zabzine, *Commun. Math. Phys*, 257:235–256, 2005. hep-th/0405085.
22. A. Alekseev and T. Strobl, *JHEP*, 03:035, 2005. hep-th/0410183.
23. D. Roytenberg, math/0203110, 2002.
24. M. Alexandrov, M. Kontsevich, A. Schwarz and O. Zaboronsky, *Int.J.Mod.Phys.*, A12:1405-1430, 1997. hep-th/9502010.
25. A. S. Cattaneo and G. Felder, *Lett. Math. Phys.*, 56:163–179, 2001. math/0102108.
26. D. Roytenberg, *Lett. Math. Phys.*, 79:143–159, 2007. hep-th/0608150.
27. N. Ikeda, *Int. J. Mod. Phys.*, A18:2689-2702, 2003. hep-th/0203043.
28. C. Hofman and J.-S. Park, preprint hep-th/0209148, 2002.
29. K. I. Izawa, *Prog. Theor. Phys.*, 103:225–228, 2000. hep-th/9910133.
30. I. Batalin and R. Marnelius, *Phys. Lett.*, B512:225–229, 2001. hep-th/0105190.
31. A. Yu. Alekseev, P. Schaller, and T. Strobl, *Phys. Rev.*, D52:7146–7160, 1995.
32. L. D. Faddeev and R. Jackiw, *Phys. Rev. Lett.*, 60:1692, 1988.
33. M. Bojowald and T. Strobl, *Rev. Math. Phys.*, 15:663–703, 2003. hep-th/0112074.

Chapter 9

Some Local and Global Aspects of the Gauge Fixing in Yang–Mills-Theories

Daniel N. Blaschke[1], François Delduc[2], François Gieres[3],
Manfred Schweda[1] and Silvio P. Sorella[4]

[1] *Institute for Theoretical Physics, Vienna University of Technology,
Wiedner Hauptstrasse 8-10, A-1040 Vienna (Austria)*

[2] *Université de Lyon, Laboratoire de Physique, UMR 5672
École Normale Supérieure de Lyon, 46 allée d'Italie,
F-69364 Lyon cedex 07 (France)*

[3] *Université de Lyon, Université Lyon 1 and CNRS/IN2P3,
Institut de Physique Nucléaire, Bat. P. Dirac,
4 rue Enrico Fermi, F - 69622 - Villeurbanne (France)*

[4] *UERJ – Universidade do Estado do Rio de Janeiro,
Instituto de Física – Departamento de Física Teórica
Rua São Francisco Xavier 524, 20550-013 Maracanã,
Rio de Janeiro, Brasil*

We provide an introduction to the global aspects of the gauge fixing procedure for Yang-Mills theories, as well as a short account of the quantization in the Landau gauge and in axial-type gauges.

We dedicate these notes to the memory of Wolfgang Kummer, the "father" of the axial gauge.[1] Wolfgang contributed novel ideas to gauge and gravitational theories with a lot of enthusiasm and energy. Those who had the chance to meet him will miss his cheerful nature, his kindness, his passion for physics and the gentle pervasiveness with which he pursued his work.

9.1. Outline

These notes are organized as follows. After recalling the general setting for Yang-Mills (YM) theories in the next section, we discuss the global aspects of the gauge fixing procedure (notably the Gribov ambiguities), i.e. issues which are important (in particular, but not only) for the nonperturbative aspects of gauge theories like the confinement of quarks. Section 9.4 deals with the BRST-quantization of Yang-Mills theories in a relativistically covariant gauge, namely the Lorenz/Landau gauge. Section 9.5 discusses the advantages and drawbacks of the quantization in noncovariant gauges, in particular in the axial gauge.

9.2. YM in d-dimensional Minkowski space

9.2.1. General setting

Let G be a compact matrix Lie group (e.g. $G = SU(N)$) and $\mathbf{g}$ its Lie algebra with basis elements $\{T^a\}_{a=1,\ldots,\dim G}$ satisfying

$$[T^a, T^b] = f^{abc} T^c, \qquad \mathrm{Tr}\,(T^a T^b) = -\delta^{ab}. \qquad (9.1)$$

Here and in section 9.3, we assume that the matrices T^a are anti-Hermitian, i.e. $(T^a)^\dagger = -T^a$.

In the following, we will consider pure YM theory with structure group G in Minkowski space-time $M = \mathbb{R}^d$ (e.g. see Ref. 2). The basic field is the $\mathbf{g}$-valued *gauge potential* $A_\mu(x) = A_\mu^a(x) T^a$ where the A_μ^a are real-valued functions on space-time. The associated *field strength* is defined by

$$F_{\mu\nu} = \partial_\mu A_\nu - \partial_\nu A_\mu + [A_\mu, A_\nu],$$

where the self-coupling constant of the gauge fields has been absorbed into A_μ. With these conventions, a *finite gauge transformation* of A_μ writes as

$$^g A_\mu = g^{-1} A_\mu g + g^{-1} \partial_\mu g, \qquad \text{with } g \equiv \mathrm{e}^\omega. \qquad (9.2)$$

Here, the $\mathbf{g}$-valued parameter $\omega(x) = \omega^a(x) T^a$ describes *infinitesimal gauge transformations*: $^g A_\mu \simeq A_\mu + \delta_\omega A_\mu$ with

$$\delta_\omega A_\mu = D_\mu \omega \equiv \partial_\mu \omega + [A_\mu, \omega]. \qquad (9.3)$$

The *gauge algebra* $\{\omega : \mathbb{R}^d \to \mathbf{g}\}$ can be identified with the (infinite dimensional) Lie algebra of the *gauge group* $\mathcal{G} \equiv \{g : \mathbb{R}^d \to G\}$.

The *dynamics* of pure YM theory is described by the classical action

$$S_{inv}[A] = \frac{1}{4} \int_{\mathbb{R}^d} d^d x\, \mathrm{Tr}\, F_{\mu\nu} F^{\mu\nu} = -\frac{1}{4} \sum_{a=1}^{\dim G} \int_{\mathbb{R}^d} d^d x\, F_{\mu\nu}^a F^{a\mu\nu}, \qquad (9.4)$$

which is invariant under gauge transformations.

9.2.2. *Geometric framework*

The space $\mathcal{A}$ of all gauge fields $A \equiv (A_\mu)$ defined on the space-time manifold $M = \mathbb{R}^d$ is referred to as the *configuration space* of the theory (in analogy to the configuration space of a system of particles in classical mechanics). The gauge transformations (9.2) define an equivalence relation in the space $\mathcal{A}$: the equivalence class $\mathcal{O}_A$ of the gauge field A consists of all potentials $^g\!A$ obtained from A by a gauge transformation $g : M \to G$, i.e.

$$\mathcal{O}_A \equiv \{a \in \mathcal{A} \,|\, a = \,^g\!A \ \text{for some } g \in \mathcal{G}\}\,.$$

This set represents a hypersurface in the space $\mathcal{A}$ which is called the *gauge orbit* of A. The quotient of the space of gauge fields by the group of gauge transformations,

$$\mathcal{M} \equiv \mathcal{A}/\mathcal{G} = \{\mathcal{O}_A \,|\, A \in \mathcal{A}\} \tag{9.5}$$

is known as the *orbit space* or the *true (or physical) configuration space* since it describes physically inequivalent gauge field configurations. The geometric set-up is schematically depicted in Figure 1 (which also involves a gauge slice to be discussed in the next paragraph).

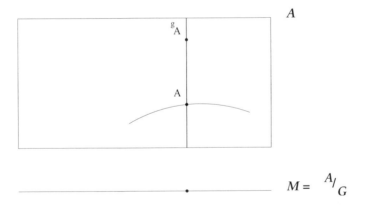

Figure 9.1 Sketch of the configuration space $\mathcal{A}$, the orbit space $\mathcal{M} \equiv \mathcal{A}/\mathcal{G}$, the gauge orbit $\mathcal{O}_A$ through $A \in \mathcal{A}$ and a local gauge slice through A.

9.2.3. *Admissibility criteria for the gauge fixing function*

A gauge fixing condition $f(A) = 0$ is given by a **g**-valued function f of the variables A_μ and their derivatives which satisfies the following two properties:

(1) The function $f(A)$ is not gauge invariant. From the geometric point of view this means that, within the space of all gauge fields, the fields satisfying $f(A) = 0$ represent a hypersurface which is 'transversal' to the gauge orbits (i.e not tangent to any of these orbits – see Fig. 1). This hypersurface is called a *gauge slice* and amounts to choosing in a continuous manner a representative in each gauge orbit.

(2) On each gauge orbit, the gauge slice $f(A) = 0$ should be reachable by means of a gauge transformation. Thus, for any gauge field A, one must be able to find a unique gauge transformation g such that $0 = f({}^gA)$, where ${}^gA_\mu = g^{-1}A_\mu g + g^{-1}\partial_\mu g$.

The most popular choice is the (relativistically invariant) **Lorenz gauge** $\partial^\mu A_\mu = 0$. We note that this gauge first appeared in print in 1867 in an article of the Danish physicist *Ludwig Valentin Lorenz*.[3] Other popular choices (which are not relativistically invariant) are the various special cases of the condition $n^\mu A_\mu = 0$ for some constant vector field (n^μ):

(General) axial gauge : $n^\mu A_\mu = 0$,   with n^μ constant.   (9.6)

The space-like axial gauge (i.e. $n^2 \equiv n^\mu n_\mu \equiv (n^0)^2 - \vec{n}^2 < 0$) has been invented by W. Kummer.[1] In gauges of this type, the Faddeev-Popov ghost fields decouple and therefore these gauges are also referred to as *ghost-free* or *'physical'* gauges.[4,5] This leads to a simplification of the quantization procedure though there also exist various drawbacks – see section 9.5.

9.3. Global issues of gauge fixing: Gribov problem

One may view the gauge fixing procedure, i.e. the choice of a gauge slice in the space $\mathcal{A}$ of all connections, as a procedure for obtaining a parametrization of the orbit space $\mathcal{M} = \mathcal{A}/\mathcal{G}$. For non-Abelian YM-theories, it was shown by Gribov that on a gauge orbit $\mathcal{O}_A$ there generally exist many potentials satisfying a given gauge fixing condition.[6] These so-called Gribov ambiguities, which we will discuss in subsection 9.3.2, imply that the orbit space has a complicated mathematical structure.

One generally assumes that Gribov's problem can be ignored in the perturbative approach to quantum field theory which is developed by choosing a local gauge slice in the space $\mathcal{A}$ of gauge fields. Yet, the work initiated by Gribov, Zwanziger and Stingl[6–8] shows that this appears to be a prejudice and that the gauge field propagator is modified if one takes into account the global aspects of the gauge fixing (an issue that we will address in subsection 9.3.3 below). The investigation of nonperturbative effects like the confinement in quantum chromodynamics also requires the study of the global aspects of gauge fixing and of the non-trivial topology of the orbit space $\mathcal{M} = \mathcal{A}/\mathcal{G}$.[9,10] A convenient and elegant approach to this study consists of formulating the gauge fixing condition in terms of a variational principle, i.e. obtaining this equation by minimizing an appropriate action functional of A.[7,11,12] We will now discuss this principle after describing the general framework.

9.3.1. *Geometric setting: Gauge fixing through minimization*

Consider a non-Abelian structure group G, e.g. $G = SU(N)$ with $N \geq 2$. Furthermore, consider the four-dimensional Euclidean space-time[9,13] and suppose that the gauge fields vanish at infinity, i.e. for $(x^0)^2 + \cdots + (x^3)^2 \to \infty$. This situation is tantamount to studying fields on compactified space $M = S^4 \simeq \mathbb{R}^4 \cup \{\infty\}$, i.e. the four-sphere $S^4 \subset \mathbb{R}^5$. The Euclidean metric of $\mathbb{R}^5$ induces a metric $(\rho_{\mu\nu}(x))$ on $M = S^4$ so that M may be viewed as a four-dimensional Riemannian manifold. One can also consider other compact 4-manifolds like the torus which corresponds to periodic boundary conditions for the gauge fields in $\mathbb{R}^4$.[7,10]

As was shown by I.M. Singer,[13] the **Gribov problem for the gauge fixing**[6] can be stated as follows. Consider a gauge fixing condition which is not algebraic, e.g. the Lorenz gauge or the Coulomb gauge, but not the axial gauge $n^\mu A_\mu = 0$ (for which the boundary conditions of gauge fields at infinity are not satisfied). Then, there generally exists more than one element of a gauge orbit which satisfies the gauge fixing condition. (One says that there exist **gauge copies** or that there are **Gribov ambiguities** or **gauge fixing degeneracies**. An explicit example is given in subsection 9.3.2 below.) Thus, the gauge fixing is not globally defined on $\mathcal{A}$, but only locally. If one wants the gauge slice to cut the gauge orbits only once, one has to restrict oneself to a subset of it which is known as a fundamental modular domain, or fundamental domain for short. This problem (which does not affect Abelian theories) is due to the non-linearities of the theory and reflects the non-trivial geometric structure of the orbit space. For a proof of Singer's theorem, we refer to the literature.[13,14] Here, we rather discuss the global properties of a gauge slice[7,9,10] by considering the gauge condition which is most natural from the geometric point of view, namely the so-called **background gauge**[a]: one chooses an arbitrary, but fixed connection $\bar{A} \in \mathcal{A}$ as background field (e.g. $\bar{A} = 0$) and one considers the gauge slice $\Gamma_{\bar{A}} \subset \mathcal{A}$ defined by

$$\Gamma_{\bar{A}} \equiv \{ A \in \mathcal{A} \,|\, D_{\mu}^{(\bar{A})}(A^{\mu} - \bar{A}^{\mu}) = 0 \} \,, \tag{9.7}$$

where $D_{\mu}^{(\bar{A})} \omega \equiv \partial_{\mu} \omega + [\bar{A}_{\mu}, \omega]$ for a Lie algebra-valued function ω. For the choice $\bar{A} = 0$, the gauge condition $D_{\mu}^{(\bar{A})}(A^{\mu} - \bar{A}^{\mu}) = 0$ reduces to the Lorenz gauge condition $\partial_{\mu} A^{\mu} = 0$. By putting $\bar{A} = 0$ and $A_0 = 0$, one obtains the Coulomb gauge condition which is discussed in Ref. 10. Although the choice of the "vacuum configuration" $\bar{A} = 0$ simplifies calculations, it entails various mathematical complications.[9] This technical point can be tackled by choosing a background connection $\bar{A}$ which has the property that it is *irreducible:* this means that the only gauge transformations $x \mapsto g(x) \in G$ which leave $\bar{A}$ invariant (i.e. $\bar{A} = {}^{g}\bar{A} \equiv g^{-1}\bar{A}g + g^{-1}dg$) are the constant transformations which commute with all elements of G. (One says that the *stabilizer* of $\bar{A}$, i.e. $\mathcal{S}_{\bar{A}} \equiv \{ g \in \mathcal{G} \,|\, {}^{g}\bar{A} = \bar{A}\}$, reduces to the center of G.) Obviously, $\bar{A} = 0$ is not an irreducible connection.

For the following considerations, we introduce the L^2 inner product of Lie algebra-valued functions B, C on $M = S^4$:

$$\langle B, C \rangle \equiv \int_{M} d^4 x \, \sqrt{\rho} \, \mathrm{Tr}\,(B^{\dagger} C) = \int_{M} d^4 x \, \sqrt{\rho} \, B^a C^a \,. \tag{9.8}$$

Here, ρ denotes the determinant of the metric tensor $(\rho_{\mu\nu})$ on M and we have used the sign conventions introduced in subsection 9.2.1. The norm associated to the inner product (9.8) is given by $\|B\| = \sqrt{\langle B, B \rangle}$ and the Lie algebra-valued functions B on M which have a finite norm define the Hilbert space $L^2(M, \mathbf{g})$.

As we will now show, the connections A satisfying the background gauge fixing condition, i.e. $A \in \Gamma_{\bar{A}}$, can be viewed as stationary points of the following functional,[7,9] which is defined on the gauge group $\mathcal{G}$ and which is referred to as *Morse functional*:

$$\phi_A \equiv \phi_{\bar{A},A} : \quad \mathcal{G} \longrightarrow [0, \infty[\tag{9.9}$$
$$g \longmapsto \phi_A[g] = \|{}^{g}A - \bar{A}\|^2$$
$$= -\int_{M} d^4 x \, \sqrt{\rho} \, \mathrm{Tr}\,({}^{g}A_{\mu} - \bar{A}_{\mu})({}^{g}A^{\mu} - \bar{A}^{\mu}) \,.$$

---

[a]We note that the background approach has originally been conceived[15] as a decomposition of a generic gauge field A into a background or "classical" part $\bar{A}$ and a "quantum" part $\hat{A}$ (i.e. $A = \bar{A} + \hat{A}$) so that the gauge fixing condition (9.7) then reads $D_{\mu}^{(\bar{A})} \hat{A}^{\mu} = 0$.

Here, ρ denotes the determinant of the metric tensor $(\rho_{\mu\nu})$ on $M = S^4$ and $\phi_A[g]$ may be viewed as the L^2-norm of A along its gauge orbit. To determine the extrema of the Morse functional as a function of $g \in \mathcal{G}$, i.e. along the gauge orbits, we proceed as follows. We consider a curve $t \mapsto g_t \equiv e^{t\omega}$ in $\mathcal{G}$ passing through the identity $g_0 = \mathbb{1} \in \mathcal{G}$, and we expand the functional $\phi_A[g_t]$ with respect to the parameter t :

$$\phi_A[g_t] = -\int_M d^4x \sqrt{\rho}\, \mathrm{Tr}\,[(A_\mu - \bar{A}_\mu) + tD_\mu^{(A)}\omega + \frac{t^2}{2}\,[D_\mu^{(A)}\omega, \omega] + \ldots]^2.$$

By comparing with the general form of the Taylor expansion,

$$\phi_A[g_t] = \phi_A[\mathbb{1}] + t\,\frac{d\phi_A[g_t]}{dt}\bigg|_{t=0} + \frac{t^2}{2}\,\frac{d^2\phi_A[g_t]}{dt^2}\bigg|_{t=0} + \ldots,$$

by using integration by parts and applying the identity

$$D_\mu^{(\bar{A})}(A^\mu - \bar{A}^\mu) = D_\mu^{(A)}(A^\mu - \bar{A}^\mu),$$

we readily find

$$\frac{d\phi_A[g_t]}{dt}\bigg|_{t=0} = \int_M d^4x \sqrt{\rho}\, \mathrm{Tr}\,[\frac{\delta\phi_A}{\delta g}\bigg|_{\mathbb{1}}\omega],$$

$$\text{with}\quad \frac{\delta\phi_A}{\delta g}\bigg|_{\mathbb{1}} = -2D_\mu^{(\bar{A})}(A^\mu - \bar{A}^\mu),$$

$$\frac{d^2\phi_A[g_t]}{dt^2}\bigg|_{t=0} = 2\int_M d^4x \sqrt{\rho}\, \mathrm{Tr}\,[\omega\,\frac{\delta^2\phi_A}{\delta g^2}\bigg|_{\mathbb{1}}\omega],$$

$$\text{with}\quad \frac{\delta^2\phi_A}{\delta g^2}\bigg|_{\mathbb{1}} = -D^{(\bar{A})\mu}D_\mu^{(A)}. \tag{9.10}$$

The critical points of the Morse functional are characterized by the condition $\frac{\delta\phi_A}{\delta g}\big|_{\mathbb{1}} = 0$. Obviously, the expression $\frac{\delta^2\phi_A}{\delta g^2}\big|_{\mathbb{1}}$, i.e. the Hessian of ϕ_A evaluated at its critical points, is nothing but the *Faddeev-Popov (FP) operator*. For any $A \in \Gamma_{\bar{A}}$, this operator is Hermitian with respect to the inner product (9.8):

$$\Delta_{\mathrm{FP}}^{(\bar{A},A)} \equiv -D^{(\bar{A})\mu}D_\mu^{(A)} = -D^{(A)\mu}D_\mu^{(\bar{A})} \qquad \text{for } A \in \Gamma_{\bar{A}}.$$

(For $\bar{A} = 0$, this operator reduces to the familiar Lorenz gauge expression $-\partial^\mu D_\mu^{(A)}$.)

We recall[b] that the operator $\Delta_{\mathrm{FP}}^{(\bar{A},A)}$ acting on the Hilbert space

---

[b]Of course a differential operator like $\Delta_{\mathrm{FP}}^{(\bar{A},A)}$ is only well defined when acting on sufficiently regular functions: here and in the following, we ignore these functional analytic details in order the simplify the presentation.

$L^2(M, \mathbf{g})$ is said to be *positive*, $\Delta_{\mathrm{FP}}^{(\bar{A},A)} \geq 0$, if its expectation values,

$$\langle \omega, \Delta_{\mathrm{FP}}^{(\bar{A},A)} \omega \rangle = - \int_M d^4x \sqrt{\rho} \, \mathrm{Tr} \, (\omega^\dagger D^{(\bar{A})\mu} D_\mu^{(A)} \omega)$$

$$= \int_M d^4x \sqrt{\rho} \, (D^{(\bar{A})\mu} \omega)^a (D_\mu^{(A)} \omega)^a \,,$$

are positive for all $\omega \in L^2(M, \mathbf{g})$. This operator is said to be *strictly positive* if $\langle \omega, \Delta_{\mathrm{FP}}^{(\bar{A},A)} \omega \rangle > 0$ for all $\omega \neq 0$. We also recall that the eigenvalues of a (strictly) positive operator are (strictly) positive.

From the results (9.10), we can now draw the following conclusions. A gauge field $A \in \Gamma_{\bar{A}}$ corresponds to a *stationnary point* of the Morse functional $g \mapsto \phi_A[g]$ (at $g = \mathbb{1} \in \mathcal{G}$). If the FP-operator $\Delta_{\mathrm{FP}}^{(\bar{A},A)}$ is strictly positive for the gauge field $A \in \Gamma_{\bar{A}}$, then this field corresponds to a *local minimum* of the Morse functional $g \mapsto \phi_A[g]$ (at $g = \mathbb{1} \in \mathcal{G}$). Conversely, for local minima of the latter functional, one has $\Delta_{\mathrm{FP}}^{(\bar{A},A)} \geq 0$.

For an irreducible background field $\bar{A}$, the *Gribov region* $\Omega_{\bar{A}} \subset \Gamma_{\bar{A}}$ is now defined to be the subset of the gauge slice $\Gamma_{\bar{A}}$ on which the FP-operator is positive:

$$\Omega_{\bar{A}} \equiv \{ A \in \Gamma_{\bar{A}} \, | \, \Delta_{\mathrm{FP}}^{(\bar{A},A)} \geq 0 \} \,. \tag{9.11}$$

Its boundary $\partial \Omega_{\bar{A}}$, which is referred to as *Gribov horizon*, is determined by the vanishing of the FP-determinant $\det \Delta_{\mathrm{FP}}^{(\bar{A},A)}$: at the Gribov horizon, the lowest eigenvalue of the FP-operator vanishes. The conclusions summarized above can now be reformulated as follows.[16] The set of all local minima of the Morse functional ϕ_A is a subset of the region $\Omega_{\bar{A}}$, and every interior point A of $\Omega_{\bar{A}}$ is necessarily a local minimum of ϕ_A.

One can show[7,17] that any gauge orbit crosses the hyperplane $\Omega_{\bar{A}}$ *at least once*, hence field configurations located on the gauge slice on the outside of $\Omega_{\bar{A}}$ are gauge copies of configurations located inside $\Omega_{\bar{A}}$. As we will discuss below, some gauge copies already appear within the Gribov region. Examples of gauge copies are obtained from the zero modes of the FP-operator, see next subsection.

The local minima of the Morse functional (corresponding to the interior points of the Gribov region $\Omega_{\bar{A}}$) may represent absolute minima or relative minima. By definition, a *fundamental domain* $\Lambda_{\bar{A}}$ (for the orbit space $\mathcal{M}$) is the subset $\Lambda_{\bar{A}} \subset \Omega_{\bar{A}}$ which consists of absolute minima of the Morse functional:

$$\Lambda_{\bar{A}} \equiv \{ A \in \Gamma_{\bar{A}} \, | \, \phi_A[g] \geq \phi_A[\mathbb{1}] \text{ for all } g \in \mathcal{G} \} \,. \tag{9.12}$$

Thus, we have the inclusion $\bar{A} \in \Lambda_{\bar{A}} \subset \Omega_{\bar{A}} \subset \Gamma_{\bar{A}}$. One can prove that the fundamental domain $\Lambda_{\bar{A}}$ is a bounded, convex set and that it is not equal to the set $\Omega_{\bar{A}}$.[9,16] Furthermore, one can show[9,12,17] that this set contains at least one representative of any gauge orbit $\mathcal{O}_A$, an issue that we will now address in more detail (cf. Figure 3).

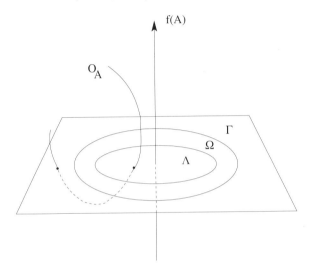

Figure 9.2[18] Here, the space $\mathcal{A}$ of gauge fields is depicted as three-dimensional, the vertical direction being labeled by the gauge fixing function $f(A) \equiv D_\mu^{(\bar{A})}(A^\mu - \bar{A}^\mu)$. The horizontal plane represents the gauge slice $\Gamma \equiv \Gamma_{\bar{A}} = \{A \in \mathcal{A} \mid f(A) = 0\}$, which contains as a subset the Gribov region $\Omega \equiv \Omega_{\bar{A}}$, the latter containing the fundamental domain $\Lambda \equiv \Lambda_{\bar{A}}$. For any $A \in \mathcal{A}$, the corresponding gauge orbit $\mathcal{O}_A$ passes only once through Λ, but more than once through the gauge slice Γ, whence the gauge copies (Gribov ambiguities).

Since the Morse functional has a unique absolute minimum in the interior of the fundamental region $\Lambda_{\bar{A}}$, gauge copies can only lie on its boundary $\partial\Lambda_{\bar{A}}$. Thus,[9,10,19] one is led to identify the irreducible connections belonging to $\partial\Lambda_{\bar{A}}$ which are gauge copies of each other: in this way, one obtains a *true fundamental domain*

$$\breve{\Lambda}_{\bar{A}} \equiv \Lambda_{\bar{A}}|_{\text{boundary identifications}} \, ,$$

i.e. a subset of the gauge slice which has the property that it contains exactly one representative of each gauge orbit. Accordingly, this domain projects down bijectively to the orbit space $\mathcal{M} = \mathcal{A}/\mathcal{G}$ and thereby allows for a description of this space.

The described identification of boundary points of the convex set $\Lambda_{\bar{A}}$ implies the existence of non-contractible loops involving the boundary points. (This is analogous to the rotation group manifold $SO(3)$, i.e. a solid ball in $\mathbb{R}^3$ of radius π for which all antipodal points on the boundary identified.) The non-trivial topology of the orbit space $\mathcal{M} = \mathcal{A}/\mathcal{G}$ thus reflects the fact that the gauge slice has been restricted to $\Lambda_{\bar{A}}$ (i.e. a non-local procedure) and that certain identifications have been made on its boundary $\partial\Lambda_{\bar{A}}$. For further details (e.g. the handling of different instanton sectors) and some explicit examples, we refer to Refs. 9, 10.

For other gauge conditions like the covariant linear gauge condition $\partial^\mu A_\mu + \lambda b = 0$ or the maximal Abelian gauge, different issues like the Gribov copies, the Morse functional or the infrared behavior of the gauge field propagator have been addressed in the literature.[20,21]

9.3.2. Determination of Gribov copies

For simplicity, we consider $\bar{A} = 0$, i.e. the Lorenz gauge condition $\partial^\mu A_\mu = 0$. The FP-operator then reduces to $\Delta_{\mathrm{FP}}^{(A)} \equiv -\partial^\mu D_\mu^{(A)}$ and its eigenvalue equation reads as

$$-\partial^\mu \left(\partial_\mu \psi_\lambda + [A_\mu, \psi_\lambda] \right) = \lambda \psi_\lambda \,,$$

where $\psi_\lambda : M \to \mathbf{g}$ and where λ depends on the gauge potential A.

By definition, the gauge fields A and ^gA are gauge copies if they satisfy, for some $g \in \mathcal{G}$, the relations

$$\begin{aligned}
^gA_\mu &= g^{-1} A_\mu g + g^{-1} \partial_\mu g \\
\partial^\mu A_\mu &= 0 = \partial^\mu \, {}^gA_\mu \,.
\end{aligned} \tag{9.13}$$

We note that the first relation can also be written as $^gA_\mu = A_\mu + g^{-1}\left(\partial_\mu g + [A_\mu, g]\right)$. Thus, by virtue of $\partial^\mu A_\mu = 0$, the condition $\partial^\mu \, {}^gA_\mu = 0$ is equivalent to

$$\partial^\mu \left[g^{-1}\left(\partial_\mu g + [A_\mu, g]\right) \right] = 0 \,.$$

Writing $g = e^\alpha \simeq 1 + \alpha$ (where α is a Lie algebra-valued function), we obtain to order α

$$0 = \partial^\mu \left(\partial_\mu \alpha + [A_\mu, \alpha] \right) \equiv \partial^\mu D_\mu^{(A)} \alpha \,. \tag{9.14}$$

Thus, gauge copies can be obtained as zero modes of the FP-operator $\Delta_{\mathrm{FP}}^{(A)} = -\partial^\mu D_\mu^{(A)}$, i.e. from eigenfunctions of $\Delta_{\mathrm{FP}}^{(A)}$ associated to the eigenvalue zero.[6]

For a generic gauge fixing condition $f(A) = 0$, we have[21]

$$0 = f(^gA) \simeq f(A + D^{(A)}\alpha) \simeq \underbrace{f(A)}_{=0} + \frac{\delta f}{\delta A_\mu} D_\mu^{(A)}\alpha = N^{(A)\mu} D_\mu^{(A)}\alpha,$$

where $N^{(A)} \equiv \frac{\delta f}{\delta A}$ is normal to the gauge fixing surface Γ. For the Lorenz or Coulomb gauge, the normal $N^{(A)}$ does not depend on A and the condition $N^{(A)\mu} D_\mu^{(A)} = 0$ (i.e. orthogonality of $N^{(A)}$ and $D^{(A)}$) describes gauge orbits that are tangent to the gauge slice Γ at some point A.[21]

The simplest example of gauge copies with axial symmetry has been given by Henyey[19,22]). In this case, one considers Euclidean $\mathbb{R}^3$, the structure group $G = SU(2)$ and a gauge field taking values in the Cartan subalgebra of $su(2)$:

$$\vec{A}(\vec{x}) = i\,\vec{a}(\vec{x})\sigma_3 \quad \text{with } \vec{\nabla} \cdot \vec{a} = 0, \tag{9.15}$$

where σ_3 denotes the third Pauli matrix. Introducing spherical coordinates (r, θ, φ) and setting

$$\vec{a} = a(r, \theta)\,\vec{e}_\varphi,$$

one can find a normalizable zero mode α of the FP-operator, i.e. a solution of

$$\vec{\nabla}^2\alpha + \sum_{i=1}^{3}[A_i, \partial_i\alpha] = 0$$

which has the form

$$\alpha(r, \theta, \varphi) = r\,b(r)\sin\theta\,[\sigma_1\cos\varphi + \sigma_2\sin\varphi].$$

Explicitly, one obtains

$$a(r, \theta) = \frac{15}{2}\frac{rr_0^2}{(r^2 + r_0^2)^2}\sin\theta,$$

$$b(r) = \frac{k}{(r^2 + r_0^2)^{3/2}}, \quad \text{with } r_0, k \in \mathbb{R},\ r_0 \neq 0. \tag{9.16}$$

This example and various others are discussed in detail in Ref. 23.

9.3.3. *Tentatives to tackle the Gribov problem in the quantization procedure*

In this subsection, we consider Euclidean $\mathbb{R}^4$, the structure group $G = SU(N)$, the gauge fixing condition $0 = \partial \cdot A \equiv \partial^\mu A_\mu$ and we spell out the gauge coupling constant g. With QCD in mind, the gauge field is also referred to as gluon field.

In order to account for the existence of gauge copies, Gribov[6] restricted the domain of integration in the Feynman path integral to the Gribov region Ω. The modified Faddeev-Popov formula for the partition function then writes

$$\mathcal{Z} = \int_\Omega \mathcal{D}A \; \delta(\partial \cdot A) \; \det(\Delta_{\mathrm{FP}}^{(A)}) \; e^{-S_{YM}} \;. \qquad (9.17)$$

Zwanziger has been able to show[24,25] that this restriction of the domain of integration can be achieved at the level of the Lagrangian by the introduction of a new term appearing in the Boltzmann weight defining the Yang-Mills measure, i.e.

$$\mathcal{Z} = \int_{\mathcal{A}} \mathcal{D}A \; \delta(\partial \cdot A) \; \det(\Delta_{\mathrm{FP}}^{(A)}) \; e^{-(S_{YM}+S_\gamma)} \;. \qquad (9.18)$$

Here, the term S_γ, which is referred to as *horizon functional*, is given by the nonlocal expression[24,25]

$$S_\gamma[A] = \gamma^4 \int_{\mathbb{R}^4} d^4x \int_{\mathbb{R}^4} d^4y \, h_A(x,y) \,,$$

$$\text{with} \quad h_A(x,y) = g^2 f^{abc} A_\mu^b(x) \left(\Delta_{\mathrm{FP}}^{(A)} \right)^{-1 \, ad}(x,y) \, f^{dec} A^{\mu \, e}(y) \,. \qquad (9.19)$$

The parameter $\gamma > 0$ has the dimension of a mass and is known as the *Gribov parameter* or *Gribov mass*. It does not represent a free parameter since it is determined in a self-consistent way as a function of the gauge coupling constant g and of the invariant scale Λ_{QCD} of Yang-Mills theory through the so-called *gap equation*[24,25]

$$\langle h_A \rangle = 4(N^2 - 1) \;. \qquad (9.20)$$

Here, $\langle h_A \rangle$ is the vacuum expectation value of h_A and the factor on the right-hand-side is the product of the dimension of space-time and of the dimension of the structure group $SU(N)$.

Very much like the Faddeev-Popov determinant, the nonlocal horizon functional, Eq. (9.19), can be cast into local form through the introduction of a set of additional auxiliary fields, so that a local action can be obtained from the restriction to the Gribov region Ω. Remarkably, this local action enjoys the important property of being multiplicatively renormalizable,[24-29] a feature which has allowed for explicit higher loop calculations.[30] It is worth mentioning here that the inclusion of the horizon functional does not alter the number of renormalization factors. Only two renormalization factors are in fact needed, which can be chosen to be those of the gauge field A and of the coupling constant g, i.e. Z_A and Z_g (see next section for the quantization in the Lorenz/Landau gauge). This follows from the fact that the nonrenormalization theorem for the gluon-ghost-antighost vertex (discussed in the next section) remains valid in the presence of the horizon function. Moreover, also the renormalization factor of the Gribov parameter γ can be proven to be expressed in terms of Z_g and Z_A.[24–29]

As for the gluon propagator, the tree level expression

$$\langle A_\mu^a(p)A_\nu^b(-p)\rangle = \delta^{ab}\left(\delta_{\mu\nu} - \frac{p_\mu p_\nu}{p^2}\right)\mathcal{D}(p^2), \qquad \text{with } \mathcal{D}(p^2) = \frac{1}{p^2},$$
(9.21)

is modified if one takes into account the Gribov problem: the tree level form factor $\mathcal{D}(p^2) = 1/p^2$ is to be replaced by the so-called *Gribov-Stingl propagator*[6,8]

$$\mathcal{D}_{GS}(p^2) = \frac{1}{p^2 + \frac{\gamma^4}{p^2+\mu^2}} = \frac{p^2+\mu^2}{p^4 + \mu^2 p^2 + \gamma^4}, \tag{9.22}$$

which involves a second mass parameter μ. The latter represents a dynamical parameter too, which reflects the nontrivial dynamics of the additional fields needed to localize the horizon functional, see Refs. 28, 29. Similarly to γ, the parameter μ can be expressed in terms of g and of Λ_{QCD}. While expression (9.22) still enjoys the usual ultraviolet behaviour $\sim 1/p^2$ at very high momenta, its behaviour in the infrared region is deeply modified due to the presence of the parameters γ, μ (which appear as a consequence of the modification (9.18) of the usual Faddeev-Popov quantization formula). In particular, expression (9.22) is suppressed at low momenta and displays violation of reflection positivity,[28,29] invalidating the usual particle interpretation. In other words, in the infrared region, the gluons cannot be considered as excitations of the physical spectrum of the theory, a feature which is interpreted as a manifestation of gluon confinement.

The Gribov phenomenon has been intensively investigated in recent years, from both analytical and numerical viewpoints. In particular, the Landau gauge has been analyzed from the numerical point of view, as it possesses a useful lattice formulation.[31] The determination of the relative minima of the Morse functional can be achieved numerically by employing suitable minimizing algorithms.[32–35] In this respect, it is worth mentioning that the most recent results obtained on huge lattices are in agreement with a gluon propagator which is strongly suppressed in the infrared region and which exhibits violation of the reflection positivity.[32–35]

As we just discussed, the restriction of the domain of integration in the Faddeev-Popov formula to the *Gribov region* can be implemented at the Lagrangian level. So far an analogous restriction could not be achieved for the *true fundamental domain* $\check{\Lambda}$. This issue is an open problem for which the numerical investigations also represent a big challenge. Indeed, the numerical characterization of field configurations corresponding to absolute minima of the Morse functional is a highly nontrivial task, as the number of variables is rapidly growing when the lattice volume is increased in order to reach the thermodynamic limit.

To conclude we mention that, for the Lorenz gauge condition $\partial^\mu A_\mu = 0$, the boundaries of the fundamental domain and of the Gribov region have some points in common.[10] Recently, it has been argued by D. Zwanziger[36] that the field configurations on this common boundary dominate the functional integral of YM-theory. He also put forward some arguments according to which the gauge copies located in the Gribov region outside of the fundamental domain do not contribute to the correlations functions. So far these issues have not yet been settled, but they currently represent the object of intensive investigations. If these conjectures are true, the restriction of the functional integral to the Gribov region would provide a satisfactory quantization for YM-theories in the Lorenz gauge.

9.4. On the quantization in the Lorenz/Landau gauge

In this section, we consider the Lorenz gauge fixing condition $\partial^\mu A_\mu = 0$ and the Landau gauge choice $\lambda = 0$ for the gauge parameter. In the literature, this set of conditions is referred to for short as the **Landau gauge** and here we will also adhere to this terminology.

The Landau gauge is one of the most popular, relativistically covariant gauges. Besides its perturbative renormalizability, it displays several interesting features. This is the case, for instance, of the so-called non-renormalization theorem of the gluon-ghost-antighost vertex, which we will discuss in this section. This result holds to all orders of perturbation theory thanks to the existence of an additional Ward identity, known as the ghost Ward identity.[37] As we already saw in the last subsection, the Landau gauge is also largely employed in the study of nonperturbative features of YM-theories, such as the Gribov phenomenon which leads to a modification of the behaviour of the correlation functions in the infrared region.

In the following, we will give a short account of the Landau gauge by considering the case of *pure Euclidean YM-theory in four dimensions*, as described by the action

$$S_{YM} = \frac{1}{4} \int_{\mathbb{R}^4} d^4x \ F_{\mu\nu}^a F_{\mu\nu}^a \ . \tag{9.23}$$

9.4.1. *Gauge fixing action*

For convenience we summarize the basic relations while spelling out all indices. In the Landau gauge

$$\partial_\mu A_\mu^a = 0 \ , \tag{9.24}$$

the gauge fixed action reads as

$$S = S_{YM} + \int_{\mathbb{R}^4} d^4x \ \left(b^a \partial_\mu A_\mu^a - \bar{c}^a \mathcal{M}^{ab} c^b \right) \ . \tag{9.25}$$

Here, b^a is the Lagrange multiplier field, $(\bar{c}^a, c^a)$ denote the anticommuting Faddeev-Popov ghosts and

$$\mathcal{M}^{ab} = -\partial_\mu \left(\partial_\mu \delta^{ab} + g f^{acb} A_\mu^c \right) \tag{9.26}$$

represents the Faddeev-Popov operator in the Landau gauge. Expression (9.25) enjoys BRST invariance, $sS = 0$, where s denotes the nilpotent BRST-operator

$$sA_\mu^a = -D_\mu^{ab} c^b \ , \qquad\qquad sc^a = \frac{1}{2} g f^{abc} c^b c^c \ ,$$
$$s\bar{c}^a = b^a \ , \qquad\qquad sb^a = 0 \ , \tag{9.27}$$

with $D_\mu^{ab} c^b \equiv \partial_\mu c^a + g f^{abd} A_\mu^b c^d$. The (tree level) gluon propagator stemming from the action (9.25) is transverse and given by expression (9.21), i.e.

$$\left\langle A_\mu^a(p) A_\nu^b(-p) \right\rangle = \delta^{ab} \left(\delta_{\mu\nu} - \frac{p_\mu p_\nu}{p^2} \right) \mathcal{D}(p^2) \ , \qquad \text{with } \mathcal{D}(p^2) = \frac{1}{p^2} \ . \tag{9.28}$$

9.4.2. *The nonrenormalization of the gluon-ghost-antighost vertex*

In order to convert the BRST invariance of the Faddeev-Popov action (9.25) into suitable Ward identities, a pair of BRST invariant external sources,

$\left(\Omega^a_\mu, L^a\right)$, coupled to the nonlinear transformations of the gauge and ghost fields has to be introduced,

$$S_{\text{ext}} = s \int d^4x \left(-\Omega^a_\mu A^a_\mu + L^a c^a\right) = \int d^4x \left(-\Omega^a_\mu D^{ab}_\mu c^b + \frac{1}{2} g L^a f^{abc} c^b c^c\right),$$
$$(9.29)$$

so that the complete action Σ

$$\Sigma = S + S_{\text{ext}},$$
$$(9.30)$$

is BRST invariant. This action enjoys the following Ward identities:

- The *Slavnov-Taylor identity*

$$\mathcal{S}(\Sigma) \equiv \int d^4x \left(\frac{\delta\Sigma}{\delta\Omega^a_\mu}\frac{\delta\Sigma}{\delta A^a_\mu} + \frac{\delta\Sigma}{\delta L^a}\frac{\delta\Sigma}{\delta c^a} + b^a\frac{\delta\Sigma}{\delta\bar{c}^a}\right) = 0.$$
$$(9.31)$$

- The *Landau gauge condition* and the *antighost equation*, given by

$$\frac{\delta\Sigma}{\delta b^a} = \partial_\mu A^a_\mu,$$
$$(9.32)$$

$$\frac{\delta\Sigma}{\delta\bar{c}^a} + \partial_\mu\frac{\delta\Sigma}{\delta\Omega^a_\mu} = 0.$$
$$(9.33)$$

- Furthermore, in the Landau gauge, it turns out that the action Σ obeys an additional integrated Ward identity, called the *ghost Ward identity*:[37]

$$\mathcal{G}^a\Sigma = \Delta^a_{\text{cl}},$$
$$(9.34)$$

with

$$\mathcal{G}^a = \int d^4x \left(\frac{\delta}{\delta c^a} + g f^{abc}\bar{c}^b\frac{\delta}{\delta b^c}\right),$$
$$(9.35)$$

and

$$\Delta^a_{\text{cl}} = g\int d^4x f^{abc}\left(\Omega^b_\mu A^c_\mu - L^b c^c\right).$$
$$(9.36)$$

Notice that the breaking term Δ^a_{cl} in the right hand side of eq.(9.34) is linear in the quantum fields. As such, Δ^a_{cl} is a classical breaking, not affected by quantum corrections.[38] When we turn to the quantum level, we can use these Ward identities to characterize the most general invariant counterterm Σ^c, which is an integrated local polynomial in the fields and sources with dimension bounded by four, and with vanishing ghost number. Following the algebraic renormalization procedure,[38] one perturbs the action Σ by considering the quantity $\Sigma + \varepsilon\Sigma^c$, where ε

stands for an infinitesimal expansion parameter, and one imposes that $\Sigma + \varepsilon \Sigma^c$ fulfills, to the first order in ε, the same set of Ward identities obeyed by Σ. This implies that the most general invariant counterterm Σ^c is constrained by the following conditions:

• The linearized Slavnov-Taylor identity:

$$\mathcal{B}_\Sigma \Sigma^c = 0 \,, \qquad (9.37)$$

where $\mathcal{B}_\Sigma$ is the nilpotent linearized Slavnov-Taylor operator

$$\mathcal{B}_\Sigma = \int d^4 x \left(\frac{\delta \Sigma}{\delta \Omega_\mu^a} \frac{\delta}{\delta A_\mu^a} + \frac{\delta \Sigma}{\delta A_\mu^a} \frac{\delta}{\delta \Omega_\mu^a} + \frac{\delta \Sigma}{\delta L^a} \frac{\delta}{\delta c^a} + \frac{\delta \Sigma}{\delta c^a} \frac{\delta}{\delta L^a} + b^a \frac{\delta}{\delta \bar{c}^a} \right) \,, \qquad (9.38)$$

with

$$\mathcal{B}_\Sigma \mathcal{B}_\Sigma = 0 \,. \qquad (9.39)$$

• The Landau gauge condition and the antighost equation:

$$\frac{\delta \Sigma^c}{\delta b^a} = 0 \,, \qquad (9.40)$$

$$\frac{\delta \Sigma^c}{\delta \bar{c}^a} + \partial_\mu \frac{\delta \Sigma^c}{\delta \Omega_\mu^a} = 0 \,.$$

• The ghost Ward identity:

$$\mathcal{G}^a \Sigma^c = 0 \,. \qquad (9.41)$$

Conditions (9.40) imply that Σ^c does not depend on the Lagrange multiplier b^a, and that the antighost $\bar{c}^a$ can enter only through the combination

$$\widetilde{\Omega}_\mu^a = \Omega_\mu^a + \partial_\mu \bar{c}^a \,.$$

Moreover, from eq.(9.41), it follows that Σ^c depends only on the differentiated ghost $\partial_\mu c^a$. Finally, from condition (9.37), it turns out that the most general counterterm fulfilling the conditions (9.37)-(9.41) contains only two arbitrary parameters, a_0, a_1, and reads

$$\Sigma^c = a_0 S_{YM} + a_1 \int d^4 x \left(A_\mu^a \frac{\delta S_{YM}}{\delta A_\mu^a} + \widetilde{\Omega}_\mu^a \partial_\mu c^a \right) \,. \qquad (9.42)$$

Once the most general counterterm has been determined, it remains to check if it can be reabsorbed through a multiplicative renormalization of the fields, sources and coupling constant, namely

$$\Sigma(g, \phi, \Phi) + \varepsilon \Sigma^c(g, \phi, \Phi) = \Sigma(g_0, \phi_0, \Phi_0) + O(\varepsilon^2) \,, \qquad (9.43)$$

with

$$g_0 = Z_g g \,,$$
$$\phi_0 = Z_\phi^{1/2} \phi \,,$$
$$\Phi_0 = Z_\Phi \Phi \,, \tag{9.44}$$

where we have set $\phi = (A_\mu^a,\, c^a,\, \bar{c}^a,\, b^a)$ for all the fields and $\Phi = (\Omega_\mu^a,\, L^a)$ for the sources. After an elementary calculation, one finds

$$Z_g = 1 + \varepsilon \frac{a_0}{2} \,,$$
$$Z_A^{1/2} = 1 + \varepsilon \left(a_1 - \frac{a_0}{2} \right) \,, \tag{9.45}$$

These are the only independent renormalization constants. For example, the Faddeev-Popov ghosts $(c^a, \bar{c}^a)$ have a common renormalization constant, determined by the renormalization constants Z_g and $Z_A^{1/2}$,

$$Z_c = Z_{\bar{c}} = 1 - \varepsilon a_1 = Z_g^{-1} Z_A^{-1/2} \,. \tag{9.46}$$

Finally, Z_b, Z_Ω and Z_L are also not independent, as they are given by

$$Z_b = Z_A^{-1} \,, \quad Z_\Omega = Z_c^{1/2} \,, \quad Z_L = Z_A^{1/2} \,. \tag{9.47}$$

Equation (9.46), usually stated as

$$Z_g Z_A^{1/2} Z_c = 1 \,, \tag{9.48}$$

is known as the *gluon-ghost-antighost nonrenormalization theorem,* expressing the fact that, due to the ghost Ward identity (9.34), the gluon-ghost-antighost vertex does not renormalize, namely

$$g_0 f^{abc} \left(\partial_\mu \bar{c}_0^a \right) A_{0\mu}^b c_0^c = g f^{abc} \left(\partial_\mu \bar{c}^a \right) A_\mu^b c^c \,. \tag{9.49}$$

Equation (9.48) holds to all orders of perturbation theory. It can be explicitly verified up to four loops by using the expressions of the gauge β-function and of the field anomalous dimensions evaluated in the $\overline{MS}$ scheme in Ref. 39. Let us also mention that, recently, this theorem has been investigated through lattice numerical simulations,[40] which have provided indications of its possible validity beyond perturbation theory.

9.5. On the quantization in axial gauges

In this section, we again consider four dimensional Minkowski space with metric tensor $(\eta_{\mu\nu}) = \mathrm{diag}\,(+1, -1, -1, -1)$.

9.5.1. *Preliminary considerations*

There are various possibilities for implementing axial-type gauges[1] in the action. Among the most common are the following gauge fixing Lagrangians

$$\mathcal{L}_{g.f.} \sim -\frac{1}{2\lambda}(nA)^2, \; bn_\mu A^\mu, \; -\frac{1}{2\lambda}(nA)\partial^2(nA),\dots, \qquad (9.50)$$

where λ is some real dimensionless parameter and n^μ is a constant vector characterizing the different axial-like gauges. It is important to remark that in the first term of Eq. (9.50) the dimension of n^μ is one, in the third term it has dimension zero and for the second term one can choose either one or zero for the dimension of n^μ, depending on the choice for the dimension of b.

As has already been mentioned concerning the axial gauge condition (9.6), different values of n^2 specify different **non-covariant gauges**:

$$n^2 < 0 : (\text{pure}) \text{ axial gauge},$$
$$n^2 = 0 : \text{light-cone gauge},$$
$$n^2 > 0 : \text{temporal gauge}.$$

The case $n^2 < 0$ is also referred to as the *planar gauge*. Note, that all choices of (n^μ) break Lorentz invariance. The first term of Eq. (9.50) for $\lambda \to 0$, as well as the second one correspond to the axial-like gauge condition

$$n^\mu A_\mu = 0 . \qquad (9.51)$$

For this choice, the first term in Eq. (9.50) for $n^2 \neq 0$ and $\lambda \neq 0$ amounts to the following massless gauge field propagator:

$$\tilde{\Delta}_{\mu\nu}(k) = \frac{-\mathrm{i}}{k^2 + \mathrm{i}\varepsilon}\left[\eta_{\mu\nu} - \frac{k_\mu n_\nu + k_\nu n_\mu}{kn} + k_\mu k_\nu \frac{n^2 + \lambda k^2}{(kn)^2}\right], \qquad (9.52)$$

whereas, for the second term of Eq. (9.50), one gets

$$\tilde{\Delta}_{\mu\nu}(k) = \frac{-\mathrm{i}}{k^2 + \mathrm{i}\varepsilon}\left[\eta_{\mu\nu} - \frac{k_\mu n_\nu + k_\nu n_\mu}{kn}\right] . \qquad (9.53)$$

The advantages of the non-covariant gauges are[4,41]:

i) The ghost fields needed to describe a non-Abelian gauge theory in a consistent manner decouple from physical S-matrix elements (see Ref. 42 and references therein).

ii) The proof of the ultraviolet finiteness of supersymmetric Yang-Mills theory becomes more transparent in the light-cone gauge.

iii) The Wess-Zumino gauge condition for supersymmetric gauge theories can be implemented in non-covariant gauges without losing the superfield formalism.

iv) The proof of the finiteness for the topological field theories becomes trivial.

v) The infrared dangerous term $1/(k^2)^2$ in the propagators of the supersymmetric gauge theories is resolved in a very elegant manner with the help of non-covariant gauges.

vi) Some aspects of string theories are more tractable in non-covariant gauges.[42]

Beside these remarkable advantages, there exist also severe disadvantages:

i) The handling of Feynman integrals is much trickier.[42]

ii) The occurrence of unphysical singularities of the form $(kn)^{-\beta}$, $\beta = 1, 2$ needs further mathematical prescriptions which are sometimes in conflict with the Wick rotation.

At this point, one has to discuss the prescriptions for the unphysical poles since they constitute a central point in the discussion of the non-covariant gauges.

9.5.2. *Prescriptions for the unphysical poles*

In a typical generic Feynman integral of the form

$$\int d^{2\bar{m}}k \; \frac{f(k_\mu, n_\mu, p_\mu)}{(k^2 + i\varepsilon)((k - p)^2 + i\varepsilon)(kn)^\beta} \; , \qquad (9.54)$$

the principal value prescription

$$\frac{1}{(kn)^\beta} \to \mathrm{PV}\frac{1}{(kn)^\beta} = \frac{1}{2}\lim_{\varepsilon \to 0}\left[\frac{1}{(kn + i\varepsilon)^\beta} + \frac{1}{(kn - i\varepsilon)^\beta}\right] \; , \qquad (9.55)$$

with $\varepsilon > 0$, seems to be an adequate possibility to regularize the spurious $(kn)^{-\beta}$ pole. Indeed, this prescription works reasonably well for axial and temporal gauges, though it is unsuitable for the light-cone gauge as will be explained later on. More precisely, let us investigate if the prescription Eq. (9.55) allows for a Wick rotation. In order to clarify this point, one observes that for $n^2 > 0$, the poles of $kn \pm i\varepsilon$ in the k^0-plane are

$$k_0^\pm = (\vec{k}\vec{n} \pm i\varepsilon)/n^0 \; , \qquad (9.56)$$

and that they are located in the first and forth quadrant, on a line parallel to the imaginary k^0-axis and this prevents us from making a Wick rotation as shown in the following Fig. 9.3.

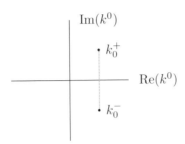

Figure 9.3 The poles for $n^2 > 0$.

In addition, there exists a further argument which also shows that it is impossible to perform a continuous transition from $n^2 > 0$ to the light-cone gauge in a typical n^μ-dependent Feynman integral. For example, one considers

$$I_\mu(p, \overline{m}, n) = \int \frac{d^{2\overline{m}} k}{(2\pi)^{2\overline{m}}} \frac{k_\mu}{(kn)} f(k^2, (pk), p^2, n^2) \ . \tag{9.57}$$

From this expression one finds

$$I_\mu(0, \overline{m}, n) = \int \frac{d^{2\overline{m}} k}{(2\pi)^{2\overline{m}}} \frac{k_\mu}{(kn)} f(k^2, 0, 0, n^2) = n_\mu S(\overline{m}, n^2) \ , \tag{9.58}$$

with

$$S(\overline{m}, n^2) = \frac{1}{n^2} \int \frac{d^{2\overline{m}} k}{(2\pi)^{2\overline{m}}} f(k^2, 0, 0, n^2) \ , \tag{9.59}$$

and one observes that the limit $n^2 \to 0$ may create singularities.

For these reasons, at the beginning of 1982, Mandelstam and Leib-brandt independently proposed two light-cone gauge prescriptions which are not of the principal value type[c]. The Mandelstam and the Leibbrandt prescriptions are based on the introduction of a further gauge direction $n^{*\mu} = (n^0, -\vec{n})$ besides $n^\mu = (n^0, \vec{n})$ as

$$nn^* = 1 \quad , \quad n^2 = (n^*)^2 = 0 \ . \tag{9.60}$$

---

[c]Mandelstam used his new prescription to demonstrate the ultraviolet finiteness of the $N = 4$ supersymmetric Yang-Mills theory.

The gauge direction $n^{*\mu}$ is sometimes called the dual gauge direction.

The Mandelstam prescription is defined by

$$M\left(\frac{1}{kn}\right) = \frac{1}{(kn) + i\varepsilon kn^*} , \qquad \varepsilon > 0 , \qquad (9.61)$$

and the Leibbrandt prescription is given by

$$L\left(\frac{1}{kn}\right) = \frac{kn^*}{(kn)(kn^*) + i\varepsilon} , \qquad \varepsilon > 0 . \qquad (9.62)$$

These two prescriptions lead to identical results, at least at the one-loop level.[43] Furthermore, it is easy to show that the Leibbrandt prescription immediately allows for a Wick rotation. Indeed, the poles of $(n^0 k^0 - \vec{k}\vec{n})(n^0 k^0 + \vec{k}\vec{n}) + i\varepsilon = 0$ are given by

$$k^0 = \pm\frac{1}{n^0}(\vec{n}\vec{k}) \mp i\varepsilon' , \quad \varepsilon' = \frac{\varepsilon}{2n^0(\vec{n}\vec{k})^2} , \qquad (9.63)$$

and are located in the second and fourth quadrant of the complex k^0-plane allowing for a Wick rotation.

Another nice feature of the Mandelstam-Leibbrandt prescription is that it can be extended outside the light-cone in a natural way, as it was first shown by P. Gaigg *et al.*[44,45] and G. Leibbrandt.[46] The generalization presented by Gaigg *et al.* is sometimes referred to as the "Vienna Prescription".

However, the existence of a further gauge direction n^* entails a more complicated tensor structure for the Feynman integrals. For example, the Yang-Mills self energy[47] calculated in the context of the dimensional regularization from

$$\tilde{\Delta}_{\mu\nu}^{ab}(k) = \frac{-i}{k^2 + i\varepsilon}\delta^{ab}\left(\eta_{\mu\nu} - \frac{k_\mu n_\nu + k_\nu n_\mu}{kn}\right) \qquad (9.64)$$

and the prescription Eq. (9.62) gives the following result for the divergent part of the one-loop contribution

$$\begin{aligned} \text{div}\,\Pi_{\mu\nu}^{ab}(p) = i\pi^2 g^2\,{}_3F_2(2-\overline{m})C_{\text{YM}}\delta^{ab}\Bigg\{ &\frac{11}{3}(p^2\eta_{\mu\nu} - p_\mu p_\nu) \\ &+ \frac{2(pn^*)}{(pn)(nn^*)}\left[2p^2 n_\mu n_\nu - (pn)(p_\mu n_\nu + p_\nu n_\mu)\right] \\ &+ \frac{2(pn)}{nn^*}(p_\mu n_\nu^* + p_\nu n_\mu^*) - \frac{2p^2}{nn^*}(n_\mu n_\nu^* + n_\nu n_\mu^*)\Bigg\} ,(9.65) \end{aligned}$$

where $f^{acd}f^{bcd} = \delta^{ab}C_{\text{YM}}$ and g denotes the coupling constant. It is now clear that with these new prescriptions, the light-cone gauge is well defined, the $\frac{1}{n^2}$ term of Eq. (9.59) being now formally replaced by $\frac{1}{nn^*}$.

Apart from the traditional transversal factor, Eq. (9.65) differs drastically from the corresponding one in the axial or temporal gauge. Indeed, the self-energy for the axial gauge $n^\mu A_\mu = 0$ ($\lambda = 0$, $n^2 < 0$) calculated with the propagator of Eq. (9.52) with the PV-prescription Eq. (9.55)

$$\tilde{\Delta}^{ab}_{\mu\nu}(k) = \frac{-\mathrm{i}}{k^2 + \mathrm{i}\varepsilon}\delta^{ab}\left(\eta_{\mu\nu} - \frac{k_\mu n_\nu + k_\nu n_\mu}{kn} + n^2\frac{k_\mu k_\nu}{(kn)^2}\right) , \qquad (9.66)$$

takes the much simpler form

$$\operatorname{div}\Pi^{ab}_{\mu\nu}(p) = \mathrm{i}\pi^2 g^2 {}_3F_2(2-\overline{m})C_{\text{YM}}\delta^{ab}\frac{11}{3}(p^2\eta_{\mu\nu} - p_\mu p_\nu) . \qquad (9.67)$$

One also has to note that the result Eq. (9.65) is not only gauge dependent and non-Lorentz covariant, but that it also contains non-local terms in the external momentum p_μ. These non-localities arise from the use of the decomposition formula

$$\frac{d^{2\overline{m}}k}{(kn)((k-p)n)} = \frac{d^{2\overline{m}}k}{(pn)}\left[\frac{1}{((k-p)n)} - \frac{1}{kn}\right] . \qquad (9.68)$$

Obviously, these non-localities are the real obstacle to the renormalization program and contradict the fact that usually the divergent parts of Feynman graphs are local functions in the external momenta. More generally, for renormalizable field models one expects the following behaviour (self-energy corrections) for an arbitrary $\Pi(p)$

$$\Pi(p) = \Pi(0) + p^\lambda\frac{\partial\Pi(0)}{\partial p^\lambda} + \frac{1}{2}p^\lambda p^\rho\frac{\partial^2\Pi(0)}{\partial p^\lambda \partial p^\rho} + \ldots . \qquad (9.69)$$

Nevertheless, it is interesting to observe that the expression Eq. (9.65) passes into equation Eq. (9.67) for $n^* \to n$. For this limit, the whole gauge dependence and the non-localities disappear.

Finally, a similar calculation for the planar gauge, with the principal value prescription Eq. (9.55), furnishes[48-50]

$$\operatorname{div}\Pi^{ab}_{\mu\nu}(p) = \mathrm{i}\pi^2 g^2 {}_3F_2(2-\overline{m})C_{\text{YM}}\delta^{ab}\left\{\frac{11}{3}(p^2\eta_{\mu\nu} - p_\mu p_\nu)\right. \qquad (9.70)$$

$$\left. -2\lambda(p_\mu p_\nu - \eta_{\mu\nu}p^2) - \frac{2\lambda}{n^2}\left[2n_\mu n_\nu p^2 - (pn)(p_\mu n_\nu + p_\nu n_\mu)\right]\right\} .$$

At this stage, the tensor structure of Eqs. (9.65) and (9.71) seems to occur by accident. However, we will see later that it can be understood as the consequence of an extended version of the BRST transformations.

9.5.3. *Extended BRST symmetry*

Let us consider the gauge fixed Yang-Mills action

$$\Gamma^{(0)} = S_{\text{inv}} + S_{\text{gf}} + S_{\phi\pi} + S_{\text{ex}}$$

$$= -\frac{1}{4g^2}\,\text{Tr}\int d^4x F_{\mu\nu}F^{\mu\nu} + \text{Tr}\int d^4x\,[bn^\mu A_\mu - \bar{c}n^\mu D_\mu c]$$

$$+ \text{Tr}\int d^4x\,[\rho^\mu sA_\mu + \sigma sc]\,, \tag{9.71}$$

where we have introduced unquantized external sources ρ^μ and σ in order to handle the non-linear pieces of the BRST symmetry

$$
\begin{aligned}
sA_\mu &= D_\mu c = \partial_\mu c + \mathrm{i}[c, A_\mu], & s\bar{c} &= b, \\
sc &= \mathrm{i}cc, & sb &= 0, \\
s\rho^\mu &= 0, & s\sigma &= 0, \\
s^2\varphi &= 0 \quad \text{for} \quad \varphi &\in \{A_\mu, c, \bar{c}, b\}\,. &&
\end{aligned}
\tag{9.72}
$$

In order to control the n^μ-dependence of the theory one observes that the classical action (9.71) fulfills

$$\frac{\partial\Gamma^{(0)}}{\partial n^\mu} = \text{Tr}\int d^4x\,[bA_\mu - \bar{c}D_\mu c]$$

$$= \text{Tr}\int d^4x\, s(\bar{c}A_\mu)\,. \tag{9.73}$$

This result motivates us to extend the action of the BRST operator to the vector (n^μ) by the relations

$$sn^\mu = \chi^\mu\,, \qquad\qquad s\chi^\mu = 0\,, \tag{9.74}$$

where χ^μ is dimensionless and has a $\phi\pi$-charge of one. Hence one has

$$\chi^\mu\frac{\partial\Gamma^{(0)}}{\partial n^\mu} = \chi^\mu\,\text{Tr}\int d^4x\, s(\bar{c}A_\mu) = \text{Tr}\int d^4x\, s(\bar{c}\chi^\mu A_\mu)\,. \tag{9.75}$$

The new action, which is now invariant under the extended BRST transformations (9.72) and (9.74), is given by

$$\Gamma^{(0)} \longrightarrow \Gamma^{(0)} - \text{Tr}\int d^4x(\bar{c}\chi^\mu A_\mu)\,. \tag{9.76}$$

The new term in expression (9.76) is $\phi\pi$-neutral and physics is defined by $\chi = 0$.

The extended BRST symmetry also leads to an enlarged Slavnov-Taylor identity for the vertex functional:

$$\mathcal{S}\left(\Gamma^{(0)}\right) \equiv \mathrm{Tr} \int d^4x \left[\frac{\delta\Gamma^{(0)}}{\delta\rho^\mu}\frac{\delta\Gamma^{(0)}}{\delta A_\mu} + \frac{\delta\Gamma^{(0)}}{\delta\sigma}\frac{\delta\Gamma^{(0)}}{\delta c} + b\frac{\delta\Gamma^{(0)}}{\delta\bar{c}} \right] + \chi^\mu\frac{\partial\Gamma^{(0)}}{\partial n^\mu} = 0. \tag{9.77}$$

9.5.4. *Stability problem in axial-like gauges*

9.5.4.1. *Non-trivial field renormalization*

So far we have dealt with an alternative formulation of pure Yang-Mills theory quantized in axial-like gauges. We have been able to characterize the model by the Slavnov-Taylor identity (9.77). It is important to note that (9.77) is invariant under the following general, local field renormalization

$$A_\mu \longrightarrow A'_\mu = (\alpha\eta_{\mu\nu} + \beta\frac{n_\mu n_\nu}{n^2})A^\nu \equiv C_{\mu\nu}A^\nu \ ,$$

$$\hat{\rho}_\mu \longrightarrow \hat{\rho}'_\mu = \alpha^{-1}(\eta_{\mu\nu} - \frac{\beta}{\alpha+\beta}\frac{n_\mu n_\nu}{n^2})\hat{\rho}^\nu \equiv C^{-1}_{\mu\nu}\hat{\rho}^\nu \ . \tag{9.78}$$

where $C_{\mu\nu}$ is an invertible matrix. Moreover, the identity (9.77) allows for a field renormalization of c and σ in the following manner:

$$c \longrightarrow c' = yc \ ,$$

$$\sigma \longrightarrow \sigma' = y^{-1}\sigma \ . \tag{9.79}$$

Here we want to stress that in our analysis, the dimension of n^μ is zero. An alternative formulation, where n^μ bears a mass dimension would lead to an infinite number of possible renormalization terms in Eq. (9.78)

$$A_\mu \longrightarrow A'_\mu = (\alpha'\eta_{\mu\nu} + \beta'\frac{n_\mu n_\nu}{n^2} + \gamma'\frac{\partial_\mu n_\nu + \partial_\nu n_\mu}{n^2} + \delta'\frac{\partial^2 n_\mu n_\nu}{(n^2)^2} + \cdots)A^\nu \ . \tag{9.80}$$

9.5.4.2. *Consequences of the extended BRST and field renormalization*

A first interesting and formulation independent consequence of the extended BRST symmetry is that, at the classical level, it forbids terms like

$$\mathrm{Tr} \int d^4x[\frac{n_\mu n_\nu}{n^2}F^{\mu\lambda}F_\lambda{}^\nu] \ , \tag{9.81}$$

which, due to its gauge invariance, would be perfectly acceptable within the usual unextended BRST symmetry. This is inferred by inspection of the right hand sides of the BRST transformations (9.72) and (9.74). One observes that Eq. (9.81) is non-trivial, i.e.

$$\mathrm{Tr} \int d^4x \, [\frac{n_\mu n_\nu}{n^2} F^{\mu\lambda} F_\lambda{}^\nu] \neq s \, \mathrm{Tr} \int d^4x \, [O(x)] \; . \tag{9.82}$$

For the ghost-dependent formulation, the field renormalizations Eqs. (9.78) and (9.79) allow for an overall renormalization for $\bar{\Gamma}$ defined in Eq. (9.79)

$$\bar{\Gamma} \longrightarrow Z(n^2) \, \bar{\Gamma}[A', \hat{\rho}', \sigma', c'; n, \chi] \; . \tag{9.83}$$

Now we have to look at the restrictions which are entailed by the extended BRST symmetry for the renormalization constants α, β, y and Z being functions depending on n^2 and the coupling constant g. A lengthy but straightforward calculation shows for example[d]

$$\frac{\partial}{\partial n^2} Z(n^2) = 0 \; . \tag{9.84}$$

The field renormalization in Eq. (9.78) has physical consequences because it alters the bilinear part of the action as follows:

$$-\frac{1}{4g^2}\mathrm{Tr}(F'_{\mu\nu} F'^{\mu\nu})_{bilin} = \frac{1}{g^2}\mathrm{Tr}\Big(A^\mu\Big\{\alpha^2(\eta_{\mu\nu}\partial^2 - \partial_\mu\partial_\nu) - \frac{\alpha\beta}{2n^2}[2n_\mu n_\nu \partial^2 -$$
$$- (n\partial)(\partial_\mu n_\nu + \partial_\nu n_\mu)] - \frac{\beta^2}{2n^2} n_\mu n_\nu (\partial^2 - \frac{(n\partial)^2}{n^2})\Big\}A^\nu\Big) \; . \tag{9.85}$$

Eq. (9.85) produces two additional possible non-transversal counterterms for the self-energy. Indeed the second term in Eq. (9.85) has been found in (9.71) by explicit evaluation of Feynman graphs. It is consistent with Eq. (9.81) which forbids such terms due to the fact that

$$\mathrm{Tr} \int d^4x \Big(n_\mu \frac{F^\mu{}_\lambda F^{\lambda\nu}}{n^2} n_\nu\Big)\Big|_{bilin} \neq \mathrm{Tr} \int d^4x (F'_{\mu\nu} F'^{\mu\nu}) \; . \tag{9.86}$$

9.5.5. *Stability problem in the light-cone gauge*

As has already been discussed in Section 9.5.2, in the light-cone gauge the LM-prescription needs the presence of another light-like vector $n^{*\mu} = (n^0, -\vec{n})$. Nevertheless, in the framework of the extended BRST symmetry where only the gauge direction n^μ is regarded as a varying gauge parameter, $sn^\mu = \chi^\mu$ (due to its occurrence in the Lagrangian density), we treat only n^μ and χ^μ as fundamental gauge variables, especially $n^{*\mu} = n^{*\mu}(n^\mu)$, $\chi^{*\mu} = \chi^{*\mu}(\chi^\mu, n^\mu)$, etc. Some useful properties between n, χ, n^* and χ^* are

---

[d]See Ref. 51 for more details.

presented in Ref. 41. The $n^{*\mu}$-dependence will then be controlled by an invariance proposed by Taylor

$$n^\mu \longrightarrow e^\theta n^\mu \ ,$$
$$n^{*\mu} \longrightarrow e^{-\theta} n^{*\mu} \ . \tag{9.87}$$

In addition one has the further restrictions

$$n\chi = n^*\chi^* = 0 \ . \tag{9.88}$$

This has the consequence that the number of n^μ and $n^{*\mu}$ must be in balance.

An analysis of the possible gauge invariant counterterms leads to the following expressions:

$$L_1 = \text{Tr} \int d^4x \left[\frac{1}{nn^*}(nFFn^*)\right] \ ,$$
$$L_2 = \text{Tr} \int d^4x \left[\frac{1}{(nn^*)^2}(nFn^*)^2\right] \ , \tag{9.89}$$

where the L_i denote integrated, local counterterms. Since power counting is not valid, there might also appear non-local, n^μ and $n^{*\mu}$-dependent counterterms containing $(nD)^{-\alpha}$. Some non-trivial expressions of this kind are

$$N_1 = \text{Tr} \int d^4x \left[(nn^*)^{-1}(nD)^{-2}(nFn^*)(nFFn)\right] \ ,$$
$$N_2 = \text{Tr} \int d^4x \left[(nn^*)^{-1}(nD)^{-1}(n^*D)(nFFn)\right] \ ,$$
$$N_3 = \text{Tr} \int d^4x \left[(nn^*)^{-1}(nD)^{-1}(DFn)(nFn^*)\right] \ ,$$
$$N_4 = \text{Tr} \int d^4x \left[(nD)^{-4}(nFFn)^2\right] \ ,\dots \tag{9.90}$$

The notation has to be understood as representing a whole class of terms rather than a certain representative, for example N_3 also stands for

$$\text{Tr} \int d^4x [(nn^*)^{-1}(DFn)\left((nD)^{-1}(nFn^*)\right)] \ , \tag{9.91}$$

as well as for

$$\text{Tr} \int d^4x \left[(nn^*)^{-1}(D^\mu nFn^*)(nD)^{-1}(F_{\mu\nu}n^\nu)\right] \ . \tag{9.92}$$

The term $(nD)^{-1}$ has to be understood perturbatively

$$Y = (nD)^{-1}X = (n\partial)^{-1}X + \text{i}(n\partial)^{-1}[nA, \ (n\partial)^{-1}X] + \cdots \ . \tag{9.93}$$

In order to explain the gauge invariant quantities in Eq. (9.89) one uses the transformation property of $M = (nD)^{-\alpha}(nFn^*)$ which transforms covariantly

$$\delta M = \mathrm{i}\big[\omega,\, (nD)^{-\alpha}(nFn^*)\big] \ . \tag{9.94}$$

In fact the situation is very similar to the case of the axial gauge. As usual, we find an invertible, non-local field renormalization, first calculated by Basetto et al.,[52] which is of the form

$$A'_\mu = \alpha\eta_{\mu\nu}A^\nu + \beta n_\mu \frac{1}{Dn}\frac{(nFn^*)}{(nn^*)} = \mathcal{R}_\mu(A) \ . \tag{9.95}$$

Equation (9.95) is compatible with the extended BRST symmetry and, at least in one-loop calculations, leads to a non-local behavior of the gluon self-energy.[47] One-loop considerations are characterized by $(nD)^{-\alpha}$ with $\alpha = 1$, and an explicit calculation shows

$$\begin{aligned}
\mathrm{Tr}F'_{\mu\nu}(A')F'^{\mu\nu}(A') = \mathrm{Tr}\,\Big\{ & 2\alpha^2 A^\mu(\eta_{\mu\nu}\partial^2 - \partial_\mu\partial_\nu)A^\nu \\
& + 2\alpha\beta A^\mu\big[(nn^*)^{-1}(n\partial)(n^*_\mu\partial_\nu + n^*_\nu\partial_\mu) \\
& - (nn^*)^{-1}\partial^2(n^*_\mu n_\nu + n_\nu n^*_\mu) \\
& + (n\partial)^{-1}(n^*\partial)(nn^*)^{-1} \\
& \cdot (2\partial^2 n_\mu n_\nu - (n\partial)(n_\mu\partial_\nu + n_\nu\partial_\mu))\big] A^\nu + \cdots \Big\} \ .
\end{aligned} \tag{9.96}$$

This result reproduces Eq. (9.65). Another consistency check at one-loop level is given by

$$\mathrm{Tr}(F'_{\mu\nu}F'^{\mu\nu}) \neq \{L_1,\ L_2,\ N_2,\ \ldots\}\ . \tag{9.97}$$

In Ref. 53, another philosophy based on a general theorem has been used to analyze the structures of counterterms. According to this theorem the counterterms at the one-loop level should only yield local vertex functions when we set $(n^*\partial) = \lambda(n\partial)$, for real λ. In this sense N_1 and N_2 are not allowed.

From the results obtained in the last two sections we thus conclude that the gauge invariant sector is completely governed by an extended BRST symmetry, at least at the one-loop level.

References

1. W. Kummer, *Acta Phys. Austriaca* **14**, 149 (1961);
   W. Kummer, *Acta Phys. Austriaca* **41**, 315 (1975);
   W. Konetschny and W. Kummer, *Nucl. Phys. B* **108**, 397 (1976).

2. A. Das, *Quantum Field Theory*, (World Scientific, 2008).
3. B. Z. Iliev, arXiv:0803.0047 [physics.hist-ph].
4. G. Leibbrandt, *Noncovariant gauges: Quantization of Yang-Mills and Chern-Simons theory in axial type gauges*, (World Scientific, 1994).
5. P. Gaigg, W. Kummer and M. Schweda, eds., *Physical and Nonstandard Gauges*, Lecture Notes in Physics, Vol. 361 (Springer, 1990).
6. V. N. Gribov, *Nucl. Phys.* B **139**, 1 (1978).
7. D. Zwanziger, *Nucl. Phys.* B **209**, 336 (1982).
8. M. Stingl, *Z. Phys.* A **353**, 423 (1996), arXiv:hep-th/9502157.
9. J. Fuchs, M.G. Schmidt and C. Schweigert, *Nucl. Phys.* B **426**, 107 (1994), arXiv:hep-th/9404059.
10. P. van Baal, *in "50 years of Yang-Mills theory"*, G. 't Hooft, ed., (World Scientific, 2005).
11. L.G. Yaffe, *Nucl. Phys.* B **151**, 247 (1979).
12. M.A. Semenov-Tyan-Shanskii and V.A. Franke, *J. Math. Sci.* **34**, 1999 (1986)
13. I.M. Singer, *Commun. Math. Phys.* **60**, 7 (1978).
14. V.P. Nair, *Quantum Field Theory: A Modern Perspective*, Graduate Texts in Contemporary Physics (Springer Verlag, 2005).
15. J. Honerkamp, *Nucl. Phys.* B **48**, 269 (1972).
16. G. Dell'Antonio and D. Zwanziger, *Nucl. Phys.* B **326**, 333 (1989).
17. G. Dell'Antonio and D. Zwanziger, *Commun. Math. Phys.* **138**, 291 (1991).
18. D. Zwanziger, *in "Fundamental Problems Of Gauge Field Theory" (Erice School, July 1985)* G. Velo and A.S. Wightman, eds., (Plenum Press, 1986).
19. P. van Baal, *Nucl. Phys.* B **369**, 259 (1992).
20. R.F. Sobreiro and S.P. Sorella, *JHEP* **0506**, 054 (2005), arXiv:hep-th/0506165.
21. F. Bruckmann, T. Heinzl, T. Tok and A. Wipf, *Nucl. Phys.* B **584**, 589 (2000), arXiv:hep-th/0001175;
    M.A.L. Capri, A.J. Gomez, V.E.R. Lemes, R.F. Sobreiro and S.P. Sorella, arXiv:0811.2760 [hep-th].
22. F.S. Henyey, *Phys. Rev.* D **20**, 1460 (1979).
23. R. F. Sobreiro and S. P. Sorella, arXiv:hep-th/0504095.
24. D. Zwanziger, *Nucl. Phys.* B **323** (1989) 513.
25. D. Zwanziger, *Nucl. Phys.* B **399**, 477 (1993).
26. N. Maggiore and M. Schaden, *Phys. Rev.* D **50**, 6616 (1994), arXiv:hep-th/9310111.
27. D. Dudal, R. F. Sobreiro, S. P. Sorella and H. Verschelde, *Phys. Rev.* D **72**, 014016 (2005), arXiv:hep-th/0502183.
28. D. Dudal, S. P. Sorella, N. Vandersickel and H. Verschelde, *Phys. Rev.* D **77**, 071501 (2008), arXiv:0711.4496 [hep-th].
29. D. Dudal, J. A. Gracey, S. P. Sorella, N. Vandersickel and H. Verschelde, *Phys. Rev.* D **78**, 065047 (2008), arXiv:0806.4348 [hep-th].
30. J. A. Gracey, *Phys. Lett.* B **632**, 282 (2006) arXiv:hep-ph/0510151.
31. K.G. Wilson, *in "Recent Developments in Gauge Theories" (Cargèse School 1979)* G. 't Hooft, et al., eds., (Plenum Press, 1980).

174   *D. N. Blaschke et al.*

32. A. Cucchieri and T. Mendes, *PoS* **LAT2007**, 297 (2007), arXiv:0710.0412 [hep-lat].
33. I. L. Bogolubsky, E. M. Ilgenfritz, M. Muller-Preussker and A. Sternbeck, *PoS* **LAT2007**, 290 (2007), arXiv:0710.1968 [hep-lat].
34. P. O. Bowman *et al.*, *Phys. Rev.* D **76**, 094505 (2007), arXiv:hep-lat/0703022.
35. A. Cucchieri and T. Mendes, *Phys. Rev. Lett.* **100**, 241601 (2008), arXiv:0712.3517 [hep-lat].
36. D. Zwanziger, *Phys. Rev.* D **69**, 016002 (2004), arXiv:hep-ph/0303028.
37. A. Blasi, O. Piguet and S. P. Sorella, *Nucl. Phys.* B **356**, 154 (1991).
38. O. Piguet and S. P. Sorella, *Algebraic Renormalization: Perturbative Renormalization, Symmetries and Anomalies*, Lecture Notes in Physics, Volume M28 (Springer, 1995).
39. K. G. Chetyrkin, *Nucl. Phys.* B **710**, 499 (2005) arXiv:hep-ph/0405193.
40. A. Cucchieri, T. Mendes and A. Mihara, *JHEP* **0412**, 012 (2004) arXiv:hep-lat/0408034.
41. A. Boresch, S. Emery, O. Moritsch, M. Schweda, T. Sommer and H. Zerrouki, *Applications of Noncovariant Gauges in the Algebraic Renormalization Procedure*, (World Scientific, 1998).
42. G. Leibbrandt, *Rev. Mod. Phys.* **59**, 1067 (1987).
43. H.C. Lee and M.S. Milgram, *Nucl. Phys.* B **268**, 543 (1986).
44. P. Gaigg, M. Kreuzer, O. Piguet and M. Schweda, *J. Math. Phys.* **28**, 2781 (1987).
45. P. Gaigg and M. Kreuzer, *Phys. Lett.* B **205**, 530 (1988).
46. G. Leibbrandt, *Nucl. Phys.* B **310**, 405 (1989).
47. G. Leibbrandt, *Phys. Rev.* D **29**, 1699 (1984).
48. A. Andrasi and J.C. Taylor, *Nucl. Phys.* B **192**, 283 (1981).
49. D.M. Capper and G. Leibbrandt, *Phys. Lett.* B **104**, 158 (1981).
50. A.I. Mil'shtein and V.S. Fadin, *Yad. Fiz.* **34**, 1403 (1981).
51. P. Gaigg, O. Piguet, A. Rebhan and M. Schweda, *Phys. Lett.* B **175**, 53 (1986).
52. A. Bassetto, M. Dalbosco and R. Soldati, *Phys. Rev.* D **36**, 3138 (1987).
53. M. Schweda and H. Skarke, *Int. J. Mod. Phys.* A **4**, 3025 (1989).

Chapter 10

Frozen Ghosts in Thermal Gauge Field Theory

P. V. Landshoff[1] and A. Rebhan[2]

[1] *Department of Applied Mathematics and Theoretical Physics,*
University of Cambridge, Cambridge CB3 0WA, UK

[2] *Institut für Theoretische Physik, Technische Universität Wien,*
A-1040 Wien, Austria

We review an alternative formulation of gauge field theories at finite temperature where unphysical degrees of freedom of gauge fields and the Faddeev-Popov ghosts are kept at zero temperature.

10.1. Introduction

Thermal gauge field theory is a combination of two difficult areas of physics, and so it is no surprise that some of its aspects are subtle. An apparently simple and rather powerful formalism is in common use,[1,2] which introduces propagators, a version of Wick's theorem, and even what appear to be quantum states, all of which are thermal generalisations of their zero-temperature counterparts. While many things may be calculated from this formalism, sometimes it does not apply,[3] and sometimes it is needlessly complicated.[4] It can be valuable, therefore, to go back to first principles, rather than making use of the conventional formalism without thinking. Doing so, we shall find that it in the real-time formalism it is possible and often advantageous to keep the tree-level propagators of ghosts and unphysical degrees of freedom free from thermal modifications in arbitrary linear gauges including covariant gauges, whereas in the standard formulation this is the case only for noncovariant gauges without propagating ghosts, such as the axial gauge.[5]

---

[1] E-mail: `pvl@damtp.cam.ac.uk` k
[2] E-mail: `rebhana@tph.tuwien.ac.at`

Already at zero temperature, there are two approaches to deriving the gauge-field-theory formalism. The first works with field operators and commutation relations, and introduces a space of kets. Some of these kets do not correspond to physical states, because the fields have unphysical degrees of freedom. It is necessary, therefore, to identify a subset of the kets corresponding to the physical states, most simply those that contain no scalar or longitudinal gauge particles. However, as was first noticed by Feynman,[6] unless one introduces additional ghost fields and the resulting kets, the probability of scattering from a physical state to an unphysical one is not zero. That is, the ghosts are needed to ensure that the S matrix is unitary within the subspace of physical states.

The other approach, which uses path integrals, does not explicitly consider states and the ghosts have to be introduced for an apparently very different reason. One can show that the two approaches are equivalent of course, but to do so is not simple: one has to introduce the BRS operator.[7,8] The operator approach is closer to the physics and so it is the one we use here.

At nonzero temperature, the propagators acquire a thermal part that has to be added to the zero-temperature Feynman propagator. As we shall discuss, there are two formalisms:

- All components of the gauge field, and the ghosts, become heated to the temperature T.

- Only the two physical degrees of freedom of the gauge field (the transverse polarisations) acquire the additional thermal propagator; the other components of the gauge field, and the ghosts, remain frozen at zero temperature. (This is for the bare propagators; self-energy insertions in the unphysical bare propagators do depend on the temperature.)

The second of these is the less commonly used, but in practice it is sometimes much simpler to apply. We shall describe it here. But before that, we go back to basics and remind ourselves of just what thermal field theory is trying to achieve.

10.2. Basics of equilibrium thermal field theory

For definiteness, consider QCD in Feynman gauge, though the discussion of any gauge theory in any covariant gauge will be similar. A system in thermal equilibrium is not in any particular quantum state; all one knows is the probability of it being in any one of a complete set of physical states. That is, one describes the system through a density matrix that expresses the knowledge that it is in thermal equilibrium:

$$\rho = Z^{-1} \mathbb{P} \exp(-H/T) \tag{10.1}$$

Here the units are such that Boltzmann's constant $k_B = 1$, H is the Heisenberg-picture Hamiltonian. $\mathbb{P}$ is a projection operator onto a complete set of physical states; we may choose to express it in terms of a complete orthonormal set of asymptotic in-states:

$$\mathbb{P} = \sum_i |i \text{ in}\rangle\langle i \text{ in}| \tag{10.2}$$

Z is called the grand partition function and is defined so as to make ρ have unit trace:

$$Z = \text{tr } \mathbb{P} \exp(-H/T) \tag{10.3}$$

A trace is invariant under a change of the basis of states used to calculate it: any complete orthonormal set of states may be used and it may or may not include unphysical states, because their contribution is removed by $\mathbb{P}$.

10.3. Freezing unphysical degrees of freedom in the real-time formalism

Thermal field theory with gauge fields is more complicated than for scalar fields largely because of the presence of $\mathbb{P}$. The theory for scalar fields relies for its comparative simplicity on the commutativity of traces, tr $AB = $ tr BA, but it is usually not true that tr $\mathbb{P}AB = $ tr $\mathbb{P}BA$.

Note that the states $|i \text{ in}\rangle$ are ordinary zero-temperature states, and the fields used to construct H are ordinary zero-temperature operators. The temperature T comes in only in that it weights the way that the states are combined together to construct the density matrix: from (10.1) and (10.2)

$$\rho = Z^{-1} \sum_i |i \text{ in}\rangle\langle i \text{ in}| \exp(-H/T) \tag{10.4}$$

Because we have chosen to express ρ in terms of asymptotic in states, to pick out a complete set of physical states we need to consider only non-interacting fields.

In the real-time formalism one can switch off interactions adiabatically at $t_i \to -\infty$, so that nonabelian gauge theories reduce to (a number of) noninteracting abelian ones. Here t_i refers to the point in time where the interaction-picture operators of perturbation theory coincides with the full Heisenberg operators.

One can then reduce the interaction-picture gauge fields to physical ones by projecting to transverse modes according to

$$A^\mu_{\text{phys.}}(k) = T^{\mu\nu}(k)A_\nu(k) \tag{10.5}$$

with

$$T^{0\mu} = 0, \quad T^{ij} = -(\delta^{ij} - \frac{k^i k^j}{\mathbf{k}^2}). \tag{10.6}$$

The unphysical fields are given by

$$A^\mu_{\text{unphys.}}(k) = (g^{\mu\nu} - T^{\mu\nu}(k))\,A_\nu(k), \tag{10.7}$$

and the ghost fields $\bar{c}$, c.

With this decomposition the corresponding parts in the free Hamiltonians commute and one can factorise

$$\sum_i \langle i \text{ in} | e^{-\beta H_{0I}} \cdots A_{\text{phys.}} \cdots A_{\text{unphys.}} \cdots \bar{c} \cdots c \cdots | i \text{ in} \rangle$$

$$= \sum_i \langle i \text{ in} | e^{-\beta H^{\text{phys.}}_{0I}} \cdots A_{\text{phys.}} \cdots | i \text{ in} \rangle$$

$$\times \langle 0 | \cdots A_{\text{unphys.}} \cdots \bar{c} \cdots c \cdots | 0 \rangle \tag{10.8}$$

where H_{0I} is the free Hamiltonian in the interaction picture and $|i \text{ in}\rangle$ are states obtained by acting exclusively with operators for the physical fields onto the vacuum state.[a] This leads to a perturbation theory where only the propagator for the physical gauge field $A^\mu_{\text{phys.}}$ is thermal and all other propagators remain as at zero temperature.

In the real-time formalism, propagators have a 2×2 matrix structure, which (with the Schwinger-Keldysh choice of complex time path) reads

$$iD^{\mu\nu}(x) = \begin{pmatrix} \langle \text{T} A^\mu(x)A^\nu(0)\rangle & \langle A^\mu(0)A^\nu(x)\rangle \\ \langle A^\mu(x)A^\nu(0)\rangle & \langle \tilde{\text{T}} A^\mu(x)A^\nu(0)\rangle \end{pmatrix} \tag{10.9}$$

where T and $\tilde{\text{T}}$ refer to time and anti-time ordering, respectively. In particular, a massless scalar (momentum-space) propagator reads

$$iD = M \begin{pmatrix} \frac{i}{k^2 - i\epsilon} & 0 \\ 0 & \frac{-i}{k^2 + i\epsilon} \end{pmatrix} M \tag{10.10}$$

---

[a]It is of course still true that there is a many-one correspondence between physical states and the kets that represent them. There is also still the issue of indefinite metrics and negative-norm states. The probability of scattering into any given unphysical state is in fact not zero, but it is cancelled by the probability of scattering into other unphysical states.

with

$$M = \sqrt{n(|k_0|)} \begin{pmatrix} e^{\beta|k_0|/2} & e^{-\beta k_0/2} \\ e^{\beta k_0/2} & e^{\beta|k_0|/2} \end{pmatrix}, \qquad (10.11)$$

where n is the Bose-Einstein distribution function. When only transverse gauge field modes have a nontrivial density matrix, this matrix structure applies only to the $T^{\mu\nu}$ projection of the gauge field propagator, whereas its complement is to be taken at zero temperature. So the latter as well as the ghost propagator involves the zero-temperature limit of the matrix M,

$$M_0 = \begin{pmatrix} 1 & \theta(-k_0) \\ \theta(k_0) & 1 \end{pmatrix}. \qquad (10.12)$$

In Feynman gauge, the gauge field propagator thus reads

$$D^{\mu\nu} = -T^{\mu\nu}D - (g^{\mu\nu} - T^{\mu\nu})D_0, \qquad (10.13)$$

which can be easily generalized[11] to arbitrary linear gauges by replacing $g^{\mu\nu}$ in the above expression by the corresponding Lorentz structure appearing in the zero-temperature propagator.

Using these propagators, one can rather easily verify explicitly the gauge fixing independence of hard thermal loops[9,10] and also of the thermodynamic potential at the multi-loop level.[4]

However, a subtlety appears in applications of the hard-thermal-loop resummation program. Upon resummation of hard thermal loops, the gauge field propagator has not only physical poles corresponding to transverse polarizations, but also a collective mode with spatially longitudinal polarization.[12] One can show[11] that after resumming the hard-thermal-loop self-energy, the spatially longitudinal propagator component acquires the usual matrix structure of a propagator at finite temperature, which is in fact necessary so that no pinch singularities appear at higher orders of the loop expansion.

References

1. J Kapusta, *Finite-temperature field theory*, Cambridge University Press (1989)
2. M Le Bellac, *Thermal field theory*, Cambridge University Press (1996)
3. P V Landshoff and J C Taylor, *Nuclear Physics B* **430** (1994) 683
4. P V Landshoff and A Rebhan, *Nuclear Physics B* **383** (1992) 607
5. W Kummer, *Acta Physica Austriaca* **14** (1961) 149; *ibid.* **41** (1975) 315
6. R P Feynman, *Acta Physica Polonica* **24** (1963) 697

7. S Weinberg, *The quantum theory of fields*, volume II, Cambridge University Press (1996)
8. H. Hata and T. Kugo, *Physical Review D* **21** (1980) 3333
9. J Frenkel and J C Taylor, *Nuclear Physics B* **334** (1990) 199
10. E Braaten and R D Pisarski, *Nuclear Physics B* **337** (1990) 569
11. P V Landshoff and A Rebhan, *Nuclear Physics B* **410** (1993) 23
12. H A Weldon, *Physical Review D* **26** (1982) 1394

PART II

Classical and Quantum Gravity

Chapter 11

Wolfgang Kummer and the Vienna School of Dilaton (Super-)Gravity

Luzi Bergamin[1] and René Meyer[2]

[1] *ESA Advanced Concepts Team, ESTEC, DG-PI,*
Keplerlaan 1, 2201 AZ Noordwijk, The Netherlands

[2] *Max-Planck-Institut für Physik (Werner-Heisenberg-Institut),*
Föhringer Ring 6, 80805 München, Germany

Wolfgang Kummer was well known for his passion for axial gauges and for the formulation of gravity in terms of Cartan variables. The combination of the two applied to two-dimensional dilaton gravity is the basis of the "Vienna School", which provided numerous significant results over the last seventeen years. In this review we trace the history of this success with particular emphasis on dilaton supergravity. We also present some previously unpublished results on the structure of non-local vertices in quantum dilaton supergravity with non-minimally coupled matter.

11.1. Historical Introduction

11.1.1. *Early Attempts to Non-Einsteinian Gravity in 2D*

The earliest works by Wolfgang Kummer connected to gravity in two dimensions[1–4] date back to the year 1991, where he realized together with D. Schwarz (cf. the article by D. Schwarz in this volume) that the Katanaev-

---

[1] Email: bergamin@tph.tuwien.ac.at
[2] Email: meyer@mppmu.mpg.de

Volovich model,[5-7a]

$$S = \int \mathrm{d}^2 x \sqrt{-g} \left[\frac{\mu^2}{2} R^2_{abcd} + \frac{\gamma^2}{2} T^2_{abc} + \lambda \right], \qquad (11.1)$$

$$= \int \mathrm{d}^2 x \sqrt{-g} \left[\frac{\mu^2}{2} R^2 - \gamma^2 \tau_a \tau^a + \lambda \right], \qquad (11.2)$$

easily can be solved in an axial gauge using light-cone variables[1,3] and applying the first order formulation of gravity theories, since the theory exhibits a (nonlinear) Yang-Mills like gauge structure. In that work, the global structure of the solutions, in particular the classification according to their singularities, was discussed as well. Certainly, this insight was based on Wolfgang Kummer's long-standing experience with noncovariant gauges in non-Abelian gauge theories (cf. the article by P.V. Landshoff in this volume).[8] Furthermore they observed the existence of two branches of classical solutions: A constant curvature branch with vanishing torsion, yielding de Sitter space in the case of (11.1), and a nontrivial branch of solutions with both curvature and torsion, which is however labeled by a conserved quantity relating the torsion scalar and the Ricci scalar in a gauge invariant manner. As we will see, this conserved quantity exists in a much larger class of two-dimensional gravity theories and has the interpretation of a quasi-local mass.

The noncovariant gauge was then put to good use[1,4] to show the renormalizability of the Katanaev-Volovich model (11.1) in a fixed background quantization around flat space, albeit its nonpolynomial interactions. Though these interactions lead to an infinite set of ultraviolet divergent one-loop graphs, only a single quantity is renormalized, such that the renormalized Greens functions reduce to the tree-level graphs. Infrared divergences were shown to be treatable by introducing a small mass regulator, and it was realized that, although no physical S-matrix exists in the model because of the lack of propagating on-shell degrees of freedom, correlators of physical observables such as the Ricci curvature scalar can still yield interesting information.

Based on the idea that the integrability of (11.1) should be due to an additional dynamical symmetry, it was then realized by Wolfgang Kummer and collaborators[9-12] that the Katanaev-Volovich model can be reformu-

---

[a]Here R_{abcd} are the components of the curvature two-form written with tangent space indices only, T_{abc} are the components of the torsion form, and μ, γ, λ are constants. In the second line, we expressed everything in terms of the Ricci scalar $R = R^{\mu\nu}{}_{\mu\nu}$ and the Hodge dual of the torsion form, $T^a_{\mu\nu} = \epsilon_{\mu\nu} \tau^a$.

lated as a $sl(2, \mathbb{R})$ gauge theory. They went on to analyze the constraint structure of the theory, and found that the secondary first class constraints form a deformed $iso(2, 1)$ algebra, i.e. a deformation of the Poincaré algebra of $(2+1)$-dimensional Minkowski space. The deformation parameter turned out to be the constant γ, and sending[b] $\gamma \to \infty$ restored an undeformed $iso(2, 1)$ symmetry. They found that the above-mentioned conserved quantity[3] was one generator of the center of the deformed symmetry algebra, while a second generator vanished on the constraint surface. Wolfgang Kummer investigated the Katanaev-Volovich model minimally coupled to real scalars or fermions[12] as well, and found the same dynamical symmetry, albeit the system in general is no longer integrable. Integrability was found to be preserved for chiral solutions, i.e. if the fermion has definite handedness, whereby the conserved quantity is just conserved in time but no longer in space, as its spatial divergence is connected to the spatial parts of the chiral current, $\chi_L^\dagger \overleftrightarrow{\partial} \chi_L$. These currents provide a central extension of the algebra of constraints. This result already hinted towards a generalized conservation law including the matter contributions. In Ref. 12 it was also mentioned that this dynamical symmetry and the classical integrability might hint towards nonperturbative quantum integrability of the Katanaev-Volovich model, a statement which is in fact correct for a large class of generalized two-dimensional dilaton gravities.

After so many interesting facets of the Katanaev-Volovich model being found, it might seem that it would already have revealed all of its mysteries. This was, however, not the case. Ikeda and Izawa[19,20] wrote (11.1) in first order form,[c]

$$S = \int_{\mathcal{M}_2} [\phi d\omega + X_a(De^a)] + \int_{\mathcal{M}_2} d^2x\sqrt{-g}\left[\lambda - \frac{\phi^2}{8\mu^2} + \frac{X_a X^a}{4\gamma^2}\right], \quad (11.3)$$

a fact which was then used by Wolfgang Kummer and Peter Widerin[22] to further analyze the symmetry structure of the theory: They found a field-

---

[b]Though the solutions of the classical equations of motion are in some sense singular in this limit,[11] it is more interesting than the de Sitter solution, as it has vanishing torsion but nonvanishing curvature (R^2 gravity without torsion[13]), as can be seen from the first order formulation (11.3). For this reason it was named "Einstein branch". Different limits of the parameters μ, γ, λ in (11.3) yield further 2d gravity models:[14] Taking $(\mu, \gamma) \to \infty$ and integrating out X_a afterwards yields 2D BH Gravity.[15,16] The Jackiw-Teitelboim model[17,18] is obtained in the same limit when rescaling the cosmological constant such that $\lambda/\gamma = \Lambda^2$ is kept fixed.

[c]Here, the dilaton ϕ and the Lagrange multipliers for torsion X^a are auxiliary fields, introduced to linearize the square terms in (11.1). The notation has been adapted to the one used in 11.1.2, $T^a = De^a = de^a + \varepsilon^a{}_b\omega \wedge e^b$ is the torsion two-form with spin connection $\omega_{ab} = \epsilon_{ab}\omega$, and according to our conventions[21] the Ricci scalar is $R = 2*d\omega$.

dependent off-shell global symmetry of (11.1) whose conserved charge is the conserved quantity mentioned above. Furthermore, they showed, driven by the desire to reinterpret the algebra of constraints in a framework not involving the Hamiltonian phase space, that the deformed $iso(2,1)$ symmetry could be re-formulated as a current algebra. The currents were readily identified with the components of the energy-momentum tensor in lightcone gauge, which were shown to fulfill (on compactified space) a Virasoro algebra, similar to the situation in string theory. As will become clear to the reader later, this first order reformulation of a much wider class of gravity theories, together with the choice of lightcone gauge, lies at the heart of the results outlined in this article.

Also the case with minimally coupled bosonic[23] and fermionic[24] matter offered some new surprises: A different lightcone gauge, defined in terms of 'lapse' and 'shift' variables, together with the first order formulation (11.3) of the gravity sector, allowed a counting of free functions in the general solution of the constraints and the equations of motion. It was found that the inclusion of scalars and fermions does spoil integrability of (11.1), but only in a mild way, since only one non-trivial first order PDE needs to be solved.

The combination of Eddington-Finkelstein gauge and first order formulation (11.3) also played an important role in the nonperturbative quantum treatment of the Katanaev-Volovich model:[25] In that work Wolfgang Kummer and Florian Haider found the correct path integral measure that allowed to integrate out both the Zweibeine as well as the spin connection. This measure differs from the one used in the perturbative approach[4] by a factor of $(-g)^{-3/4}$, arising from the Gaussian path integration over the fields ϕ, X^a in lightcone gauge. As different path integral measures correspond in a renormalizable theory to different renormalization schemes with different counterterms, the nonperturbatively useful measure differed from the perturbative one only by a shift in the counterterm present in the quantum effective action.[4] Besides reading off the correct measure just from the symplectic structure which can be found directly from the first order form (11.3), they also derived the same result from a canonical BRST analysis. We will see in section 11.1.3 that this method reveals best the underlying symmetry structure of the quantum theory, being in essence a nonlinear Yang-Mills theory. Imposing again lightcone gauge the quantum effective action may be calculated exactly and they found that all local quantum effects disappear right away, i.e. the counterterms vanish. They also discuss that the same result can be obtained by careful renormalization in the fixed-background quantization of Ref. 4, as expected for a gauge-invariant quantum effective action. The Katanaev-Volovich model thus shows no local quantum effects, and all Green functions are determined by their tree-level contributions. Nonetheless, the theory is neither classically nor quantum mechanically trivial since there exists a global degree of freedom,[26] namely the quasilocal mass (conserved quantity). Its role becomes even more important when considering spacetimes with boundary (see Sect. 11.3).

11.1.2. *First Order Formulation of Generalized 2D Dilaton Gravity*

Ref. 25 was the last in a series of papers[1-4,9-12,22,24,25] devoted to "Non-Einsteinian Gravity in $d = 2$", i.e. the Katanaev-Volovich model (11.1). In the subsequent works Wolfgang Kummer and his collaborators[27-32] realized that a much larger class of gravity theories in two dimensions can be treated along similar lines. This step was possible thanks to a generalization of the first order action (11.3) to

$$\mathcal{S}_{\text{FOG}} = \int_{\mathcal{M}} \left(\phi \mathrm{d}\omega + X^a D e_a + \epsilon \mathcal{V}(\phi, Y) \right) , \qquad (11.4)$$

with $Y = X^a X_a / 2$, which first appeared in Ref. 33. These models, called First Order Gravities (FOGs), in general do not permit a formulation exclusively in terms of the geometric quantities (curvature and torsion), since the possibility to eliminate the fields X and X^a (which actually represents a Legendre transformation[34]) is not guaranteed. However, already in Ref. 30 it was then realized that it is preferable to eliminate the torsion dependent part of the spin-connection instead of the dilaton. Remarkably this procedure only involves linear and algebraic equations for any potential $\mathcal{V}$, which thus may be reinserted into the action without restrictions.[35] In general, this yields higher derivative theories of gravity. Still, if one restricts to models of type

$$\mathcal{V}(\phi, Y) = V(\phi) + Y U(\phi) \qquad (11.5)$$

the ensuing second order action are Generalized Dilaton Theories (GDTs)

$$\mathcal{S}_{\text{GDT}} = \frac{1}{2} \int \mathrm{d}^2 x \sqrt{-g} \left[\phi R - U(\phi)(\nabla\phi)^2 + 2V(\phi) \right] . \qquad (11.6)$$

The classical equivalence also holds on the quantum level,[36] as long as possible additional matter fields do not couple to torsion, which is the case for scalars and fermions in two dimensions, but for example not for four-dimensional spherically reduced fermions.[37]

Analogous to the Katanaev-Volovich model, all matterless FOGs with potential (11.5) are classically integrable. Their equations of motion[d]

$$0 = \mathrm{d}\phi + X^- e^+ - X^+ e^- \,, \qquad 0 = (\mathrm{d} \pm \omega)X^\pm \mp \mathcal{V}e^\pm + W^\pm \,, \qquad (11.7)$$

$$0 = \mathrm{d}\omega + \epsilon \partial \mathcal{V}\phi + W \,, \qquad 0 = (\partial \pm \omega)e^\pm + \epsilon \partial \mathcal{V}X^\mp \,, \qquad (11.8)$$

in the matter free case, $W = \frac{\delta L^{(m)}}{\delta \phi} = 0$, $W^\pm = \frac{\delta L^{(m)}}{\delta e^\mp} = 0$, can be solved just by form manipulations. In a patch with $X^+ \neq 0$, the solution is

$$\mathrm{d}s^2 = 2e^+ \otimes e^- = 2\mathrm{d}f\mathrm{d}\tilde\phi + \xi(\tilde\phi)\mathrm{d}f^2 \,, \qquad (11.9)$$

$$Q(\phi) = \int^\phi U(y)\mathrm{d}y \,, \quad w(\phi) = \int^\phi e^{Q(y)}V(y)\mathrm{d}y \,, \qquad (11.10)$$

$$\xi(\tilde\phi) = 2e^Q(C - w)\big|_{\phi=\phi(\tilde\phi)} \,, \quad \mathrm{d}\tilde\phi = \mathrm{d}\phi e^Q \,, \qquad (11.11)$$

$$C = e^{Q(\phi)}Y + w(\phi) \,, \qquad \mathrm{d}C = 0 \,. \qquad (11.12)$$

Here f is a free function on two-dimensional space-time. By using f and $\tilde\phi$ directly as coordinates the line element (11.9) is found naturally in Eddington-Finkelstein gauge, although no diffeomorphism gauge fixing was employed. The conserved quantity (11.12), which enters the Killing norm (11.11), is exactly the nontrivial integral of motion found in the Katanaev-Volovich model (11.1). The solutions (11.9) can have many Killing horizons, and in fact can be extended globally.[21,29,30,38,39] As for the Katanaev-Volovich model the integrability extends to cases with special "chiral" matter.[21]

FOG also allows for a convenient reformulation as a so-called Poisson-Sigma-Model (PSM),[40–42] illuminating some of its structure more clearly: Grouping together the target-space coordinates $X^I = (\phi, X^a)$ and the gauge fields $A_I = (\omega, e_a)$, FOG (11.4) can be written as

$$\mathcal{S}_{\mathrm{PSM}} = \int_{\mathcal{M}_2} \left[\mathrm{d}X^I \wedge A_I + \frac{1}{2}P^{IJ}A_J \wedge A_I\right] \,, \qquad (11.13)$$

where $P^{IJ} = \{X^I, X^J\}$ is a Poisson tensor related to a Schouten-Nijenhuis bracket defined on the manifold. Since the Poisson tensor must have a vanishing Nijenhuis tensor with respect to this bracket,

$$P^{IL}\partial_L P^{JK} + perm\,(IJK) = 0 \,, \qquad (11.14)$$

[d]The $\pm$-indices are lightcone indices in tangent space, cf. App. A.

the PSM action is invariant under the symmetries

$$\delta X^I = P^{IJ} \varepsilon_J , \qquad \delta A_I = -\mathrm{d}\varepsilon_I - \left(\partial_I P^{JK}\right) \varepsilon_K A_J . \qquad (11.15)$$

For P^{IJ} linear in X^I the symmetries constitute a linear Lie algebra and (11.14) reduces to the Jacobi identity for the structure constants of a Lie group. Finally, we mention that the variation of A_I and X^I in (11.13) yields the PSM field equations

$$\mathrm{d}X^I + P^{IJ} A_J = 0 , \qquad \mathrm{d}A_I + \frac{1}{2}(\partial_I P^{JK})A_K A_J = 0 . \qquad (11.16)$$

All PSMs are in essence topological theories of gauge fields on a by itself dynamical target space. They also allow for the straightforward inclusion of other topological degrees of freedom (such as gauge fields in two dimensions). The central object in (11.13) is the Poisson tensor P^{IJ}, which for bosonic FOG (11.4) reads

$$P^{\phi\pm} = \pm X^\pm , \quad P^{+-} = X^+ X^- U(\phi) + V(\phi) , \quad P^{IJ} = -P^{JI} . \qquad (11.17)$$

The PSM formulation of FOG is very useful in the context of supergravity, see Sect. 11.2. Since the Poisson tensor (11.17) has odd dimension it cannot have full rank. Therefore there appears at least one "Casimir function" defined by the condition

$$\{X^I, C\} = P^{IJ} \frac{\partial C}{\partial X^J} = 0 . \qquad (11.18)$$

The conserved quantities are thus central elements of the algebra of target space coordinates under the Schouten-Nijenhuis bracket. It is straightforward to show the conservedness by using the equations of motion (11.16), as well as to reproduce the form of (11.12).

11.1.3. *Exact Path Integral Quantization*

In the years 1997-1999 Wolfgang Kummer, Herbert Liebl and Dimitri Vassilevich[36,43,44] found that an exact path integral quantization is possible for the first order formulation (11.4), (11.6). Even more, if (11.4) is coupled to matter,[36,43–51] the geometric sector (ω, e^a, X^a, ϕ) can still be integrated out exactly, yielding a nonlocal and nonpolynomial effective action for the matter fields, which then can be treated perturbatively. In this section, we will shortly review the most important steps in the path integral quantization, more details are explained in the context of supergravity in Sect. 11.2 or can be found e.g. in the Review 21.

The goal is to evaluate the path integral

$$\mathcal{Z} = \int \mathcal{D}(\Phi, \phi, X^a, \omega_\mu, e^a_\mu) e^{iS_{\text{FOG}}[\phi, X^a, \omega, e^a] + iS_{\text{mat}}[\phi, e^a, \Phi]} \qquad (11.19)$$

where Φ denote all kinds of matter fields, which should not couple directly to the spin connection if the equivalence with generalized dilaton theories (11.6) shall not be lost. Since the model can be formulated as a nonlinear gauge theory with diffeomorphisms and local Lorentz invariance as symmetries, the first step is to construct an appropriate gauge fixed action. To this end, one analyzes the constraints in the theory, by first noting that (11.4) furnishes a natural symplectic structure with canonical variables[e] $q^I = (\phi, X^a)$, $p_I = (\omega_1, e^a_1)$, and $\bar{p}_I = (\omega_0, e^a_0)$. The second set of momenta $\bar{p}_I$ does not appear with time derivatives in the Lagrangian and thus its conjugate coordinates are constrained to zero, $\bar{q}^I \approx 0$, which constitutes three primary first class constraints. These constraints give rise to three secondary first class constraints $G_I = \{\bar{p}_I, \mathcal{H}\}$. If the matter extension yields additional second class constraints, as e.g. in the case of fermions,[48] the Poisson bracket should be replaced by a the Dirac bracket. The three constraints G_I form a closed nonlinear algebra

$$\{G_I, G'_J\} = G^K f_K{}^{IJ} \delta(x - x'), \qquad (11.20)$$

with structure functions $f_K{}^{IJ}$. A Virasoro-like algebra closing on derivatives of δ-functions typical for 2D gravity models can be recovered from linear combinations of the G^I.[22,23] As expected for a theory of gravity the Hamiltonian density vanishes on the constraint surface,

$$\mathcal{H} = \dot{q}^I p_I - L = -G^I \bar{p}_I. \qquad (11.21)$$

From this knowledge it is now straightforward to construct the gauge fixed action with ghosts, using the BVF formalism:[54–56] Introducing one pair of ghosts and antighosts for each secondary first class constraint, (c_I, p^I_c),[f] one finds the nilpotent BRST charge

$$\Omega = G^I c_I + \frac{1}{2} p^K_c f_K{}^{IJ} c_J c_I, \qquad \{\Omega, \Omega\} = 0, \qquad (11.22)$$

---

[e]Note that these conventions accord with most literature on supergravity[52,53] but differ from those in Ref. 21, where the components of the gauge fields were chosen as momenta rather than canonical coordinates.

[f]Strictly speaking it would also be necessary to introduce (anti)ghosts for the primary constraints. The procedure is straightforward, as described in chapter 7 of Ref. 21, and does not yield much new insight.

having the same structure as in ordinary Yang-Mills theory, albeit the field-dependence in $f_K{}^{IJ}$. At this point it is important to follow Wolfgang Kummer's concept of temporal gauges and to fix e.g. $(\omega_0, e_0^-, e_0^+) = (0, 1, 0)$, which again establishes Eddington-Finkelstein gauge for the metric. This can be implemented by a simple multiplier gauge $\Psi = p_c^+$,[57] another possibility was used in Ref. 36. Now the gauge fixed Hamiltonian $\mathcal{H}_{gf} = \{\Omega, \Psi\}$, via Legendre transformation[g] yields the gauge fixed Lagrangian,

$$\mathcal{L}_{gf} = \dot{\bar{q}}^I \bar{p}_I + \dot{q}^I p_I + G^+ + p_c^K (\partial_0 \delta_K^L + f_K{}^{+L}) c_L \,. \tag{11.23}$$

The path integral (11.19) can now be evaluated as follows: First all (anti)ghosts are integrated out yielding the Faddeev-Popov determinant $\mathrm{Det}\Delta_{FP} = \mathrm{Det}(\partial_0 \delta_K^L + f_K{}^{+L})$. Now an important observation is made: Eq. (11.23) only depends linearly on p_I, while $\mathrm{Det}\Delta_{FP}$ is independent thereof. After eventual integration of matter momenta the geometric fields p_I thus can be integrated. This generates three "functional δ functions", whose arguments contain parts of the equations of motion and imply in the matterless case (j^I are sources for the p_I)

$$0 = \partial_0 \phi - X^+ - j^\phi \,, \qquad 0 = \partial_0 X^+ - j^+ \,, \tag{11.24}$$

$$0 = (\partial_0 + X^+ U(\phi)) X^- + V(\phi) - j^- \,. \tag{11.25}$$

The final integration of $X^I = q^I$ sets these fields to the formal solution of the above equations given in terms of Green functions for ∂_0 acting on the sources. Note that these solutions depend nonlocally on the sources, and in particular the solution of (11.25) is also nonpolynomial. In defining the Green functions the asymptotic values of X^I are fixed, which also fixes the quasilocal mass. For this reason, the path integral as it is yields local physics. It is a question worth investigating whether in the path integral formalism an "integration over masses" is possible — see Sect. 11.3 for some further comments on this topic. Finally, it should be mentioned that the integration over q^I cancels exactly the Faddeev-Popov determinant.

In the absence of matter fields the story ends here since the path integral is fully evaluated. If an integration over matter fields is left, the so-far obtained effective action is a complicated, nonlocal and nonpolynomial expression in the matter fields. Still, it is possible to derive in a systematic way non-local vertices of the interaction of matter with the quantized gravitational background. Therefrom, higher loops[43] and scattering processes

---

[g]Of course the Legendre transform also has to be done w.r.t. possible matter canonical variables present. We tacitly ignored this here, as we do not want to specify the matter content yet, but describe the overall structure.

can be calculated.[45] For a real scalar field unitarity, i.e. absence of information loss, and CPT invariance of the tree-level four-particle S-matrix element was found. The specific heat of the Witten black hole was also shown to be positive[58] once loop corrections are taken into account.

A phenomenon worth mentioning is the "virtual black hole (VBH)":[46,58–60] Some of the nonlocal interaction geometries resemble black holes, although being off-shell entities. The curvature scalar of a VBH has a δ-peak at some point and is discontinuous up to that peak (see the Penrose diagram in Fig. 2 of ref. 61, an effective line element can also be found there). Although this interpretation is a feature of the chosen gauge, in the calculation of (gauge invariant) S-matrix elements one has to integrate over all such VBHs, which leads to the idea that nonperturbative and nonlocal excitations such as VBHs might also play an important role in higher-dimensional quantum gravity.

11.2. Two-Dimensional Dilaton Supergravity

In the following we will leave the field of bosonic dilaton gravity, since many results thereof have been summarized in various reviews.[21,61] Instead we will concentrate on dilaton supergravity in two dimensions, a field in which Wolfgang Kummer made substantial contributions as well but which is not covered exhaustively by the existing reviews.

Of course, dilaton supergravity in two dimensions existed long before Wolfgang Kummer entered the field. Many early attempts were based on superspace techniques,[62] which led to a generalized dilaton supergravity action of the form[63]

$$\mathcal{S} = \int \mathrm{d}^2 x \mathrm{d}^2 \theta E \big(\Phi S - \frac{1}{4} U(\Phi) D^\alpha \Phi D_\alpha \Phi + \frac{1}{2} u(\Phi) \big) , \qquad (11.26)$$

where S and Φ are the supergravity and dilaton superfields, resp, and $U(\Phi)$ and $u(\Phi)$ two dilaton dependent functions (potentials) defining the model.[h] Despite its successes and advantages,[63,65,66] in particular the straightforward way to couple additional gauge or matter fields, the superspace formulation shares its drawbacks with the purely bosonic second order action: It does not include bosonic torsion, exact solutions are not easy to obtain although the matterless model is integrable, and a nonperturbative quantization is cumbersome. Some attempts to relax the standard condition of

---

[h]In Ref. 63 a third potential $J(\Phi)S$ was introduced in the first term. If $J(\Phi)$ is not invertible, the analysis performed here applies locally, only. However, such models are of limited interest.[34,64]

vanishing bosonic torsion led to a mathematically complex formalism[67,68] and thus were not pursued further.

11.2.1. *2D Dilaton Supergravity from Graded PSMs*

In this situation it appeared promising to extend FOG (11.4) to dilaton supergravity. Already in Refs. 69, 70 gauge theoretic methods were used to find supersymmetric extensions of some dilaton gravity models. However, this approach is limited to models representable as a linear gauge theory, in particular the Jackiw-Teitelboim model.[17,18] Its extension to nonlinear gauge theories showed that generalized dilaton supergravity can be treated in this framework.[40,71] This led to a first order action for dilaton supergravity using free differential algebras,[72] which however is restricted to vanishing bosonic torsion. This model, which is classically equivalent to the action (11.26) for $U = 0$, then was shown to be a special case of a graded PSM.[73]

Wolfgang Kummer, Martin Ertl and Thomas Strobl then showed[35] that the use of graded PSMs (gPSMs) provides a simpler and more systematic tool to find supergravity extensions of FOG. To this end one replaces the target space (Poisson manifold) of the action (11.13) by a graded Poisson manifold.[i] On this manifold we use the coordinates (scalar fields) $X^I(x) = (X^i(x), X^\alpha(x))$, where X^i refers to bosonic (commuting) coordinates, while X^α are fermions (anti-commuting coordinates.) All equations (11.13)–(11.18) have been displayed in such a way that they hold for gPSMs as well, if one replaces the permutations in (11.14) by graded permutations and keeps in mind that P^{IJ} is now graded anti-symmetric.

As in the non-supersymmetric case, not every gPSM describes a supergravity model, but certain additional structures are needed. Firstly this concerns the choice of target space and gauge fields. In the application to two-dimensional $N = (1,1)$ supergravity, the bosonic fields in (11.4) are complemented by two Majorana spinors, ψ_α ("gravitino") and χ^α ("dilatino"):

$$A_I = (\omega, e_a, \psi_\alpha) \qquad X^I = (\phi, X^a, \chi^\alpha) \qquad (11.27)$$

Secondly, local Lorentz invariance determines the ϕ-components of the

---

[i]Instead of a graded target space, supersymmetric PSMs can be obtained by grading the world-sheet manifold. While such models attracted interest in the context of string theory,[74,75] they are less useful to derive supergravity actions.

Poisson tensor

$$P^{a\phi} = X^b \epsilon_b{}^a \,, \qquad\qquad P^{\alpha\phi} = -\frac{1}{2}\chi^\beta \gamma^3{}_\beta{}^\alpha \,. \qquad (11.28)$$

Finally it should be noted that the existence of a tanget space metric η_{ab} is assumed as an extra structure.[j] Remarkably it was possible to solve the nonlinear Jacobi identity (11.14) explicitly with just these three assumptions.[35,68] Having defined the fermionic extension, all components of the graded Poisson tensor may be written as a systematic expansion in these fields, restricted by local Lorentz invariance encoded in the ε_ϕ component in (11.15). As an example P^{ab} must be of the form $P^{ab} = \mathcal{V}(\phi, Y) + \chi^2 v_2(\phi, Y)$, where $\mathcal{V}(\phi, Y)$ is the bosonic potential (11.5) of the model. Then it is a straightforward but tedious calculation to solve the condition (11.14) order by order in the fermionic fields. Though a purely algebraic solution was found,[35,68] this result was not yet satisfactory, as it depends besides the bosonic potential $\mathcal{V}(\phi, Y)$ on five arbitrary Lorentz covariant functions. However, many choices of these five five functions impose new singularities and obstructions in the variables ϕ and Y in points where the bosonic potential remains regular. In extreme cases this prohibits any supersymmetrization at all.

From this result it is obvious that not all Lorentz covariant gPSMs permit an interpretation as dilaton supergravity, but a suitable implementation of supersymmetry transformations is needed.[k] This can be achieved by restricting the fermionic extension $P^{\alpha\beta}$ to the form

$$P^{\alpha\beta} = -2i X^a \gamma_a^{\alpha\beta} + Z(\phi, Y, \chi^2)\gamma^{3\,\alpha\beta} \,, \qquad (11.29)$$

where the first term generates supersymmetry transformations of the form $\{Q^\alpha, Q^\beta\} = -2ip^{\alpha\beta}$ in the commutator of two symmetry transformations (11.15). The solution of the Jacobi identities now proved a unique class of $N = (1,1)$ dilaton supergravity models with Poisson tensor[l]

$$P^{ab} = \left(V + YU - \frac{1}{2}\chi^2 \left(\frac{VU + V'}{2u} + \frac{2V^2}{u^3} \right) \right) \epsilon^{ab} \,, \qquad (11.30)$$

$$P^{\alpha b} = \frac{U}{4}X^a(\chi\gamma_a\gamma^b\gamma^3)^\alpha + \frac{iV}{u}(\chi\gamma^b)^\alpha \,, \qquad (11.31)$$

$$P^{\alpha\beta} = -2i X^c \gamma_c^{\alpha\beta} + \left(u + \frac{U}{8}\chi^2 \right)\gamma^{3\,\alpha\beta} \,, \qquad (11.32)$$

---

[j]Attempts to relax some of these conditions have been presented in Refs. 76, 77.

[k]This procedure is similar to the choice of "consistent" supergravity in the standard formulation.[78–81]

[l]Notice that the potential $U(\phi)$, which determines the torsion part of the spin connection, in most of the literature on supergravity is denoted by $Z(\phi)$.

whereby primes indicate a derivative with respect to the dilaton field ϕ and $u(\phi)$ acts as a prepotential for the bosonic potential $V(\phi)$,

$$V(\phi) = -\frac{1}{8}\left((u^2)' + u^2 U(\phi)\right).$$ (11.33)

First, this result was established from symmetry arguments[52] by constructing a deformed superconformal algebra as previously done for FOG.[23] Then it turned out that the supergravity action from (11.28) and (11.30)–(11.32),

$$\mathcal{S} = \int_{\mathcal{M}} \left(\phi \mathrm{d}\omega + X^a D e_a + \chi^\alpha D\psi_\alpha + \epsilon\left(V + YU - \frac{1}{2}\chi^2\left(\frac{VU + V'}{2u} + \frac{2V^2}{u^3}\right)\right)\right.$$
$$+ \frac{U}{4}X^a(\chi\gamma_a\gamma^b e_b\gamma^3\psi) + \frac{iV}{u}(\chi\gamma^a e_a\psi)$$
$$\left. + iX^a(\bar{\psi}\gamma_a\psi) - \frac{1}{2}\left(u + \frac{U}{8}\chi^2\right)(\psi\gamma_3\psi)\right),$$ (11.34)

after elimination of the auxiliary fields X^a and ω is equivalent to the action (11.26) after integrating out superspace.[82] As in FOG there exists at least one Casimir function, which receives additional fermionic contributions:

$$C = e^Q\left(Y - \frac{1}{8}u^2 + \frac{1}{16}\chi^2(u' + \frac{1}{2}uU)\right).$$ (11.35)

The Casimir function plays a crucial role in the discussion of BPS states.[83] From the symmetry transformation of the dilatino in Eq. (11.15) it follows that the fermionic part of the Poisson tensor for BPS solutions does not have full rank. For purely bosonic states this implies $\mathrm{Det}P^{\alpha\beta} = 8Y - u^2 = 0$ and thus $C = 0$. Furthermore, an additional fermionic Casimir function exists and it can be checked[35,82] that the quantity

$$\tilde{c} = e^{\frac{1}{2}Q}\sqrt{|X^{++}|}\left(\chi^- - \frac{\chi^+}{2\sqrt{2}X^{++}}\right)$$ (11.36)

for $C = 0$ is a Casimir function as defined in (11.18), which corresponds to the unbroken supersymmetry of the BPS state.[83] Besides the purely bosonic states there also exist BPS states with non-vanishing fermionic fields, which however must have a trivial bosonic background.[83] Nonetheless, some of these states exhibit the interesting feature that $\chi^2(u' + \frac{1}{2}uU) \neq 0$ and thus the bosonic part of the Casimir vanishes, while its fermionic extension (its soul) is non-zero.

Also the complete classical solution now is straightforward to obtain.[35,68,82] The bosonic line element again may be written in the form (11.9). From this result the extremality of BPS Killing horizons[84,85] is immediate since by virtue of the constraint (11.33) the conformally invariant potential acquires the form $w(\phi) = -e^Q u^2/8$. Thus, the Killing norm ξ in Eq. (11.11) for BPS solutions is positive semi-definite, an eventual horizon must be extremal and represents a ground state.[83]

Coupling of additional fields (e.g. gauge fields or matter fields) is straightforward for FOG, but in supergravity this task becomes considerably more complicated since a correct supersymmetry transformation of the new fields must be ensured. Since it is known that all genuine gPSM supergravity models are classically equivalent to superspace models, the coupling of matter fields can be inferred therefrom. As an example, it has been demonstrated in Ref. 83 that conformal matter fields can be coupled non-minimally to the general action (11.34). The matter action for a chiral multiplet with real scalar field f and Majorana fermion λ_α and with coupling function $K(\phi)$ for second order supergravity is obtained from standard superspace techniques as

$$
\mathcal{S}_{\text{matter}} = \int d^2x e \left[K\left(\frac{1}{2}(\partial^m f \partial_m f + i\lambda\gamma^m \partial_m \lambda) + i(\psi_n \gamma^m \gamma^n \lambda)\partial_m f \right.\right.
$$
$$
\left. + \frac{1}{4}(\psi^n \gamma_m \gamma_n \psi^m)\lambda^2 \right) + \frac{u}{8}K'\lambda^2 - \frac{1}{4}K'(\chi\gamma^3\gamma^m\lambda)\partial_m f
$$
$$
\left. - \frac{1}{32}\left(K'' - \frac{1}{2}\frac{[K']^2}{K}\right)\chi^2\lambda^2 \right] . \quad (11.37)
$$

Without further changes this action provides the correct extension of genuine gPSM supergravity by a conformal matter field. Still, the supersymmetry transformations of some of the gPSM fields acquire additional terms from the matter action.[83]

The gPSM approach to two-dimensional dilaton supergravity is not restricted to the above example of $N = (1,1)$ supergravity. As an example the gPSM version of the model of Ref. 86 has been presented in Refs. 87, 88. This model deals with chiral or twisted-chiral $N = (2,2)$ supergravity. Thus besides a pair of complex Dirac dilatini/gravitini an additional graviphoton gauge field is needed in the supergravity multiplet, while the (twisted-)chiral dilaton multiplet is complemented by an additional Lorentz scalar π:

$$
X^I = (\phi, \pi, X^a, \chi^\alpha, \bar{\chi}^\alpha) , \qquad A_I = (\omega, B, e_a, \psi_\alpha, \bar{\psi}_\alpha) . \quad (11.38)
$$

Again symmetry constraints fix certain components of the Poisson tensor. Local Lorentz invariance and supersymmetry are implemented analogously to (11.28) and (11.29) and supersymmetry is encoded as $P^{\alpha\bar{\beta}} =$

$-2iX^c\gamma_c^{\alpha\beta}$ + terms $\propto \gamma^3$. The additional B gauge symmetry imposes (here for chiral supergravity)

$$P^{a\pi} = 0, \qquad P^{\alpha\pi} = -\frac{i}{2}\chi^\beta\gamma^3{}_\beta{}^\alpha , \qquad P^{\bar{\alpha}\pi} = \frac{i}{2}\bar{\chi}^\beta\gamma^3{}_\beta{}^\alpha . \qquad (11.39)$$

Starting with these restrictions it is again possible to solve the non-linear Jacobi identity (11.14) and to derive the complete action of extended supergravity.[87] Again, once reformulated as a gPSM the model can be solved explicitly, whereby some subtleties arise as a consequence of the grading of the Poisson manifold.[88] The model is found to have at least two Casimir functions, one being the $N = (2,2)$ extension of (11.12), the second one is the charge with respect to the graviphoton field. This new conserved quantity diverges in the limit $C \to 0$ unless all fermionic contributions vanish. Nevertheless, after integrating all equations of motion it is found that there exist well-defined solutions at $C = 0$ with non-vanishing fermion fields. However, the remaining conserved quantity (the charge with respect to the graviphoton) can no longer be expressed entirely in terms of the target space variables, i.e. for those special solutions this conserved quantity is not a Casimir function of the graded Poisson manifold. This is a unique feature of graded PSMs, since for purely bosonic PSMs the existence of Casimir-Darboux coordinates excludes such a behavior.

11.2.2. *Quantization of 2D Supergravity*

Having formulated dilaton supergravity as a gPSM it is self-evident to try to extend the nonperturbative quantization outlined in Sect. 11.1.3 to dilaton supergravity.[47,53]

11.2.2.1. *Quantization Without Matter*

Let us first consider the dilaton supergravity action (11.34) without matter couplings. As in Sect. 11.1.3 the canonical coordinates are chosen as $q^I = X^I$, $p_I = A_{1I}$, $\bar{p}_I = A_{0I}$ and $\bar{q}^I \approx 0$ are the primary first class constraints. In supergravity these coordinates obey the graded Poisson bracket[m]

$$\{q^I, p'_J\} = (-1)^{I \cdot J+1}\{p'^J, q_I\} = (-1)^I \delta_I^J \delta(x - x') . \qquad (11.40)$$

The secondary constraints,

$$G^I = \partial_1 q^I + P^{IJ} p_J , \qquad (11.41)$$

---

[m]This bracket should not be confused with the Schouten bracket, associated to the Poisson tensor P^{IJ} in (11.13).

despite the new fermionic constraints formally obey the same constraint algebra (11.20) as in the purely bosonic case. The structure functions in the matterless case are simply $f_K{}^{IJ} = -\partial_K P^{IJ}$ with P^{IJ} as given in (11.28) and (11.30)–(11.32), and with fermionic derivatives being left-derivatives. The ensuing Hamiltonian (11.21) again is quantized by introducing ghosts, which now obey the graded commutation rules $[c_I, p_c^J] = -(-1)^{(I+1)(J+1)}[p_c^J, c_I] = \delta_I^J$. This yields an additional sign in the BRST charge[53]

$$\Omega = G^I c_I + \frac{1}{2}(-1)^I p_c^K f_K{}^{IJ} c_J c_I \ . \tag{11.42}$$

The BRST charge (11.42) is found to be nilpotent as a consequence of the graded Jacobi identity of P^{IJ} (11.14). This characteristic does not depend on the specific form of the Poisson tensor as found in (11.30)–(11.32) and thus also applies to graded dilaton models not related to the supergravity action (11.26). Also, any gPSM gravity model without matter is free of ordering problems although the constraints are nonlinear in the fields.[53]

Imposing again a multiplier gauge $\bar{p}_I = a_I$ with[n] $a_I = -i\delta_I^{++}$ one arrives at the gauge fixed Lagrangian

$$L^0_{\text{g.f.}} = \dot{q}^I p_I + p_c^I \dot{c}_I - iP^{++|J} p_J \ . \tag{11.43}$$

The evaluation of the path integral now follows exactly the same steps as in the purely bosonic case. Having integrated out all ghosts the effective Lagrangian including sources for p_I and q^I becomes

$$L^0_{\text{eff}} = \dot{q}^I p_I - iP^{++|J} p_J + q^I j_{qI} + j_p^I p_I \tag{11.44}$$

which again is linear in all p_I. Integration of the momenta thus yields five functional δ functions, whose arguments imply (for the Poisson tensor (11.28), (11.30)-(11.32))

$$\dot{q}^\phi = -iq^{++} - j_p^\phi \ , \qquad \dot{q}^{++} = -j_p^{++} \ , \tag{11.45}$$

$$\dot{q}^{--} = i\Big(e^{-Q}w' + q^{++}q^{--}U + \frac{1}{2\sqrt{2}}q^-q^+e^{-Q/2}\big(\sqrt{-w}\big)''\Big) - j_p^{--} \ , \tag{11.46}$$

$$\dot{q}^+ = -j_p^+ \ , \qquad \dot{q}^- = -i\big(e^{-Q/2}(\sqrt{-w})'q^+ - \frac{1}{2}Uq^{++}q^-\big) - j_p^- \ . \tag{11.47}$$

These functional δ functions may be used to evaluate the remaining q^I integration. As expected, this exactly cancels the super-determinant from

[n]Notice that the elements of the spin-tensor decomposition are related to lightcone indices as $v^{++} = iv^\oplus$, $v^{--} = -iv^\ominus$, cf. Eq. (A.8). Thus $\bar{p}_{++}$ is imaginary.

the integration of the ghosts. The final generating functional becomes

$$\mathcal{W}[j_p^I, j_{qI}] = \exp i L_{\text{eff}}^0 , \quad L_{\text{eff}}^0 = \int \mathrm{d}^2 x \left(B^I j_{qI} + \tilde{L}^0(j_p^I, B^I) \right) , \quad (11.48)$$

where B^I are the solutions of Eqs. (11.45)–(11.47) and $\tilde{L}^0$ are the so-called "ambiguous terms".[21,25,36] An expression of this type is generated by the integration constants $G_I(x^1)$ from the term $\int \mathrm{d} x^0 \int \mathrm{d} y^0 (\partial_0^{-1} A^I) j_{qI}$.

From this result it is straightforward to calculate the quantum effective action

$$\Gamma \left(\langle q^I \rangle, \langle p_I \rangle \right) := L_{\text{eff}}^0 \left(j_{qI}, j_p^I \right) - \int \mathrm{d}^2 x \left(\langle q^I \rangle j_{qI} + j_p^I \langle p_I \rangle \right) , \quad (11.49)$$

in terms of the mean fields

$$\langle p_I \rangle := \left. \frac{\overrightarrow{\delta}}{\delta j_p^I} L_{\text{eff}}^0 \right|_{j=0} , \quad \langle q^I \rangle := \left. L_{\text{eff}}^0 \frac{\overleftarrow{\delta}}{\delta j_{qI}} \right|_{j=0} = \left. B^I \right|_{j_p=0} . \quad (11.50)$$

By re-expressing all sources in terms of the (classical) target space coordinates it is found[53] that—up to boundary terms—the quantum effective action is nothing but the gauge fixed classical action. Thus similar to FOG[36] two-dimensional dilaton supergravity exhibits local quantum triviality. Still, the quantization procedure is not completely trivial as the generating functional (11.48) is essentially non-local.

11.2.2.2. *Quantization Including Matter Fields*

Local quantum triviality provides an important consistency check, the main purpose of the nonperturbative quantization procedure is the straightforward way to couple matter fields perturbatively to the fully quantized geometric background.[21,43,44,57] While an immediate extension to supergravity could be expected for the matterless theory on general grounds, this came as a surprise for the theory including matter fields:[47,53] The matter action (11.37) is obtained from the superspace action by a non-trivial integration of auxiliary fields and exactly this step, in the second order formalism, generates quartic ghost terms, which would spoil the nonperturbative quantization procedure. In the first order formulation this does not happen, which makes the full integration over geometry possible in the first place.

From the matter action (11.37) together with the matter fields and momenta, $q = f$, $\mathfrak{p} = \partial L_{(m)}/\partial \dot{q}$ and $\mathfrak{q}^\alpha = \lambda^\alpha$, $\mathfrak{p}_\alpha = \partial L_{(m)}/\partial \dot{\mathfrak{q}}^\alpha$, the Hamiltonian density follows as $H_{(m)} = \dot{q}\mathfrak{p} + \dot{\mathfrak{q}}^+ \mathfrak{p}_+ + \dot{\mathfrak{q}}^- \mathfrak{p}_- - L_{(m)}$. The total Hamiltonian density is the sum of this contribution and (11.21). For the

Poisson bracket of two matter field monomials one finds $\{\mathfrak{q}, \mathfrak{p}'\} = \delta(x - x')$ and $\{\mathfrak{q}^\alpha, \mathfrak{p}'_\beta\} = -\delta^\alpha_\beta \delta(x - x')$. We do not provide the explicit form of the matter Hamiltonian, as it can again be written in terms of secondary constraints, $H = G^I \bar{p}_I$. The constraints itself divide into matter and geometry, $G^I = G^I_{(g)} + G^I_{(m)}$, where $G^I_{(g)}$ has been derived in (11.41) and the matter part reads ($\partial = \partial_1$):

$$
\begin{aligned}
G^{++}_{(m)} &= -\frac{K}{4p_{++}}(\partial\mathfrak{q} - \frac{1}{K}\mathfrak{p})^2 + i(\partial\mathfrak{q} - \frac{1}{K}\mathfrak{p})(\frac{K}{p_{++}}p_+\mathfrak{q}^+ - \frac{K'}{4\sqrt{2}}\frac{p_{--}}{p_{++}}q^- q^-) \\
&\quad + \frac{iK'}{4\sqrt{2}}(\partial\mathfrak{q} + \frac{1}{K}\mathfrak{p})q^+ \mathfrak{q}^+ + \frac{K}{\sqrt{2}}\mathfrak{q}^+\partial\mathfrak{q}^+ - \frac{K'}{2\sqrt{2}}\frac{p_{--}}{p_{++}}p_+ q^- q^- \mathfrak{q}^+ \\
&\quad - p_{--}\Big(\frac{uK'}{4} - \frac{1}{8}(K'' - \frac{K'^2}{K})q^- q^+\Big)q^- \mathfrak{q}^+ ,
\end{aligned}
$$

$$(11.51)$$

$$
\begin{aligned}
G^{--}_{(m)} &= \frac{K}{4p_{--}}(\partial\mathfrak{q} + \frac{1}{K}\mathfrak{p})^2 - i(\partial\mathfrak{q} + \frac{1}{K}\mathfrak{p})(\frac{K}{p_{--}}p_- q^- + \frac{K'}{4\sqrt{2}}\frac{p_{++}}{p_{--}}q^+ \mathfrak{q}^+) \\
&\quad + \frac{iK'}{4\sqrt{2}}(\partial\mathfrak{q} - \frac{1}{K}\mathfrak{p})q^- \mathfrak{q}^- - \frac{K}{\sqrt{2}}\mathfrak{q}^-\partial\mathfrak{q}^- + \frac{K'}{2\sqrt{2}}\frac{p_{++}}{p_{--}}p_- q^+ \mathfrak{q}^- \mathfrak{q}^+ \\
&\quad + p_{++}\Big(\frac{uK'}{4} - \frac{1}{8}(K'' - \frac{K'^2}{K})q^- q^+\Big)q^- \mathfrak{q}^+ ,
\end{aligned}
$$

$$(11.52)$$

$$
G^+_{(m)} = iK(\partial\mathfrak{q} - \frac{1}{K}\mathfrak{p})\mathfrak{q}^+ - \frac{K'}{2\sqrt{2}}p_{--}q^- \mathfrak{q}^- \mathfrak{q}^+ ,
$$

$$(11.53)$$

$$
G^-_{(m)} = -iK(\partial\mathfrak{q} + \frac{1}{K}\mathfrak{p})\mathfrak{q}^- + \frac{K'}{2\sqrt{2}}p_{++}\mathfrak{q}^+ q^- \mathfrak{q}^+ .
$$

$$(11.54)$$

As the kinetic term of the matter fermion λ is first order only, this part of the action leads to constraints as well. From $\mathfrak{p}_+ = -Kp_{++}\mathfrak{q}^+/\sqrt{2}$ and $\mathfrak{p}_- = Kp_{--}\mathfrak{q}^-/\sqrt{2}$ the usual primary second-class constraints are deduced:

$$
\Psi_+ = \mathfrak{p}_+ + \frac{K}{\sqrt{2}}p_{++}\mathfrak{q}^+ \approx 0, \qquad \Psi_- = \mathfrak{p}_- - \frac{K}{\sqrt{2}}p_{--}\mathfrak{q}^- \approx 0. \qquad (11.55)
$$

These second class constraints are treated by substituting the Poisson bracket by the "Dirac bracket"[89] $\{f, g\}^* = \{f, g\} - \{f, \Psi_\alpha\}C^{\alpha\beta}\{\Psi_\beta, g\}$, where $C^{\alpha\beta}C_{\beta\gamma} = \delta^\alpha_\gamma$ and $C_{\alpha\beta} = \{\Psi_\alpha, \Psi_\beta\}$. Despite the complexity of these expressions it can be shown[47] that the secondary constraints with respect to the Dirac bracket obey an algebra of the type (11.20). In the case of minimal coupling, $K(\phi) = 1$, the structure functions of the matterless theory (11.20) are reproduced,[53] while in the generic case additional matter contributions to the structure functions arise, explicit expression can be found in Ref. 47.

The quantization follows the same steps as in the matterless case. As most important observation a lengthy calculation unravels that the homological perturbation theory again stops at first order,[47,53] with the BRST charge as given in Eq. (11.42). Furthermore it is again found that despite the different non-linearities the theory does not exhibit ordering problems.[47,53] Thanks to this unexpectedly simple result we can pursue to formulate a path integral along the same lines as for the matterless case. In the same temporal gauge as used there the gauge fixed Lagrangian becomes

$$
\begin{aligned}
L_{\text{g.f.}} = {}& \dot{q}^I p_I + \dot{\mathsf{q}}\mathsf{p} + \dot{\mathsf{q}}^\alpha \mathsf{p}_\alpha + p_c^I \dot{c}_I - i P^{++|J} p_J - i(-1)^K p_c^I C_I^{++|K} c_K \\
& + \frac{i}{4}\frac{K}{p_{++}}(\partial\mathsf{q} - \frac{1}{K}\mathsf{p})^2 + (\partial\mathsf{q} - \frac{1}{K}\mathsf{p})(\frac{K}{p_{++}}p_+\mathsf{q}^+ - \frac{K'}{4\sqrt{2}}\frac{p_{--}}{p_{++}}q^- q^-) \\
& + \frac{K'}{4\sqrt{2}}(\partial\mathsf{q} + \frac{1}{K}\mathsf{p})q^+\mathsf{q}^+ - \frac{i}{\sqrt{2}}K\mathsf{q}^+\partial\mathsf{q}^+ + \frac{i}{2\sqrt{2}}K'\frac{p_{--}}{p_{++}}p_+ q^- q^- \mathsf{q}^+ + \\
& + i p_{--}\Big(\frac{uK'}{4} - \frac{1}{8}(K'' - \frac{K'^2}{K})\Big)q^- q^+\Big)\mathsf{q}^- \mathsf{q}^+ . \quad (11.56)
\end{aligned}
$$

As in the matterless case the path integrals of $\bar{q}^I$, $\bar{p}_I$ are trivial and the ghosts just yield the super-determinant $\operatorname{sdet} M_I{}^J \doteq \operatorname{sdet}(\delta_I{}^J \partial_0 + i f_I^{++|J})$. The fermionic momenta p_α by means of the constraint (11.55) are integrated trivially as well, while this is possible for p after a quadratic completion. The ensuing determinant can be absorbed by the re-definition of the path-integral measure of q and q^α with correct superconformal properties.[90,91] This yields the effective matter Lagrangian

$$
\begin{aligned}
L_{(m)} = {}& iKp_{++}\dot{\mathsf{q}}^2 - \frac{K}{\sqrt{2}}p_{++}\dot{\mathsf{q}}^+\mathsf{q}^+ + \frac{K}{\sqrt{2}}p_{--}\dot{\mathsf{q}}^-\mathsf{q}^- \\
& + \dot{\mathsf{q}}\Big(K\partial\mathsf{q} - 2iKp_+\mathsf{q}^+ + \frac{i}{2\sqrt{2}}K'(p_{--}q^- q^- + p_{++}q^+\mathsf{q}^+)\Big) \\
& + \frac{K'}{2\sqrt{2}}\partial\mathsf{q}q^+\mathsf{q}^+ - \frac{i}{\sqrt{2}}K\mathsf{q}^+\partial\mathsf{q}^+ \\
& + i p_{--}\Big(\frac{uK'}{4} - \frac{1}{8}(K'' - \frac{1}{2}\frac{K'^2}{K})q^- q^+\Big)\mathsf{q}^- \mathsf{q}^+ . \quad (11.57)
\end{aligned}
$$

Since this expression is again linear in p_I all geometric variables can be integrated out and one is left with the integration of the matter variables, which must be treated perturbatively. It should be noted that the above Lagrangian explicitly depends on the prepotential u as a consequence of the elimination of (superspace) auxiliary fields. This is different than in all bosonic models, where a strict separation of the potentials appearing in the geometric part (V and U) and the one of the matter extension, K, occurs.

Having performed the remaining integrals of all geometric quantities one is left with a path integral in the matter fields q and q^α:

$$\mathcal{W}[\mathcal{J}] = \int \mathcal{D}(\mathsf{q}, \mathsf{q}^\alpha) \exp\left[i \int \mathrm{d}^2 x \left(K(\dot{\mathsf{q}}\partial\mathsf{q} - \frac{i}{\sqrt{2}}\mathsf{q}^+\partial\mathsf{q}^+) \right. \right.$$
$$\left. \left. + \frac{K'}{2\sqrt{2}}\partial\mathsf{q}\mathsf{q}^+\partial\mathsf{q}^+ + B^I j_{qI} + \tilde{L}(j_p^I, B^I) + \mathsf{q}J + \mathsf{q}^\alpha J_\alpha \right) \right] . \quad (11.58)$$

Here, the B^I are solutions to the functional δ functions

$$\dot{q}^\phi = q_{(g)}^\phi , \qquad \dot{q}^{++} = q_{(g)}^{++} - iK\dot{\mathsf{q}}^2 + \frac{K}{\sqrt{2}}\dot{\mathsf{q}}^+\mathsf{q}^+ - i\frac{K'}{2\sqrt{2}}\dot{\mathsf{q}}\mathsf{q}^+\mathsf{q}^+ , \quad (11.59)$$

$$\dot{q}^{--} = q_{(g)}^{--} - \frac{K}{\sqrt{2}}\dot{\mathsf{q}}^-\mathsf{q}^- - i\frac{K'}{2\sqrt{2}}\dot{\mathsf{q}}\mathsf{q}^-\mathsf{q}^-$$
$$\qquad\qquad - i\left(\frac{uK'}{4} - \frac{1}{8}\left(K'' - \frac{1}{2}\frac{K'^2}{K} \right) q^- q^+ \right) q^- \mathsf{q}^+ \qquad\qquad (11.60)$$

$$\dot{q}^+ = q_{(g)}^+ - 2iK\dot{\mathsf{q}}\mathsf{q}^+ , \qquad \dot{q}^- = q_{(g)}^- , \qquad (11.61)$$

with the $q_{(g)}^I$ being the right hand sides of Eqs. (11.45)–(11.47).

11.2.2.3. *Four-Point Vertices*

From the matter Lagrangian (11.57) and the differential equations (11.59)–(11.61) it is possible to derive the non-local vertices of matter to lowest order (tree level.) These results were presented in Ref. 53 for minimal coupling, $K(\phi) = 1$, here we derive the more general result for non-minimal couplings. In this calculation the concept of localized matter[44,58,59] is used, here we follow the notation of Ref. 53 and define

$$\Phi_i(x) = \frac{1}{2}\partial_i\mathsf{q}(x)\partial_0\mathsf{q}(x) \Rightarrow a_i\delta^2(x - y) , \quad (11.62)$$

$$\Psi_i^{\pm\pm}(x) = \frac{1}{2}\partial_i\mathsf{q}^\pm(x)\mathsf{q}^\pm(x) \Rightarrow b_i^{\pm\pm}\delta^2(x - y) , \quad (11.63)$$

$$\Pi_i^\pm(x) = \partial_i\mathsf{q}(x)\mathsf{q}^\pm(x) \Rightarrow c_i^\pm\delta^2(x - y) , \quad (11.64)$$

$$\Lambda(x) = \mathsf{q}^-\mathsf{q}^+(x) \Rightarrow e\delta^2(x - y) . \quad (11.65)$$

Notice that with our choice of gauge only Ψ_1^{++} and Π_1^+, but no terms in Ψ_1^{--} or Π_1^- appear in the interaction. Furthermore, due to supersymmetry a_0 and b_0^{++} only appear in the linear combination

$$A_0 = 2ia_0 - \sqrt{2}b_0^{++} , \quad (11.66)$$

which thus will be used as abbreviation in the following. Thanks to the local quantum triviality of the matterless theory, the lowest order vertices can be determined from the matter interaction terms in the gauge-fixed Lagrangian (11.57) by solving to first order in localized matter the *classical* equations of motion of the geometrical variables involved.[21,44,58] To this end the asymptotic integration constants must be chosen in a convenient way. Following the calculations of the purely bosonic case[44,58] $q^\phi(x^0 \to \infty) = x^0$, which implies $q^{++}(x^0 \to \infty) = i$. In addition $p_{--}(x^0 \to \infty) = ie^Q$ may now be imposed. Finally we have to fix the asymptotic value of the Casimir function (11.35), $C(x^0 \to \infty) = C_\infty$. Due to the matter interactions $dC \neq 0$, but the conservation law receives contributions from the matter fields as well.[83] Finally, considering the asymptotic values of the fermions q^+ and q^- we follow Ref. 53 and set $q^+_\infty = q^-_\infty = 0$. This considerably reduces the complexity of the equations of motion since all contributions quadratic in the fermions vanish as they are second order in localized matter.

It turns out that the gauge fixed equations with the above choice of the asymptotic values can be solved explicitly to first order in localized matter. Therefore all non-local four-point vertices can be evaluated exactly to lowest order. To economize writing of the subsequent non-local quantities we introduce the notations

$$[fg]_{x^0} = f(x^0)g(x^0) \,, \qquad [fg]_{x^0 \pm y^0} = [fg]_{x^0} \pm [fg]_{y^0} \,. \qquad (11.67)$$

Furthermore, the abbreviation $h_{xy} = \theta(y^0 - x^0)\delta(x^1 - y^1)$ is used. We do not repeat the complete solution here, as an example the Casimir function gets the new contributions

$$C = -m_\infty + \left[iA_0\left(m_\infty[K]_{y^0} + [Kw]_{y^0}\right) \right.$$
$$\left. + \sqrt{2}i[e^Q K]_{y^0}b_0^{--} - \frac{1}{4}[e^Q uK]_{y^0}e\right]h_{xy} \,. \qquad (11.68)$$

m_∞ is the integration constant of $q^{--} = ie^{-Q}m_\infty + \ldots$, which, however, turns out to be equivalent to the asymptotic value $-C_\infty$. Notice that all contributions except this integration constant are proportional to h and thus to first order in localized matter all geometric variables may be replaced by their asymptotic values in that equation. Of course, this is equivalent to the statement that C is constant in the absence of matter fields.

For the explicit expressions of the vertices one derives the relevant interaction terms from (11.57) as

$$L_{(m)} = (2i\Phi_0 - \sqrt{2}\Psi_0^{++})Kp_{++} + \sqrt{2}Kp_{--}\Psi_0^{--} + 2K\Phi_1$$
$$+ (\frac{i}{2\sqrt{2}}K'p_{++}q^+ - 2iKp_+)\Psi_0^+ + \frac{i}{2\sqrt{2}}K'p_{--}q^-\Psi_0^-$$
$$+ \frac{K'}{2\sqrt{2}}q^+\Pi_1^+ + \sqrt{2}iK\Psi_1^{++} + \frac{i}{4}uK'p_{--}\Lambda \ . \quad (11.69)$$

With the solution obtained one now finds that the vertices depend on seven different functions:

$$V_1(x,y) = -K(x^0)K(y^0)\left(2[w]_{x^0-y^0} - (x^0-y^0)\left([w']_{x^0+y^0}\right.\right.$$
$$\left.\left. + \left[\frac{K'}{K}(w+m_\infty)\right]_{x^0+y^0}\right)\right)h_{xy} \ , \quad (11.70)$$

$$V_2(x,y) = -\sqrt{2}iK(x^0)K(y^0)\left(\left[\sqrt{-w'}\right]_{x^0-y^0}\left[\sqrt{-w}\right]_{x^0-y^0}\right.$$
$$\left. - \frac{1}{2}\left[\frac{K'}{K}(w+m_\infty)\right]_{x^0+y^0}\right)h_{xy} \ , \quad (11.71)$$

$$V_3(x,y) = 2iK(x^0)K'(y^0)|x^0-y^0|\delta(x^1-y^1) \ , \quad (11.72)$$

$$V_4(x,y) = -\frac{i}{4}K(x^0)\left[e^Q uK'\right]'_{y^0}|x^0-y^0|\delta(x^1-y^1) \ , \quad (11.73)$$

$$V_5(x,y) = \sqrt{2}K(x^0)\left[e^Q K'\right]'_{y^0}|x^0-y^0|\delta(x^1-y^1) \ , \quad (11.74)$$

$$V_6(x,y) = -\frac{1}{\sqrt{2}}K(x^0)\left[\sqrt{-w}\right]_{x^0-y^0}\left[e^{\frac{Q}{2}}K'\right]_{y^0}(h_{xy}-h_{yx}) \ , \quad (11.75)$$

$$V_7(x,y) = -\frac{i}{\sqrt{2}}K(x^0)K'(y^0)(h_{xy}-h_{yx}) \ . \quad (11.76)$$

The full interaction vertices from these functions are obtained as[21,53,58]

$$\mathcal{V} = \int_x \int_y \Xi_i(x)\Xi_j(y)V_{ij}(x,y) \ , \quad (11.77)$$

where Ξ_i is one of the contributions from localized matter according to (11.62)–(11.65) and V_{ij} is—up to eventual constants—the vertex function from (11.70)–(11.76) that describes the interaction between Ξ_i and Ξ_j.

V_1 determines the interaction of

$$\dot{q}\dot{q}(x) \to \dot{q}\dot{q}(y) = -4V_1(x,y) \;, \tag{11.78}$$

$$\dot{q}^+q^+(x) \to \dot{q}^+q^+(y) = 2V_1(x,y) \;, \tag{11.79}$$

$$\dot{q}\dot{q}(x) \to \dot{q}^+q^+(y) = -2\sqrt{2}iV_1(x,y) \;, \tag{11.80}$$

while V_2 yields

$$\dot{q}q^+(x) \to \dot{q}q^+(y) = V_2(x,y) \;. \tag{11.81}$$

These are the only vertices that do not vanish for minimal coupling,[53] $K = 1$. Both of them are conformally invariant, but while while V_1 vanishes at $x^0 = y^0$, V_2 does not unless $K = 1$, which thus yields a local four-point interaction for non-minimal coupling. The vertex function V_1 with minimal coupling vanishes for models with $w \propto \phi$, in particular for the CGHS model.[92] Since V_1 is the only vertex function of bosonic models with minimal couplings, the CGHS model exhibits scattering triviality to this order. This does not apply to the supersymmetric extension of the CGHS model[63] since V_2 does not vanish here.[53]

V_3 appears in two different interaction terms in (11.69) leading to the four vertices

$$\dot{q}\dot{q}(x) \to \partial q\dot{q}(y) = 2iV_3(x,y) \;, \qquad \dot{q}^+q^+(x) \to \partial q\dot{q}(y) = -\sqrt{2}V_3(x,y) \;, \tag{11.82}$$

$$\dot{q}\dot{q}(x) \to \partial q^+q^+ = 2\sqrt{2}V_3(x,y) \;, \qquad \dot{q}^+q^+(x) \to \partial q^+q^+ = 2iV_3(x,y) \;. \tag{11.83}$$

The remaining two interactions with $(2i\Phi_0 - \sqrt{2}\Psi_0^{++})$ as initial or final state are

$$\dot{q}\dot{q}(x) \to q^-q^+(y) = 2iV_4(x,y) \;, \qquad \dot{q}^+q^+(x) \to q^-q^+(y) = -\sqrt{2}V_4(x,y) \;, \tag{11.84}$$

$$\dot{q}\dot{q}(x) \to \dot{q}^-q^-(y) = 2iV_5(x,y) \;, \qquad \dot{q}^+q^+(x) \to \dot{q}^-q^-(y) = -\sqrt{2}V_5(x,y) \;. \tag{11.85}$$

Notice that these vertices are not conformally invariant° as they cannot be written exclusively in terms of the conformally invariant potential w.

°It might come as a surprise that a conformally invariant model generates vertices which are not invariant under this transformation. Nonetheless, it should be remembered that conformal invariance applies to the complete Lagrangian, here (11.69), including the (asymptotic) matter states.[48,61]

Finally, for non-minimal coupling besides (11.81) there exist two additional vertices with mixed bosonic/fermionic initial and final states,

$$\dot{\mathsf{q}}\mathsf{q}^+(x) \to \dot{\mathsf{q}}\mathsf{q}^-(y) = V_6(x,y) \,, \tag{11.86}$$

$$\dot{\mathsf{q}}\mathsf{q}^+(x) \to \partial\mathsf{q}\mathsf{q}^-(y) = V_7(x,y) \,. \tag{11.87}$$

V_6 is not conformally invariant but zero for $x^0 = y^0$, while V_7 exactly behaves the other way around.

11.3. Two-Dimensional Dilaton Gravity with Boundaries

In all calculations of the previous sections boundary terms were assumed to vanish and asymptotically the values of the target space variables ϕ and X^I were fixed, which removes eventual boundary degrees of freedom. However, in many applications this setup is not suitable and a careful treatment of boundary terms is necessary. Already in Ref. 32 it was found by Wolfgang Kummer and Stephen Lau that the action (11.4) should be complemented by the boundary term

$$\mathcal{S}_{\text{boundary}} = \int_{\partial\mathcal{M}} \left(X\omega + \frac{1}{2}X\mathrm{d}\ln\left|\frac{e_{\parallel}^+}{e_{\parallel}^-}\right| \right) \,. \tag{11.88}$$

in order to make the theory globally equivalent to the model (11.6) complemented with the standard York-Gibbons-Hawking boundary term.[93,94] If Dirichlet boundary conditions are chosen for X, $e_{\parallel}^+$ and $e_{\parallel}^-$ the second term in Eq. (11.88) is formulated exclusively in terms of fields fixed at the boundary. However, this term is essential to restore invariance under unrestricted Lorentz transformations.[p]

In one of his last publications[97] Wolfgang Kummer resumed the discussion of boundary terms in FOG from a quite different point of view. This work was motivated by results from Refs. 98, 99 where it was argued that black hole entropy should emerge from Goldstone-like degrees of freedom that emerge from a symmetry breaking in the presence of a horizon.[q] In these works a stretched horizon was imposed as a boundary, implemented

---

[p]In order to ensure a well-definedness of the semiclassical approximation and thus of thermodynamics of black hole spacetimes, further boundary counterterms, solely depending on the boundary values of the fields held fixed there, can be important. In the Euclidean approach these counterterms have been discussed in Refs. 95, 96.

[q]This is thought to be a spontaneous symmetry breaking happening in the full dynamical theory. However, because of the inability to treat the fully dynamical picture, the investigation of Carlip's idea is done by implementing the symmetry breaking explicitly through boundary constraints.

by suitable boundary constraints. It was then found that these constraints break parts of the symmetry, which allowed to deduce the correct entropy by means of the Cardy formula.[100] Since the formalism of Refs. 98, 99 does not allow to impose sharp horizon constraints, it however remained open in which sense the stretched horizon really is a special choice of a boundary. In FOG the Eddington-Finkelstein type solutions are not singular at the horizon and thus the first order formulation provides the possibility to replace the stretched horizon in Refs. 98, 99 by a true horizon. This led to the idea to study FOG with boundaries, once chosen as a generic boundary and once chosen as a horizon, and to compare these two situations.

In Ref. 97 the boundary was considered at a fixed value of x^1, whereby x^0 represents Hamiltonian time. This choice allowed to implement the specific values of the fields fixed at the boundary as boundary constraints, which then were introduced in the Hamiltonian analysis. Since the dilaton is constant at the horizon, the first boundary constraint was chosen as $\hat{B}_1[\eta] = (p_1 - p_1^b)\eta|_{\partial\mathcal{M}}$. A generic boundary is determined by the two additional constraints (η is a smearing function)

$$\hat{B}_2[\eta] = (\bar{q}_2 - E_0^-(x^0))\eta|_{\partial\mathcal{M}} , \quad \hat{B}_3[\eta] = (\bar{q}_3 - E_0^+(x^0))\eta|_{\partial\mathcal{M}} , \quad (11.89)$$

which turns all secondary constraints G^I into second class constraints. On-shell (11.89) with the choice $E_0^-(x^0) \equiv 0$ could be used to fix the boundary to be a horizon. However, off-shell this choice is problematic since the Killing norm expressed in terms of target space variables, Eq. (11.11), need not vanish and not surprisingly it was found that the Hamiltonian treatment of (11.89) becomes singular at the horizon. Still, the first order formulation offers a different set of horizon constraints, namely

$$B_2[\eta] = \bar{q}_2\eta|_{\partial\mathcal{M}} , \qquad B_3[\eta] = p_3\eta|_{\partial\mathcal{M}} , \qquad (11.90)$$

which removes all the problems encountered with (11.89). As an important difference to the generic boundary it is now found that two of the three secondary constraints, namely the Lorentz constraints and diffeomorphisms along the boundary, remain first class. This picture was confirmed by constructing the reduced phase space. Both situations have zero physical degrees of freedom in the bulk, but while a generic boundary exhibits one pair of boundary degrees of freedom (which could be related to mass and proper time as previously found by Kuchař[101]), no boundary degrees of freedom are left at the horizon. This suggests a quite different picture of black hole entropy:[97,102] The physical degrees of freedom present on a generic boundary are converted into gauge degrees of freedom on a horizon and entropy arises because approaching the black hole horizon does not commute with constructing the physical phase space.

Already during the preparation of Ref. 97 it was realized by Wolfgang Kummer and his collaborators that it could be advantageous to choose the boundary at constant value of Hamiltonian time. Indeed, the extremely complex constraint algebras emerging from the calculations above destroyed any hopes to quantize the model along the lines of Sect. 11.1.3, not to mention the impossibility to couple matter fields. Nonetheless, if in the Hamiltonian picture the boundary is rather chosen as initial or final values, the canonical formalism is not affected at all. However, since for a spacelike boundary the boundary values are no longer fixed via boundary constraints, the necessary restrictions should be obtained from the "lost constraints,"[r] which turned out to be difficult to tackle. Thus this line of investigations was given up in favor of the picture presented in Ref. 97. Only recently, this unfinished work was continued[103] and it was shown that for the matterless theory the result expected from previous works,[97,101] namely the existence of one boundary degree of freedom – the mass – was obtained. In particular, fixing $e_1^{\pm}$ at the boundary instead of $X^{\pm}$ implies for the arguments of the functional δ functions (11.24)-(11.25) not to evaluate the path integral completely, but rather leaving an integration over field boundary values unevaluated. Furthermore, additional contributions from the York-Gibbons-Hawking boundary term made the evaluation the of quantum equations of motion and the identification of the additional boundary degree of freedom possible. The formalism presented in Ref. 103 thus may provide a way to finally do a path integration over the remaining boundary degree of freedom – a real "sum over boundary conditions" labeled by the mass of the spacetime. Though relevant questions regarding the path integral remained open this work, and will hopefully addressed in the future, it shows that Wolfgang Kummer's philosophy of the "Vienna School of dilaton gravity" remains a powerful formalism which also in the future will provide deeper insight into important questions of classical and quantum gravity.

---

[r]Integrating out the p_I reproduces only the equations of motion (11.24)-(11.25). There is another set of these equations with spatial rather than time-derivatives, which should fix the remaining freedom in the choice of certain integration functions (depending on x^1). These are the "lost constraints", which play the role of Ward identities for the diffeomorphism and local Lorentz invariance. See also Ref. 57 and references therein.

Appendix A. Notations and conventions

Most of the notation follows the one used in Refs. 35, 68, which should be consulted for further explanations.

For indices of target-space coordinates and gauge fields the notation

$$X^I = (X^i, X^\alpha) = (X^\phi, X^a, X^\alpha) = (\phi, X^a, \chi^\alpha) \,, \tag{A.1}$$

$$A_I = (A_i, A_\alpha) = (A_\phi, A_a, A_\alpha) = (\omega, e_a, \psi_\alpha) \,, \tag{A.2}$$

i.e. capital Latin indices are generic, $i, j, k \ldots$ are bosonic, $a, b, c \ldots$ denote the anholonomic coordinates and Greek indices are fermionic. The summation convention is always $NW \to SE$, e.g. for a fermion χ: $\chi^2 = \chi^\alpha \chi_\alpha$. Our conventions are arranged in such a way that almost every bosonic expression is transformed trivially to the graded case when using this summation convention and replacing commuting indices by general ones. This is possible together with exterior derivatives acting *from the right*, only. Thus the graded Leibniz rule is given by

$$\mathrm{d}\,(AB) = A\mathrm{d}B + (-1)^B (\mathrm{d}A)B \,. \tag{A.3}$$

In terms of anholonomic indices the metric and the symplectic 2×2 tensor are defined as

$$\eta_{ab} = \begin{pmatrix} 1 & 0 \\ 0 & -1 \end{pmatrix} \,, \quad \epsilon_{ab} = -\epsilon^{ab} = \begin{pmatrix} 0 & 1 \\ -1 & 0 \end{pmatrix} \,, \quad \epsilon_{\alpha\beta} = \epsilon^{\alpha\beta} = \begin{pmatrix} 0 & 1 \\ -1 & 0 \end{pmatrix} \,. \tag{A.4}$$

The metric in terms of holonomic indices is obtained by $g_{mn} = e_n^b e_m^a \eta_{ab}$ and for the determinant the standard expression $e = \det e_m^a = \sqrt{-\det g_{mn}}$ is used. The volume form reads $\epsilon = \frac{1}{2}\epsilon^{ab} e_b \wedge e_a$; by definition $*\epsilon = 1$.

The γ-matrices used are in a chiral representation:

$$\gamma^0{}_\alpha{}^\beta = \begin{pmatrix} 0 & 1 \\ 1 & 0 \end{pmatrix} \quad \gamma^1{}_\alpha{}^\beta = \begin{pmatrix} 0 & 1 \\ -1 & 0 \end{pmatrix} \quad \gamma^3{}_\alpha{}^\beta = (\gamma^1\gamma^0)_\alpha{}^\beta = \begin{pmatrix} 1 & 0 \\ 0 & -1 \end{pmatrix} \tag{A.5}$$

Covariant derivatives of anholonomic indices with respect to the geometric variables $e_a = \mathrm{d}x^m e_{am}$ and $\psi_\alpha = \mathrm{d}x^m \psi_{\alpha m}$ include the two-dimensional spin-connection one form $\omega^{ab} = \omega\epsilon^{ab}$. When acting on lower indices the explicit expressions read ($\frac{1}{2}\gamma^3$ is the generator of Lorentz transformations in spinor space):

$$(De)_a = \mathrm{d}e_a + \omega\epsilon_a{}^b e_b \qquad (D\psi)_\alpha = \mathrm{d}\psi_\alpha - \frac{1}{2}\omega\gamma^3{}_\alpha{}^\beta \psi_\beta \tag{A.6}$$

For Majorana spinors in chiral representation,

$$\chi^\alpha = (\chi^+, \chi^-) \,, \qquad\qquad \chi_\alpha = \begin{pmatrix} \chi_+ \\ \chi_- \end{pmatrix} \,, \qquad (A.7)$$

upper and lower chiral components are related by $\chi^+ = \chi_-$, $\chi^- = -\chi_+$, $\chi^2 = \chi^\alpha \chi_\alpha = 2\chi_- \chi_+$. Vectors conveniently are used in the spin tensor decomposition $v^{\alpha\beta} = \frac{i}{\sqrt{2}} v^c \gamma_c^{\alpha\beta}$. Due to the additional factor i the spin tensor components are related to standard light cone components as

$$v^{++} = iv^\oplus \,, \qquad\qquad v^{--} = -iv^\ominus \,, \qquad (A.8)$$

in particular the spin tensor components of a real vector are imaginary. This notation implies that $\eta_{++|--} = 1$, $\epsilon_{--|++} = -\epsilon_{++|--} = 1$ and for the γ matrices one finds

$$(\gamma^{++})_\alpha{}^\beta = \sqrt{2}i \begin{pmatrix} 0 & 1 \\ 0 & 0 \end{pmatrix} \,, \qquad (\gamma^{--})_\alpha{}^\beta = -\sqrt{2}i \begin{pmatrix} 0 & 0 \\ 1 & 0 \end{pmatrix} \,. \qquad (A.9)$$

References

1. W. Kummer and D. J. Schwarz, *Czech. J. Phys.* **41**, 13–22, (1991).
2. W. Kummer and D. J. Schwarz. NonEinsteinian gravity with torsion at d = 2. In *Strings and Symmetries, 1991: SUNY Stony Brook, 20-25 May 1991*, pp. 168–169. World Scientific, (1992).
3. W. Kummer and D. J. Schwarz, *Phys. Rev.* **D45**, 3628–3635, (1992).
4. W. Kummer and D. J. Schwarz, *Nucl. Phys.* **B382**, 171–186, (1992).
5. M. O. Katanaev and I. V. Volovich, *Phys. Lett.* **B175**, 413–416, (1986).
6. M. O. Katanaev and I. V. Volovich, *Ann. Phys.* **197**, 1, (1990).
7. M. O. Katanaev, *J. Math. Phys.* **31**, 882, (1990).
8. W. Kummer, *Acta Phys. Austriaca* **41**, 315–334, (1975).
9. H. Grosse, W. Kummer, P. Presnajder, and D. J. Schwarz, *Czech. J. Phys.* **42**, 1325–1329, (1992).
10. W. Kummer and D. J. Schwarz, *Class. Quant. Grav.* **10**, S235–S238, (1993).
11. H. Grosse, W. Kummer, P. Presnajder, and D. J. Schwarz, *J. Math. Phys.* **33**, 3892–3900, (1992).
12. W. Kummer. Deformed ISO(2,1) symmetry and non-Einsteinian 2d-gravity with matter. In eds. D. Bruncko and J. Urban, *Hadron Structure '92*, (1992). Stara Lesna, Czechoslovakia.
13. H. Kawai and R. Nakayama, *Phys. Lett.* **B306**, 224–232, (1993).
14. T. Strobl, *Int. J. Mod. Phys.* **D3**, 281–284, (1994).
15. D. Cangemi and R. Jackiw, *Phys. Rev. Lett.* **69**, 233–236, (1992).
16. H. Verlinde. Black holes and strings in two dimensions. In *Trieste Spring School on Strings and Quantum Gravity*, pp. 178–207 (April, 1991).
17. C. Teitelboim, *Phys. Lett.* **B126**, 41, (1983).

18. R. Jackiw. Liouville field theory: a two-dimensional model for gravity. In ed. S. Christensen, *Quantum theory of gravity : essays in honor of the 60th birthday of Bryce S.DeWitt*, pp. 327–344, Bristol, (1984). Hilger.

19. N. Ikeda and K. I. Izawa, *Prog. Theor. Phys.* **89**, 223–230, (1993).

20. N. Ikeda and K. I. Izawa, *Prog. Theor. Phys.* **89**, 1077–1086, (1993).

21. D. Grumiller, W. Kummer, and D. V. Vassilevich, *Phys. Rept.* **369**, 327, (2002).

22. W. Kummer and P. Widerin, *Mod. Phys. Lett.* **A9**, 1407–1414, (1994).

23. M. O. Katanaev, *Nucl. Phys.* **B416**, 563–605, (1994).

24. W. Kummer. Exact classical and quantum integrability of R**2 + T**2 gravity in (1+1) dimensions. In eds. J. Carr and M. Perrottet, *International Europhysics Conference On High-Energy Physics (HEP93)*. Editions Frontieres, (1993).

25. F. Haider and W. Kummer, *Int. J. Mod. Phys.* **A9**, 207–220, (1994).

26. P. Schaller and T. Strobl, *Class. Quant. Grav.* **11**, 331–346, (1994).

27. W. Kummer. Unified treatment of all 1+1 dimensional gravitation models. In eds. J. Lemonne, C. Vander Velde, and F. Verbeure, *International Europhysics Conference On High Energy Physics (HEP 95)*. World Scientific, (1995).

28. W. Kummer and P. Widerin, *Phys. Rev.* **D52**, 6965–6975, (1995).

29. W. Kummer. General treatment of all 2d covariant models. In ed. S. Moskaliuk, *12th Hutsulian Workshop On Methods Of Mathematical Physics*. Hadronic Press, (1995).

30. M. O. Katanaev, W. Kummer, and H. Liebl, *Phys. Rev.* **D53**, 5609–5618, (1996).

31. M. O. Katanaev, W. Kummer, and H. Liebl, *Nucl. Phys.* **B486**, 353–370, (1997).

32. W. Kummer and S. R. Lau, *Annals Phys.* **258**, 37–80, (1997).

33. T. Strobl. *Poisson structure induced field theories and models of 1+1 dimensional gravity*. PhD thesis, Technische Universität Wien, (1994).

34. T. Strobl. Gravity in two spacetime dimensions. Habilitation thesis, (1999).

35. M. Ertl, W. Kummer, and T. Strobl, *JHEP.* **01**, 042, (2001).

36. W. Kummer, H. Liebl, and D. V. Vassilevich, *Nucl. Phys.* **B493**, 491–502, (1997).

37. H. Balasin, C. G. Boehmer, and D. Grumiller, *Gen. Rel. Grav.* **37**, 1435–1482, (2005).

38. T. Klösch and T. Strobl, *Class. Quant. Grav.* **13**, 965–984, (1996).

39. T. Klösch and T. Strobl, *Class. Quant. Grav.* **13**, 2395–2422, (1996).

40. N. Ikeda, *Ann. Phys.* **235**, 435–464, (1994).

41. P. Schaller and T. Strobl. Poisson sigma models: A generalization of 2-d gravity Yang-Mills systems. In *Finite dimensional integrable systems*, pp. 181–190, (1994). Dubna.

42. P. Schaller and T. Strobl, *Mod. Phys. Lett.* **A9**, 3129–3136, (1994).

43. W. Kummer, H. Liebl, and D. V. Vassilevich, *Nucl. Phys.* **B513**, 723–734, (1998).

44. W. Kummer, H. Liebl, and D. V. Vassilevich, *Nucl. Phys.* **B544**, 403–431, (1999).
45. P. Fischer, D. Grumiller, W. Kummer, and D. V. Vassilevich, *Phys. Lett.* **B521**, 357–363, (2001). Erratum ibid. **B532** (2002) 373.
46. D. Grumiller, *Class. Quant. Grav.* **19**, 997–1009, (2002).
47. L. Bergamin. Quantum dilaton supergravity in 2d with non-minimally coupled matter. In eds. P. Fiziev and M. Todorov, *Gravity, Astrophysics, and Strings @ the Black Sea*, pp. 17–28, Sofia, (2005). St.Kliment Ohridski University Press.
48. D. Grumiller and R. Meyer, *Class. Quant. Grav.* **23**, 6435–6458, (2006).
49. R. Meyer. Constraints in two-dimensional dilaton gravity with fermions. To appear in the proceedings of International V.A. Fock School of Advances of Physics (IFSAP 2005), St. Petersburg, Russia, 21-27 Nov 2005., (2005).
50. R. Meyer. Classical and quantum dilaton gravity in two dimensions with fermions. Master's thesis, (2006).
51. R. Meyer. Quantizing two-dimensional dilaton gravity with fermions: The Vienna way. In eds. H. Kleinert and R. Jantzen, *The eleventh Marcel Grossman meeting.* pp. 2698–2700. World Scientific, (2008).
52. L. Bergamin and W. Kummer, *JHEP.* **05**, 074, (2003).
53. L. Bergamin, D. Grumiller, and W. Kummer, *JHEP.* **05**, 060, (2004).
54. E. S. Fradkin and G. A. Vilkovisky, *Phys. Lett.* **B55**, 224, (1975).
55. E. S. Fradkin and T. E. Fradkina, *Phys. Lett.* **B72**, 343, (1978).
56. I. A. Batalin and G. A. Vilkovisky, *Phys. Lett.* **B69**, 309–312, (1977).
57. D. Grumiller. *Quantum dilaton gravity in two dimensions with matter.* PhD thesis, Technische Universität Wien, (2001).
58. D. Grumiller, W. Kummer, and D. V. Vassilevich, *Eur. Phys. J.* **C30**, 135–143, (2003).
59. D. Grumiller, W. Kummer, and D. V. Vassilevich, *Nucl. Phys.* **B580**, 438–456, (2000).
60. D. Grumiller, *Int. J. Mod. Phys.* **D13**, 1973–2002, (2004).
61. D. Grumiller and R. Meyer, *Turk. J. Phys.* **30**, 349–378, (2006).
62. P. S. Howe, *J. Phys.* **A12**, 393–402, (1979).
63. Y.-C. Park and A. Strominger, *Phys. Rev.* **D47**, 1569–1575, (1993).
64. D. Grumiller. Three functions in dilaton gravity: The good, the bad and the muggy. Lectures given at 14th International Hutsulian Workshop on Mathematical Theories and their Physical and Technical Applications (Timpani - Mathyphys 2002), Chernivtsi, Ukraine, (2002).
65. A. Bilal, *Phys. Rev.* **D48**, 1665–1678, (1993).
66. S. Nojiri and I. Oda, *Mod. Phys. Lett.* **A8**, 53–62, (1993).
67. M. F. Ertl, M. O. Katanaev, and W. Kummer, *Nucl. Phys.* **B530**, 457–486, (1998).
68. M. Ertl. *Supergravity in two spacetime dimensions.* PhD thesis, Technische Universität Wien, (2001).
69. V. O. Rivelles, *Phys. Lett.* **B321**, 189–192, (1994).
70. D. Cangemi and M. Leblanc, *Nucl. Phys.* **B420**, 363–378, (1994).
71. N. Ikeda, *Int. J. Mod. Phys.* **A9**, 1137–1152, (1994).

72. J. M. Izquierdo, *Phys. Rev.* **D59**, 084017, (1999).
73. T. Strobl, *Phys. Lett.* **B460**, 87–93, (1999).
74. L. Bergamin, *Mod. Phys. Lett.* **A20**, 985–996, (2005).
75. I. Calvo, *Lett. Math. Phys.* **77**, 53–62, (2006).
76. T. Strobl, *Commun. Math. Phys.* **246**, 475–502, (2004).
77. M. Adak and D. Grumiller, *Class. Quant. Grav.* **24**, F65, (2007).
78. D. Z. Freedman, P. van Nieuwenhuizen, and S. Ferrara, *Phys. Rev.* **D13**, 3214–3218, (1976).
79. D. Z. Freedman and P. van Nieuwenhuizen, *Phys. Rev.* **D14**, 912, (1976).
80. S. Deser and B. Zumino, *Phys. Lett.* **B62**, 335, (1976).
81. R. Grimm, J. Wess, and B. Zumino, *Phys. Lett.* **B73**, 415, (1978).
82. L. Bergamin and W. Kummer, *Phys. Rev.* **D68**, 104005, (2003).
83. L. Bergamin, D. Grumiller, and W. Kummer, *J. Phys.* **A37**, 3881–3901, (2004).
84. G. W. Gibbons and C. M. Hull, *Phys. Lett.* **B109**, 190, (1982).
85. K. P. Tod, *Phys. Lett.* **B121**, 241–244, (1983).
86. W. M. Nelson and Y. Park, *Phys. Rev.* **D48**, 4708–4712, (1993).
87. L. Bergamin and W. Kummer, *Eur. Phys. J.* **C39**, S41–S52, (2005).
88. L. Bergamin and W. Kummer, *Eur. Phys. J.* **C39**, S53–S63, (2005).
89. P. A. M. Dirac, *Lectures on Quantum Mechanics.* (Belfer Graduate School of Science, Yeshiva University, New York, 1996).
90. M. Rocek, P. van Nieuwenhuizen, and S. C. Zhang, *Ann. Phys.* **172**, 348, (1986).
91. U. Lindstrom, N. K. Nielsen, M. Rocek, and P. van Nieuwenhuizen, *Phys. Rev.* **D37**, 3588, (1988).
92. J. C. G. Callan, S. B. Giddings, J. A. Harvey, and A. Strominger, *Phys. Rev.* **D45**, 1005–1009, (1992).
93. J. W. J. York, *Phys. Rev. Lett.* **28**, 1082–1085, (1972).
94. G. W. Gibbons and S. W. Hawking, *Phys. Rev.* **D15**, 2752–2756, (1977).
95. D. Grumiller and R. McNees, *JHEP.* **04**, 074, (2007).
96. L. Bergamin, D. Grumiller, R. McNees, and R. Meyer, *J. Phys.* **A41**, 164068, (2008).
97. L. Bergamin, D. Grumiller, W. Kummer, and D. V. Vassilevich, *Class. Quant. Grav.* **23**, 3075–3101, (2006).
98. S. Carlip, *Class. Quant. Grav.* **22**, 1303–1312, (2005).
99. S. Carlip, *Int. J. Theor. Phys.* **46** 2192–2203, (2007).
100. J. L. Cardy, *Nucl. Phys.* **B270**, 186–204, (1986).
101. K. V. Kuchař, *Phys. Rev.* **D50**, 3961–3981, (1994).
102. L. Bergamin and D. Grumiller, *Int. J. Mod. Phys.* **D15**, 2279–2284, (2006).
103. L. Bergamin and R. Meyer. Two-dimensional quantum gravity with boundary. In eds. P. Fiziev and M. Todorov, *Gravity, Astrophysics, and Strings @ the Black Sea.* St.Kliment Ohridski University Press, (2008).

Chapter 12

Order and Chaos in Two-Dimensional Gravity

Robert B. Mann

Dept. of Physics & Astronomy,
University of Waterloo,
Waterloo, Ontario, Canada N2L 3G1
E-mail: rbmann@sciborg.uwaterloo.ca

The N-body problem in two-dimensional gravity affords a unique opportunity to explore the interplay between relativistic gravity and chaos. I review the main achievements in the field to date, concentrating on the 2, 3 and N body problems.

12.1. Introduction

For more than 25 years two-dimensional gravity has been a fruitful arena for furthering our understanding of classical and quantum gravity. An extensive body of work[1] has been built up over this time period covering a diverse range of topics including cosmology, black holes, quantum gravity, conformal field theory, string theory, quantum information, the N-body problem, and more. What unifies these diverse topics is the common expectation that the simplicity of two-dimensional gravity will afford significantly more progress in difficult technical matters than is possible in the four-dimensional case. Such progress comes at a price insofar as some key features of four-dimensional gravity – gravitational waves and rotation for example – are lost in stepping down to two dimensions. Theoretical investigations in this subject are pinned on the hope that key conceptual features in many of these problems will carry over into the four-dimensional world, making the price worth paying.

The N-body problem – that of determining the motion of a system of N particles mutually interacting through specified forces – is a case in point. One of the oldest problems in physics, it remains of key importance in nuclear physics, atomic physics, stellar dynamics, and cosmology.[2] For gravitational interactions this problem is notoriously vexing: although an exact solution for pure Newtonian gravity is known in three spatial dimensions for $N = 2$ case, no corresponding solution exists in the general-relativistic case. Energy dissipation via gravitational radiation forces one to resort to approximation schemes. When $N \geq 3$ there are unsolved problems even in the Newtonian case, and consideration of the OGS (One-spatial-dimension self-Gravitating System) has been under investigation for nearly 40 years.[3] Its statistical properties and ergodic behaviour are still not fully understood, the circumstances (if any) under which equipartition of energy can be attained are not clear, it is not known if an OGS can attain true equilibrium state from arbitrary initial conditions, and the appearance of fractal behaviour in such systems came as a surprise.[4] Furthermore, such systems approximate the behaviour of some physical systems in 3 spatial dimensions. For example, long-lived core-halo configurations, resembling structures observed in globular clusters, exist in the OGS phase space,[2] modelling a dense massive core in near-equilibrium, surrounded by a halo of high kinetic energy stars that interact only weakly with the core. Other examples include collisions of flat parallel domain walls moving in directions orthogonal to their surfaces and stellar dynamics in a direction orthogonal to the plane of a highly flattened galaxy.

The relativistic OGS, or ROGS, has been an active area in my research group for the past 10 years[5–22] . Although a body of literature exists with regard to dynamical systems in cosmology, comparatively little is known about N-body chaotic behaviour in general relativity. The ROGS has furthered insight into this problem, providing a unique window into the study of chaos in relativistic systems. In this article I provide a brief review of this body of research. Beginning with the foundations of two-dimensional gravity, I go on to illustrate the relationship between the ROGS and the OGS. I then summarize the main results obtained for the relativistic 2-body, 3-body, and N-body problems. A number of new and unique solutions have been obtained, including a new solution to the static balance problem and a new equilibrium solution for circular topology.

Wolfgang Kummer believed that by pushing our models of two-dimensional gravity to encompass a broader range of theoretical structures, we could make progress in understanding the behaviour of quantum gravity. By pushing our models to include a larger number of physical interactions via the ROGS, we likewise can make progress in understanding the many-body problem. It is in this spirit that I dedicate this article to his memory.

12.2. Foundations of Two-Dimensional Gravity

To obtain a ROGS one must first have an action principle that sensibly couples particles to gravity in two spacetime dimensions. While there are many ways to do this using a dilaton (whose presence is unavoidable since the Einstein-Hilbert action is a topological invariant[5]), for the ROGS one would like to impose constraints that ensure the system under study resembles that in three space dimensions as closely as possible. In particular only the stress-energy of the N bodies should source the spacetime curvature. In the non-relativistic limit the system should reduce to the well-known OGS. These constraints require one to adopt a particular kind of dilaton gravity[6] with dilaton Ψ, metric $g_{\mu\nu}$, and covariant derivative ∇_μ, whose action is

$$I = \int d^2x \left[\frac{\sqrt{-g}}{2\kappa} \left(\Psi R + \frac{1}{2} (\nabla\Psi)^2 + \Lambda \right) - \sum_{a=1}^{N} \int d\tau_a m_a \mathcal{S}_a \delta^2(x - z_a(\tau_a)) \right] \tag{12.1}$$

where the arc length $\mathcal{S}_a = \sqrt{-g_{\mu\nu}\frac{dz_a^\mu}{d\tau_a}\frac{dz_a^\nu}{d\tau_a}}$ of the a-th particle minimally couples it to gravity, with $\kappa = 8\pi G/c^4$, and $R = g^{\mu\nu}R_{\mu\nu}$ the Ricci scalar.

Ignoring boundary-term issues, applying the variational principle to (12.1) yields after some manipulation the field equations[5]

$$R - \Lambda = \kappa T^{P\mu}_{\ \ \mu} \qquad \frac{d}{d\tau_a}\left\{ \frac{dz_a^\nu}{d\tau_a} \right\} + \Gamma^\nu_{\alpha\beta}(z_a)\frac{dz_a^\alpha}{d\tau_a}\frac{dz_a^\beta}{d\tau_a} = 0 \tag{12.2}$$

$$\frac{1}{2}\nabla_\mu\Psi\nabla_\nu\Psi - g_{\mu\nu}\left(\frac{1}{4}\nabla^\lambda\Psi\nabla_\lambda\Psi - \nabla^2\Psi \right) - \nabla_\mu\nabla_\nu\Psi = \kappa T^P_{\mu\nu} + \frac{\Lambda}{2}g_{\mu\nu} \tag{12.3}$$

which form a closed sytem of equations for the ROGS. Here $T_{\mu\nu} = \sum_a m_a \int d\tau_a \frac{1}{\sqrt{-g}} g_{\mu\sigma} g_{\nu\rho} \frac{dz_a^\sigma}{d\tau_a}\frac{dz_a^\rho}{d\tau_a}\delta^2(x - z_a(\tau_a))$ is the stress-energy of the N-body system, which (along with the cosmological constant Λ) from (12.2) sources the gravitational field as required.

Solution of these equations is most fruitfully attained by writing the metric as $ds^2 = -N_0^2(x,t)\,dt^2 + \gamma\left(dx + \frac{N_1}{\gamma}dt \right)^2$ so that the action becomes[7]

$$I = \int dx^2 \left\{ \sum_{a=1}^{N} p_a \dot{z}_a \delta(x - z_a(x^0)) + \pi\dot{\gamma} + \Pi\dot{\Psi} + N_0 R^0 + N_1 R^1 \right\} \tag{12.4}$$

where $\{\pi, \Pi, p_a\}$ are the respective conjugate momenta to $\{\gamma, \Psi, z_a\}$, and overdot denotes $\frac{d}{dt}$. Choosing coordinate conditions $\gamma = 1 \quad \Pi = 0$ yields the canonically reduced Hamiltonian[7]

$$H = -\frac{1}{\kappa} \int dx \Psi'' \tag{12.5}$$

where the constraint equations $R_0 = 0$, $R_1 = 0$ imply

$$\Psi'' - \frac{1}{4}(\Psi')^2 + \kappa^2 \pi^2 - \frac{\Lambda}{2} + \kappa \sum_a \sqrt{p_a^2 + m_a^2} \delta(x - z_a) = 0 \tag{12.6}$$

$$2\pi' + \sum_a p_a \delta(x - z_a) = 0 \tag{12.7}$$

with prime denoting an x derivative. In physical terms eqs. (12.6,12.7) describe the respective energy and momentum balance between the gravitational field and the particles. The consistency of this canonical reduction can be demonstrated[7] by showing that the full set of canonical equations of motion (not shown) derived from eq. (12.4) are identical with the field equations eqs.(12.2,12.3).

The procedure for obtaining the behaviour of the ROGS is straightforward: one chooses a value of N, solves the constraints eqs. (12.6,12.7) for Ψ, and then obtains H as a function of the canonical variables (z_a, p_a) of the particles, with boundary conditions chosen to ensure $H(z_a, p_a)$ is finite. The behaviour of the N-body system, whose equations

$$\dot{z}_a = \frac{\partial H}{\partial p_a} \qquad \dot{p}_a = -\frac{\partial H}{\partial z_a} \tag{12.8}$$

follow from Hamilton's principle, is governed by this Hamiltonian.

The Hamiltonian H is conserved and its constant value corresponds to the energy E of the system. A more careful treatment that takes boundary terms into account[11] yields for a one-dimensional space-like region D with boundary points P_1 and P_2 the definition

$$E_D(\xi) = \mathcal{U}(\xi)|_{P_2} - \mathcal{U}(\xi)|_{P_1} \tag{12.9}$$

where ξ is a time-like vector field, $(g(x), \psi(x))$ a solution of the field equations (relative to some background $(\bar{g}(x), \bar{\psi}(x))$), and

$$\mathcal{U}(\xi) = \frac{\sqrt{g}}{2\kappa} \left\{ 2 \nabla_\beta \psi \, g^{\beta\alpha} \, \xi^\sigma - \psi \, g^{\alpha\mu} \, \nabla_\mu \xi^\sigma + \psi \, g^{\mu\nu} \, (w_{\mu\nu}^\sigma - \bar{w}_{\mu\nu}^\sigma)\xi^\alpha \right\} \epsilon_{\alpha\sigma}$$

$$- \frac{\sqrt{\bar{g}}}{2\kappa} \left\{ 2 \bar{\nabla}_\beta \bar{\psi} \, \bar{g}^{\beta\alpha} \, \xi^\sigma - \bar{\psi} \, \bar{g}^{\alpha\mu} \, \bar{\nabla}_\mu \xi^\sigma \right\} \epsilon_{\alpha\sigma} \tag{12.10}$$

is the superpotential corresponding to the *density* of energy. Here $w_{\beta\mu}^\alpha \equiv \Gamma_{\beta\mu}^\alpha - \delta_{(\beta}^\alpha \Gamma_{\mu)\nu}^\nu$ This definition is the covariant counterpart of the ADM formulation (12.5), reducing to it for the ROGS described by eq. (12.1).

12.3. The Non-Relativistic Limit

The ROGS Hamiltonian (12.5) (or action (12.1)) is the natural relativistic version of the Newtonian OGS. Reintroducing the speed of light c by rescaling $p_a \to p_a c$ and $m_a \to m_a c^2$ into eqs. (12.6,12.7), one makes this comparison[7,19] by expanding the theory powers of $\frac{1}{c}$, noting that p_a^2/m_a^2 and $\sqrt{\kappa}$ are each of order of c^{-2}. Iteratively computing H in (12.5) yields

$$
H_{pN} = \sum_a \left\{ \frac{\tilde{p}_a^2}{2m_a} + \pi G \sum_b m_a m_b |z_{ab}| - \frac{\tilde{p}_a^4}{8m_a^3 c^2} + \frac{\pi G}{c^2} \sum_b m_a \frac{\tilde{p}_b^2}{m_b} |z_{ab}| \right.
$$
$$
\left. - \frac{\pi G}{c^2} \sum_b \tilde{p}_a \tilde{p}_b |z_{ab}| + \left(\frac{\pi G}{c} \right)^2 \sum_{b,c} m_a m_b m_c \left\{ |z_{ab}||z_{ac}| - z_{ab} z_{ac} \right\} \right\}
$$
$$
(12.11)
$$

where $z_{ab} = z_a - z_b$, $\tilde{p}_a = p_a - 2\pi G \sum_b m_a m_b(z_{ab})$, and I have set $\Lambda = 0$.

The quantity $H_{pN} = H - \sum_a m_a c^2$ is the post-Newtonian Hamiltonian. Its leading term is governs the OGS, whose gravitational potential grows linearly particle separation as is clear from the second term in (12.11). The force between each body is constant, changing only with their relative distribution as they pass through one another. However the ROGS Hamiltonian (12.5) is a highly non-linear function of the canonical degrees of freedom of the particles. Yet in between the particles the spacetime has constant curvature. The non-linear interactions between the particles (the first few terms of which appear in (12.11) are the result of matching the constraint equations (12.6,12.7) consistently across each particle with appropriate boundary conditions at infinity. In this sense all the physics comes from global effects.

12.4. 2-Body Motion

The first non-perturbative relativistic curved-spacetime treatment of the two-body problem[8-10] was obtained by solving eqs. (12.6,12.7) for $N = 2$. The full solution is a function only of $r = z_{12}$ and $p = p_1 = -p_2$, since the canonical equations imply conservation of the total momenta $p_1 + p_2$. Solving first (12.7) for π and then (12.6) for Ψ, eq. (12.5) yields

$$
\tanh(\frac{\kappa \mathcal{J}}{8} |r|) = \frac{\mathcal{J}(h_1 + h_2)}{\mathcal{J}^2 + h_1 h_2} \tag{12.12}
$$

where $h_{1,2} = H - 2\sqrt{p^2 + m_{1,2}^2}$, $\mathcal{J}^2 = (\sqrt{H^2 + 8\Lambda/\kappa^2} - 2\epsilon \tilde{p})^2 - 8\Lambda/\kappa^2$ and $\tilde{p} = p|r|/r$. For a given $\Lambda \geq -(\kappa H)^2/8$, equation (12.12) describes the surface in (r, p, H) space of all allowed phase-space trajectories. It is easy to verify by t-differentiation of (12.12) that H is a constant of the motion.

Consider first the equal-mass case. Setting $H = H_0$, Hamilton's equations (12.8) yield $\dot{r}(t)$ and $\dot{p}(t)$ as a function of H_0, Λ and $m = m_1 = m_2$. Superficial singularities appear in these equations, since t is a coordinate time. The proper time τ_a for each particle is

$$d\tau = \frac{(H_0 - 2\tilde{p})m}{\left\{2 - \frac{\kappa r}{4}(H_0 - 2\tilde{p})\right\}\left(\sqrt{p^2 + m^2} - \tilde{p}\right)\sqrt{p^2 + m^2}} dt \qquad (12.13)$$

and the canonical equations (12.8) have the exact solution

$$p(\tau) = \frac{|r|m}{2rf(\tau)}\left(f^2(\tau) - 1\right) \qquad (12.14)$$

where

$$f(\tau) = \begin{cases} \dfrac{\frac{H_0}{m}\left(1 + \sqrt{\gamma_H}\right)\left\{1 - \eta\, e^{\frac{\epsilon\kappa m}{4}\sqrt{\gamma_m}(\tau - \tau_0)}\right\}}{1 + \sqrt{\gamma_m} + \left(\sqrt{\gamma_m} - 1\right)\eta\, e^{\frac{\epsilon\kappa m}{4}\sqrt{\gamma_m}(\tau - \tau_0)}} & \gamma_m > 0, \\[4mm] \dfrac{1 + \sqrt{\gamma_H}}{\frac{m}{H_0} + \frac{\sigma}{m - \sigma\frac{\epsilon\kappa H_0}{8}(\tau - \tau_0)}} & \gamma_m = 0, \\[4mm] \dfrac{\frac{H_0}{m}\left(1 + \sqrt{\gamma_H}\right)}{1 + \sqrt{-\gamma_m}\frac{\sigma + \frac{m^2}{H_0}\sqrt{-\gamma_m}\tan\left[\frac{\epsilon\kappa m}{8}\sqrt{-\gamma_m}(\tau - \tau_0)\right]}{\frac{m^2}{H_0}\sqrt{-\gamma_m} - \sigma\tan\left[\frac{\epsilon\kappa m}{8}\sqrt{-\gamma_m}(\tau - \tau_0)\right]}} & \gamma_m < 0, \end{cases} \qquad (12.15)$$

with p_0 being the initial momentum at $\tau = \tau_0$, and where

$$\gamma_H = 1 + \frac{8\Lambda}{\kappa^2 H_0^2}, \qquad \gamma_m = 1 + \frac{8\Lambda}{\kappa^2 m^2},$$
$$\eta = \frac{\sigma - \frac{m^2}{H_0}\sqrt{\gamma_m}}{\sigma + \frac{m^2}{H_0}\sqrt{\gamma_m}}, \qquad \sigma = (1 + \sqrt{\gamma_H})(\sqrt{p_0^2 + m^2} - \frac{|r|}{r}p_0) - \frac{m^2}{H_0}. \qquad (12.16)$$

Note that there are three possibilities, depending on the relative size and sign of Λ compared to m and H_0. For each of these there are four possible solutions for the proper separation r between the particles, given by

$$|r(\tau)|^A_\pm = \frac{16\tanh^{-1}\left[\left(\frac{\kappa\left(H_0 - m|f(\tau) + \frac{1}{f(\tau)}|\right)}{\sqrt{\left(\sqrt{\kappa^2 H_0^2 + 8\Lambda} - m\kappa(f(\tau) - \frac{1}{f(\tau)})\right)^2 - 8\Lambda}}\right)^{\pm 1}\right]}{\sqrt{\left(\sqrt{\kappa^2 H_0^2 + 8\Lambda} - m\kappa(f(\tau) - \frac{1}{f(\tau)})\right)^2 - 8\Lambda}} \qquad (12.17)$$

or, depending on the relative size and sign of $\mathcal{J}(\tau)$, by

$$|r(\tau)|_{\pm}^{B} = \frac{16\left(\tan^{-1}\left[\left(\frac{\kappa\left(m\left|f(\tau)+\frac{1}{f(\tau)}\right|-H_0\right)}{\sqrt{8\Lambda-\left(\sqrt{\kappa^2 H_0^2+8\Lambda}-m\kappa(f(\tau)-\frac{1}{f(\tau)})\right)^2}}\right)^{\pm 1}\right]+n\pi\right)}{\sqrt{8\Lambda-\left(\sqrt{\kappa^2 H_0^2+8\Lambda}-m\kappa(f(\tau)-\frac{1}{f(\tau)})\right)^2}}.$$

(12.18)

If $\Lambda = 0$ then only bounded periodic motions exist. All solutions have $H_0 > 2m$. As H_0 increases, the amplitude and period of the bounded motion become large, and the shape of the separation trajectory $r(\tau)$ deforms from the parabolic Newtonian motion. The maximal separation of the particles is much smaller than its non-relativistic counterpart and is achieved far more quickly. After maximal separation, the particles move toward each other at a slower velocity until they are within 10% of their maximal separation, and then rapidly accelerate toward the same point, after which the motion repeats with the particles interchanged.

If $\Lambda \neq 0$ the possible variety of motions that can emerge is rich and varied.[9,10] Type B motions are always unbounded, whereas type-A motions may or may not be bounded depending on the relative values of the parameters. As Λ becomes increasingly negative the particles do not achieve as wide a proper separation, and the oscilliation frequency is more rapid. As Λ becomes increasingly positive the oscillation frequency decreases and the particles more widely separate until eventually there is no bound motion and they fall out of causal contact with one another.

For the unequal masses these same features recur. However the equations must be solved numerically since the asymmetry between the masses obstructs obtention of an analytic solution.

It is also possible to extend the ROGS to include electromagnetism,[12,13] with each particle having a different charge. An even more varied array of possible motions ensues. The quantity $\mathcal{J} \to (\sqrt{H^2 + 8\Lambda/\kappa^2} - 2\epsilon\tilde{p})^2 - 8(\Lambda + e_1 e_2)/\kappa^2$ where e_i is the charge of the i-th particle.

The defining equation (12.12) for the Hamiltonian is correspondingly modified. Some qualitatively new features emerge, including the existence of unbounded motion for $H_0 < 2m$, provided charge repulsion is sufficiently strong.[12] Another is the existence of a new solution to the long-standing problem of static balance,[13] in which an equilibrium solution for 2 or more bodies is attained by offsetting gravitational attraction with another force. The only known solution in general relativity[23] has $e_i = \pm\sqrt{4\pi G m_i}$, which is much stronger than the corresponding non-relativistic condition $e_1 e_2 = 4\pi G m_1 m_2$. However for the 2-body ROGS, setting $\frac{\partial H}{\partial r} = 0$ yields

$$p = p_s = \frac{\left| \left(\frac{\kappa}{2}\right)^2 m_1^2 m_2^2 - e_1^2 e_2^2 \right|}{\sqrt{2\kappa e_1 e_2} \sqrt{\left(\frac{\kappa}{2} m_1^2 + e_1 e_2\right) \left(\frac{\kappa}{2} m_2^2 + e_1 e_2\right)}} \qquad (12.19)$$

which is a new relativistic force-balance condition[13] where the two particles move with constant velocity. If the particles are at rest then $p_s = 0$ and the non-relativistic static balance condition holds even for this relativistic case. This suggests that analogous equilibrium states might also exist in four-dimensional general relativity.

12.5. 3-Body Motion

There is a substantive qualitative difference between 2-body and 3-body motion. The high degree of regularity in the 2-body system is replaced by a much richer variety of motions that are either quasiperiodic or chaotic. The applications of the 3-body system are also much broader. The $N = 2$ OGS is equivalent to a single body falling in a hole drilled through the centre of a large spherically symmetric body. However the $N = 3$ system models perfectly elastic collisions of a particle with a wedge in a uniform gravitational field,[24] two elastically colliding billiard balls in a uniform gravitational field[25] and a bound state of three quarks to form a "linear baryon".[26] A much richer dynamical system, the 3-body ROGS is the first to be studied in which gravitational back-reaction is fully accounted for.[15,18–20,22]

A number of intriguing relationships exist between the relativistic and non-relativistic 3-body ROGS. Tightly bound states of two bodies undergoing a low-frequency oscillation with the third exhibit features similar to the 2-body ROGS discussed in the previous section, whereas the corresponding motions in the non-relativistic case have the expected parabolic behaviour.[15,18] Bound-state oscillations in the relativistic system have a higher frequency and cover a smaller region of the position part of the phase space than its non-relativistic counterpart does at the same energy.

To see these features explicitly one must first solve (12.6,12.7) for $N = 3$. This is algebraically straightforward, though considerably more tedious, and yields the implicit equation[18–20]

$$L_1 L_2 L_3 = \frac{1}{2} \sum_{ijk} \left| \epsilon^{ijk} \right| \mathcal{M}_{ij} \mathcal{M}_{ji} L_k^* e^{\frac{\kappa}{4} s_{ij} [(L_i + \mathcal{M}_{ij}) z_{ik} - (L_j + \mathcal{M}_{ji}) z_{jk}]} \qquad (12.20)$$

where

$$\mathcal{M}_{ij} = M_i - p_i s_{ij}, \qquad\qquad M_i = \sqrt{p_i^2 + m_i^2} \qquad (12.21)$$

$$L_i = H - M_i - \left(\sum_j p_j s_{ji}\right) \qquad L_i^* = \left(1 - \prod_{j<k\neq i} s_{ij} s_{ik}\right) M_i + L_i$$

$$(12.22)$$

with $z_{ij} = (z_i - z_j)$, $s_{ij} = sgn(z_{ij})$, and ϵ^{ijk} is the 3-dimensional Levi-Civita tensor. Solving (12.20) for H as a function of the canonical variables in closed form is not possible, though one can obtain Hamilton's equations (12.8) by straightforward differentiation.

The following canonical transformation

$$\rho = \frac{1}{\sqrt{2}}(z_1 - z_2) \qquad \lambda = \frac{1}{\sqrt{6}}(z_1 + z_2 - 2z_3) \qquad Z = z_1 + z_2 + z_3$$

$$p_\rho = \frac{1}{\sqrt{2}}(p_1 - p_2) \qquad p_\lambda = \frac{1}{\sqrt{6}}(p_1 + p_2 - 2p_3) \qquad p_Z = \frac{1}{3}(p_1 + p_2 + p_3)$$

$$(12.23)$$

shows that there are only four canonical degrees of freedom, since nothing in eq. (12.20) depends on Z or p_Z. For the OGS this transformation renders the system isomorphic to the motion of a single particle with coordinates (ρ, λ) in a linear, hexagonal well potential. An analogous hexagonal potential exists for the ROGS with the sides of the well no longer increasing linearly with increasing particle separation. When the relative masses of the particles are changed, the shape of the hexagonal cross-section expands orthogonal to one of the lines connecting opposite vertices.

Using this single-particle representation the global structure of phase space can be probed using Poincare sections. The bisectors of the hexagonal cross-section correspond to points where two of the bodies are coincident, dividing the ρ-λ plane into 6 sextants corresponding to the 6 different configurations the 3 particles can assume. As the "hex-particle" moves from one sextant to the next, this corresponds to two particles passing through each other. In the equal mass case, all 6 sextants are equivalent, and in the unequal mass case, opposite sextants correspond to the opposite configuration of particles (*i.e.* $(1, 2, 3) \rightarrow (3, 2, 1)$); if two particles have the same mass further symmetries exist.

The global structure of this phase space is best studied by constructing Poincaré sections. The motion is confined to a 3 dimensional hypersurface in phase space because the total energy of the system is constant as a consequence of the time-independence of the Hamiltonian. Further reduction to 2 dimensions can be obtained by plotting the hex-particle's radial momentum p_R and squared angular momentum p_θ^2 each time it crosses one of the bisectors,[18,24] yielding a surface of section, or Poincaré section. For equal masses all bisectors are equivalent, yielding one surface of section for a given energy, whereas for unequal masses one must distinguish between the different bisectors and the directions in which they cross.

A thorough numerical study has been carried out for various types of 3-body systems of equal and unequal masses, and extended to include both cosmological constant and electromagnetic charge. A number of interesting results concerning the dynamics of the 3-body ROGS have emerged from this study, which are summarized here.

(I) Despite the high degree of non-linearlity in the ROGS the qualitative features of its Poincare sections are the same as the OGS for all values of the energy studied. The regions of chaos in the ROGS do not increase in size but instead develops an asymmetric distortion that increases with increasing energy because of a weaker symmetry in its Hamiltonian.[15]

(II) Both the ROGS and the OGS exhibit three types of motion:[18,19] annulus (no two particles ever cross twice in succession), pretzel (a pair of particles cross each other at least twice before either crosses the third) and chaotic. Annulus orbits can be either periodic, quasi-periodic, or densely filled, whereas pretzels have the latter two features. Stable bound subsystems of two particles exist for each of these two cases. Chaotic orbits occur when the hex-particle crosses the origin. All three types of motion model that seen in galactic dynamics.[27]

(III) Differences between the OGS and the ROGS become more pronounced with increasing energy, with orbits in the latter having higher frequency and covering a smaller region in (ρ, λ) space at the same energy. ROGS pretzel orbits develop an hourglass shape not seen in the OGS system.[18]

(IV) Additional chaotic regions in the unequal mass case appear that are not present in the equal mass space at the same energy in both the ROGS and the OGS.[19] These additional chaotic regions appear within the pretzel regions of the corresponding equal mass surface of section. This indicates that the unequal case is not simply a deformation of its equal mass counterpart but instead contains novel dynamics whose origin is currently unknown.

(V) Unlike the ROGS or the OGS, a similar study of the equal-mass post-Newtonian system (12.11) indicates that at energies larger than $\approx$

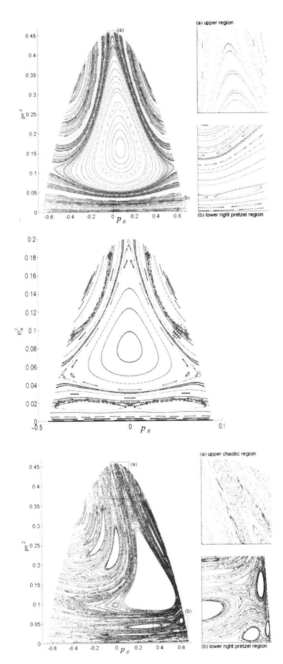

Figure 12.1   A comparison of Poincaré sections. Top: $H_0 = 1.5, \Lambda = 0.02$ for the ROGS. Middle: $H_0 = 1.3$ for the OGS. Bottom: $H_0 = 1.5, \Lambda = -0.25$, just above its critical value. The insets illustrate the increase (decrease) in chaos for $\Lambda > 0$ ($\Lambda < 0$).

$1.26Mc^2$ the chaotic regions grow,[18] eventually becoming connected areas on the Poincare section, a phenomenon called a Kolmogorov, Arnold and Moser (KAM) transition.[28]

(VI) When $\Lambda < 0$ is introduced in the ROGS there is a rapid decrease in the amount of chaos for all values of the energy.[20] Despite its high degree of non-linearity, the size of the chaotic regions are even smaller than in the OGS (for which nonzero Λ has no meaning). As $\Lambda \to -\frac{H_0^2 \kappa^2}{8}$ the chaotic regions nearly vanish (see fig 12.1). Since the area of the chaotic regions is roughly proportional to $\left| \frac{8\Lambda}{H_0^2 \kappa^2} \right|$ this suggests that this occurs for arbitrarily large H_0.

(VII) For $\Lambda > 0$ the chaotic regions increase in area,[20] both in the regions between the annulus and pretzel orbits, and within the regions corresponding to the pretzel orbits. The lines in the pretzel regions start to thicken as Λ increases and small regions of chaos appear between the group of ellipses, reminicent of the preliminary stages of KAM breakdown[28] (see 12.1).

(VIII) Stable orbits remain stable for $\Lambda < 0$ but can become chaotic for $\Lambda > 0$. The point of transition in the latter situation depends on the initial conditions. These results are consistent with the higher-dimensional understanding of a negative cosmological constant providing stronger gravitational binding, leading to an increase of the integrability of the dynamics and thus an increase in the stability of trajectories.[20]

(IX) A study of the charged ROGS indicates that electromagnetic coupling significantly modifies the chaotic properties in markedly different ways depending on the relative signs of the charges of the particles.[22] If all charges are positive, similiar features to $\Lambda > 0$ appear but with a less rapid transition to KAM breakdown. The overall magnitude of the charges is more effective at inducing chaos than a change in relative magnitudes of the charges for a given energy. If one particle is neutral or negatively charged then chaotic and pretzel regions increase in areas a U-shaped area around the dissolving annulus orbits forms. A new hourglass-shaped class of chaotic orbits emerge surrounding the U-shaped areas. The existence of both localized positive vacuum energy and of potential difference as the arrangment of the charges change when crossing each other drives the hex particle to accelerate, destroying all annulus orbits for higher value of the charges. If one particle is neutral, and the other two have opposite charge the properties of the U-shaped areas remain generally the same apart from a considerable increase in pretzel areas caused by negative localized vaccum energy, leaving no annulus orbits for higher value of the charges.

12.6. N-bodies

A method exists for determining H for the N-body ROGS[21] , though no further detailed study has been carried out. Despite this, a number of interesting results have been obtained.

Consider a relativistic gas of N gravitating particles.[14] For a canonical ensemble, all phase-space averages are carried out with a weighting function $\exp[-\beta H]$, where $k_B T = \beta^{-1}$ is the temperature multiplied by Boltzmann's constant k_B. Setting $\bar{p} = \frac{1}{N} \sum_{a=1}^{N} p_a$ and $\bar{z} \equiv \frac{1}{M} \sum_{a=1}^{N} m_a z_a$, with $M = \sum_{a=1}^{N} m_a$ the partition function

$$\mathcal{Z} = \int \int \frac{d\mathbf{p}d\mathbf{z}}{N!} \delta\left(\bar{p}\right) \delta\left(\bar{z}\right) \exp\left(-\beta H\right) \qquad (12.24)$$

to leading order in $1/c$ is[14]

$$\mathcal{Z} = \frac{\exp\left[-\beta Mc^2 - \frac{3(N-1)\ln\left(\beta mc^2\right)}{2} - \left\{\frac{(5N+3)(N-1)+8Nq(N)}{8N\beta mc^2}\right\}\right]}{\sqrt{N}\left(\sqrt{2\pi}G/c^3\right)^{(N-1)}\left[(N-1)!\right]^2} \qquad (12.25)$$

using (12.11). The average energy is[14]

$$\langle E \rangle = -\frac{\partial \ln \mathcal{Z}}{\partial \beta} = Mc^2 + \frac{3(N-1)}{2\beta} - \frac{(5N+3)(N-1)+8Nq(N)}{8Mc^2\beta^2} \qquad (12.26)$$

where $q(N) = \sum_{k=1}^{N-1} \sum_{l=k+1}^{N-1} \frac{(l-k)}{l(N-k)}$. For fixed $M = Nm$ the relativistic correction grows quadratically with N is negative, indicating that relativistic effects cool the gas down: at a given energy, the ROGS temperature is smaller than its OGS counterpart.

The canonical ensemble is somewhat unrealistic in that the system is kept at constant temperature $T = \beta^{-1}$ via coupling to a heat bath, with the energy undergoing fluctuations of order $k_B T$. In a more realistic astrophysical case the ROGS is isolated and its total energy is conserved. The microcanonical ensemble, in which phase space integrations are carried out by constraining the total energy to be E, properly describes this situation. The canonical partition function $\mathcal{Z}$ in (12.24) is replaced with

$$\Omega = \frac{1}{N!} \int \int d\mathbf{p}d\mathbf{z}\delta\left(\bar{p}\right) \delta\left(\bar{z}\right) \delta\left(E - H\right) \qquad (12.27)$$

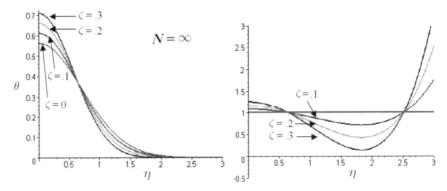

Figure 12.2   Momentum density ϑ as a function of $\eta = \frac{3p}{4mc\sqrt{\zeta}}$ as $N \to \infty$ for various values of $\zeta = E/Mc^2 - 1$.

and the microcanonical single-particle distribution function is

$$f(p, z) = \int \int \frac{dpdz}{N\Omega N!} \delta\left(\overline{p}\right) \delta\left(\overline{z}\right) \delta\left(E - H\right) \sum_a \delta\left(z - z_a\right) \delta\left(p - p_a\right) \quad (12.28)$$

from which the distributions $\rho(z) = \int_{-\infty}^{\infty} dp\, f(p, z)$ for density and $\vartheta(p) = \int_{-\infty}^{\infty} dz\, f(p, z)$ for momentum are obtained.

For statistical systems a key quantity of interest is the large N limit, in E and $M = Nm$ are fixed. For the OGS, $f(p, z)$ approaches the isothermal solution of the Vlasov equations.[3] Relativistic corrections tend to yield more sharply-peaked density distributions,[14] with $\rho_{\text{OGS}} > \rho_{\text{ROGS}}$ for sufficiently large z, due to the quadratic ROGS potential in (12.11) which yields a tighter binding also seen in the 2 and 3 body cases. For momentum densities, $\vartheta_{\text{OGS}} > \vartheta_{\text{ROGS}}$ for intermediate values of p, but for large p this inequality is reversed[14] due to the p^4 corrections in (12.11). See fig. 12.2 for a comparison of the momentum density at different values of the energy.

An exact equilibrium solution on a circle of coordinate length $2L$ is the other main achievement for the N-body case.[16,17] This solution is also obtained using canonical methods by setting the extrinsic curvature $K = \gamma\kappa(\pi - \Pi/\gamma) = \tau$, where τ is a time coordinate. Solving the canonical equations with equilibrium conditions $\dot{z}_a = 0$, $p_a = 0$, and $m_a = m$ yields

$$\sqrt{\gamma} = \frac{N}{L\mathcal{T}} \text{arctanh}\left(\frac{\xi\sqrt{\frac{\kappa^2 m^2}{16} + (\xi^2 - 1)\mathcal{T}^2} - \frac{\kappa m}{4}}{\frac{\kappa^2 m^2}{16} + \xi^2\mathcal{T}^2}\mathcal{T}\right) \quad (12.29)$$

where ξ is an integration constant and $\mathcal{T}^2 = \tau^2 - \Lambda/2$. The lapse N_0

and shift N_1 can also be explicitly obtained. Even though the particles are in gravitational equilibrium, spacetime itself can expand or contract depending on the value of Λ, counterbalancing the mass attraction. An electromagnetic generalization of this solution for N charged particles was obtained using similar methods.[17]

12.7. Summary and Acknowledgments

There is a dynamical richness to lower dimensional N-body dynamics, with much left to explore. So far all work on the 3-body ROGS has been restricted to $H_0 \leq 2Mc^2$; more sophisticated numerical techniques will be needed to probe the dynamics of the system at high energies and larger values of Λ and charge. The development of a ROGS where particles elastically collide instead of passing through each other would allow one to see if the increased chaos found in the OGS[24] has an analogue in the ROGS. New analytic methods will be needed to extend the statistical work beyond the post-Newtonian approximation, and to obtain more general solutions for circular topology.

A new frontier for exploration is the quantum N-body problem. Preliminary work[29] on the quantization of H_{pN} for $N = 2$ in (12.11) has shown that its ground state energy is lower than in the non-relativistic case, while all other relativistic states have higher energies than their Newtonian counterparts. Numerical work has indicated that the expectation values $\langle \hat{r}^2 \rangle_{\psi_n^{OGS}} < \langle \hat{r}^2 \rangle_{\psi_n^{ROGS}}$ and $\langle \hat{p}^2 \rangle_{\psi_n^{OGS}} > \langle \hat{p}^2 \rangle_{\psi_n^{ROGS}}$ – quantum-relativistically, the two particles are less tightly bound. A full quantization of the action (12.1) remains an interesting problem for future study.

I have enjoyed my collaborations with the many students that have worked with me on this program and with my longtime collaborator Tadayuki Ohta. This work was supported by the Natural Sciences and Engineering Research Council of Canada.

References

1. For a review of recent developments see D. Grumiller, W. Kummer and D. V. Vassilevich, *Phys. Rept.* **369**, 327 (2002) .
2. See B.N. Miller and P. Youngkins, *Phys. Rev. Lett.* **81** 4794 (1998); K.R. Yawn and B.N. Miller, *Phys. Rev. Lett.* **79** 3561 (1997) and references therein.
3. G. Rybicki, *Astrophys. Space. Sci* **14** 56 (1971); H.L. Wright, B.N. Miller, and W.E. Stein, *Astrophys. Space. Sci.* **84, 2** 421 (1982) and references therein.

4. H. Koyama and T. Kinoshi, *Phys.Lett.* **A279** 226 (2001); H. Koyama and T. Kinoshi, *Europhys.Lett.***58** 356 (2002).

5. R.B. Mann, A. Shiekh and L. Tarasov, *Nuc. Phys.* **B341** 134 (1990); A.E. Sikkemma and R.B. Mann, *Class. Quant. Grav.* **8** 219 (1991); R.B. Mann, S.M. Morsink, A.E. Sikkema and T.G. Steele, *Phys. Rev.* **D43** 3948 (1991); R.B. Mann, *Found. Phys. Lett.* **4** 425 (1991); *Gen. Rel. Grav.* **24** 433 (1992).

6. S.F.J. Chan and R.B. Mann, *Class. Quant. Grav.* **12** 351 (1995).

7. T. Ohta and R.B. Mann, *Class. Quant. Grav.* **13** 2585 (1996).

8. R.B. Mann and T. Ohta, *Phys. Rev.* **D57** 4723 (1997); *Class. Quant. Grav.* **14** 1259 (1997).

9. R.B. Mann, D. Robbins and T. Ohta, *Phys. Rev. Lett.* **82** 3738 (1999).

10. R.B. Mann, D. Robbins and T. Ohta, *Phys. Rev.* **D60** 104048 (1999).

11. R.B. Mann, G. Potvin and M. Raiteri, *Class. Quant. Grav.* **17** 4941 (2000).

12. R.B. Mann, D. Robbins, T. Ohta and M. Trott, *Nucl. Phys.* **B590** 367 (2000).

13. R.B. Mann and T. Ohta, *Class. Quant. Grav.* **17** 4059 (2000).

14. R.B. Mann and P. Chak, *Phys. Rev.* **E65** (2002) 026128.

15. F. Burnell, R.B. Mann and T. Ohta, *Phys. Rev. Lett.* **90** 134101 (2003).

16. R. Kerner and R.B. Mann, *Class. Quant. Grav.* **20** 133 (2003).

17. R. Kerner and R.B. Mann, *Class. Quant. Grav.* **21** 5789 (2004).

18. F.J. Burnell, J.J. Malecki, R.B. Mann, and T. Ohta, *Phys. Rev.* **E69** 016214 (2004).

19. J. J, Malecki and R. B. Mann, *Phys. Rev.* **E69** 066208 (2004) .

20. S. Bachmann, M.J. Koop and R.B. Mann, *Phys. Rev.* **D76** 104051 (2007).

21. P.S. Farrugia, R.B. Mann and T.C. Scott *Class. Quant. Grav.* **24** 4647 (2007).

22. M.J. Koop, R.B. Mann and M. Rohanizadegan, University of Waterloo preprint (2008).

23. S.D. Majumdar, *Phys. Rev.* **72** 390 (1947); A. Papapetrou, *Proc. Roy. Irish. Acad.* **A51** 191 (1947).

24. H.E. Lehtihet and B.N. Miller, *Physica* **21D** 93 (1986).

25. N.D. Whelan, D. Goodings and J.K. Cannnizzo, *Phys. Rev.* **A42** 742 (1990).

26. D. Bukta, G. Karl and B. Nickel, *Can. J. Phys.* **78** 449 (2000).

27. D. Merritt *Science* **271** 337 (1996).

28. V. I. Arnold and A. Avez, *Ergodic Problems of Classical Mechanics* (Springer, New York, 1968); A. N. Kolmogorov *Dokl. Akad. Nauk SSSR* **98** 525 (1954); V.I. Arnold *Russ. Math. Surveys* **18** 85 (1963); J. Moser, *Nachr. Akad. Wiss. Goett. II, Math.-Phys. K1.* **K1** 1 (1962).

29. R.B. Mann and M. Young *Class. Quant. Grav.* **24** 951 (2007).

Chapter 13

2-D Midisuperspace Models for Quantum Black Holes

Jack Gegenberg[1] and Gabor Kunstatter[2]

[1] *Department of Mathematics and Statistics,*
University of New Brunswick, Fredericton,
New Brunsick E3B 5A3, Canada

[2] *Physics Department, University of Winnipeg,*
and Winnipeg Institute for Theoretical Physics,
Winnipeg, Manitoba R3B 2E9, Canada

Dimensionally reduced spherically symmetric gravity and its generalization, generic 2-D dilaton gravity, provide ideal theoretical laboratories
for the study of black hole quantum mechanics and thermodynamics.
They are sufficiently simple to be tractable but contain enough structure to allow the study of many deep issues in quantum gravity, such as
the endpoint of Hawking radiation and the source of black hole entropy.
This article reviews recent progress in a particular geometrical approach
to the study of quantum black holes in generic 2-d dilaton gravity.

13.1. Introduction

Despite significant progress in recent years, particularly in the context of
string theory[1] and loop quantum gravity,[2] the nature of quantum gravity
remains the great unsolved mystery of modern theoretical physics. With the
possible exception of some brane-world scenarios, most conceivable versions
of a quantum theory of gravity will likely only be relevant at energy and
distance scales that will remain experimentally inaccessible in the forseeable
future. Most research in the field consists of an examination of the internal
self-consistency of given candidate theories as well as their potential for

---

[1] E-mail: `lenin@math.unb.ca`
Affiliate Member, Perimeter Institute for Theoretical Physics.
[2] E-mail: `g.kunstatter@uwinnipeg.ca`
Affiliate Member, Perimeter Institute for Theoretical Physics.

solving the outstanding theoretical problems in classical and semi-classical general relativity. These theoretical problems include resolution of the singularities predicted by general relativity as well as the black hole entropy and information loss puzzles.

In order to make any theory of quantum gravity amenable to rigorous quantitative analysis, it is necessary to simplify the equations by focussing on a sector of the theory that is thought to contain enough structure to access the relevant issues. In this regard, the most promising arenas of study are cosmology, in which homegeneity and isotropy are good approximate symmetries, and black holes for which the no-hair theorem suggests that spherical symmetry provides a reasonable starting point[a]. This review will deal only with black holes, whose thermodynamic properties suggest that they contain valuable clues about the underlying microscopic theory of quantum gravity.

One particular model that has been extensively studied in the context of quantum black holes is dimensionally reduced spherically symmetric gravity.[3,4] The Birkhoff theorem guarantees that the vacuum theory is dynamically trivial (there are no propagating spherically symmetric graviton modes) but the model is diffeormorphism invariant and hence maintains the essential kinematical features of the full theory, therebye providing a valuable "midi-superspace model" for quantization.

The class of models that we will describe generalizes spherically symmetric gravity to a large class of diffeomorphism invariant theories in two space-time dimensions. These theories all satisfy a Birkhoff theorem and allow for the existence of black hole solutions with the same thermodynamic properties (entropy, semi-classical temperature) as physical black holes in higher dimensions. They are collectively known as generic 2-D dilaton gravity.

Few researchers have made more substantial and important contributions to the study of generic 2-D dilaton gravity than Professor Kummer and his collaborators. An excellent description of these contributions is contained in the very thorough review written by D. Grumiller, W. Kummer and D.V. Vassilevich.[5] The present article cannot hope to compete with the completeness and depth of that review. Instead, we will concentrate on the small part of the field with which we are most familiar. In particular, we will focus on analyses in terms of the geometrodynamical variables of the theory: the metric and the dilaton field. In the context of spherically

---

[a]Strictly speaking, one should consider axial symmetry in this context, but for in many cases, zero angular momentum should be a good approximation.

symmetric gravity, the dilaton has a geometrical interpretation as the area of the Killing sphere at fixed radius from the origin. If one is considering a theory that cannot be derived from higher dimensional gravity, the dilaton still plays a crucial role in determining the thermodynamic properties of black holes in the generic theory. This will be described in more detail below.

The paper is organized as follows: In the next section we will provide a brief introduction to the class of models that we are considering, including the action and general solution. Section III will describe how the physical observables and thermodynamic properties can be extracted in a very simple, diffeomorphism invariant manner. Section IV will present the Hamiltonian analysis, including the complete reduction to the physical phase space, which will be a precursor to the study in Section V of its semi-classical quantum properties. Section VI will describe the Dirac quantization of the theory, showing that the Hamiltonian constraint can be solved to reveal interesting quantum structure. Finally, we close with conclusions and prospects for future work, particularly the possibility of incorporating matter so as to provide a self-consistent treatment of the quantum dynamics of gravitational collapse in the generic theory.

13.2. Generic 2-D Dilaton Gravity: Action and Solutions

The gravitational action we wish to consider is:

$$S_G[g, \phi] = \frac{1}{2G} \int dx dt \sqrt{-g} \left(\phi R(g) + \frac{V(\phi)}{l^2} \right), \qquad (13.1)$$

where l is a parameter with dimensions of length which is generally taken to be the Planck length in the theory. Note that the generic theory is completely specified by the form of the dilaton potential $V(\phi)$. This action is the most general diffeomorphism invariant action in two space-time dimensions that has at most second derivatives of the fields.[6] Note that there is no kinetic term for the scalar field in (13.1). Had we chosen to add such a term, it could always be removed by a conformal reparametrization of the metric of the general form:

$$g_{\mu\nu} \rightarrow \Omega(\phi) g_{\mu\nu} \qquad (13.2)$$

Such reparametrizations leave the conformal structure of the geometry invariant, but do affect geodesics of test particles. In the cases of interest to be described below, the physical metric is in fact related to $g_{\mu\nu}$ above by precisely such a conformal reparametrization, one that is regular everywhere except at the curvature singularity of the physical metric.

As expected from spherically symmetric gravity, the generic vacuum theory has no propagating modes. There is a one parameter family of classical solutions with a single Killing vector.[7] In Schwarzschild-like coordinates in which the dilaton is used as the spatial coordinate the solution takes the form:

$$\phi = x/l,$$
$$ds^2 = -(j(\phi) - 2GlM)dt^2 + (j(\phi) - 2GlM)^{-1}d\phi^2, \qquad (13.3)$$

where

$$j(\phi) := \int_0^\phi d\tilde{\phi}V(\tilde{\phi}). \qquad (13.4)$$

As will be discussed in the next section, the existence of black hole solutions in the theory and their corresponding conformal properties depend on the form of the dilaton potential $V(\phi)$ and its first integral $j(\phi)$.

Specific dilaton gravity theories with action equivalent (up to local reparametrizations) to S_G were considered extensively in the past. One of the earliest was proposed in 1984 by Jackiw and Teitelboim.[9] The Jackiw-Teitelboim theory has dilaton potential $V(\phi) = \lambda\phi$. This theory came into further prominence when it was realized that it is equivalent to the cylindrically symmetric dimensionally reduced 2+1 gravity with cosmological constant λ, which also lacks local propagating modes but nevertheless has interesting black hole solutions.[10] The vacuum theory can be reduced via dimensional reduction to J-T coupled to an Abelian gauge field theory.

Another theory that received a great deal of attention in the early 1990's because of its connection to string theory is the 2-d vacuum dilatonic black hole (the so-called Witten black hole[11] in the string motivated CGHS model.[12] A thorough analysis of the thermodynamic properties of the Witten black holes can be found in Bose et al.[13] As stated above, this theory is exactly solvable both classically and quantum mechanically, and there was hope that it would provide clues about the back reaction and end-point of the collapse/radiation process (see for example Mann[14] and Bose et al.[15] and references therein.).

13.3. Classical Vacuum Theory: Observables and Thermodynamics

The beauty of 2-d dilaton gravity is that it is simple enough to be tackled generically. A systematic analysis of the generic theory was undertaken in the early to mid 1990's by several groups using a variety of techniques. One can prove a Birkhoff theorem[7] for arbitrary potential $V(\phi)$, i.e. for the most general theory, and explicitly write down all the solutions in terms of a single physical parameter, which can be interpreted as the total energy. The physical observables and thermodynamics properties associated which such black hole solutions were initially derived for various specific 2-d theories,[14,16] but it is possible to do a completely general analysis for the generic action in Eq. (13.1) as well.[17,19] The basis for such an analysis is existence of a global Killing vector that can be written down in covariant form for the generic theory in terms of the dilaton[17]:

$$k^{\mu} = \eta^{\mu\nu}\phi_{,\nu}. \tag{13.5}$$

The vanishing of the norm of this Killing vector signals as usual the presence of a Killing horizon. When the spatial slice $\phi = 0$ is excluded from the spacetime this Killing horizon is indeed an event horizon that provides a boundary between the interior of the black hole and the asymptotic region. In the context of the metric $g_{\mu\nu}$ it is not obvious that the surface $\phi = 0$ has a curvature singularity and indeed for some solutions this surface is completely regular. In this regard, one observes that the metric (13.3) is not generically asymptotically flat, but if $j(\phi)$ diverges for large ϕ, one can define a physical metric by a conformal rescaling:

$$ds^2_{phys} = \frac{1}{j(\phi)}ds^2 = -(1 - 2GMl/j(\phi))dt^2 + (1 - 2GMl/j(\phi))^{-1}\left(\frac{d\phi}{j(\phi)}\right)^2. \tag{13.6}$$

The physical metric (13.6) is asymptotically flat and has a singularity at $\phi = 0$. When the dilaton gravity theory corresponds to the spherical reduction of D-dimensional gravity, this physical metric is precisely the radial part of the higher dimensional metric, as can be verified by changing coordinates to $\phi = r^{D-2}$ so that $j \propto r^{D-3}$. This correspondence also points out that the dilaton has a geometrical interpretation as the area of a sphere at fixed r. More generally, Cadoni[8]showed for power law potentials $V \sim \phi^{-b}$ that black hole solutions exist providing $-1 \leq b < 1$. The dilaton potential for spherically symmetric gravity in D spacetime dimensions is of this form with $b = 1/(D-2)$.

The energy of black hole solutions can be written in covariant form using the Killing field (13.5). In particular, the mass observable M:

$$2M = j(\phi) - |\nabla k|^2, \tag{13.7}$$

is constant on-shell and a Hamiltonian analysis[19] confirms that it corresponds to the ADM mass.

From the expression for the mass observable (13.7), one can readily derive the thermodynamic properties such as the temperature, surface gravity and entropy. Clearly the horizon location is a surface of constant dilaton field $\phi = \phi_h$ given by:

$$j(\phi_h) = 2M. \tag{13.8}$$

Variation of the above gives the analogue of the "first law of black hole mechanics" in this simple context:

$$\delta M = J_{,\phi}(\phi_h)\delta\phi_h = V(\phi_h)\delta\phi_h, \tag{13.9}$$

where V is the dilaton potential. A direct calculation of the black hole surface gravity gives:

$$\kappa \propto V(\phi_h). \tag{13.10}$$

From this one can identify the value of the dilaton at the horizon as the analogue of the black hole entropy. This is to be expected from the fact that in dimensionally reduced spherically symmetric gravity $\phi = r^{D-2}$ is the area of a surface of constant radius. It can also be verified directly from the 2-d theory by using Wald's method.[22]

The analysis above shows the beauty, simplicity and universality of the formalism: the thermodynamic properties of black holes are fundamental and completely generic[b]. The diffeomorphism invariance allows for the possibility that generic dilaton gravity may help to provide a deeper understanding of the microscopic source of the thermodynamics, which is likely rooted in the quantum structure of the diffeomorphism group. We will now show that the subtle blend of simplicity and underlying complexity is also manifest in the quantum theory.

13.4. Hamiltonian Structure and Reduced Theory

In the following, we present a summary of the canonical quantization of generic dilaton gravity in terms of geometric variables following Louis-Martinez et al.[19] We again emphasize that this class of theories has been quantized by a variety of authors. The Vienna group used the very elegant Poisson-Sigma Model approach to perform a complete classification of the

---

[b]For a recent review of the thermodynamics of 2-d black holes see Grumiller and McNees.[23]

classical solutions of the model coupled to a Yang-Mills field.[42] They were able to classify all the global solutions for the generic model and determine physical quantum states, determining the mass spectra in some cases. Interesting results have also been obtained via path integral methods (see for example the references cited in Ref. 43).

The first step is a Hamiltonian analysis, which has been well documented in the literature, so we give only essential details. The metric is first parametrized in ADM form:

$$ds^2 = e^{2\rho} \left(-\sigma^2 dt^2 + (dx + Ndt)^2 \right).$$ (13.11)

which leads to the action:

$$I[\rho, \phi] = \int d^2x \left(\Pi_\rho \dot{\rho} + \Pi_\phi \dot{\phi} - (\sigma \mathcal{G} + N \mathcal{F}) - H_B \right),$$ (13.12)

where H_B is the boundary term needed to make the variational derivative of the action well defined, the lapse σ and shift N are Lagrange multipliers that enforce the Hamiltonian and diffeomorphism constraints, respectively:

$$\mathcal{G} := \frac{\phi''}{G} - \frac{\phi' \rho'}{G} - G \Pi_\phi \Pi_\rho - \frac{e^{2\rho}}{2G} \frac{V(\phi)}{l^2} \approx 0,$$ (13.13)

$$\mathcal{F} := \rho' \Pi_\rho - \Pi_\rho' + \phi' \Pi_\phi + \psi' \Pi_\psi \approx 0.$$ (13.14)

The presence of the two Lagrange multipliers and two first class constraints means that the number of degrees of freedom are (heuristically): 3 metric components + 1 dilaton - 2 lagrange multipliers - 2 constraints = 0. This only applies to the field theoretic degrees of freedom. It can be shown that the mass function:

$$\mathcal{M} = \frac{l}{2G} \left(e^{-2\rho}((G\Pi_\rho)^2 - (\phi')^2) + \frac{j(\phi)}{l^2} \right),$$ (13.15)

commutes with the constraints and hence is a physical observable in the Dirac sense. It is spatially constant on the constraint surface and is equal to the boundary term H_B which is the ADM mass of the system.

The momentum canonically conjugate to the mass observable M (13.7) can be written covariantly[17] as an integral over the Killing vector field k^μ, and corresponds generically to the Schwarzschild time separation at infinity as shown[20] in the case of spherically symmetric 4-D gravity[c]. The complete phase space is therefore two dimensional and can be coordinatized by the

---

[c]An earlier derivation of the phase space observables for spherically symmetric gravity was given by Thiemann and Kastrup[21] in terms of Ashtekar variables.

mass, M and its canonical conjugate, P_M. For suitably chosen boundary conditions, the fully reduced action given in terms of M and P_M is simply:

$$I_{red} = \int dt \left(P_M \dot{M} - M \right).$$ (13.16)

The resulting equations of motion imply that M is time independent and that P_M is equal to the time coordinate. This elegant coordinatization, while geometrically motivated, nonetheless makes it difficult to extract further information about the system without further assumptions.

13.5. Semi-Classical Arguments and the Area Spectrum

We now show that it is possible to use the laws of black hole mechanics/thermodynamics to make very intriguing general arguments about the generic semi-classical black hole mass/area spectrum in terms of the fully reduced phase space variables. Bekenstein[25] and then Bekenstein and Mukhanov[26] conjectured that the area of 4-D black holes is an adiabatic invariant whose semi-classical spectrum must, by the Bohr-Sommerfeld quantization condition, be equally spaced. A modern, completely general, rationale (it is not quite a proof) for this claim goes like this: Suppose that there is a natural frequency, ω, associated with the dynamics of Schwarzschild-like black holes, which by definition have a single horizon completely parametrized by a single dimensionful variable that can without loss of generality be taken as the mass or energy, E. Although at first glance the notion of oscillatory motion of event horizons may seem far fetched since one normally thinks of black holes as static with no dynamics whatsoever, we will see below that there do exist candidates for such black hole vibrational frequencies. Moreover, as argued above, the frequency $\omega(E)$ is a function of at most the energy of the black hole.

Generally, a dynamical system with an energy dependent natural frequency $\omega(E)$ has an associated adiabatic invariant that is given up to an additive constant by the indefinite integral:

$$I = \int \frac{dE}{\omega(E)}.$$ (13.17)

By virtue of the Bohr-Sommerfeld quantization condition, the semi-classical energy spectrum is given by :

$$I = nh, \qquad n >> 1.$$ (13.18)

Incidentally, the above argument is equivalent to the Bohr correspondence principle which states that for large quantum numbers the classical frequency is proportional to the change in energy due to a quantum transition between adjacent states: $\Delta E = \hbar\omega(E)\Delta n$. For large n, $\Delta n = 1$ can be treated as infinitesmal which immediately implies the differential forms of (13.17) and (13.18).

In the mid-nineties, it was noticed by two groups[27,28] that there is a natural candidate for such an oscillation frequency, namely:

$$\omega(E) = \frac{k}{\hbar} T_{BH}.$$ (13.19)

This frequency corresponds to the inverse of the period in imaginary time of the Euclidean Gibbons-Perry instanton solution. An application of the semi-classical argument above[27] or a direct quantization of the reduced Hamiltonian in Euclidean time[28] (13.16) both yield an equally spaced area/entropy spectrum:

$$A = 2\pi(n + 1/2)\hbar G.$$ (13.20)

Analoguous arguments were later applied to the quantization of charged[29] and rotating[30] black holes.

An intriguing proposal for the vibrational frequency of black holes came from Hod[32] in 1998 who argued that the frequency of the highly damped quasinormal modes of black holes are the vibrational frequencies that determine the semi-classical quantum spectrum for black holes. This proposal not only gave an equally spaced area spectrum, but the spacing that results from this choice has tantalizing consequences: the Bekenstein-Hawking entropy takes the form $S = k\ln(3^n)$, which is consistent the statistical mechanical entropy that one might associate with an event horizon built out of n elements of area, each with three allowed microscopic states. Hod's argument went more or less unheeded until Dreyer[33] showed that such a spectrum with precisely this spacing followed from LQG. Hod's argument is even more compelling in the light of the elegant calculations of Motl and Neitzke.[37] Using WKB methods, they were able to obtain analytic expressions for the highly damped QNM's of Schwarzschild black holes in higher dimensions that were consistent with a generalization of Hod's conjecture.[38] This lead to a veritable cottage industry in highly damped QNM calculations for a large variety of black holes that were designed to test the universal applicability of area spectrum derived by Hod. The results were somewhat mixed. In addition, it turned out that Dreyer's analysis was based on a incorrect expression for the entropy of LQG black holes[34] that has since been corrected.[35] The new expression seems on the face of it inconsistent with Hod's conjecture, but a slightly different interpretation of what one means by spherically symmetric black holes in the LQG context can be used to bring the two expressions back in line.[36]

The argument for the universal applicability of Hod's conjecture was revitalized by very recent paper of Maggiore,[39] who used an analogy with damped harmonic oscillators to argue that the physical black hole frequency is not the real part of the QNM frequency. Instead, the physically relevant frequency is given by:

$$\omega_0 = \sqrt{\omega_R^2 + \omega_I^2} \,, \qquad (13.21)$$

where ω_R (ω_I) are the real and imaginary parts, respectively. In this case, the imaginary part, not the real part, dominates the expression in the high damping limit. Moreover, ω_I appears to be more or less universal, since it is determined by the periodicity of the Euclidean instanton solution. The spectrum that results from Maggiore's reinterpretation is precisely (13.20).

The unifying theme that has emerged is that all semi-classical quantization schemes appear to give an equally spaced area spectrum for Schwarzschild-like black holes in the semi-classical limit, albeit with different spacings. The apparent universality can be understood at a very basic level by noting that for Schwarschild-like black holes that are completely specified by a single dimensionful parameter (the mass/horizon radius/surface gravity), the only natural time scale is the time for light to cross the horizon, or $2GM/c^3$, whose inverse gives a frequency that is proportional to $kT/\hbar$. This fact, plus the first law of black hole thermodynamics, invariably produces an equally spaced area spectrum[d]. A strong hint that this argument is physically relevant (despite our uncertainty about what precisely constitutes the physical vibration frequency of a black hole) comes from the recent work[40] wherein the LQG spectrum is calculated. They find a periodic structure, not unlike interference fringes, in the spectrum. This periodic structure is highly suggestive of an equally spaced area spectrum, and provides startling evidence for the relevance of the above semi-classical arguments.

---

[d]In a somewhat different vein, Louko and Makela[31] performed a rigorous quantization of a reduced theory in which the radius of the throat of the Einstein-Rosen bridge provided the physical observable. Although strictly speaking there was no periodic motion in this model, they also obtained an equally spaced area spectrum, presumably for reasons that could be traced to the dimensional arguments above.

13.6. Dirac Quantization

We have shown that despite the lack of local degrees of freedom, the completely reduced theory can contain some interesting semi-classical information about the area/entropy spectrum of black holes. However, to gain insight into the microscopic source of black hole entropy it is necessary to go deeper. By going to the completely reduced theory, one might be throwing out the baby with the bathwater. Carlip has argued[46] that the entropy of black holes can be understood as a consequence of boundary conditions at the horizon which effectively break diffeomorphism invariance and result in "would-be gauge degrees of freedom" becoming physical. A few years ago, he examined this issue in the context of generic 2-d dilaton gravity and found[46] that indeed the black hole boundary conditions at the horizon resulted in a modified (anomalous) constraint algebra which, when quantized, provided the right number of microstates to account for the black hole entropy.

Remarkably, Professor Kummer's last paper on the subject of 2-D dilaton gravity[47] shed a different and interesting new light on this type of analysis. In this paper it is shown that physical degrees of freedom on the horizon can, by imposition of horizon constraints, be converted to gauge degrees of freedom, in agreement with a conjecture by 't Hooft[48] and in apparent contradiction with the results of Carlip. As argued by Bergamin *et al.*,[49] the two sets of results can in some sense be viewed as complementary descriptions of entropy in terms of inaccessible states, but clearly more work is required to understand this issue fully. It is therefore useful to explore the quantum behaviour of unreduced generic 2-D dilaton gravity.

13.6.1. *Exact Dirac Wave Functionals*

It turns out to be possible in the generic theory to write down candidates for exact mass eigenstates which solve the quantized constraints.[18] These solutions were first found in the specific case of Jackiw-Teitelboim gravity by Henneaux.[41]

In terms of the ADM parametrization given in Eq. (13.11), Dirac quantization in the Schrodinger representation entails a search for functionals $\Psi_M[\phi, \rho]$ that satisfy the quantized version of the diffeomorphism and Hamiltonian constraints. Following Henneaux[41] one first solves the two constraints classically to obtain an expression for the momenta in terms of

ρ and ϕ:

$$\Pi_\phi = \frac{g[\phi, \rho]}{Q[\phi, \rho; M]},$$

$$\Pi_\rho = Q[\phi, \rho; M], \tag{13.22}$$

where

$$g[\phi, \rho] = 4\phi'' - 4\phi'\rho' + 2e^{2\rho}V(\phi)$$

$$Q[\phi, \rho] = 2\sqrt{(\phi')^2 + (2M - j(\phi))e^{2\rho}} . \tag{13.23}$$

M is a constant of integration that corresponds, as the notations suggests, to the black hole mass. If one replaces the conjugate momenta by the standard operators:

$$\hat{\Pi}_\phi = -i\hbar \frac{\delta}{\delta\phi(x)},$$

$$\hat{\Pi}_\rho = -i\hbar \frac{\delta}{\delta\rho(x)}. \tag{13.24}$$

it is straightforward to integrate the quantum version of (13.22). The result is:

$$\Psi_M[\phi, \rho; M] = \exp\left(\frac{i}{\hbar}S[\phi, \rho; M]\right), \tag{13.25}$$

where:

$$S[\phi, \rho; M] = \int dx \left[Q + \phi' \ln\left(\frac{2\phi' - Q}{2\phi' + Q}\right)\right]. \tag{13.26}$$

The wave-functional Ψ_M satisfies the quantum diffeomorphism and Hamiltonian constraints with a particular (non-standard) choice of factor ordering and is an eigenstate of the quantum version of the mass function, again with a particular factor ordering.

Although the interpretation of Ψ_M as an exact physical mass eigenstate has difficulties related to the choice of functional measure and self-adjointness of the relevant operators, the phase S has a natural and un-ambiguous interpretation as the Hamilton-Jacobi function for the classical theory that derives from the classical constraints. Thus, the wave-functional (13.25) is at least correct to lowest order in the WKB approximation and has some interesting quantum properties: the phase is imaginary in the classically forbidden regions, $Q^2 < 0$ and $4\phi'^2 - Q^2 < 0$. The latter corresponds to the region below the horizon, again forbidden along a Schwarzschild slice.

These imaginary parts were given an interesting, albeit speculative, interpretation[18] for classically forbidden configurations of ρ that correspond to Schwarzschild slices with a mass parameter, m, different from the mass eigenvalue M of the wave functional. By defining the probability amplitude for the black hole in an eigenstate of the mass function with eigenvalue M to have mass $m \neq M$ as: $P[M] \propto |\psi_M[m; M]|^2$, it was found that the relative probability of having mass $m = M$ to having no mass $m = 0$ was:

$$\frac{P[m = M]}{P[m = 0]} = \exp 2\pi \frac{M^2}{m_{pl}^2}. \tag{13.27}$$

This expression can be interpreted as the inverse of the tunnelling probability from a state with mass M to the vacuum state. Remarkably, it is proportional to the exponential of the black hole entropy (up to a factor of two), so that this interpretation is consistent with the fact that in statistical mechanics the exponential of the entropy is equal the number of accessible microstates.

13.6.2. *Partially Reduced Theory*

It is possible to implement a procedure that is part way between complete gauge fixing at the classical level and the Dirac quantization of the completely unreduced theory. Since one expects the resolution of many key issues in quantum gravity to reside in the Hamiltonian constraint, it makes sense to choose a partial gauge fixing which eliminates only the diffeomorphism constraint and leaves the Hamiltonian constraint to be implemented as an operator constraint via the Dirac prescription. A few years ago Husain and Winkler[52] started a program designed to formulate the quantum dynamics of black hole formation in four dimensions. They partially fixed the gauge so as to allow slicings that were regular across the horizon. Their boundary conditions were consistent with the so-called "flat slice" or Painleve-Gullstrand (PG) coordinates. A similar program was initiated[51] for the generic theory in which the analogue of PG coordinates takes the form:

$$ds^2 = j(\phi) \left(-dt^2 + (dx + \sqrt{\frac{2GMl}{j}} dt)^2 \right). \tag{13.28}$$

The partial gauge choice was therefore $\phi'(x) = j(\phi)/\ell$, which when the diffeomorphism constraint is imposed strongly and the corresponding Lagrange multiplier, i.e. the shift function, is fixed so as to preserve the gauge fixing condition, leaves a partially reduced action of the form:

$$I = \int dx P \dot{X} - \int dx \left(-\frac{\sigma X^2}{2j(\phi)} \mathcal{M}' \right) + \int dx (\frac{\sigma X^2}{j(\phi)} \mathcal{M})',$$

$$(13.29)$$

where we have done a canonical transformation from ρ, Π_ρ to $X = e^\rho$ and its conjugate P, and M is again the mass function, which in this class of partially-fixed gauges is:

$$\mathcal{M} := \frac{l}{2G} \left(P^2 - \frac{j(\phi)^2}{X^2} + \frac{j(\phi)}{l^2} \right).$$

$$(13.30)$$

One can now satisfy the Hamiltonian constraint quantum mechanically by finding eigenstates of the mass function. Remarkably, the chosen partial gauge fixing results in a mass function that no longer couples different spatial points, so that the eigenvalue problem reduces to a set of decoupled quantum mechanical systems, each of which corresponds to that of a particle moving in an attractive $1/X^2$ potential. The quantization of the $1/X^2$ potential has been extensively studied in part because of the scale invariance and $SO(2,1)$ symmetry algebra that are broken at the quantum level.

The eigenstates of the mass function were found using two distinct quantization schemes, with interesting, but somewhat distinct results: Bohr, or polymer quantization[51] for fixed mass M forced a non-trivial discretization of the spatial slice: $j(\phi)$ can take on only a countable infinity of discrete values. Schrödinger quantization,[50] on the other hand, yielded solutions to the quantum Hamiltonian constraint in terms of (generalized) eigenstates of the ADM mass operator and allowed the specification of a physical inner product in such a way as to guarantee self-adjointness of the time operator affinely conjugate to the ADM mass. The interesting result there was that regularity of the time operator across the horizon gave rise to a factor ordering term that distinguished the future and past horizons, and gave rise to a quantum correction to the black hole surface gravity.

13.7. Conclusion

Our discussion so far has dealt with vacuum 2-d dilaton gravity. Despite its underlying simplicity, it has the potential to yield significant insights into the underlying microscopic quantum theory. Of course, in order to examine important issues such as Hawking radiation, the end-point of gravitational collapse and the quantum dynamics of black hole formation it is necessary to add matter. This is a difficult problem for arbitrary matter couplings but is tractable in the case of conformal coupling.[12] A first step in this direction for general matter couplings in the generic theory has been taken recently[52,53] by deriving the gravity-matter Hamiltonian for a massless scalar field with partial gauge fixing $\phi' = j(\phi)/\ell$. This Hamiltonian takes a rather simple and suggestive form:

$$
H(X, P, \psi, \Pi_\psi) = \int dx \left(-\frac{\sigma X^2}{j(\phi)}\mathcal{M}' + \sigma\mathcal{G}_M + \sigma l\frac{X P \psi' \Pi_\psi}{j(\phi)} \right)
$$
$$
+ \int dx (\frac{\sigma X^2}{j(\phi)}\mathcal{M})', \qquad (13.31)
$$

where $\mathcal{G}_\mathcal{M}$ is the matter energy density:

$$
\mathcal{G}_M := \frac{1}{2} \left(\frac{\Pi_\psi^2}{h(\phi)} + h(\phi)(\psi')^2 \right). \qquad (13.32)
$$

There is no space to describe this model in detail, but this Hamiltonian has some potentially useful properties. There is a clean separation between the pure gravitational sector (the first term), the matter sector (second term) and a quartic interaction (third term). This rather simple form is a direct consequence of our partial gauge fixing: the function $\phi(x)$ is no longer dynamical but a fixed function of the spatial coordinates.

One possible approach for solving the Hamiltonian constraint may be as follows: one can take as a hopefully complete basis the eigenstates of the mass function found via Bohr quantization in Ref. 51 or via Schrödinger quantization in Ref. 50. In addition, one can use standard techniques to find a complete basis of states for the scalar field using just the matter term in the Hamiltonian constraint. The interaction term can then be formally expressed in the corresponding direct product basis, allowing the Hamiltonian constraint to be solved using perturbative techniques. Perturbation theory will likely not be valid near the singularity but may be relevant near the horizon of macroscopic black holes. One can thus hope to address interesting questions related to Hawking radiation, including the emergence of the standard semi-classical approximation and quantum corrections to geometrical quantities such as surface gravity.

Acknowledgments

This work was supported in part by the Natural Sciences and Engineering Research Council of Canada.

References

1. For a review of black holes in string theory, see S.D. Mathur, Class. Qu. Gravity **23** (2006) R115.
2. T. Thiemann, Modern Canonical Quantum General Relativity (Cambridge University Press, Cambridge, 2007).
3. B Berger, D.M. Chitre, V.E. Moncrief and Y. Nutka, Phys. Rev. **D8** (1973) 3247.
4. W.G. Unruh, Phys. Rev. **D14** (1976) 870.
5. D. Grumiller, W. Kummer and D.V. Vassilevich, Phys. Rep. **369** (2002) 327.
6. T. Banks and M. O'Loughlin, Nucl. Phys. **B362** (1991) 649.
7. D. Louis-Martinez and G. Kunstatter, Physical Review **D49** (1994) 5227-5230.
8. M. Cadoni, Phys. Rev. **D53** (1991) 4413.
9. R. Jackiw, in *Quantum Theory of Gravity*, edited by S. Christensen (Hilger, Bristol, 1984), p. 403; C. Teitelboim, *ibid* p. 327.
10. M. Bañados, C. Teitelboim, J. Zanelli, Phys. Rev. Lett. 69 (1992) 1849-1851.
11. E. Witten, Phys. Rev. **D44** (1991) 314-324.
12. C.G. Callan, S.B. Giddings, J.A. Harvey and A. Strominger, Phys. Rev. D45, R1005 (1992). See also S. Hawking, Phys. Rev. Lett. 69, 406 (1992).
13. S. Bose, J. Louko, L. Parker and Y. Peleg, Phys. Rev. D53 (1996) 5708-5716.
14. R.B. Mann, Nucl. Phys. **B418** (1994) 231-256.
15. S. Bose, L. Parker and Y. Peleg, Phys. Rev. D52 (1995) 3512-3517.
16. J.P. Lemos and P.M. Sa, Phys. Rev. D49 (1994) 2897-2908; Erratum-ibid. D51 (1995) 5967-5968.
17. J. Gegenberg, G. Kunstatter and D. Louis-Martinez, Phys. Rev. D51 (1995) 1781-1786.
18. J. Gegenberg and G. Kunstatter, Phys. Rev. D47 (1993) 4192-4195.
19. D. Louis-Martinez, J. Gegenberg and G. Kunstatter, Phys. Lett. B321 (1994) 193-198.
20. K. Kuchar, Phys. Rev. D50 (1994) 3961-3981.
21. T. Thiemann and H.A. Kastrup, Nucl. Phys. **B399** (1993) 211-258;Nucl. Phys. B425 (1994) 665-686.
22. R.M. Wald, Phys. Rev **D48** (1993) R3427.
23. D. Grumiller, R. McNees, JHEP 0704:074 (2007).
24. T. Strobl, Phys. Rev. D50 (1994) 7346-7350.
25. J. Bekenstein, Lett. Nuovo Cim. 11:467,1974.
26. J.D. Bekenstein and V.F. Mukhanov, Phys. Lett. **B360** (1995) 7-12.
27. H.A. Kastrup, Phys. Lett. B385 (1996) 75-80.
28. A. Barvinsky and G. Kunstatter, Phys. Lett. B389 (1996) 231-237.
29. A. Barvinsky, S. Das, G. Kunstatter, Found. Phys. 32 (2002) 1851-1862.
30. G. Gour, A.J.M. Medved, Class. Quant. Grav. 20 (2003) 1661-1672.
31. J. Louko, J. Makela, Phys. Rev. D54 (1996) 4982-4996.
32. S. Hod, Phys. Rev. Lett. 81 (1998) 4293.
33. Olaf Dreyer, Phys. Rev. Lett. 90 (2003) 081301.

34. A. Ashtekar, J. Baez, A. Corichi, K. Krasnov, Phys. Rev. Lett. 80 (1998) 904-907.
35. M. Domagala, J. Lewandowski, Class. Qu. Grav. 21 (2004) 5233-5244.
36. O. Dreyer, F. Markopoulou, L. Smolin, Nucl. Phys. B744 (2006) 1-13.
37. L. Motl, Adv. Theor. Math. Phys. 6 (2003) 1135-1162; L. Motl and A. Neitzke, Adv. Theor. Math. Phys. 7 (2003) 307-330.
38. G. Kunstatter, Phys. Rev. Lett. 90 (2003) 161301.
39. M. Maggiore, Phys. Rev. Lett. **100** (2008), 141301.
40. A. Corichi, J Diaz-Polo, E. Fernandez-Borja, Phys. Rev. Lett. 98, 181301 (2007); I. Agullo, J. Fernando Barbero, J. Diaz-Polo, E. Fernandez-Borja, E. J. S. Villaseñor, Phys. Rev. Lett. 100 (2008) 211301.
41. M. Henneaux, Phys. Rev. Lett. **54** (1985) 959.
42. T. Klosch, P. Schaller and T. Strobl, Helv. Phys. Acta **69** (1996) 305-308.
43. A. Chamseddine, Phys. Lett. B256 (1991), 379; B258 (1991), 97. S.P. de Alwis, Phys. Lett. B289 (1992), 278.E. Elizalde, S. Naftulin, S.D. Odintsov, Int. J. Mod. Phys. A9 (1994) 933-952. K. Kirsten, S. Odintsov, Mod. Phys. Lett. A9 (1994) 2761-2766. D. Grumiller, W. Kummer and D.V. Vassilevich, JHEP 0307 (2003), 009.
44. D. Louis-Martinez and G. Kunstatter, Phys. Rev. D52 (1995) 3494-3505.
45. R.B. Mann, Phys. Rev. D47 (1993) 4438-4442.
46. S. Carlip, Class. Quant. Grav. 22 (2005) 1303-1312.
47. L. Bergamin, D. Grumiller, W. Kummer and D.V. Vassilevich, Class. Quant. Grav. **23** (2006) 3075-3101.
48. G. 't Hooft, "Horizons", gr-qc/0401027 (2004).
49. L. Bergamin and D. Grumiller, Int. J. Mod. Phys. **D15** (2006) 2279-2284.
50. G. Kunstatter and J. Louko, Phys. Rev. D75 (2007) 024036.
51. J. Gegenberg, G. Kunstatter and D. Small, Class. Quant. Grav. 23 (2006) 6087-6100.
52. V. Husain, O. Winkler, Phys. Rev. D71 (2005) 104001; Phys. Rev. D73 (2006) 124007.
53. R. Daghigh, J. Gegenberg and G. Kunstatter, Class. Qu. Grav. 24 (2007) 2099-2107.

Chapter 14

Global Solutions in Gravity. Euclidean Signature

Michael O. Katanaev

Steklov Mathematical Institute,
ul. Gubkina, 8, Moscow, 117966, Russia
E-mail: katanaev@mi.ras.ru

We consider a wide class of two-dimensional metrics having one Killing vector. The method is proposed for the construction of maximally extended surfaces with the given Riemannian metric which is the analog of the conformal block method for two-dimensional Lorentzian signature metrics. The Schwarzschild solution is considered as an example.

14.1. Introduction

For physical interpretation of solutions in different gravity models, one has not only to find the metric as a solution of the equations of motion but also to analyze the behavior of extremals (geodesics) corresponding, in particular, to trajectories of point particles. Therefore the construction of global solutions is of uttermost significance. By global solution we mean a pair $(\mathcal{M}, g)$ where $\mathcal{M}$ is a manifold and $g = \{g_{\alpha\beta}\}$ is a metric given on $\mathcal{M}$. The metric is to be found as a solution to some system of the Euler–Lagrange equations and a manifold is supposed to be maximally extended. The last requirement means that any extremal can be prolonged either to infinite value of the canonical parameter in both directions or it ends up at a singular point at a finite value of the canonical parameter where at least one of the geometric invariants is infinite or not defined. The well known example is the Kruskal–Szekerez extension of the Schwarzschild solution.[1,2]

In general, this problem is very complicated because it requires an exact solution of the equations of motion as well as the analysis of extremals. The case of spherically symmetric solutions in general relativity was analyzed by Carter in Ref. 3. The method of conformal blocks for constructions of global solutions for a wide class of two-dimensional metrics having one

Killing vector was proposed in Ref. 4. This method was developed as a result of construction and classification of all global solutions[5-7] in two-dimensional gravity with torsion.[8-10] The method of conformal blocks was also used for construction of global solutions in many two-dimensional dilaton gravity models[11,12] and for complete classification of global vacuum solutions in general relativity with a cosmological constant assuming that the four-dimensional space-time is a warped product of two surfaces with a block diagonal metric.[13] In the last case, all solutions were known locally. The analysis of global properties gave physical interpretation of many solutions which were earlier known only locally. Besides the black hole solutions, the vacuum Einstein equations have solutions describing cosmic strings, domain walls of curvature singularities, cosmic strings surrounded by domain walls, and other physically interesting solutions.

The method of conformal blocks is applicable for metrics of Lorentzian signature. At the same time, the Euclidean formulation of the theory plays important role in quantum field theory and statistical mechanics and allows in some cases to avoid difficulties related to the indefinite signature of the metric. In the present paper, the method of conformal blocks is generalized to a wide class of two-dimensional metrics of Euclidean signature having one Killing vector. We prove that there is the global solution for each conformal block with positive or negative definite metric. This method was applied earlier to two-dimensional gravity with torsion[14] and in the analysis of hyperbolically symmetric solutions in general relativity.[13]

14.2. Local form of the metric

We start with the Lorentz case to explain the choice of the Riemannian metric for the present paper. At first glance, the choice of Lorentzian metric may seem artificial, but many exact solutions of general relativity depending essentially on two coordinates can be written in this way. Besides, a general solution of the equations of motion of two-dimensional gravity has exactly this form in the conformal gauge.[15]

We consider a plane $\mathbb{R}^2$ with Cartesian coordinates $x^\alpha = \{\tau, \sigma\}$, $\alpha = 1, 2$. Two-dimensional metrics of constant curvature as well as many solutions of general relativity and other gravity models can be written in the form[4]

$$ds^2 = |N(q)|(d\tau^2 - d\sigma^2).$$ (14.1)

where the conformal factor $N(q) \in \mathcal{C}^l$, $l \geq 2$, is l times continuously differentiable function of one variable $q \in \mathbb{R}$ except finite number of singularities. Let variable q depend only on one coordinate, the dependence being given by the ordinary differential equation

$$\left|\frac{dq}{d\zeta}\right| = \pm N(q), \qquad (14.2)$$

with the following sign rule

$$\begin{aligned} N > 0: &\quad \zeta = \sigma, \quad + \text{ sign (static solution)}, \\ N < 0: &\quad \zeta = \tau, \quad - \text{ sign (homogeneous solution)}. \end{aligned} \qquad (14.3)$$

Equations (14.1) and (14.2) define four different metrics due to the modulus and $\pm$ signs in Eq. (14.2). There are two surfaces with static metrics and two surfaces with homogeneous metrics which differ by the sign of the derivative $dq/d\zeta$. Let us denote these domains by the Roman numbers,

$$\begin{aligned} \text{I}: &\quad N > 0, \quad dq/d\sigma > 0, \\ \text{II}: &\quad N < 0, \quad dq/d\tau < 0, \\ \text{III}: &\quad N > 0, \quad dq/d\sigma < 0, \\ \text{IV}: &\quad N < 0, \quad dq/d\tau > 0. \end{aligned} \qquad (14.4)$$

To clarify the form of the metric (14.1), we note that in domain I there are Schwarzschild like coordinates τ, q in which the metric becomes

$$ds^2 = N(q)d\tau^2 - \frac{dq^2}{N(q)}.$$

So the variable q can be interpreted there as the radius and the conformal factor as the g_{00} component of the metric.

To transform Lorentzian metric (14.1) to the Euclidean signature metric, we perform the rotation in the complex plane of the coordinate on which the conformal factor does not depend. The corresponding Riemannian metric is the solution of the same system of the Euler–Lagrange equations as the original Lorentzian metric (14.1) because the conformal factor does not depend on this coordinate. This transformation is given by the coordinate changes $\tau = i\rho$ and $\sigma = i\rho$ in domains I,III and II,IV, respectively. As a result, we obtain the metric

$$ds^2 = -N(q)(d\sigma^2 + d\rho^2), \qquad (14.5)$$

where we changed notations $\tau \to \sigma$ in domains II and IV. The sign of the conformal factor $N(q)$ is not fixed, and we consider both positive and negative definite metrics. After the transformation the variable q depends only on σ, this dependence being given by the ordinary differential equation

$$\left|\frac{dq}{d\sigma}\right| = |N(q)|. \tag{14.6}$$

The modulus signs are opened in the following way

$$
\begin{aligned}
\text{I}: &\quad N > 0, &\quad dq/d\sigma > 0, &\quad \operatorname{sign} g_{\alpha\beta} = (--), \\
\text{II}: &\quad N < 0, &\quad dq/d\sigma < 0, &\quad \operatorname{sign} g_{\alpha\beta} = (++), \\
\text{III}: &\quad N > 0, &\quad dq/d\sigma < 0, &\quad \operatorname{sign} g_{\alpha\beta} = (--), \\
\text{IV}: &\quad N < 0, &\quad dq/d\sigma > 0, &\quad \operatorname{sign} g_{\alpha\beta} = (++),
\end{aligned} \tag{14.7}
$$

where the signature of the metric (14.5) is shown in the last column. Metrics in domains I and III as well as in domains II and IV are essentially the same because they are related by the transformation $\sigma \to -\sigma$.

Riemannian metric defined by Eqs. (14.5) and (14.6) is the subject of the present paper. We admit the conformal factor to have zeroes and singularities at a finite number of points q_i, $i = 1, \ldots, k$. Infinite points $q_1 = -\infty$ and $q_k = \infty$ are included in this sequence. In this way the real line q is divided on intervals by points q_i in which the conformal factor is either strictly positive or negative. We consider power behavior of the conformal factor near the boundary points q_i:

$$|q_i| < \infty: \quad N(q) \sim |q - q_i|^m, \tag{14.8}$$

$$|q_i| = \infty: \quad N(q) \sim |q|^m. \tag{14.9}$$

For finite q_i, the exponent differs from zero, $m \neq 0$, because the conformal factor either equals zero or singular by assumption. In a general case, the exponent m can be an arbitrary real number. Horizons of the space-time correspond to zeroes of the conformal factor at finite points $|q_i| < \infty$ in the Lorentzian case.[5]

Metric (14.5) has at least one Killing vector $K = \partial_\rho$, its length being $-N(q)$.

Christoffel's symbols are defined by the metric,

$$\Gamma_{\alpha\beta}{}^\gamma = g^{\gamma\delta}(\partial_\alpha g_{\beta\delta} + \partial_\beta g_{\alpha\delta} - \partial_\delta g_{\alpha\beta}),$$

and have the following nontrivial components

$$\text{I, II}: \quad \Gamma_{\sigma\sigma}{}^\sigma = \Gamma_{\sigma\rho}{}^\rho = \Gamma_{\rho\sigma}{}^\rho = -\Gamma_{\rho\rho}{}^\sigma = \frac{1}{2}N', \tag{14.10}$$

$$\text{III, IV}: \quad \Gamma_{\sigma\sigma}{}^\sigma = \Gamma_{\sigma\rho}{}^\rho = \Gamma_{\rho\sigma}{}^\rho = -\Gamma_{\rho\rho}{}^\sigma = -\frac{1}{2}N', \tag{14.11}$$

where prime denotes the derivative with respect to the argument, $N' = dN/dq$, and there is no summation over indices ρ and σ. The curvature tensor in our notations is

$$R_{\alpha\beta\gamma}{}^{\delta} = \partial_\alpha \Gamma_{\beta\gamma}{}^{\delta} - \Gamma_{\alpha\gamma}{}^{\epsilon}\Gamma_{\beta\epsilon}{}^{\delta} - (\alpha \leftrightarrow \beta).$$

It is the same in all domains and has only one independent component

$$R_{\sigma\rho\sigma}{}^{\rho} = \frac{1}{2}NN'' \tag{14.12}$$

Nonzero components of the Ricci tensor and the scalar curvature are

$$R_{\sigma\sigma} = R_{\rho\rho} = \frac{1}{2}NN'', \tag{14.13}$$

$$R = -N''. \tag{14.14}$$

As the consequence, the conformal factor N is the second power polynomial for constant curvature surfaces $R = \mathsf{const}$.

According to Eq. (14.14), the scalar curvature is singular at q_i for the following exponents in the power behavior in Eq. (14.8):

$$|q_i| < \infty: \qquad m < 0, \;\; 0 < m < 1, \;\; 1 < m < 2, \tag{14.15}$$

$$|q_i| = \infty: \qquad m > 2. \tag{14.16}$$

At infinite boundary points $q_i = \pm\infty$ the scalar curvature goes to nonzero constant for $m = 2$ and to zero for $m < 0$. Note that the nonzero value of the scalar curvature at finite points $|q_i| < \infty$ can occur also for $m = 1$ due to the next power corrections in expansion (14.8).

The range of definition for metric (14.5) on the σ, ρ plane depends on the conformal factor. Coordinate $\rho \in \mathbb{R}$ runs through all real line because nothing depends on it, but the range of definition of coordinate σ is defined by Eq. (14.6). We have finite, semifinite or infinite interval for coordinate σ in the each interval $q \in (q_i, q_{i+1})$ depending on whether the integral

$$\sigma \sim \int^{q_i, q_{i+1}} \frac{dq}{N(q)} \tag{14.17}$$

converge or diverge at the boundary points. Depending on the exponent m we have

$$\begin{aligned} |q_i| < \infty: & \quad \begin{cases} m < 1, & \text{converge}, \\ m \geq 1, & \text{diverge}, \end{cases} \\ |q_i| = \infty: & \quad \begin{cases} m \leq 1, & \text{diverge}, \\ m > 1, & \text{converge}, \end{cases} \end{aligned} \tag{14.18}$$

Coordinate σ runs through all real line, $\sigma \in (-\infty, \infty)$ if the integral diverge at both ends of the interval (q_i, q_{i+1}), and the metric is defined on the whole plane $\sigma, \rho \in \mathbb{R}^2$. If at one boundary point q_{i+1} or q_i the integral converge, then the metric is defined on the half plane $\sigma \in (-\infty, \sigma_{i+1})$ or $\sigma \in (\sigma_i, \infty)$, respectively. The choice of boundary points σ_{i+1} and σ_i is arbitrary, and without loss of generality, we can put $\sigma_{i,i+1} = 0$. If the integral converges at both boundary points, then the solution is defined on the strip $\sigma \in (\sigma_i, \sigma_{i+1})$, and only one of the ends of the interval can be set to zero.

To construct the maximally extended surface in the Lorentzian case, we attribute the conformal block of definite shape to each interval (q_i, q_{i+1}) and then glue them together. For the Euclidean signature metric, there is no need for this procedure because "lightlike" extremals with asymptotics $\rho = \pm\sigma + \mathsf{const}$ are shown to be absent.

The value of variable q and hence the scalar curvature are constant along Killing trajectories $\sigma = \mathsf{const}$. Variable q is monotonically increasing on σ in domains I,IV and monotonically decreasing in domains II,III according to the definition of the domains (14.7).

14.3. Extremals

To describe maximally extended surface for metric (14.5) we must analyze the behavior of extremals $x^\alpha(t) = \{\sigma(t), \rho(t)\}$, $t \in \mathbb{R}$ given by the system of ordinary differential equations

$$\ddot{x}^\alpha = -\Gamma_{\beta\gamma}{}^\alpha \dot{x}^\beta \dot{x}^\gamma,$$

where dot denotes the derivative with respect to the canonical parameter t which is defined up to a linear transformation. For definiteness, consider domain I. Expressions for Christoffel's symbols (14.10) yield the system of equations for extremals

$$\ddot{\sigma} = \frac{1}{2}N'(\dot{\rho}^2 - \dot{\sigma}^2), \qquad (14.19)$$

$$\ddot{\rho} = -N'\dot{\sigma}\dot{\rho}. \qquad (14.20)$$

It has two first integrals

$$-N(\dot{\sigma}^2 + \dot{\rho}^2) = C_0 = \mathsf{const}, \qquad (14.21)$$

$$-N\dot{\rho} = C_1 = \mathsf{const}. \qquad (14.22)$$

The existence of the integral (14.21) allows one to choose the length of the extremal as a canonical parameter. The second integral (14.22) is connected to the existence of the Killing vector. Now we formulate the theorem determining extremals.

Theorem 14.1. *Any extremal in domain I belongs to one of the four classes.*
1. Straight extremals of the form (the analog of lightlike extremals)

$$\rho = \pm\sigma + \text{const}, \tag{14.23}$$

exist only for the Euclidean metric $N = \text{const}$ and the canonical parameter can be chosen as $\sigma = t$.
2. General type extremals which form is defined by equation

$$\frac{d\rho}{d\sigma} = \pm\frac{1}{\sqrt{-1 - C_2 N}}, \tag{14.24}$$

where C_2 is a negative constant. Canonical parameter is defined by any of the equations

$$\dot{\sigma} = \pm\frac{\sqrt{-1 - C_2 N}}{N}, \tag{14.25}$$

$$\dot{\rho} = \frac{1}{N}. \tag{14.26}$$

The signs plus or minus in Eqs. (14.24) and (14.25) must be chosen simultaneously.
3. Straight extremals parallel to axis σ and going through each point $\rho = \text{const}$. Canonical parameter is defined by equation

$$\dot{\sigma} = \frac{1}{\sqrt{N}}. \tag{14.27}$$

4. Straight degenerate extremals parallel to axis ρ and going through the points $\sigma_0 = \text{const}$ in which

$$N'(\sigma_0) = 0. \tag{14.28}$$

Canonical parameter can be chosen as

$$t = \rho. \tag{14.29}$$

The proof of this theorem repeats almost word by word the proof of the corresponding theorem in the Lorentzian case[4] and is not given here. Let us remind that constant C_2 is defined by the integrals (14.21) and (14.22)

$$C_2 = \frac{C_0}{C_1^2}. \tag{14.30}$$

Equation (14.24) shows that constant C_2 parameterizes the angle at which an extremal of general type goes through a given point.

Behavior of extremals for metrics with Euclidean signature is essentially different from that for Lorentzian metrics. First, the analog of lightlike extremals is absent for Riemannian metrics. This is important because lightlike extremals are incomplete on horizons and must be continued. This problem is absent in the Riemannian case. Second, Eqs. (14.24) and (14.25) differ from the corresponding equations in the Lorentzian case by the sign before the unity inside the square root. At first glance insignificant, this difference leads to the absence of "lightlike" asymptotics for extremals of general type as $q \to q_i$ near zeroes of conformal factor $N(q_i) = 0$ which define horizons.

Qualitative behavior of extremals of general type is easily analyzed. We consider domain I for definiteness. The conformal factor is positive in domain I, and extremals of a general type exist only for negative values of the constant C_2 because otherwise the right hand side of Eq. (14.26) becomes imaginary. The inequality

$$N(q) \geq -1/C_2.$$

must hold. For sufficiently large values of the modulus of C_2, this inequality determines the range of $q \in (q', q'')$ where the boundary points q' and q'' are given by equations $N(q) = -1/C_2$. This range definits points σ' and σ'' which correspond to q' and q''. The extremal of general type can not go out of the strip $\sigma \in (\sigma', \sigma'')$, $\rho \in (-\infty, \infty)$. Simple analysis of Eq. (14.26) shows that extremals of general type oscillate between the values σ' and σ'' as shown in Fig. 14.1a_3, b_3. Oscillating extremals of general type are always complete because the right hand side of Eq. (14.26) is bounded from the top by $1/N(q')$ or $1/N(q'')$ and bottom by $1/\max N(q)$.

If the conformal factor is equal to infinity in point q_{i+1} as shown in Fig. 14.1b, then an extremal of general type can start and end at the singular boundary. This boundary corresponds to finite value σ_{i+1} for $|q_{i+1}| < \infty$, $m < 1$ and $|q_{i+1}| = \infty$, $m > 1$. All these extremals approach the singular boundary at right angle because the right hand side of Eq. (14.24) goes to zero. Completeness of extremals of general type going to the singular boundary is defined by the integral

$$\lim_{q \to q_{i+1}} t \to \int^{q_{i+1}} \frac{dq}{\sqrt{N}}, \tag{14.31}$$

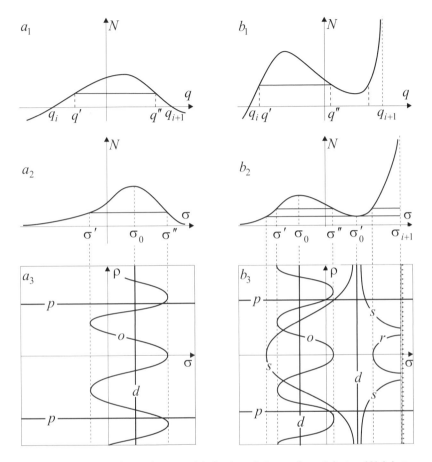

Figure 14.1   Top row (a_1, b_1): typical behavior of the conformal factor $N(q)$ between two zeroes and a zero and singularity. Middle row (a_2, b_2): dependence of the conformal factor $N(\sigma)$ on coordinate σ. Point σ_{i+1} can be either at finite or infinite distance from the origin. Bottom row (a_3, b_3): extremals of general type (o) for different values of C_2 oscillate between σ' and σ'' near local maximum. There is one degenerate extremal (d) which goes through every local extremum. There are extremals (s) that asymptotically approach the degenerate extremal at local minimum σ_0' and can end at the singular boundary σ_{i+1}. Extremals of general type (r) of finite length start and end at the singular boundary. All extremals can be arbitrary moved along axis ρ. Straight extremals (p) parallel to axis σ are going through every point ρ.

as the consequence of Eqs. (14.25) and (14.6). So completeness of extremals of general type which approach the singular boundary is the same as completeness of straight extremals parallel to axis σ (14.27). They are complete only for $|q_{i+1}| = \infty$ and $1 < m \leq 2$.

Typical behavior of the conformal factor between two zeroes and with two local extrema between zero and singularity is shown in the top row of Fig. 14.1, (a_1, b_1). In case $1b$ we assume $|q_{i+1}| < \infty$ and $m < 1$ or $1 < m < 2$ corresponding to the curvature singularity. In the middle row (a_2, b_2), we shaw the dependence of the conformal factors on coordinate σ. The value of coordinate σ_{i+1} is finite for $m < 1$ and infinite for $1 < m < 2$ in Fig. 14.1,b_2. Qualitative behavior of extremals on the σ, ρ plane is shown in the bottom row (a_3, b_3). Extremals of general type (o) oscillate near local maximum $N(\sigma_0)$ between σ' and σ'' which are determined by the value of constant C_2. These extremals (r) can also start and end at the singular boundary σ_{i+1} having a finite length. Degenerate extremals (d) go through every extremum of the conformal factor. There are also extremals of general type (s) which asymptotically approach the degenerate extremal going through local minimum σ'_0 when $\rho \to \pm\infty$. Part of these extremals end at the singular boundary σ_{i+1} at a finite value of the canonical parameter. All extremals can be shifted arbitrary along axis ρ. Straight extremals (p) parallel to axis σ go through every point ρ.

If both boundary points q_i and q_{i+1} of the interval are zeroes, then the conformal factor N has at least one maximum as the consequence of continuity through which goes the degenerate extremal (d). Degenerate extremals are always complete, i.e. have infinite length, because the canonical parameter coincides with coordinate ρ (14.29).

The above analysis shows that incompleteness of extremals in strips $\sigma \in (\sigma_i, \sigma_{i+1})$ and $\rho \in (-\infty, \infty)$ is defined entirely by completeness of straight extremals parallel to axis σ when they approach boundary points σ_i, σ_{i+1}. In its turn, this is defined by the convergence of the integral (14.31). These extremals are incomplete at finite points $|q_i| < \infty$ for $m < 2$. At infinite points $|q_i| = \infty$ they are incomplete for $m > 2$ and complete in all other cases. Extension of the surface is necessary only for the conformal factor which has simple zero $m = 1$ at a finite point q_i because we must extend the surface only for nonsingular curvature. Note that a simple zero in the Lorentzian case corresponds to a horizon, and this is the unique case when four conformal blocks meet at a saddle point. The Carter–Penrose diagram for the Schwarzschild solution has exactly this form.

14.4. Global solutions

To construct maximally extended surfaces with metric (14.5), we perform the following procedure which is useful also for visualization of the surfaces.

We identify points ρ and $\rho + L$ where L is an arbitrary positive constant which is always possible because nothing depends on coordinate ρ. After this identification the plane σ, ρ becomes a cylinder. The circumference of the directing circle is equal to

$$P = LN(q) \to \begin{cases} 0, & N(q_i) = 0, \\ \infty & N(q_i) = \infty. \end{cases} \tag{14.32}$$

The plane σ, ρ is the universal covering space for this cylinder.

We summarize properties of the boundary points q_i in Table 14.1. Depending on the value of the exponent m, we show there the values of

Table 14.1   Properties of boundary points depending on the exponent m. The symbol const in the rows for the scalar curvature denotes a nonzero constant.

			$|q_i| < \infty$			
	$m < 0$	$0 < m < 1$	$m = 1$	$1 < m < 2$	$m = 2$	$m > 2$
R	∞	∞	const	∞	const	0
σ_i	const	const	∞	∞	∞	∞
P	∞	0	0	0	0	0
Completeness	$-$	$-$	$-$	$-$	$+$	$+$

			$|q_i| = \infty$			
	$m < 0$	$m = 0$	$0 < m \leq 1$	$1 < m < 2$	$m = 2$	$m > 2$
R	0	0	0	0	const	∞
σ_i	∞	∞	∞	const	const	const
P	0	const	∞	∞	∞	∞
Completeness	$+$	$+$	$+$	$+$	$+$	$-$

scalar curvature R on the corresponding surface, finiteness of coordinate σ_i at the point q_i, circumference of directing circles for cylinders P and completeness of extremals which are parallel to axis σ.

Forms of the surfaces near boundary points q_{i+1} after the identification $\rho \sim \rho + L$ are shown in Fig. 14.2. Surfaces near points q_i have similar form but are turned in the opposite direction. The surface corresponding to the whole interval (q_i, q_{i+1}) is obtained by gluing two such surfaces for boundary points q_i and q_{i+1} together. So extremals must be continued only near the boundary point $|q_i| < \infty$ for $m = 1$. We call this point horizon because it corresponds to a horizon in the Lorentzian case. The continuation at a horizon is performed as follows. First of all note that a horizon in the Euclidean case is itself a point because the length of the directing circle goes to zero. Next, this "infinite" point in the ρ, σ plane lays, in fact, at a finite distance because all extremals reach this point at a finite value of the canonical parameter. Up to higher order terms, the conformal

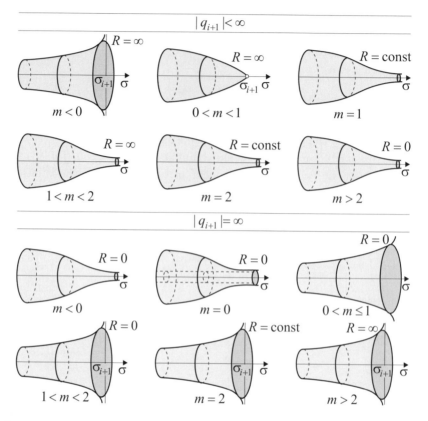

Figure 14.2   Forms of the surfaces near boundary points q_{i+1} after the identification $\rho \sim \rho + L$. By assumption, the coordinate σ grows from left to right and $\sigma_{i+1} > \sigma_i$. Surfaces near points q_i have a similar form but are turned in the opposite direction. The surface for the whole interval q_i, q_{i+1} is obtained by gluing two such surfaces for boundary points q_i q_{i+1} together.

factor in the neighborhood of the point q_i has the form

$$N(q) = N_i'(q - q_i), \tag{14.33}$$

where $N_i' = N'(q_i) = \mathsf{const} \neq 0$. In domain I for $N_i' > 0$ and $q > q_i$, Eq. (14.6) is easily integrated

$$q - q_i = e^{N_i'\sigma},$$

where we dropped an insignificant constant of integration related to a shift of σ. Thus the boundary point $|q_i| < \infty$ is reached for $\sigma \to -\infty$. Metric (14.5) in coordinates q, ρ which play the role of Schwarzschild coordinates takes the form

$$-ds^2 = \frac{dq^2}{N_i'(q - q_i)} + N_i'(q - q_i)d\rho^2. \tag{14.34}$$

· In polar coordinates r, φ, defined by the transformation

$$q - q_i = \frac{N_i'}{4}r^2, \qquad \rho = \frac{2}{N_i'}\varphi, \tag{14.35}$$

the metric becomes Euclidean

$$-ds^2 = dr^2 + r^2 d\varphi^2.$$

Here the polar angle varies within the interval $\varphi \in (0, LN_i'/2)$ and the radius r is defined in the neighborhood of the boundary point q_i by Eq. (14.35). This coordinate transformation maps the "infinite" line $\sigma_i = -\infty$, $\rho \in \mathbb{R}$ into the origin of Euclidean plane. Conical singularity can appear at the origin because the polar angle varies within the interval which in general differs from $(0, 2\pi)$. The corresponding deficit angle is

$$2\pi\theta = \frac{LN_i'}{2} - 2\pi. \tag{14.36}$$

Thus we get the Euclidean metric on a plane $\mathbb{R}^2$ with a conical singularity at the origin. The deficit angle is zero for $L = 4\pi/N_i'$, conical singularity is absent, and we are left with the flat Euclidean metric which is evidently smooth at the origin. In general, conformal factor (14.33) has corrections of higher order near the boundary point q_i and transformation to polar coordinates yields the metric of the same differentiability as the conformal factor.

In general continuation of the solution through the point $|q_i| < \infty$ for $m = 1$ has no meaning because this point correspond to a conical singularity. We assume that this point as well as any other singular point does not belong to a manifold. Therefore the plane σ, ρ or its part is the universal covering space for the surface with metric (14.5). Continuation is necessary only in the absence of conical singularity $L = 4\pi/N_i'$. In this case, straight extremals parallel to axis σ and going through points ρ and $\rho + L/2$ become two halves of the same extremal shown in Fig. 14.3. The fundamental group is trivial in the absence of a conical singularity, and therefore the corresponding surface is itself a universal covering space.

If the conformal factor has asymptotics $N \sim (q - q_i)$ and $N \sim (q_{i+1} - q)$ on both sides of the interval (q_i, q_{i+1}), then the surface must be continued in both points $\sigma = \pm\infty$. In general, after the identification $\rho \sim \rho + L$

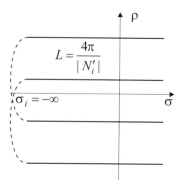

Figure 14.3   Continuation of straight extremals going through ρ and $\rho + L/2$ in the absence of conical singularity $L = 4\pi/|N'_i|$. The identification is performed at $\sigma_i = -\infty$.

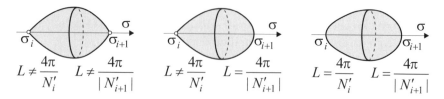

Figure 14.4   Three types of possible maximally extended surfaces, corresponding to the interval (q_i, q_{i+1}) when the conformal factor has asymptotics $N \sim (q - q_{i,i+1})$ in boundary points. In general, the surface has two conical singularities after the identification $\rho \sim \rho + L$. For $L|N'_{i,i+1}| = 4\pi$ conical singularities are absent.

the surface has conical singularities in both points. For $LN'_i = 4\pi$ and $LN'_{i+1} = 4\pi$ conical singularities are absent. There are three types of global surfaces shown in Fig. 14.4 according to the number of conical singularities. These surfaces have topology of a cylinder, plane or sphere, respectively.

The rules for construction of maximally extended surfaces with metric (14.5) are as follows.

(1) The maximally extended surface corresponds to each interval (q_i, q_{i+1}) after the identification $\rho \sim \rho + L$ and is obtained by gluing of two surfaces shown in Fig. 14.2 corresponding to the boundary points q_i and q_{i+1}.

(2) In all cases except the absence of the conical singularity, $|q_i| < \infty$, $LN'_i \neq 4\pi$ or $|q_{i+1}| < \infty$, $L|N'_{i+1}| \neq 4\pi$ the strip $\sigma \in (\sigma_i, \sigma_{i+1})$, $\rho \in \mathbb{R}$ with metric (14.5) is the universal covering space for the corresponding maximally extended surface.

(3) In the absence of one of conical singularities, $|q_i| < \infty$, $LN_i' = 4\pi$ or $|q_{i+1}| < \infty$, $L|N_{i+1}'| = 4\pi$, the surface obtained from the plane σ, ρ by identification $\rho \sim \rho + L$ is itself the maximally extended surface with trivial fundamental group.

Note that we do not glue together coordinate charts. It means that the corresponding surfaces are smooth $\mathcal{C}^\infty$. In the absence of conical singularity, the transformation to polar coordinates (14.35) does not use the explicit form of the conformal factor, and the resulting surface is also smooth as the Euclidean plane. So we have proved the statement.

Theorem 14.2. *The universal covering space constructing according to the rules 1–3 is the maximally extended smooth surface, $\mathcal{C}^\infty$, with Riemannian $\mathcal{C}^l$, $l \geq 2$, metric such that every point not lying on a horizon has a neighborhood isometric to some domain with metric (14.5).*

The universal covering space is known to be unique and all other maximally extended surfaces are obtained as quotient spaces of the universal covering space by the transformation group which acts freely and properly discontinuous.[16]

14.5. Schwarzschild solution

We consider the Schwarzschild solution as an example. Look for spherically symmetric solutions of vacuum Einstein's equations in the form

$$ds^2 = g_{\mu\nu}dx^\mu dx^\nu = f(d\tau^2 - d\sigma^2) - m(d\theta^2 + \sin^2\theta d\varphi^2),$$

where $f(\tau, \sigma)$ and $m(\tau, \sigma)$ are two unknown functions on time τ and radius σ, $\mu, \nu = 0, 1, 2, 3$. Then, as the consequence of the equations of motion, a solution depends only on one of the coordinates (Birkhoff's theorem) and has the form

$$ds^2 = |N(q)|(d\tau^2 - d\sigma^2) - q^2(d\theta^2 + \sin^2\theta d\varphi^2), \quad q > 0, \qquad (14.37)$$

where

$$N(q) = 1 - \frac{2M}{q}, \qquad q = \begin{cases} r, & N(q) > 0 \\ t, & N(q) < 0. \end{cases}$$

Detailed calculations are given in Ref. 13. The variable q is related to one of the coordinates τ or σ by differential Eq. (14.2). So the time-radial part of the Schwarzschild metric coincides exactly with metric (14.1), (14.2). The

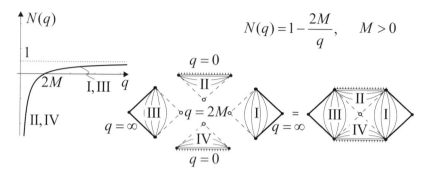

Figure 14.5   Construction of the Carter–Penrose diagram for the Schwarzschild solution by the method of conformal blocks.

appearance of the modulus and ± signs is the consequence of Einstein's equations. Maximally extended Schwarzschild solution is well known[1,2] and was obtained by introducing global coordinates. In general, introduction of such coordinates is not necessary and can be very complicated. The method of conformal blocks for construction of global solutions for two-dimensional metrics (14.2) was proposed in Ref. 4. Topologically, the maximally extended Schwarzschild solution is the direct product of two surfaces $\mathbb{U} \times \mathbb{S}^2$ where $\mathbb{S}^2$ is a sphere and $\mathbb{U}$ is the two-dimensional Lorentzian surface which is represented by the Carter–Penrose diagram shown in Fig. 14.5. The dependence of the conformal factor N on q is shown in the figure. It has one simple zero at $q = 2M$. Thus the interval $(0, \infty)$ is divided on two intervals $(0, 2M)$ and $(2M, \infty)$ where the conformal factor is negative (homogeneous solutions) and positive (static solutions). Two triangular conformal blocks II, IV and two square conformal blocks I, III correspond to intervals $(0, 2M)$ and $(2M, \infty)$, respectively. Their unique gluing yields the Carter–Penrose diagram for the Schwarzschild solution and is shown in Fig. 14.5. The rules for construction of conformal blocks, their gluing procedure, and the proof of the differentiability of the metric on the glued boundaries are given in Ref. 4.

The Carter–Penrose diagram after changing the signature from Lorentzian to Euclidean breaks into four disconnected between themselves surfaces which are shown in Fig. 14.6. Two Riemannian surfaces I and III with negative definite metric correspond to the region outside the black hole. For these surfaces the total metric has the signature $\mathrm{sign}\, g_{\mu\nu} = (- - --)$. Negative definiteness of the metric is due to the choice of the signature for the Schwarzschild solution (14.37) and can be

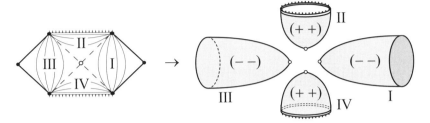

Figure 14.6   The Carter–Penrose diagram for the Schwarzschild solution in the Euclidean case breaks into four disconnected between themselves surfaces. There are two surfaces I, III with negative definite metric and two surfaces II, IV with positive definite metric corresponding to the regions outside and inside the horizon, respectively.

easily changed. This solution is usually considered as the Euclidean version of the Schwarzschild solution. Two surfaces II and IV with positive definite metric correspond to regions inside the horizon of the black hole. The signature of the total metric is $\operatorname{sign} g_{\mu\nu} = (+ + - -)$. This spherically symmetric solution is usually considered as unphysical though there is no mathematical reason to discard it.

14.6. Conclusion

We formulated the method of construction of global (maximally extended) surfaces for a large class of two-dimensional Riemannian metrics having one Killing vector. The smoothness of surfaces and differentiability of metrics are proved. This method is the counterpart of the method of conformal blocks for metrics with Lorentzian signature. Almost anytime each conformal block is the universal covering space for the maximally extended surface. The exceptions are surfaces with a horizon which is a point in the Euclidean case. If there is no conical singularity at this point then the universal covering space appears after the necessary identification of some points of the conformal block.

The transformation from Euclidean signature of the metric to Lorentzian one is interesting. If the Lorentzian surface is represented by the Carter–Penrose diagram with horizons corresponding to zeroes of the conformal factor, then, after the rotation to the Euclidean signature, the Carter–Penrose diagram breaks into disconnected surfaces with positive and negative definite metrics. The change in the signature of the Riemannian metric corresponds to crossing the horizon in the Lorentzian case.

Two-dimensional metrics considered in the present paper arise not only in two-dimensional models but also in higher dimensional gravity when solutions depend essentially only on two of the coordinates. The method can be useful in this case as well.

The author is very grateful to Prof. W. Kummer for numerous and fruitful discussions during which the idea of this work was born in the middle of 90-ties.

References

1. M. D. Kruskal. *Phys. Rev.*, 119(5):1743–1745, 1960.
2. Szekeres G. *Publ. Mat. Debrecen.* 1960. V. 7, N 1–4. P. 285–301.
3. B. Carter. Black hole equilibrium states. In C. DeWitt and B. C. DeWitt, editors, *Black Holes*, pages 58–214, New York, 1973. Gordon & Breach.
4. M. O. Katanaev. *Proc. Steklov Inst. Math.*, 228:158–183, 2000.
5. M. O. Katanaev. *J. Math. Phys.*, 31(4):882–891, 1990.
6. M. O. Katanaev. *J. Math. Phys.*, 32(9):2483–2496, 1991.
7. M. O. Katanaev. *J. Math. Phys.*, 34(2):700–736, 1993.
8. I. V. Volovich and M. O. Katanaev. *JETP Lett.*, 43(5):267–269, 1986.
9. M. O. Katanaev and I. V. Volovich. *Phys. Lett.*, 175B(4):413–416, 1986.
10. M. O. Katanaev and I. V. Volovich. *Ann. Phys.*, 197(1):1–32, 1990.
11. M. O. Katanaev, W. Kummer, and H. Liebl. *Phys. Rev. D*, 53(10):5609–5618, 1996.
12. M. O. Katanaev, W. Kummer, and H. Liebl. *Nucl. Phys.*, B486:353–370, 1997.
13. M. O. Katanaev, T. Klösch, and W. Kummer. *Ann. Phys.*, 276:191–222, 1999.
14. M. O. Katanaev. *J. Math. Phys.*, 38(3):946–980, 1997.
15. M. O. Katanaev. *Ann. Phys.*, 296(1):1–50, 2002.
16. S. Kobayashi and K. Nomizu. *Foundations of differential geometry*, volume 1, 2. Interscience publishers, New York – London, 1963, 1969.

Chapter 15

Thoughts on the Cosmological Principle

Dominik J. Schwarz

Fakultät für Physik, Universität Bielefeld, Postfach 100131,
33501Bielefeld, Germany
E-mail: dschwarz@physik.uni-bielefeld.de

15.1. Introduction

Wolfgang Kummer was a great teacher and mentor to me. Although Wolfgang never published any research on cosmology himself, he had a great interest in that field and supported me in my decision to get engaged with cosmological issues. Thus, I decided to describe current ideas and some of my own thoughts on one of the corner stones of modern cosmology — the cosmological principle. This principle says that the Universe is spatially homogeneous and isotropic. It predicts, among other phenomena, the cosmic redshift of light, the Hubble law and the black body shape of the cosmic background radiation spectrum. Nevertheless, the existence of structure in the Universe violates the (exact) cosmological principle. A more precise formulation of the cosmological principle must allow for the formation of structure and must therefore incorporate probability distributions. Below, I discuss how to formulate a new version of the cosmological principle, how to test it, and how to possibly justify it by fundamental physics. But let me, before doing so, describe in short some of my memories of Wolfgang.

15.2. Tribute to Wolfgang Kummer

My first contact with theoretical physics was with Wolfgang. He taught a course on "Methods in Theoretical Physics", which was compulsory for all physics students at the Vienna University of Technology (TU Wien) during their second year. The exercises accompanying that course were demanding and I learned how to handle complicated calculations. Later on I enjoyed his excellent lectures on "Particle Physics", which triggered my decision to devote my studies to fundamental issues in physics and to seek for a possibility to become part of his research group at the Institute for Theoretical Physics (ITP).

Luckily, during the third year of my study, the position of a library assistant was vacant in the ITP and Wolfgang was looking for a student interested in that position. So I became a member of Wolfgang's group well before I started my diploma project. The duties of a library assistant occupied not more than one or two hours a day and I was able to concentrate on my studies and research. Consequently, Wolfgang became my diploma advisor. The diploma research project was on two-dimensional gravity in the context of Riemann-Cartan geometry. Wolfgang had already done some preliminary, unpublished work, which provided a good starting point. At the same time, there appeared a very interesting work by Katanaev and Volovich, which opened up interesting perspectives. Wolfgang's style to approach a new problem and his attitude to meet his students at an equal level impressed me very much and is still influencing me in the way I try to deal with my students. We met every week to develop new ideas and to check all calculations step by step and soon managed to quantise the system and to find all its classical solutions. (For the scientific aspects of our work, see the contribution of L. Bergamin and R. Meyer to this volume.) This work resulted in my diploma thesis, three publications and several proceeding articles.

After my diploma thesis, I decided that I would like to devote my research towards a topic closer to "experiment" and chose cosmology. My first contact with modern cosmology was probably in the weekly theory seminar of Wolfgang's group, when we worked through the book of Kolb and Turner. Wolfgang took care that all graduate students of the ITP would have the chance to participate at international conferences and workshops. During one of those, which I was lucky to attend during my PhD studies, the COBE discovery of cosmic temperature anisotropies was presented for the first time in Europe. A big tradition at the ITP was and is the study of gauge theories (starting off from Wolfgangs important contributions on axial gauges, see the contributions by P. Landshoff, D. Blaschke et al., and P. Landshoff and A. Rebhan in this volume), and in the 1980s there was some confusion in the cosmology community on the issue of whether and how to use "gauges" in cosmological perturbation theory. Anton Rebhan picked up that topic and combined the cosmological perturbation theory with techniques and methods from finite temperature field theory, which attracted me to become Toni's PhD student. To my surprise, Wolfgang supported my decision to change the field, while he continued the study of two-dimensional physics. He became my mentor and continued to support my career. Besides our scientific connections, Wolfgang and Lore Kummer have been good friends to me and my family. I am very thankful for his support and will keep Wolfgang in good memory.

15.3. Modern Cosmology

Although questions of cosmology have been an issue for thousands of years, only the 20th (Christian) century saw cosmology turning into a physical science. Einstein's general relativity allows us to talk about the space-time of the Universe and to formulate dynamical laws for its geometry. Light is our most important source of information to learn about the evolution and state of the Universe. The advent of quantum mechanics, atomic, nuclear and particle physics enabled us to understand the mechanisms of light emission and absorption. At the same time, astronomical observations became sensitive, numerous and precise enough to study the global properties of the Universe.

Inspired by the ideas of Mach, Einstein decided to select very special conditions for a model of the Universe. His first attempt was to find solutions to his equations that allow for a static and spatially homogeneous and isotropic space-time. In order to achieve that, he had to introduce an additional term to his equations — the cosmological constant. With Hubble's discovery of cosmic expansion Einstein realised that the Universe was not static and he dropped the cosmological constant. This opened the way for the success of the Friedmann-Lamaître models, which are characterised by spatial homogeneity and isotropy. Milne coined the name "Cosmological Principle" for the statement that these symmetries are realised (at least approximately) in the Universe (see Peeble's book[1] for a more detailed description of the history of these ideas). Today, we have reached a high level of precision and as the cosmological principle is at its best an approximate statement about Nature, it is timely to think about possible refinements, especially in the light of the recently discovered cosmic acceleration of the Hubble expansion. In that context it has been proposed that the apparent cosmic acceleration might be an inappropriate interpretation of the data, due to our ignorance with respect to the effect of averaging over cosmic distances.[2]

15.4. Observational Facts

Typical cosmic photons belong to the cosmic microwave background (CMB) radiation. The observation of these microwave photons provides the basis of modern cosmology. The CMB radiation is well described by a black body at temperature $T_0 = 2.7$ K, is almost perfectly isotropic over the full sky and almost unpolarised. A small dipole anisotropy $\Delta T/T \simeq 10^{-3}$ is interpreted to be due to the motion of the Solar System barycentre with respect to this cosmic heat bath. At smaller angular scales cosmic temperature anisotropies are tiny, $\Delta T/T \simeq 10^{-5}$.

A high degree of isotropy is actually observed at all explored frequencies of the electromagnetic radiation (if one disregards nearby objects). Not only does the CMB have this property, but even the angular distributions of astrophysical objects on the sky at the extreme ends of the electromagnetic spectrum, radio galaxies and gamma-ray bursts, are isotropic.

This suggests that the distribution of light in the Universe is statistically isotropic. This could imply, that the probabilities to see a supernova, to find a radio galaxy or to measure a certain amount of CMB polarisation are distributed uniformly on the sky. However, this statement is obviously violated by several local phenomena, like day and night, or the Milky Way. A potentially true statement is:

Proposition 15.1 (Statistical Isotropy). *Apart from anisotropies of local origin, the distribution of light in the Universe is statistically isotropic with respect to the barycentre of the Solar system.*[a]

Local origins of anisotropy are, e.g. the Zone of Avoidance caused by the Milky Way, or the motion of the Solar system barycentre with respect to the CMB.

Causality is a fundamental principle of modern physics. However, it does not play any role when discussing the issue of statistical isotropy. This is no longer the case when we discus the question of spatial homogeneity. Our observations allow us to estimate distances of objects that are located on our backward light cone. Thus looking at distant objects means that we are also looking back in time. This is a substantial complication, as it means that we cannot study the issue of spatial distributions without a model of cosmic evolution.

---

[a]We could also refer to the barycentre of the Milky Way or of the Local Group, but those are less well known and it would not solve the problem that there might still be unresolved local effects.

The three dimensional distribution of matter in the Universe is observed by means of redshift surveys. Studying their distribution, we first of all find that galaxies come in groups, clusters and super-clusters. There exist big voids surrounded by filaments and sheets of structure. The largest object found in the Universe so far is the Sloan Great Wall, which extends over a few 100 Mpc.[3]

In a static or stationary Universe we would expect homogeneity in redshift space, but as we know that there is evolution in the Universe (e.g., the ratio of ellipticals to spirals changes as a function of redshift, the ionisation of the intergalactic medium changes at a redshift of $z \simeq 6$, ...), there cannot be homogeneity in redshift space. However, in an evolving Universe it does make sense to study the distribution of matter on spatial hypersurfaces, their definition being observer-dependent.

It seems useful to talk about the spatial hypersurface that is defined by a real astronomer. We might correct for some well understood effects, like the motion of Earth in the solar system. The astronomer can define her unique **comoving spatial hypersurface**. Let me also note that the word comoving obviously has to refer to the motion of atomic matter here. In general relativity one usually defines a class of **comoving observers**, which means that they are comoving with some form of matter. It seems feasible to define the class of **atomic/baryonic comoving observers** (as it is possible to receive information from them, while I don't know a way to receive information from an observer made out of dark matter). In the following we will refer always to them.

A perfectly homogeneous distribution is characterised by a well defined mean density (one-point correlation) and the vanishing of the (reduced) higher n-point correlation functions. A volume independent mean density seems to exist on scales larger 100 Mpc,[4] but this issue remains controversial.[5] The vanishing of the two-point correlation at scales much larger 100 Mpc is best seen by means of quasar redshift catalogues.[6] Although it is not clear if statistical homogeneity does hold, we formulate

Proposition 15.2 (Statistical Homogeneity). *The spatial distribution of visible matter in the Universe on scales larger than a homogeneity scale r_h is statistically homogeneous.*

In the following we will always assume that proposition 1.1 holds true and investigate its implications.

It is important to realise that the existence of a globally defined **cosmic time** is closely related to the large scale homogeneity (proposition 1.2) of the Universe. It implies that observers at different places in the Universe probe just different realisations of the same distribution of light and matter, which can be parametrized as a function of cosmic time. If statistical homogeneity does not hold, different observers might experience very different histories of the Universe.

15.5. Formulation(s) of the Cosmological Principle

These observationally motivated propositions are usually combined with a statement that seems to be a logic continuation of Bruno's and Copernicus' insight that we do not live at the centre of the world. Let me formulate two different versions:

Principle 15.1 (Weak Copernican Principle). *We are typical.*

Principle 15.2 (Strong Copernican Principle). *We are not distinguished.*

The strong version is more radical. The weak version implies that typical observers, wherever they are and whoever they are, observe the same distributions.

The strong version allows for different classes of observers, like there are different species of monkeys, none of them is distinguished. It is not a priori obvious that there couldn't be several species of observers, e.g. those living in a spiral galaxy and those in an elliptical, or observers in a filament and observers in a void. These observers could observe statistically different distributions.

What I call the weak Copernican principle is the commonly adopted textbook version. However, that we are made out of atomic matter, while the dominant mass/energy of the cosmic substratum seems to be non-atomic, questions the validity of the weak version. If we do not know these 95% of the Universe, how can we claim that we are typical?

We can now proceed to the formulation of a cosmological principle. At that point one usually lifts the statistical isotropy and statistical homogeneity on sufficiently large scales to an exact isotropy and homogeneity of space-time itself. The justification is that the isotropy of the CMB is almost exact and that one can assume exact isotropy as a starting point for a theory of structure formation. Combining the exact isotropy around one point with the weak Copernican principle, one concludes that every observer sees an isotropic sky. Together with some technical assumptions on the smoothness of the space-time metric, the homogeneity follows.[7]

Principle 15.3 (Cosmological Principle). *All physical quantities measured by a comoving observer are spatially homogeneous and isotropic.*

This formulation leads us to the class of Friedmann-Lemaître models, which are successful in describing the cosmic expansion and the thermal history of the Universe, especially primordial nucleosynthesis and the decoupling of light. However, these models do not explain how structure forms. In order to do so, we have to introduce cosmological perturbations, which violate the cosmological principle.

Note that the cosmological principle as usually stated is much stronger than what we can possibly establish by means of observations. At best, it is only the statistical distribution of matter and light that appears to be homogeneous and isotropic, not its actual realisation. I thus favour an alternative formulation of the cosmological principle.

Principle 15.4 (Statistical Cosmological Principle). *The distribution of light and matter in the Universe is statistically isotropic around any point, apart from anisotropies of local origin.*

The observed isotropic distribution of light (and matter) together with the weak Copernican principle implies the statistical isotropy around every point. It seems to me, that this implies statistical homogeneity, however, I am not aware of a rigorous proof of that statement. However, perhaps we should use the strong Copernican principle and then we cannot conclude that homogeneity holds true. In that case we could only state a.

Principle 15.5 (Minimal Cosmological Principle). *There exists a class of observers that see a statistically isotropic Universe, apart from anisotropies of local origin.*

This is a very interesting possibility, as this is the minimal version that seems to be justified by experiment. I think that the study of it's implications would be very interesting and could lead us to conclusions that differ significantly form today's textbook cosmology.

15.6. Testing the Cosmological Principle

As an approximation, the cosmological principle is very useful, but strictly speaking it is wrong. With respect to isotropy, we know that the violation is small, however with respect to homogeneity the case remains unclear.

This actually might be at the reason for the current crisis that we are facing in cosmology: we claim that we know with high precision that we only understand 5% of the Universe.[8] But we have no direct evidence for the existence of dark matter or energy.

This is one of the reasons why many groups started to investigate the idea of cosmic backreaction as an alternative to the existence of dark energy. Instead of looking at statistical distributions, we can average over regions of space-time and study the properties of these estimators. These regions might be one, two or three dimensional. Due to the non-linearity of gravity, it is obvious that these estimates of physical quantities and the evolution of physical observables do not necessarily commute. This could give rise to a misinterpretation of the data and thus the cosmic acceleration could be an illusion.[2]

This finally leads us to the question how one could test the statistical cosmological principle. There are some indications that statistical isotropy is violated at the largest scales on the CMB,[9] but it remains to be seen if that will eventually turn out as a Solar system contamination or a systematic effect. I mentioned already that the statistical homogeneity has not been firmly established so far.

While all observations are consistent with the strong Copernican principle, its weak version is contradicted by our claim that the Universe is dominated by non-atomic stuff. This might be an irrelevant detail, thus several tests of the weak Copernican principle have been proposed.[11]

15.7. Cosmological Inflation and Quantum Gravity

Can we justify the statistical cosmological principle? The scenario of cosmological inflation is certainly an important step towards a possible justification. In the context of eternal inflation,[10] the classical version of the cosmological principle fails miserably at super-large scales, as the Universe is extremely inhomogeneous at these scales. In any case, it fails at scales larger than the particle horizon, which are enormously bigger than what we can observe and will ever observe. But, we can hope to justify the statistical cosmological principle for regions smaller than the particle horizon.

To sum up, the historically important formulation of the cosmological principle has no justification in modern cosmology, as the quantum fluctuations during inflation spoil it. However, turning it into a statement on the statistical distribution of light and matter seems to be a logic consequence of the very same quantum fluctuations. Unless a consistent formulation of quantum gravity is available, it seems that a cosmological principle of some form is still required.

The promises of quantum gravity to eventually predict the statistical cosmological principle also provide a link to Wolfgang's dedication to fundamental science— the understanding of the quantum effects of space-time.

Acknowledgment

I would like to thank the editors Daniel Grumiller, Toni Rebhan and Dimitri Vassilevich for their effort to put together this memorial volume.

References

1. P. J. E. Peebles, *The Large Scale Structure of the Universe* (Princeton University Press, Princeton, 1980).
2. T. Buchert, "Dark Energy from Structure - A Status Report," Gen. Rel. Grav. **40** (2008) 467.
3. J. R. I. Gott et al., "A Map of the Universe," Astrophys. J. **624** (2005) 463.
4. D. W. Hogg et al., "Cosmic homogeneity demonstrated with luminous red galaxies," Astrophys. J. **624** (2005) 54;
   R. Thieberger and M. N. Celerier, "Scaling Regimes as obtained from the DR5 Sloan Digital Sky Survey," arXiv:0802.0464 [astro-ph].
5. F. S. Labini et al., "The large scale inhomogeneity of the galaxy distribution," arXiv:0805.1132 [astro-ph].
6. S. M. Croom et al., "The 2dF QSO Redshift Survey - XIV. Structure and evolution from the two-point correlation function," Mon. Not. Roy. Astron. Soc. **356** (2005) 415.
7. W. Thirring, *Lehrbuch der Mathematischen Physik, 2. Klassische Feldtheorie* (Springer, Wien, 1990).
8. E. Komatsu *et al.* [WMAP Collaboration], "Five-year Wilkinson Microwave Anisotropy Probe (WMAP) observations: cosmological interpretation," Astrophys. J. Suppl. **180** (2009) 330.
9. A. de Oliveira-Costa et al., "The significance of the largest scale CMB fluctuations in WMAP," Phys. Rev. D **69** (2004) 063516;
   D. J. Schwarz et al., "Is the low-l microwave background cosmic?," Phys. Rev. Lett. **93** (2004) 221301.
   C. J. Copi *et al.*, "No large-angle correlations on the non-Galactic mic rowave sky," arXiv:0808.3767 [astro-ph].

10. A. Vilenkin, "The Birth Of Inflationary Universes," Phys. Rev. D **27** (1983) 2848;

    A. D. Linde, "Eternally Existing Selfreproducing Chaotic Inflationary Universe," Phys. Lett. B **175** (1986) 395;

    A. D. Linde, D. A. Linde and A. Mezhlumian, "From the Big Bang theory to the theory of a stationary universe," Phys. Rev. D **49**, (1994) 1783.

11. J. Goodman, "Geocentrism reexamined," Phys. Rev. D **52**, 1821 (1995);

    R. R. Caldwell and A. Stebbins, "A Test of the Copernican Principle," Phys. Rev. Lett. **100** (2008) 191302;

    J. P. Uzan, C. Clarkson and G. F. R. Ellis, "Time drift of cosmological redshifts as a test of the Copernican principle," Phys. Rev. Lett. **100** (2008) 191303;

    C. Clarkson, B. Bassett and T. C. Lu, "A general test of the Copernican Principle," Phys. Rev. Lett. **101** (2008) 011301.

Chapter 16

When Time Emerges

Cornelia Faustmann[1], Helmut Neufeld[2] and Walter Thirring[3]

Fakultät für Physik, Universität Wien,
Boltzmanngasse 5, A-1090 Wien, Austria

We study the geodesics in a space-time where a metric with Minkowski type signature is continued across a singularity to a region with a different space-time signature. Contrary to naive expectation, the motion of test particles does not end up in a catastrophe. Depending on the form of the signature change and the initial conditions, timelike (or lightlike) geodesics are either reflected at the singularity or even leave the Minkowski sector and use the region with the flipped metric for a time travel. An observer in the Minkowski sector interprets this as the creation of a particle and an antiparticle at the singularity.

16.1. Introduction

In Einstein's theory of gravitation, the local Minkowski structure of space-time is a separate axiom. If this assumption is relaxed, one may envisage a scenario where space-time undergoes some kind of phase transition to a sector with a different signature.

Stimulated by investigations[1,2] in the context of quantum gravity, the question of a change of signature in a classical space-time has been discussed quite extensively in the literature (see for instance Refs. 3–7 and the citations therein). The simplest prototype of a space-time with signature change is obtained by cutting an S^4 along its equator and joining it to the corresponding half of a de Sitter space.

In the present work, we study a somewhat different situation where a singularity of space-time is associated with a signature change. Our main

[1] E-mail: cornelia.faustmann@univie.ac.at
[2] E-mail: helmut.neufeld@univie.ac.at
[3] E-mail: walter.thirring@univie.ac.at

interest is the behaviour of the corresponding geodesics which can be interpreted as the reaction of a test particle to a signature change in space-time. Starting with simple two-dimensional examples, we show how these ideas can be extended to a four-dimensional model with a signature changing variant of the Kasner metric.[8]

16.2. Geodesics

The geodesics of a general metric $ds^2 = g_{\mu\nu}(x)dx^\mu dx^\nu$ can be obtained from the variational principle

$$\delta \int d\lambda\, L = 0, \tag{16.1}$$

with

$$L = \frac{1}{2}g_{\mu\nu}\dot{x}^\mu \dot{x}^\nu, \quad \dot{x}^\mu = \frac{dx^\mu(\lambda)}{d\lambda}. \tag{16.2}$$

As the Lagrangean L does not depend explicitly on the parameter λ, the associated Hamiltonian $H \equiv L$ is a constant under this evolution. By rescaling λ, we can give L the values $+1/2$, 0, $-1/2$.

If $L = +1/2$ (i.e. $ds^2 \geq 0$), the parameter λ can be identified with the proper time s (in a Minkowski region). As originally proposed by Stueckelberg[9] and Feynman,[10] the corresponding worldline belongs to a particle if $\dot{t} > 0$ and to an antiparticle if $\dot{t} < 0$. The case $L = 0$ $(ds^2 = 0)$ belongs to a massless particle (light) and $L = -1/2$ $(ds^2 < 0)$ is associated with a tachyon.

The geodesics are differentiable curves $\lambda \to x^\mu(\lambda)$ and we will encounter the situation where $t(\lambda)$ is not monotonic. In such a case, there must be a λ_0 with $\dot{t}(\lambda_0) = 0$ and this is interpreted as the creation of a particle-antiparticle pair.

16.3. Two-dimensional metric with spacelike singularity

Metric A is defined by

$$ds^2 = dt^2 - \frac{dx^2}{t}. \tag{16.3}$$

For this metric we have $g = -1/t$ and the upper half plane of space-time $(t > 0)$ has a Minkowski structure and the lower part $(t < 0)$ is Euclidean. Thus time emerges at $t = 0$. The corresponding Lagrangean is given by

$$L_A = \frac{1}{2}\left(\dot{t}^2 - \frac{\dot{x}^2}{t}\right). \tag{16.4}$$

The geodesics are determined by

$$\dot{x} = ct, \quad \dot{t}^2 - c^2 t = 2L_A, \quad c = \text{const.} \tag{16.5}$$

The function $U(t) = -c^2 t$ acts like a potential for the one-dimensional "motion" $\lambda \to t(\lambda)$. In this space, only timelike geodesics can enter the Euclidean region. Lightlike geodesics have their turning point at $t = 0$ and tachyons cannot leave the Minkowski sector. From (16.5) we infer $\ddot{t} = c^2/2$, thus $t(\lambda)$ is strictly convex, and the general solution of the geodesic equations is given in terms of polynomials in the parameter λ. Expressed through the integration constants L_A, $c \neq 0$ and $x(0)$, one obtains

$$t(\lambda) = \frac{c^2}{4}\lambda^2 - \frac{2L_A}{c^2}, \quad x(\lambda) = \frac{c^3}{12}\lambda^3 - \frac{2L_A}{c}\lambda + x(0), \tag{16.6}$$

where we have used the freedom of a parameter shift $\lambda \to \lambda - \lambda_0$ such that $\dot{t}(0) = 0$. For $c = 0$ (and $L_A \geq 0$), the (trivial) solution is given by

$$t(\lambda) = \pm\sqrt{2L_A}\,\lambda + t(0), \quad x(\lambda) = x(0). \tag{16.7}$$

Figure 16.1 shows space-time diagrams of the various types of geodesics.

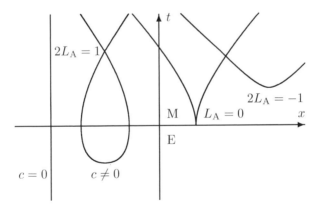

Figure 16.1   Space-time diagram of the geodesics of metric A.

For $2L_A = 1$ and $c \neq 0$, the worldline

$$t(s) = \frac{c^2}{4}s^2 - \frac{1}{c^2}, \quad x(s) = \frac{c^3}{12}s^3 - \frac{1}{c}s + x(0) \tag{16.8}$$

starts in the Minkowski sector. It enters the Euclidean region where $\dot{x} = ct$ changes sign, reaches its turning point at $t_0 = -1/c^2$ and finally turns back to the Minkowski region where the original sign of $\dot{x}$ is resumed. The worldline has an intersection at $t = -2t_0 = 2/c^2$. An observer in the

Minkowski sector sees this as the production of a particle and an antiparticle (with masses $m \neq 0$) at $t = 0$ at two different values of x. They approach each other with increasing velocities, meet at the time $t = 2/c^2$, and finally move away in opposite directions.

A lightlike geodesic,

$$t(\lambda) = \frac{c^2}{4}\lambda^2, \quad x(\lambda) = \frac{c^3}{12}\lambda^3 + x(0), \tag{16.9}$$

cannot leave the Minkowski sector. It describes the creation of a particle-antiparticle pair (mass $m = 0$) at $t = 0$.

We add a few remarks to elucidate the mathematical structure of the dynamics. Since the curvature scalar $R \sim 1/t^2$ of metric A is singular at $t = 0$, there cannot be a coordinate system where all Christoffel symbols and their derivatives remain finite for $t = 0$. It is all the more surprising that the geodesic flow in cotangent space is as well behaved as one can wish. The reason is that all factors $1/t$ are accompanied by $\dot{x}$ and this product remains finite when the trajectory passes the hypersurface $t = 0$. The underlying structure can be analysed even more clearly by passing from the Lagrangean to the Hamiltonian

$$H_A = \frac{1}{2}\left(p_t^2 - tp_x^2\right), \tag{16.10}$$

where

$$p_t = \frac{\partial H_A}{\partial \dot{t}} = \dot{t}, \quad p_x = \frac{\partial H_A}{\partial \dot{x}} = -\frac{\dot{x}}{t}. \tag{16.11}$$

H_A is regular for $t = 0$ and so are the solutions of the Hamilton equations:

$$t(\lambda) = \frac{p_x(0)^2}{4}\lambda^2 + p_t(0)\lambda + t(0),$$

$$x(\lambda) = -\frac{p_x(0)^3}{12}\lambda^3 - \frac{p_t(0)p_x(0)}{2}\lambda^2 - p_x(0)t(0)\lambda + x(0),$$

$$p_t(\lambda) = \frac{p_x(0)^2}{2}\lambda + p_t(0),$$

$$p_x(\lambda) = p_x(0). \tag{16.12}$$

Using (16.11) and setting $p_x(0) = -c$, one recovers the solutions in the previously discussed form.

The solutions (16.12) define a one-parameter group of diffeomorphisms $\tau(\lambda) : \mathbb{R}^4 \to \mathbb{R}^4$. To see this more explicitly, we start from the observation that the points in phase space with $p_x = 0$ form an invariant submanifold.

The group property of the corresponding solution (16.12) with $p_x(0) = 0$ is obvious. For the remaining subspace with $p_x \neq 0$, we define

$$a = \begin{pmatrix} 2x/p_x^2 \\ -2t/p_x \\ -2p_t/p_x \\ -p_x \end{pmatrix}.$$ (16.13)

The geodesic equations can now be rewritten in the form

$$\frac{d}{d\lambda} a(\lambda) = N a(\lambda),$$ (16.14)

where N is the nilpotent matrix

$$N = \begin{pmatrix} 0 & 1 & 0 & 0 \\ 0 & 0 & 1 & 0 \\ 0 & 0 & 0 & 1 \\ 0 & 0 & 0 & 0 \end{pmatrix}, \qquad N^4 = 0.$$ (16.15)

The solution of (16.14),

$$a(\lambda) = \exp(\lambda N) a(0),$$ (16.16)

guarantees the group property. The special form of the matrix N (with $N^4 = 0$) allows the peaceful coexistence of the group property with the polynomial structure of the solutions (16.12). This implies that $\tau(\lambda)$ is everywhere defined and invertible. Since the mappings are all analytic, the uniqueness of the continuation is guaranteed. The only unusual feature is that $\lambda \to t(\lambda)$ is not a bijection, thus $\lambda(t)$ does not exist as a map. Therefore one cannot define a trajectory in the form $x(t) = x(\lambda(t))$. Of course, as a map of sets, $t(\lambda)$ has an inverse and this expression defines $x(t)$ as a map of sets. However, for some t it may map into the empty set and for some other t it may map into two values of x. In terms of physics, this means that at some time there was no particle, at other times there are two, shortly there is some pair creation.

16.4. Two-dimensional metric with timelike singularity

The second two-dimensional metric we wish to consider is given by

$$ds^2 = \frac{dt^2}{x} - dx^2,$$ (16.17)

corresponding essentially to an interchange of space and time coordinates compared to the previous metric. In this case, $g = -1/x$ and the right half plane ($x > 0$) is of Minkowski type and the left half plane ($x < 0$) is Euclidean. The associated Lagrangean is given by

$$L_B = \frac{1}{2}\left(\frac{\dot{t}^2}{x} - \dot{x}^2\right).$$ (16.18)

In this case, particles (or antiparticles) stay away from the Euclidean region ($x < 0$) as for them L_B has to be positive. Massless particles ($L_B = 0$) are also confined in the Minkowski part but they can touch the boundary at $x = 0$. Tachyons ($L_B < 0$) can live in the Euclidean as well as the Minkowski region.

$\partial L_B/\partial t = 0$ implies that $c = \dot{t}/x$ stays constant and the dynamics is governed by the equations

$$\dot{t} = cx, \quad \dot{x}^2 - c^2x = -2L_B.$$ (16.19)

The potential for the one-dimensial motion $\lambda \to x(\lambda)$ is now given by $-c^2x$ and $\ddot{x} = c^2/2 \geq 0$. The present case (metric B) can be obtained from the previous one (metric A) by the substitutions $t \leftrightarrow x$, $L_A \to -L_B$ (interchange of the rôle of massive particles and tachyons).

The explicit solution for this case can be copied from (16.6) and (16.7). After applying the necessary changes, one finds

$$t(\lambda) = \frac{c^3}{12}\lambda^3 + \frac{2L_B}{c}\lambda + t(0), \quad x(\lambda) = \frac{c^2}{4}\lambda^2 + \frac{2L_B}{c^2},$$ (16.20)

and

$$t(\lambda) = t(0), \quad x(\lambda) = \pm\sqrt{2L_B}\,\lambda + x(0),$$ (16.21)

for $c = 0$ (and $L_B \leq 0$). The possible types of solutions are shown in Fig. 16.2.

The motion of massive particles ($2L_B = 1, \dot{t} > 0$) or antiparticles ($2L_B = 1, \dot{t} < 0$) is restricted to $x_0 \leq x \leq +\infty$ with the turning point $x_0 = 1/c^2 > 0$.

Massless particles (photons) are reflected at the time-like singularity ($x = 0$).

Tachyons can enter the Euclidean region, but they do not like it there and they come back quickly, in fact earlier than they entered.

16.5. Asymptotically flat space

We define metric C by

$$ds^2 = dt^2 - \frac{dx^2}{\tanh t}.$$ (16.22)

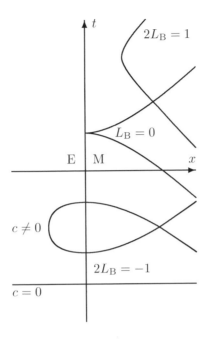

Figure 16.2   Space-time diagrams of the geodesics of metric B.

As in the case of metric A, the upper half plane of space-time is Minkowskian and the lower part is Euclidean. However, this metric becomes asymptotically flat for $t \to \pm\infty$. Its Lagrangean is given by

$$L_C = \frac{1}{2}\left(\dot{t}^2 - \frac{\dot{x}^2}{\tanh t}\right). \tag{16.23}$$

The geodesics are determined by

$$\dot{x} = c \tanh t, \quad \dot{t}^2 - c^2 \tanh t = 2L_C. \tag{16.24}$$

For particles ($2L_C = 1$, $\dot{t} > 0$), the velocity v is given by

$$v = \frac{\dot{x}}{\dot{t}} = \frac{c \tanh t}{\sqrt{1 + c^2 \tanh t}} \xrightarrow{t \to \infty} \frac{c}{\sqrt{1 + c^2}} = \frac{mc}{\sqrt{m^2 + (mc)^2}}, \tag{16.25}$$

which shows that $p = mc$ is the asymptotic momentum of a particle with mass m (corresponding to an energy $E = \sqrt{m^2 + m^2 c^2}$), giving the physical interpretation of the integration constant c.

The solutions can still be given in closed form. In terms of the variable

$$\xi = \sqrt{c^2 \tanh t + 2L}, \tag{16.26}$$

the differential equation for the trajectories of the particle (antiparticle) takes the simple form

$$\frac{dx}{d\xi} = \pm c \left(\frac{1}{2L_C + c^2 - \xi^2} + \frac{1}{2L_C - c^2 - \xi^2} \right). \tag{16.27}$$

The creation of a massive particle-antiparticle pair occurs for $2L_C = 1$ and $c^2 > 1$, which corresponds to the case of an asymptotic energy $E > 2m$. The worldline has a turning point at $t_0 = \operatorname{arctanh}(-1/c^2) < 0$ in the Euclidean region. The trajectory of the particle (antiparticle) for $t > t_0$ is described by

$$x_{p,a}(t) = \pm c \left(\frac{\operatorname{arctanh}\sqrt{\frac{1+c^2 \tanh t}{c^2+1}}}{\sqrt{c^2+1}} - \frac{\arctan\sqrt{\frac{1+c^2 \tanh t}{c^2-1}}}{\sqrt{c^2-1}} \right) + x(t_0). \tag{16.28}$$

For $c^2 \leq 1$, the geodesics with $2L_C = 1$ extend from $t = -\infty$ to $t = +\infty$ without any turning point. In other words, a massive particle (or antiparticle) with asymptotic energy $E < 2m$ has existed already in the Euclidean sector and an observer in the Minkowski part of space-time sees its appearance at the "beginning of the universe" $(t = 0)$. In this case, the trajectories are given by

$$x_{p,a}(t) = \pm c \left(\frac{\operatorname{arctanh}\sqrt{\frac{1+c^2 \tanh t}{1+c^2}}}{\sqrt{1+c^2}} + \frac{\operatorname{arccoth}\sqrt{\frac{1+c^2 \tanh t}{1-c^2}}}{\sqrt{1-c^2}} \right) + \text{const.}$$

$$\tag{16.29}$$

A massless particle $(L_C = 0)$ is always confined in its local light-cone and cannot enter the Euclidean region. Its wordline has a turning point at $t = 0$. An observer in the Minkowski sector interprets this as pair production at $t = 0$. The solutions take the form

$$x_{p,a}(t) = \pm \frac{c}{|c|} \left(\operatorname{arctanh}\sqrt{\tanh t} - \arctan\sqrt{\tanh t} \right) + x(0). \tag{16.30}$$

Tachyons are only allowed for $c^2 > 1$ with a turning point at $t_0 = \operatorname{arctanh}(1/c^2) > 0$ in the Minkowski sector. The solutions for $t > t_0$ are given by

$$x(t) = \pm c \left(\frac{\operatorname{arctanh}\sqrt{\frac{c^2 \tanh t - 1}{c^2-1}}}{\sqrt{c^2-1}} - \frac{\arctan\sqrt{\frac{c^2 \tanh t - 1}{c^2+1}}}{\sqrt{c^2+1}} \right) + x(t_0). \tag{16.31}$$

16.6. Kasner metric

The Kasner metric[8]

$$ds^2 = dt^2 - t^{2p_1} dx^2 - t^{2p_2} dy^2 - t^{2p_3} dz^2, \qquad (16.32)$$

is a solution of the free Einstein equations ($R_{\mu\nu} = 0$) for $t > 0$ if the coefficients $p_{1,2,3}$ fulfil the two relations

$$p_1 + p_2 + p_3 = 1, \quad p_1^2 + p_2^2 + p_3^2 = 1, \qquad (16.33)$$

which implies $g = \det(g_{\mu\nu}) = -t^2$. With the convention $p_1 \leq p_2 \leq p_3$, the values of these coefficients lie in the intervalls

$$-1/3 \leq p_1 \leq 0, \quad 0 \leq p_2 \leq 2/3, \quad 2/3 \leq p_3 \leq 1. \qquad (16.34)$$

The space described by (16.32) is homogenous but not isotropic. With increasing t, all volumes are also increasing like t. Distances on the y- and z-axis are increasing while those on the x-axis are decreasing. The curvature of the four-dimensional space can be characterized by the invariant

$$R_{\alpha\beta\gamma\delta} R^{\alpha\beta\gamma\delta} = \frac{4}{t^4} \left(\sum_{i=1}^{3} p_i^2 (1 - p_i)^2 + p_1^2 p_2^2 + p_1^2 p_3^2 + p_2^2 p_3^2 \right), \qquad (16.35)$$

which is nonvanishing (with a singularity at $t = 0$), except for $p_1 = p_2 = 0$, $p_3 = 1$. In the latter case, (16.32) is just a funny way of describing flat Minkowski space which can also be seen by performing the coordinate transformation $t \sinh z = \zeta$, $t \cosh z = \tau$.

As a next step, we want to extend (16.32) also to negative values of t allowing at the same time for a possible change of signature. To this end we consider metrics of the form

$$ds^2 = \begin{cases} dt^2 - t^{2p_1} dx^2 - t^{2p_2} dy^2 - t^{2p_3} dz^2, & t > 0, \\ dt^2 - \sum_{i=1}^{3} \sigma_i |t|^{2p_i} (dx^i)^2, & t < 0, \end{cases} \qquad (16.36)$$

where the signs $\sigma_i = \pm 1$ can be chosen in all possible combinations describing different types of signature change. The coefficients p_i are still restricted by (16.33).

All metrics described by (16.36) have the following properties:

- The free Einstein equations $R_{\mu\nu} = 0$ are fulfilled for $t > 0$ and $t < 0$ (independently of the choice of the σ_i).
- The four-dimensional volume element is given by $d^4x \sqrt{|g|} = dt\, dx\, dy\, dz\, |t|$ and vanishes at $t = 0$.

- The curvature invariant $R_{\alpha\beta\gamma\delta}R^{\alpha\beta\gamma\delta}$ is given by (16.35) also for $t < 0$.

If a signature change of the form proposed in (16.36) could occur in reality cannot be decided within the framework of classical (non-quantized) gravitation theory. Here, we explore the following, more modest question: Assuming that a metric like (16.36) gives an (at least approximate) description of the gravitational background near the beginning of the universe, how do test particles react to the associated change of signature?

16.7. Signature change and particle creation

To give a specific example, we consider metric (16.36) with the special values $p_1 = -1/3$, $p_2 = p_3 = 2/3$ and $\sigma_1 = -1$, $\sigma_2 = \sigma_3 = 1$:

$$ds^2 = dt^2 - \epsilon(t)|t|^{-2/3}dx^2 - |t|^{4/3}(dy^2 + dz^2). \tag{16.37}$$

In this case, we have two temporal and two spatial dimensions for $t < 0$.

The associated Lagrangean

$$L = \frac{1}{2}\left(\dot{t}^2 - \epsilon(t)|t|^{-2/3}\dot{x}^2 - |t|^{4/3}(\dot{y}^2 + \dot{z}^2)\right) \tag{16.38}$$

does not depend on x, y, z and thus, in addition to L, also

$$c_x = \epsilon(t)|t|^{-2/3}\dot{x}, \; c_y = |t|^{4/3}\dot{y}, \; c_z = |t|^{4/3}\dot{z} \tag{16.39}$$

remain constant. Combining (16.38) with (16.39), the geodesics are determined by the following system of differential equations:

$$\dot{x} = c_x\epsilon(t)|t|^{2/3}, \; \dot{y} = c_y|t|^{-4/3}, \; \dot{z} = c_z|t|^{-4/3}, \tag{16.40}$$

$$\dot{t}^2 - c_x^2\epsilon(t)|t|^{2/3} - (c_y^2 + c_z^2)|t|^{-4/3} = 2L \tag{16.41}$$

Lines with $c_x = c_y = c_z = 0$ are trivial solutions of the geodesic equations. Equation (16.41) allows a simple qualitative discussion of the nontrivial solutions. The function

$$U(t) = -c_x^2\epsilon(t)|t|^{2/3} - (c_y^2 + c_z^2)|t|^{-4/3} \tag{16.42}$$

acts as potential for the one-dimensional motion $\lambda \to t(\lambda)$:

$$\dot{t}^2 + U(t) = 2L. \tag{16.43}$$

Note that the special case $c_x \neq 0$, $c_y^2 + c_z^2 = 0$ is rather similar to the motion in metric A discussed in Sec. 16.3. The general case ($c_x \neq 0$, $c_y^2 + c_z^2 \neq 0$) is shown in Fig. 16.3.

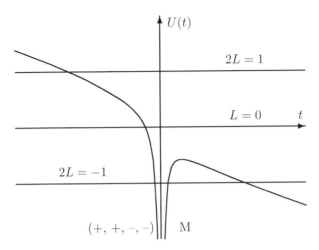

Figure 16.3 Potential $U(t)$ for the geodesics of metric (16.37) where a space with signature $(+, +, -, -)$ turns into Minkowski space-time (M) at $t = 0$.

Timelike $(2L = 1)$ and lightlike $(L = 0)$ geodesics enter the region with signature $(+, +, -, -)$. They have a turning point $\dot{t}(s_0) = 0$ at a time $t(s_0) = t_0 < 0$ characterized by $U(t_0) = 2L$. For instance, the solution of $U(t_0) = 0$ (lightlike case) is given by

$$t_0 = -\sqrt{c_y^2 + c_z^2}/|c_x|. \tag{16.44}$$

Finally the geodesics turn back to Minkowski space. We interpret such solutions as the creation of a particle-antiparticle pair in the region with the flipped metric. An observer in the Minkowski sector sees this as the production of a particle and an antiparticle at $t = 0$ and two different points of the space coordinates.

For tachyons $(L < 0)$, turning points $\dot{t}(\lambda_0) = 0$ are also possible but this does not allow an interpretation in terms of particle-antiparticle production.

Note that the qualitative behaviour of the geodesics does not change if the general form of the metric with signature $(+, +, -, -)$ for $t < 0$,

$$ds^2 = dt^2 - \epsilon(t)|t|^{2p_1} dx^2 - |t|^{2p_2} dy^2 - |t|^{2p_3} dz^2, \tag{16.45}$$

is considered.

16.8. Different types of signature change

The qualitative behaviour of the geodesics of all the metrics in (16.36) with signatures $(+1, -\sigma_1, -\sigma_2, -\sigma_3)$ for $t < 0$ can be found quite analogously by inspecting the pertinent potential

$$U(t) = \begin{cases} -c_x^2 t^{-2p_1} - c_y^2 t^{-2p_2} - c_z^2 t^{-2p_3}, & t > 0, \\ -\sigma_1 c_x^2 t^{-2p_1} - \sigma_2 c_y^2 t^{-2p_2} - \sigma_3 c_z^2 t^{-2p_3}, & t < 0. \end{cases} \qquad (16.46)$$

In the following discussion, we shall concentrate on the typical case with $(c_x, c_y, c_z) \neq (0, 0, 0)$.

Recalling (16.34), the geodesics of all metrics with $\sigma_3 = -1$ cannot cross the infinitely high potential barrier

$$c_z^2 \epsilon(t) |t|^{-2p_3} + \dots$$

at $t = 0$. A typical example is given by

$$ds^2 = dt^2 - \epsilon(t) \big(|t|^{-2/3} dx^2 + |t|^{4/3} (dy^2 + dz^2) \big), \qquad (16.47)$$

describing the transition from a Euclidean sector $(t < 0)$ to a Minkowski region $(t > 0)$. The associated potential,

$$U(t) = -\epsilon(t) \big(c_x^2 |t|^{2/3} + (c_y^2 + c_z^2)|t|^{-4/3} \big) \qquad (16.48)$$

is shown in Fig. 16.4. In this case, geodesics starting in the Minkowski region cannot leave this sector. For $L \geq 0$ a particle-antiparticle pair is produced at $t = 0$. Geodesics (with $L > 0$) can exist in the Euclidean but they are repelled by the potential barrier at $t = 0$ and have to stay in the region $t < 0$.

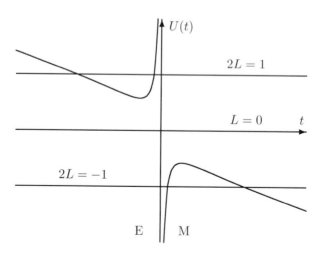

Figure 16.4   Potential $U(t)$ for the geodesics of metric (16.47) with a transition from a Euclidean sector (E) to Minkowsi space-time (M).

For $\sigma_3 = +1$, the potential (16.46) exhibits an attractive singularity at $t = 0$ and geodesics starting in the Minkowski sector can enter the region $t < 0$. This case includes the previously discussed example (16.45) and also the metric with signature $(+, +, +, -)$ for $t < 0$.

With a signature $(+, -, +, -)$ for $t < 0$, the behaviour of timelike and lightlike geodesics depends on the choice of the constants c_i^2. They either extend from $t = -\infty$ to $t = +\infty$ or they decay into unconnected segments for $t < 0$ and $t > 0$, respectively.

Finally, the case $\sigma_1 = \sigma_2 = \sigma_3 = 1$ corresponds to a metric without signature change. The geodesics pass the singularity at $t = 0$ without any problems. Timelike and lightlike geodesics extend from $t = -\infty$ to $t = +\infty$. Note that in spite of the singular behaviour of $\dot{y} = c_y |t|^{-2p_2}$ and $\dot{z} = c_z |t|^{-2p_3}$, the (squared) velocity

$$v^2 = \sum_{i=1}^{3} |t|^{2p_i} \left(\frac{dx^i}{dt} \right)^2 \leq 1 \tag{16.49}$$

stays finite.

16.9. Reflection of particles

A further solution of the free field equations can be obtained from (16.32) by an interchange of the coordinates t and x together with an appropriate change of signs:

$$ds^2 = x^{2p_1} dt^2 - dx^2 - x^{2p_2} dy^2 - x^{2p_3} dz^2. \tag{16.50}$$

This metric can now be continued to negative values of x with several possibilities of a change of signature. We pick out the particular example

$$ds^2 = \epsilon(x)|x|^{-2/3} dt^2 - dx^2 - |x|^{4/3}(dy^2 + dz^2). \tag{16.51}$$

In this case, a space-time of Minkowski type exists only for $x > 0$. At $x = 0$ it turns into a space with Euclidean signature. The Lagrangean associated with (16.51) reads

$$L = \frac{1}{2} \left(\epsilon(x)|x|^{-2/3} \dot{t}^2 - \dot{x}^2 - |x|^{4/3}(\dot{y}^2 + \dot{z}^2) \right). \tag{16.52}$$

It is independentof t, y and z,

$$c_t = \epsilon(x)|x|^{-2/3} \dot{t}, \ c_y = |x|^{4/3} \dot{y}, \ c_z = |x|^{4/3} \dot{z} \tag{16.53}$$

stay constant and we obtain

$$\dot{x}^2 - c_t^2 \epsilon(x)|x|^{2/3} + (c_y^2 + c_z^2)|x|^{-4/3} = -2L. \tag{16.54}$$

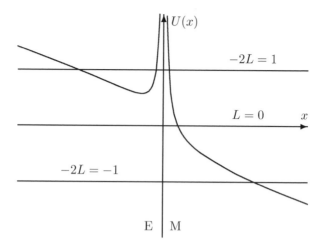

Figure 16.5　Potential $U(x)$ for the geodesics of metric (16.51).

The corresponding potential

$$U(x) = -c_t^2\epsilon(x)|x|^{2/3} + |x|^{-4/3}(c_y^2 + c_z^2) \qquad (16.55)$$

is shown in Fig. 16.5. Now, timelike and lightlike geodesics cannot leave the Minkowski sector ($x > 0$). They are reflected at $x_0 > 0$ determined by $U(x_0) = 0$ in the case of massless particles and by $U(x_0) = -1$ for massive ones, respectively. No particle-antiparticle pairs are created in this case.

Also tachyons ($2L = -1$) are repelled by the repulsive part of the potential $(c_y^2 + c_z^2)|x|^{-4/3}$ at $x = 0$, their worldlines decay into two disconnected parts. One in Minkowski space-time and another one (for suitable values of c_t^2 and $c_y^2 + c_z^2$) in the Euclidean.

16.10. Conclusions

We have proposed a scenario where a metric with Minkowski type signature is continued across a singularity to a region with a different space-time signature. One might be afraid that a signature change wrecks physics. To explore this question, we have studied the reaction of test particles to such a space-time background based on the Kasner metric and simple two-dimensional metrics, respectively. Although the considered spaces exhibit a singularity, many trajectories can traverse it quite happily. The most spectacular situation occurs for certain types of signature change when the particles use the non-Minkowski region for a time travel. Potential acausality problems can be avoided by a reinterpretation of the corresponding geodesics as the creation of particle-antiparticle pairs. Based on the present purely classical investigation, we do not see any reason why a signature change should be excluded.

Acknowledgments

We are grateful to P.C. Aichelburg, G.F.R. Ellis, F. Embacher, P.G.O. Freund, W. Rindler, H. Urbantke, and J. Yngvason for discussions, useful comments, and suggestions.

References

1. A. Vilenkin, *Phys. Rev. D* **27**, 2848 (1983); *ibid.* **30**, 509 (1984).
2. J. B. Hartle, S. W. Hawking, *Phys. Rev. D* **28**, 2960 (1983); S. W. Hawking, *Nucl. Phys. B* **239**, 257 (1983).
3. G. Ellis, A. Sumeruk, D. Coule, C. Hellaby, *Class. Quant. Grav.* **9**, 1535 (1992); G. Ellis, K. Piotrkowska, *Int. J. Mod. Phys. D* **3**, 49 (1994); M. Carfora, G. Ellis, *Int. J. Mod. Phys. D* **4**, 175 (1994).
4. S. A. Hayward, *Class. Quant. Grav.* **9**, 1851 (1992).
5. M. Kossowski, M. Kriele, *Class. Quant. Grav.* **10**, 2363 (1993); M. Kriele, J. Martin, *Class. Quant. Grav.* **12**, 503 (1995).
6. F. Embacher, *Phys. Rev. D* **51**, 6764 (1995); *ibid.* **52**, 2150 (1995); *Class. Quant. Grav.* **12**, 1723 (1995); *ibid.* **13**, 92 (1996).
7. M. Mars, J. M. M. Senovilla, R. Vera, *Phys. Rev. Lett.* **86**, 4219 (2001); G. W. Gibbons, A. Ishibashi, *Class. Quant. Grav.* **21**, 2919 (2004).
8. E. Kasner, *Amer. J. Math.* **43**, 217 (1921).
9. E. C. G. Stueckelberg, *Helv. Phys. Acta* **14**, 588 (1941).
10. R. P. Feynman, *Phys. Rev.* **74**, 939 (1948).

Chapter 17

Towards Noncommutative Gravity

Dmitri Vassilevich

CMCC, Universidade Federal do ABC, Santo André, SP, Brazil
Department of Theoretical Physics, St.Petersburg University, Russia
E-mail: dmitry@dfn.if.usp.br

In this short article accessible for non-experts I discuss possible ways of constructing a non-commutative gravity paying special attention to possibilities of realizing the full diffeomorphism symmetry and to relations with $2D$ gravities.

17.1. Preliminaries

For the first time I met Wolfgang Kummer in 1992. It happened on my way back from Italy to St.Petersburg. At that time, a hundred of US dollars was a fortune in Russia. Therefore, to save money I took a train going through Vienna, and not a plane flying over it. The most natural decision was to stop in Vienna for a couple of days and give a seminar at TU. This is how one of the most fruitful and exciting collaborations in my life started, and this is also a very rare example of a positive effect of severe financial difficulties.

The Vienna School of 2D gravity was an amazingly successful project, see Ref. 1. To keep it running, new interesting directions of research were always needed. About 2005 I told Wolfgang about my recent work on noncommutative (NC) gravity in two dimensions[2] which almost literally repeated some of the steps done previously in the commutative case. We decided to return to this after completing our current work. Unfortunately, deteriorating health did not allow Wolfgang to take up this job. This short article is a kind of a proposal for a "Vienna-style" NC gravity. This is not a (mini)review, with most visible consequence that the literature is incomplete. I am asking all authors whose papers will not be mentioned for understanding. For a systematic overview of NC gravities the reader may consult the paper by Szabo.[3]

Generally speaking, the desire to construct an NC gravity is very natural. One of the main arguments in favor of noncommutativity comes from gravity.[4] Particular ways to realize noncommutativity differ much from model to model.

To stay closer to Vienna, whenever possible, I will discuss noncommutative counterparts of dilaton gravities in two dimensions (see Ref. 5 for a review). In the commutative case, the classical first-order action reads

$$S = \int_M [X_a De^a + X d\omega + \epsilon \left(U(X) X^a X_a/2 + V(X) \right)] , \qquad (17.1)$$

where $a = 0, 1$ is a Lorentz index, e^a and ω are the zweibein and connection one-forms respectively, ϵ is a volume two-form, X is the dilaton, and X^a is an auxiliary field which generates the torsion constraint. $De^a = de^a + \varepsilon^a{}_b \omega \wedge e^b$, where ε^{ab} is the Christoffel symbol. $U(X)$ and $V(X)$ are two arbitrary functions called the dilaton potentials. With the choice $U(X) = 0$, $V(X) \propto X$ one obtains the Jackiw-Teitelboim model.[6] Other choices reproduce all gravity models in two dimensions, see Ref. 7.

17.2. What can we call a noncommutative gravity?

In principle, any theory containing some effects of noncommutativity of the coordinates and looking more or less like a gravity theory may be called a noncommutative gravity. The problem is that the people working on a particular approach are (naturally) more enthusiastic about it than the rest of the community. Therefore, I asked myself, what kind of noncommutative gravity theory could have a chance to satisfy Wolfgang? An answer to this question seems to be a rather strict point of view on NC gravity.

To construct a gravity one first needs a manifold. NC manifolds may be understood through the Gelfand-Naimark duality. To a manifold M one can associate a commutative associative algebra $C^\infty(M)$ of smooth functions. Under certain restrictions, each commutative associative algebra is an algebra of smooth functions on some manifold. In this sense, an algebra A, which is a noncommutative associative deformation of $C^\infty(M)$ defines an NC deformation of M. Most conveniently the deformation is done by replacing the point-wise product $f_1 \cdot f_2$ by a noncommutative star product $f_1 \star f_2$, which can be presented as

$$f_1 \star f_2 = f_1 \cdot f_2 + \frac{i}{2} \theta^{\mu\nu}(x) \partial_\mu f_1 \cdot \partial_\nu f_2 + O(\theta^2) . \qquad (17.2)$$

Because of the associativity, $\theta^{\mu\nu}$ is a Poisson bivector, i.e. it has to satisfy the Jacobi identity. Note, that in two dimensions the Jacobi identity is satisfied any antisymmetric tensor $\theta^{\mu\nu}(x)$.

For a constant θ there exists a simple (Moyal) formula for the star product

$$(f_1 \star_M f_2)(x) = \exp\left(\frac{i}{2}\theta^{\mu\nu}\partial_\mu^x \partial_\nu^y\right) f_1(x)f_2(y)|_{y=x} . \quad (17.3)$$

Next, one has to satisfy the relativity principle, i.e., one should realize the group of diffeomorphisms (or a deformation of this group) on an NC manifold. Then one has to construct invariants which in the commutative limit $\theta \to 0$ reproduce the Einstein-Hilbert action coupled to matter fields. This program, upon completion, should give an NC gravity.

None of the existing approaches to the NC gravity fulfills strictly all the requirements formulated above, but we still can learn a lot from each of them.

17.2.1. *Minimalistic approaches*

These are approaches which are not even trying to construct a full NC gravity but instead focus on some selected features of NC theories. For example, in one of such approaches, reviewed in Ref. 8, the nonlocality, which is a characteristic feature of NC theories, is modelled by delocalization of sources in otherwise commutative theories. Such approaches are very useful in one wishes to understand what kind of physical effects may follow from the noncommutativity, but they are not designed to check theoretical consistency.

17.2.2. *Seiberg-Witten map*

In 1999 Seiberg and Witten[9] discovered a map between commutative and noncommutative gauge theories. Due to this map, gauge symmetries, including diffeomorphisms, can be realized by standard commutative transformations on commutative fields. The NC fields are expressed through power series in θ with growing number of commutative fields and their derivatives. This map was applied also to gravity, and even some physical effects were studied, see e.g. Ref. 10. With higher orders of θ technical difficulties in applying the Seiberg-Witten map grow fast, so that no one was able to go beyond the second order. Because of this, this method can hardly be considered as an ultimate solution of the problem of constructing

296 D. V. Vassilevich

an NC gravity, but it gives a very valuable information: the statement that such a theory does exist at least in the form of power series.

17.2.3. *Gauging symplectic diffeomorphisms*

Looking at the formula (17.3) one immediately sees a source of the problems with the diffeomorphisms: $\theta^{\mu\nu}$ looks as a tensor, but the formula (17.3) is not tensorial. Then, it is natural to assume that the things become easier with the part of the diffeomorphisms group which does not change θ. For a non-degenerate $\theta^{\mu\nu}$ these diffeomorphisms (symplectomorphisms) are generated by vector fields of the form

$$\xi^\mu(x) = \theta^{\mu\nu}\partial_\nu f(x)\,. \tag{17.4}$$

Such diffeomorphisms preserve also the volume element, and thus we are dealing with unimodular gravity theories. NC theories based on gauging symplectic diffeomorphisms were indeed constricted[11] and gave rise to many interesting results. Though in our rather strict approach to NC gravities this group looks too small, we again receive an important message that a consistent NC theory may be constructed at least with this small part.

17.2.4. *Gravity through Yang-Mills type symmetries*

The action of a Yang-Mills gauge transformation can easily be extended to a noncommutative case. Let in a commutative theory $\delta_\alpha \phi = \rho(\alpha)\cdot\phi$, where ϕ is a field transformed according to a finite dimensional representation ρ of the symmetry algebra. Then in an NC case one can define $\delta^\star_\alpha \phi = \rho(\alpha)\star\phi$. A problem appears with commutators. Let T^A be a basis in the Lie algebra taken in the representation ρ. Then

$$\delta^\star_\alpha \delta^\star_\beta - \delta^\star_\beta \delta^\star_\alpha = \delta^\star_{[\alpha,\beta]_\star}$$

$$[\alpha,\beta]_\star = \frac{1}{2}[T^A,T^B](\alpha_A \star \beta_B + \beta_B \star \alpha_A)$$

$$+\frac{1}{2}\{T^A,T^B\}(\alpha_A \star \beta_B - \beta_B \star \alpha_A)\,.$$

The expression on the right hand side of the last line is a gauge generator if both commutator $[T^A,T^B]$ and anticommutator $\{T^A,T^B\}$ belong to the Lie algebra. This imposes severe restrictions on possible gauge groups and their representations.[12] For example, $su(n)$ cannot be extended to NC spaces, while $u(n)$ can.

One can demonstrate, that with the choice of the potentials $U(X) = 0$, $V(X) \propto X$ corresponding to the Jackiw-Teitelboim model[6] is equivalent to an $su(1,1)$ BF theory. Consequently, extending this symmetry to an NC $u(1,1)$ one can construct an NC version of the JT gravity.[13] The model appears to be both classical[13] and quantum[2] integrable. Of course, by extending the gauge symmetry one introduces a new gauge field, which, however, decouples in the commutative limit and does not lead to any contradictions. However, there is a different problem with this approach. One cannot deform the linear dilaton potential $V(X)$ by adding higher powers of the dilaton and preserving the number of NC gauge symmetries.[14] This means that other interesting dilaton gravity models cannot be constructed in this approach.

17.2.5. *Twisted symmetries*

Practically all symmetries of commutative theories can be realized on a noncommutative space as *twisted* symmetries. The twisting is based on an observation that the Moyal product (17.3) can be represented as a composition of the point-wise product and a Drinfeld twist. Indeed, the point-wise product $\mu : A \otimes A \to A$, $\mu(f_1 \otimes f_2) = f_1 \cdot f_2$ and the Moyal product $\mu_\star : A \otimes A \to A$, $\mu_\star(f_1 \otimes f_2) = f_1 \star_M f_2$ are related through $\mu_\star = \mu \circ \mathcal{F}^{-1}$, where

$$\mathcal{F} = \exp \mathcal{P}, \qquad \mathcal{P} = -\frac{i}{2}\theta^{\mu\nu}\partial_\mu \otimes \partial_\nu \tag{17.5}$$

is a twist.

The way how the symmetry generators act on tensor products is defined by the coproduct Δ. In commutative field theories one uses a primitive coproduct $\Delta_0(\alpha) = \alpha \otimes 1 + 1 \otimes \alpha$, so that we have the usual Leibniz rule

$$\alpha(\phi_1 \otimes \phi_2) = \Delta_0(\alpha)(\phi_1 \otimes \phi_2) = (\alpha\phi_1) \otimes \phi_2 + \phi_1 \otimes (\alpha\phi_2). \tag{17.6}$$

We may define another (twisted) coproduct

$$\Delta_{\mathcal{F}} = \mathcal{F}\Delta\mathcal{F}^{-1} \tag{17.7}$$

The action of a generator α on the star-product of fields is defined as follows

$$\alpha(\phi_1 \star_M \phi_2) = \mu_\star(\Delta_{\mathcal{F}}(\alpha)\phi_1 \otimes \phi_2) = \mu \circ \mathcal{F}^{-1}(\Delta_{\mathcal{F}}(\alpha)\phi_1 \otimes \phi_2). \tag{17.8}$$

Twisting, in a sense, pushes the symmetry generator through the star product. This makes it possible to define symmetry transformations without transforming the star product. In algebraic language, we have a Hopf algebra symmetry instead of a Lie algebra one.

The literature on twisted symmetries is very large. We like to mention an early paper by Oeckl.[15] The symmetries relevant for our discussion are the Poincare symmetry[16] (this was the first symmetry to be twisted), diffeomorphisms,[17] and gauge symmetries.[18] Moreover, the twist interpretation may be given to some star products other than the Moyal one.

Twisting the diffeomorphism transformations allowed to define a model of NC gravity[17] invariant under the full diffeomorphism algebra, though this invariance is realized in a non-standard way[a].

The twisted symmetries are not *bona fide* physical symmetries. One cannot use them, for example, to gauge away any degrees of freedom. The problem of proper interpretation of twisted local symmetries remains. One possible interpretation is as follows.[20] Let us replace the partial derivatives ∂ in (17.3) and (17.5) with covariant derivatives ∇ with a trivial connection (pure gauge). Since ∇_μ commute, the new star product will be again associative. If the original theory were twisted gauge invariant, the theory with this new star product will be both twisted gauge invariant and gauge invariant in the ordinary sense. To return back, one has to fix the gauge $\nabla = \partial$. Therefore, twisted gauge invariance is a remnant of ordinary gauge invariance after fixing the gauge by imposing a condition on gauge-trivial covariant derivatives appearing inside the star product.

17.2.6. NC geometry and spectral action

A unifying approach to describe *any* NC geometry was introduced by Connes[21] (see also Ref. 22 for a recent overview). It is based on the notion of a spectral triple (A, H, D) consisting of an associative algebra A represented by bounded operators on a Hilbert space H and a Dirac operator D acting on H. These three object satisfy certain relations and restrictions. As soon as a spectral triple is defined, the corresponding classical action follows from the so-called spectral action principle[23]

$$S = \operatorname{Tr} \Phi(D/\Lambda), \tag{17.9}$$

where Φ is a positive even function, and Λ is a scale parameter. All unitary symmetries of the operator D are inherited by the spectral action. As an expansion in Λ the action (17.9) may be calculated by the heat kernel

---

[a]There are also critics of twisting local symmetries, see Ref. 19.

methods. On Moyal spaces such methods are rather well developed.[24] The problem is "only" to find a corresponding spectral triple.

A similar idea, that the NC gravity may be induced is explored in the emergent gravity approach, see Ref. 25 and references therein.

17.3. The star products

As we have seen above, rigidity of $\theta^{\mu\nu}$ under the diffeomorphism transformations creates a lot of problems. It may be a good idea to transform both $\theta^{\mu\nu}$ and the star product under the diffeomorphisms. To this end, we need general star products.

The modern history of deformation quantization started with the papers.[26] The main part of the deformation-quantization program is a construction of a star product for a given Poisson structure $\theta^{\mu\nu}(x)$. For symplectic manifolds (non-degenerate $\theta^{\mu\nu}$) the existence of a start product was demonstrated by De Wilde and Lecomte,[27] and a very elegant construction was given by Fedosov.[28] For generic Poisson structure the existence of a star product was demonstrated by Kontsevich[29] who also gave an explicit formula (which is, however, too complicated to be used for actual calculations of higher orders in the star product). Such orders of the star product were computed by using the Weyl map and a representation of noncommutative coordinates in the form of differential operators.[30]

A very promising non-perturbative formula for the star product was suggested by Cattaneo and Felder.[31] They took a Poisson sigma model with the action

$$S_{\text{PSM}} = \int \left[A_\mu dX^\mu + \frac{1}{2}\theta^{\mu\nu}(X)A_\mu \wedge A_\nu \right] \qquad (17.10)$$

defined on a two-dimensional manifold. X and A are the fields on this manifold, which are a zero-form taking values in a Poisson manifold and a one-form with values in the cotangent space to this manifold, respectively. The two-dimensional world-sheet is supposed to be a disc (with suitable boundary conditions imposed on A). Three distinct points on the boundary of the disc are selected, denoted 0, 1, and ∞. The star product is then given by a correlation function

$$f \star g(x) = \int dA \, dX \, f(X(0))g(X(1)) \, e^{iS_{\text{PSM}}}, \qquad (17.11)$$

where the integration is restricted by the condition $X^\mu(\infty) = x^\mu$. The main advantage of this formula is that it does not imply any expansion in θ.

What is then the relation to two-dimensional dilaton gravities? The point is that the Poisson sigma models were originally introduced[32,33] as generalizations of the dilaton gravity action (17.1). Indeed, by identifying X, X^a with X^μ, and ω, e^a with A_μ and making a suitable choice of $\theta^{\mu\nu}(X)$ one can reduce (17.10) to (17.1). In the context of two-dimensional gravities rather powerful methods of calculation of the path integral were developed.[34] At least some of these methods work also for generic Poisson sigma models.[35] The approach[34] was specially tailored to study quantum gravity phenomena, like virtual black holes, and not the correlation functions of the type (17.11). However, some steps to adjust that methods to the new tasks have already been done. For example, inclusion of boundaries was considered in a paper,[36] which was the last publication of Wolfgang Kummer.

17.4. Conclusions

As we have seen, there are many rather successful approaches to NC gravity. One can be optimistic, that soon an NC gravity satisfying our (perhaps, too strict) criteria will be formulated. It is likely, that 2D dilaton gravities will play a prominent role in this process.

Acknowledgment

This work was supported in part by CNPq (Brazil).

References

1. L. Bergamin and R. Meyer, "Wolfgang Kummer and the Vienna School of Dilaton (Super-)Gravity," arXiv:0809.2245 [hep-th], in this volume.
2. D. V. Vassilevich, "Quantum noncommutative gravity in two dimensions," Nucl. Phys. B **715**, 695 (2005) [arXiv:hep-th/0406163].
3. R. J. Szabo, "Symmetry, gravity and noncommutativity," Class. Quant. Grav. **23**, R199 (2006) [arXiv:hep-th/0606233].
4. S. Doplicher, K. Fredenhagen and J. E. Roberts, "The Quantum structure of space-time at the Planck scale and quantum fields," Commun. Math. Phys. **172**, 187 (1995) [arXiv:hep-th/0303037].
5. D. Grumiller, W. Kummer and D. V. Vassilevich, "Dilaton gravity in two dimensions," Phys. Rept. **369**, 327 (2002) [arXiv:hep-th/0204253].
6. R. Jackiw, "Liouville Field Theory: A Two-Dimensional Model for Gravity?" in Quantum Theory Of Gravity, p. 403-420, S. Christensen (ed.)

(Adam Hilger, Bristol, 1983); C. Teitelboim, "Gravitation and Hamiltonian Structure in Two Space-Time Dimensions", Phys.Lett.B **126**, 41 (1983);

7. D. Grumiller and R. Meyer, "Ramifications of lineland," Turk. J. Phys. **30**, 349 (2006) [arXiv:hep-th/0604049].

8. P. Nicolini, "Noncommutative Black Holes, The Final Appeal To Quantum Gravity: A Review," arXiv:0807.1939 [hep-th].

9. N. Seiberg and E. Witten, "String theory and noncommutative geometry," JHEP **9909**, 032 (1999) [arXiv:hep-th/9908142].

10. A. H. Chamseddine, "Deforming Einstein's gravity," Phys. Lett. B **504**, 33 (2001) [arXiv:hep-th/0009153]. M. Chaichian, A. Tureanu and G. Zet, "Corrections to Schwarzschild Solution in Noncommutative Gauge Theory of Gravity," Phys. Lett. B **660**, 573 (2008) [arXiv:0710.2075 [hep-th]]. S. Mar-culescu and F. Ruiz Ruiz, "Seiberg–Witten maps for $SO(1,3)$ gauge in-variance and deformations of gravity," Phys. Rev. D **79**, 025004 (2009) [arXiv:0808.2066 [hep-th]].

11. X. Calmet and A. Kobakhidze, "Noncommutative general relativity," Phys. Rev. D **72**, 045010 (2005) [arXiv:hep-th/0506157].

12. M. Chaichian, P. Presnajder, M. M. Sheikh-Jabbari and A. Tureanu, "Non-commutative gauge field theories: A no-go theorem," Phys. Lett. B **526**, 132 (2002) [arXiv:hep-th/0107037].

13. S. Cacciatori, A. H. Chamseddine, D. Klemm, L. Martucci, W. A. Sabra and D. Zanon, "Noncommutative gravity in two dimensions," Class. Quant. Grav. **19**, 4029 (2002) [arXiv:hep-th/0203038].

14. D. V. Vassilevich, R. Fresneda and D. M. Gitman, "Stability of a noncommu-tative Jackiw-Teitelboim gravity," Eur. Phys. J. C **47**, 235 (2006) [arXiv:hep-th/0602095].

15. R. Oeckl, "Untwisting noncommutative R^d and the equivalence of quantum field theories," Nucl. Phys. B **581**, 559 (2000) [arXiv:hep-th/0003018].

16. M. Chaichian, P. P. Kulish, K. Nishijima and A. Tureanu, "On a Lorentz-invariant interpretation of noncommutative space-time and its implica-tions on noncommutative QFT," Phys. Lett. B **604**, 98 (2004) [arXiv:hep-th/0408069]; J. Wess, "Deformed coordinate spaces: Derivatives," arXiv:hep-th/0408080; M. Chaichian, P. Presnajder and A. Tureanu, "New concept of relativistic invariance in NC space-time: Twisted Poincare symmetry and its implications," Phys. Rev. Lett. **94**, 151602 (2005) [arXiv:hep-th/0409096].

17. P. Aschieri, C. Blohmann, M. Dimitrijevic, F. Meyer, P. Schupp and J. Wess, "A gravity theory on noncommutative spaces," Class. Quant. Grav. **22**, 3511 (2005) [arXiv:hep-th/0504183]; P. Aschieri, M. Dimitrijevic, F. Meyer and J. Wess, "Noncommutative geometry and gravity," Class. Quant. Grav. **23**, 1883 (2006) [arXiv:hep-th/0510059].

18. D. V. Vassilevich, "Twist to close," Mod. Phys. Lett. A **21**, 1279 (2006) [arXiv:hep-th/0602185]; P. Aschieri, M. Dimitrijevic, F. Meyer, S. Schraml and J. Wess, "Twisted gauge theories," Lett. Math. Phys. **78**, 61 (2006) [arXiv:hep-th/0603024].

19. M. Chaichian, A. Tureanu and G. Zet, "Twist as a Symmetry Principle and the Noncommutative Gauge Theory Formulation," Phys. Lett. B **651**, 319

(2007) [arXiv:hep-th/0607179].

20. D. V. Vassilevich, "Symmetries in noncommutative field theories: Hopf versus Lie," to appear in São Paulo J. Math. Sci. [arXiv:0711.4091 [hep-th]].

21. A. Connes, "Noncommutative geometry", Academic Press, 1994.

22. F. Muller-Hoissen, "Noncommutative Geometries and Gravity," AIP Conf. Proc. **977**, 12 (2008) [arXiv:0710.4418 [gr-qc]].

23. A. H. Chamseddine and A. Connes, "The spectral action principle," Commun. Math. Phys. **186**, 731 (1997) [arXiv:hep-th/9606001].

24. D. V. Vassilevich, "Heat Trace Asymptotics on Noncommutative Spaces," SIGMA **3**, 093 (2007) [arXiv:0708.4209 [hep-th]].

25. H. Steinacker, "Emergent Gravity from Noncommutative Gauge Theory," JHEP **0712**, 049 (2007) [arXiv:0708.2426 [hep-th]].

26. F. Bayen, M. Flato, C. Fronsdal, A. Lichnerowicz and D. Sternheimer, "Deformation Theory And Quantization. 1. and 2" Annals Phys. **111**, 61 (1978); **111**, 111 (1978).

27. M. De Wilde and P. B. A. Lecomte, "Existence of star-products and of formal deformations of the Poisson Lie algebra of arbitrary symplectic manifolds", Lett. Math. Phys. **7**, 487 (1983).

28. B. V. Fedosov, "A Simple Geometrical Construction Of Deformation Quantization," J. Diff. Geom. **40** (1994) 213.

29. M. Kontsevich, "Deformation quantization of Poisson manifolds, I," Lett. Math. Phys. **66**, 157 (2003) [arXiv:q-alg/9709040].

30. V. G. Kupriyanov and D. V. Vassilevich, "Star products made (somewhat) easier," Eur. Phys. J. C **58**, 627 (2008) [arXiv:0806.4615 [hep-th]].

31. A. S. Cattaneo and G. Felder, "A path integral approach to the Kontsevich quantization formula," Commun. Math. Phys. **212** (2000) 591 [arXiv:math/9902090]; "Poisson sigma models and deformation quantization," Mod. Phys. Lett. A **16**, 179 (2001) [arXiv:hep-th/0102208].

32. P. Schaller and T. Strobl, "Poisson structure induced (topological) field theories," Mod. Phys. Lett. A **9**, 3129 (1994) [arXiv:hep-th/9405110].

33. N. Ikeda, "Two-dimensional gravity and nonlinear gauge theory," Annals Phys. **235**, 435 (1994) [arXiv:hep-th/9312059].

34. W. Kummer, H. Liebl and D. V. Vassilevich, "Exact path integral quantization of generic 2-D dilaton gravity," Nucl. Phys. B **493**, 491 (1997) [arXiv:gr-qc/9612012]; "Integrating geometry in general 2D dilaton gravity with matter," Nucl. Phys. B **544**, 403 (1999) [arXiv:hep-th/9809168].

35. A. C. Hirshfeld and T. Schwarzweller, "Path integral quantization of the Poisson-sigma model," Annalen Phys. **9**, 83 (2000) [arXiv:hep-th/9910178].

36. L. Bergamin, D. Grumiller, W. Kummer and D. V. Vassilevich, "Physics-to-gauge conversion at black hole horizons," Class. Quant. Grav. **23**, 3075 (2006) [arXiv:hep-th/0512230].

Chapter 18

Superembedding Approach to Superstring in $AdS_5 \times S^5$ Superspace

Igor A. Bandos

Ikerbasque, the Basque Science Foundation, and
Department of Theoretical Physics and History of Sciences,
University of The Basque Country (EHU/UPV),
P.O. Box 644, 48080 Bilbao, Spain
and
Institute for Theoretical Physics,
NSC Kharkov Institute of Physics & Technology,
UA 61108, Kharkov, Ukraine
E-mail: `igor_bandos@ehu.es, bandos@ific.uv.es`

We review the spinor moving frame formulations and generalized action principle for super-p-branes, describe in detail the superembedding approach to superstring in general type IIB supergravity background and present the complete superembedding description of type IIB superstring in the $AdS_5 \times S^5$ superspace.

This contribution is devoted to the memory of Wolfgang Kummer who untimely left us in 2007. We collaborated with him several years beginning, in 1996, by studying gravity induced on the worldvolume of a brane;[11] this was one of the pre-Rundall-Sundrum Brane World scenarios (see also Ref. 5). Search for its supersymmetric generalizations led us to thinking on a new form of Dp-brane actions[12] and to studying the super-D9-brane dynamics.[2] This line was then continued by attacking the problem of supersymmetric Lagrangian description of the interacting superbrane systems[13] which, in my opinion, still remains open as far as the commonly accepted candidate action for coincident Dp–branes[50] does not possess neither supersymmetry nor Lorentz symmetry.

Among the main tools in our studies were embedding and superembedding approaches to bosonic and supersymmetric branes. This is why I decided to chose for my contribution the present manuscript containing a review of the superembedding approach and its specific application for the case of superstring in $AdS_5 \times S^5$ superspace (see Refs. 47, 48 for Green–Schwarz superstring action in this superspace).

Notice that superstring in $AdS_5 \times S^5$ superspace is often called $AdS_5 \times S^5$ *superstring* (see Ref. 3 and refs therein). However, in our opinion, this name might produce an erroneous impression that the model is essentially different from the Green–Schwarz (GS) superstring. Such a confusion might be further enlarged by an accent which is made in the literature on the fact that $AdS_5 \times S^5$ superspace (the superspace with bosonic body $AdS_5 \times S^5$) is a coset of $SU(2,2|4)$ supergroup. Although important, this does not change the fact that this '$AdS_5 \times S^5$ superstring' *is just a particular case of the GS superstring in a curved superspace*[37]. So is its type IIA counterpart, '$AdS_4 \times \mathbb{CP}^3$ superstring', which attracted recently much attention, but is not a model on a coset of supergroup, just because the type IIA supergravity superspace with the bosonic body $AdS_4 \times \mathbb{CP}^3$ is not a coset.[35]

Thus we prefer to formulate our problem as superembedding description of the GS superstring model in $AdS_5 \times S^5$ superspace. On one hand, this formalism can be applied to study the N=16 two dimensional supergravity induced on the worldsheet superspace of the superstring moving in AdS superspace. And, in this respect, it is proper for the present volume because two dimensional gravity and supergravity model were always in the center of Wolfgang's interests, see *e.g.* Refs. 22, 38, 44.

On the other hand, the results of this manuscript can be useful in further study of classical and quantum $AdS_5 \times S^5$ superstring, which is of current interest for the applications of AdS/CFT correspondence.[a]

18.1. Introduction

The standard GS superstring action[36] is based on embedding of a bosonic surface W^2 in the target superspace $\Sigma^{(D|n)}$ ($D = 3,4,6,10$, $n = 2(D-2)$ for heterotic and type I and $n = 4(D-2)$ for type II superstrings). This embedding is described by the bosonic and fermionic coordinate functions

$$ W^{p+1} \in \Sigma^{(D|n)} \ : \quad \hat{Z}^M(\xi) = \left(\hat{x}^\mu(\xi), \hat{\theta}^{\check\alpha}(\xi)\right), \quad \begin{smallmatrix} \mu=0,1,\ldots,(D-1), \\ \check\alpha=1,\ldots,n , \end{smallmatrix} \quad (18.1) $$

where $p = 1$ and $\xi^m = (\tau, \sigma)$ are local coordinates on W^2. The more 'ancient' Ramond–Neveu–Schwarz (RNS) or spinning string, which becomes

---

[a]See, for instance, Refs. 21, 27 where the $AdS_5 \times S^5$ superstring was used to reveal the mysterious dual superconformal symmetry of the N=4 SYM amplitudes.

equivalent to the GS sigma model on the quantum level and after imposing the so–called GSO projection (see, however, Refs. 58, 63 and more
recent Ref. 59), corresponds to an embedding of the worldsheet superspace
$W^{(2|1+1)}$ into the spacetime $M^D = \Sigma^{(D|0)}$, described by D bosonic super-
fields $\hat{X}^{\underline{m}}(\xi, \eta, \bar{\eta}) = \hat{x}^m(\xi) + i\eta\psi^{\underline{m}}(\xi) + i\bar{\eta}\bar{\psi}^{\underline{m}}(\xi) + \ldots$ depending on two
bosonic (ξ^m) and complex fermionic coordinate η (or real fermionic coor-
dinate in the case of heterotic string). There are some known obstacles for
extending such a description to supermembrane and other branes.[25,43]

Following Ref. 57, the *superembedding approach*, developed in Ref. 16
for 10D superstrings and 11D supermembrane, and applied in the first
studies of dynamics of Dirichlet p-branes (Dp–branes) and M-theory 5-
brane (M5-brane) in seminal papers Refs. 41 and 42, describes strings and
branes by *embedding of a worldvolume superspace $W^{(p+1|n/2)}$ into the target
superspace $\Sigma^{(D|n)}$*.

Let us denote the $d = p + 1 \leq D$ local bosonic coordinates and $n/2$
fermionic coordinates of $W^{(p+1|n/2)}$ by $\zeta^{\mathcal{M}} = (\xi^m, \eta^{\breve{q}})$. Then the embedding
of $W^{(p+1|n/2)}$ into the tangent superspace $\Sigma^{(D|n)}$ with coordinates $Z^M = (x^\mu, \theta^{\breve{\alpha}})$ can be described parametrically by specifying the set of coordinate
super-functions, the *worldvolume superfields* $\hat{Z}^M(\zeta) = \hat{Z}^M(\xi^m, \eta^{\breve{q}})$

$$W^{(p+1|n/2)} \in \Sigma^{(D|n)} \ : \quad Z^M = \hat{Z}^M(\zeta) = (\hat{x}^\mu(\xi, \eta), \hat{\theta}^{\breve{\alpha}}(\xi, \eta)) \ . \quad (18.2)$$

Here, $\mu = 0, 1, \ldots, (D-1)$, $\breve{\alpha} = 1, \ldots n$, $m = 0, 1, \ldots, p$ and $\breve{q} = 1, \ldots, \frac{n}{2}$.
Notice that the number of fermionic 'directions' $\eta^{\breve{q}}$ of the worldvolume
superspace are usually chosen to be one–half of the number of fermionic
dimensions of the target superspace.[b] This is proper to replace *all* the κ–
symmetries[31,36,54] of the *standard*, Dirac–Nambu–Goto type super-p-brane
actions[1,24] by the local worldvolume supersymmetry,[c] thus realizing the
idea developed for $D = 3, 4$ superparticle in Ref. 57.[d]

---

[b] For $\mathcal{N} = 1$ $n = \delta^{\breve{\alpha}}_{\breve{\alpha}}$ is the number of values of the minimal D–dimensional spinor index;
for $D \neq 2$ (mod 8) this is $n = 2^{[D/2]}$, where is $[D/2]$ is the integer part of $D/2$; and for
$D = 2$ (mod 8) it is $n = 2^{[D/2]-1}$.

[c] Under the standard super-p-branes we mean supersymmetric extended objects the
ground state of which are 1/2 BPS states, *i.e.* preserve 1/2 of the tangent space supersymmetry reflected by n/2 parametric κ–symmetry of their worldvolume actions.
See Ref. 14 as well as Refs. 7, 9, 65 for the actions in enlarged (tensorial) superspaces
with additional tensorial coordinates (see Refs. 14, 30, 60 and refs therein) describing the
excitations of $k/32$ BPS states, including the $k = 31$ models possessing the properties of
BPS preons.[8]

[d] See Refs. 49, 58 and references therein for formulations of superbranes in the worldvolume superspaces with less than $n/2$ fermionic 'directions'.

18.1.1. Superembedding equation

For all presently known superbranes the embedding (18.2) of their maximal worldvolume superspace $W^{(p+1|\frac{n}{2})}$ into the target superspace $\Sigma^{(D|n)}$ obeys the *superembedding equation*. For D=3,4 superparticle this was obtained in Ref. 57 by varying a superfield action (called STV action in nineties). To write its most general and universal form for a super–p–brane in D–dimensional supergravity background, let us denote the supervielbein of the worldvolume superspace $W^{(p+1|16)}$ by

$$e^A = d\zeta^{\mathcal{M}} e_{\mathcal{M}}{}^A(\zeta) = (e^a, e^q), \quad a = 0,1,\ldots,p, \quad q = 1,\ldots,n/2, \quad (18.3)$$

and decompose the pull–back $\hat{E}^{\underline{A}} := E^{\underline{A}}(\hat{Z}) = d\hat{Z}^M E_M{}^{\underline{A}}(\hat{Z})$ of the supervielbein of the target superspace, $E^{\underline{A}} = dZ^M E_M{}^{\underline{A}}(Z) = (E^{\underline{a}}, E^{\alpha})$ ($\underline{a} = 0,1,\ldots(D-1)$, $\alpha = 1,\ldots,n$), on the basis of (18.3). In general, such a decomposition reads

$$\hat{E}^{\underline{A}} := E^{\underline{A}}(\hat{Z}) = d\hat{Z}^M E_M{}^{\underline{A}}(\hat{Z}) = e^b \hat{E}_b^{\underline{A}} + e^q \hat{E}_q^{\underline{A}}, \quad (18.4)$$

where $\hat{E}_b^{\underline{A}} := e_b^{\mathcal{M}} \partial_{\mathcal{M}} \hat{Z}^M E_M{}^{\underline{A}}(\hat{Z})$ and $\hat{E}_q^{\underline{A}} := e_q^{\mathcal{M}} \partial_{\mathcal{M}} \hat{Z}^M E_M{}^{\underline{A}}(\hat{Z})$ are, respectively, bosonic and fermionic components of the pull–back the supervielbein form. The superembedding equations states that the fermionic component of the pull–back of the bosonic supervielbein form vanishes,

$$\boxed{\hat{E}_q^{\underline{a}} := \nabla_q \hat{Z}^M E_M{}^{\underline{a}}(\hat{Z}) = 0}, \quad \nabla_q := e_q^{\mathcal{M}}(\zeta)\partial_{\mathcal{M}}. \quad (18.5)$$

For higher dimensional superbranes of sufficiently large co-dimensions the superembedding equation contains equations of motion among their consequences. This was shown for M2-brane and D=10 type II superstring in Ref. 16, for M5-brane in Ref. 42 and for Dp–branes with $p \leq 5$ in Ref. 41 (the 'boundary' $p \leq 5$ was established in Ref. 28). Hence, in these cases, the description of the classical super-p-brane dynamics by this equation is complete. Moreover, if several types of D–dimensional p-branes exist, the superembedding equation provides their universal description (see Ref. 6 for such a universal description of fundamental type IIB superstring and D1–brane and Ref. 23 for the $SL(2)$ covariant formulation providing a unified descriptions of all the actions of p–branes related to the Dp-brane by SL(2) transformations).

On the other hand, this on-shell nature of the superembedding equation prevents from the constructing the complete worldvolume superfield action of the STV type (see Refs. 57 and 56 for the review and further references). A universal although non-standard Lagrangian framework for the superembedding approach is provided by the generalized action principle, proposed in Ref. 18 for superstrings and $D = 11$ supermembrane and in Ref. 17 for the case of super-Dp-branes. This produces the superembedding equation in its equivalent form

$$\hat{E}^i(\zeta) := d\hat{Z}^M(\zeta) \, E_M{}^{\underline{b}}(\hat{Z}(\zeta)) \, u_{\underline{b}}{}^i(\zeta) = 0 \,, \qquad (18.6)$$

where $u_{\underline{b}}{}^i$ are $(D - p - 1)$ vectors orthogonal to the worldsheet superspace. These *moving frame variables* or Lorentz harmonics (vector harmonics) will be the subject of the next section.

18.2. Spinor moving frame formulation and generalized action principle for super-p-branes

18.2.1. *Vector harmonics as moving frame adapted to (super)embedding*

The standard formulations of superstring,[36] M2-brane (supermembrane)[24] and super-p-branes[1] is based on embedding (18.1) of the bosonic worldvolume W^{p+1} into the tangent superspace $\Sigma^{(D|n)}$.

If the worldvolume W^{p+1} is flat, one always can chose a special Lorentz frame with $p + 1$ vectors being tangential and the remaining $D - p - 1$ vectors - orthogonal to W^{p+1}. In general this also can be done, but locally. It is convenient to use the dual language of the differential forms and to consider the pull–back

$$\hat{E}^{\underline{a}} := E^{\underline{a}}(\hat{Z}) = d\hat{Z}^M(\xi) E_M{}^{\underline{a}}(\hat{Z}) = d\xi^m \partial_m \hat{Z}^M E_M{}^{\underline{a}}(\hat{Z}) =: d\xi^m \hat{E}_m^{\underline{a}} \quad (18.7)$$

of the bosonic supervielbein of the target superspace $E^{\underline{a}} := dZ^M E_M{}^{\underline{a}}(Z)$ to the worldvolume W^{p+1} with local coordinates ξ^m, $m = 0, 1, \ldots, p$. Only $(p + 1)$ of the D one–forms $\hat{E}^{\underline{a}}$ may be independent on W^{p+1}. This is tantamount to saying that there exist $(D - p - 1)$ linear combinations of $\hat{E}^{\underline{a}}$ that vanish on W^{p+1}. We can express the above statement by the following *embedding equation*

$$\hat{E}^i(\xi) := \hat{E}^{\underline{b}} u_{\underline{b}}{}^i(\xi) = 0 \,, \qquad i = 1, \ldots, (D - p - 1) \,, \qquad (18.8)$$

where $u_{\underline{b}}{}^i(\xi)$ are some coefficient dependent on the point of W^{p+1}. They define $(D - p - 1)$ vectors which are linear independent and orthogonal to the worldvolume W^{p+1}. Thus one may chose them orthogonal one to another and normalized (on -1 as the vectors are spacelike and we are working with 'mostly minus' metric conventions)

$$u^{\underline{a}i} u_{\underline{a}}^j = -\delta^{ij} \,. \qquad (18.9)$$

One can complete the set of the $(D-p-1)$ vectors $u_{\underline{a}}{}^i$ orthogonal to the worldvolume by the set of the (p+1) vectors $u_{\underline{a}}{}^b$ tangential to W^{p+1} (also orthogonal among themselves and normalized). Then the $D \times D$ *moving frame matrix* constructed from $u_{\underline{a}}{}^b$ and $u_{\underline{a}}{}^j$ obeys $U\eta U^T = \eta$,

$$U_{\underline{a}}^{(\underline{b})} := \begin{pmatrix} u_{\underline{a}}{}^b \\ u_{\underline{a}}{}^j \end{pmatrix}, \qquad U^T \eta U = \eta \quad \Leftrightarrow \quad \begin{cases} u_{\underline{a}}^{\underline{c}a} u_{\underline{c}}{}^b = \eta^{ab}, \\ u^{\underline{a}a} u_{\underline{a}}{}^j = 0, \\ u^{\underline{a}i} u_{\underline{a}}{}^j = -\delta^{ij}, \end{cases} \qquad (18.10)$$

by construction, and, hence, belongs to the fundamental representation of the Lorentz group $SO(1, D-1)$,

$$U_{\underline{a}}^{(\underline{b})} := \begin{pmatrix} u_{\underline{c}}{}^b \\ u_{\underline{c}}{}^j \end{pmatrix} \in SO(1, D-1). \qquad (18.11)$$

The splitting of the $D \times D$ matrix U on the $D\times$(p+1) and $D\times$(D$-$p$-$1) blocks (18.11) is invariant under the (right multiplication by the matrix from the) $SO(1,p) \otimes SO(D-p-1)$ subgroup of the Lorentz group $SO(1,D-1)$. In the Lorentz harmonic approach of Refs. 19, 20[e] this gauge invariance is usually considered as an identification relation on the set of moving frame variables making possible to consider them as 'homogeneous' coordinate for the coset

$$\left\{ u_{\underline{a}}{}^b, u_{\underline{a}}{}^j \right\} = \frac{SO(1, D-1)}{SO(1, p) \otimes SO(D-p-1)}. \qquad (18.12)$$

This was the reason to call these moving frame variables *Lorentz harmonics*,[4,19] following the spirit of Ref. 34 where the notion of harmonic variables was introduced to construct the unconstrained superfield formulation of the $\mathcal{N} = 2$ supersymmetric theories.

Reordering the line of arguments one can start from (18.12) and notice that $SO(1,D-1)$ group valued moving frame matrix U (18.11) can be used to define, starting from $\hat{E}^{(\underline{a})}$, another vielbein attached to the worldvolume,

$$\hat{E}^{(\underline{a})} := \hat{E}^{\underline{b}} U_{\underline{b}}^{(\underline{a})} =: (\hat{E}^a, \hat{E}^i). \qquad (18.13)$$

This vielbein is adapted to the embedding of W^{p+1} into the D–dimensional spacetime if the pull–back of $D - p - 1$ 'orthogonal' forms

$$\hat{E}^i := \hat{E}^{\underline{a}} u_{\underline{a}}{}^i \qquad (18.14)$$

---

[e]See Refs. 55, 52, 45, 64, 4, 32 and 33 for earlier works.

vanishes, *i.e.* if embedding equation (18.8) is valid. The $(p+1)$ 'tangential' forms $\hat{E}^a$ defined with the use of the 'parallel' vector harmonics $u_{\underline{b}}^a$

$$\hat{E}^a := \hat{E}^{\underline{b}} u_{\underline{b}}^a \qquad (18.15)$$

can be used as a vielbein on W^{p+1};

$$e^a = \hat{E}^a := \hat{E}^{\underline{b}} u_{\underline{b}}^a . \qquad (18.16)$$

One says that this vielbein is induced by the embedding.

Now one sees that the superembedding equation (18.6) is just the straightforward supersymmetric generalization of the above embedding equation (18.8). However, in contrast to (18.8), the superembedding equation cannot be derived by imposing a conventional orientation conditions, and in this sense is nontrivial.

18.2.2. *Action of the moving frame formulation*

This is the induced vielbein (18.16) which can be understood as a square root from the induced metric

$$g_{mn}(\xi) = \hat{E}_m^{\underline{a}} \hat{E}_{n\,\underline{a}} \qquad (18.17)$$

provided the embedding equation (18.8) holds,

$$E^i = 0 \quad \Rightarrow \quad g_{mn}(\xi) = \hat{E}_m^{\underline{a}} \hat{E}_{n\,\underline{a}} = \hat{E}_m^a \hat{E}_{n\,a} = e_m^{\,a} e_{n\,a} . \qquad (18.18)$$

As a result, the invariant volume element on W^{p+1}, this is to say the Nambu–Goto term for a (super)–p–brane,

$$S_p^{N-G} := \int d^{p+1}\xi \sqrt{|g|} := \int d^{p+1}\xi \sqrt{|det(\hat{E}_m^{\underline{a}} \hat{E}_{n\,\underline{a}})|} \qquad (18.19)$$

can be equivalently presented in terms of $e^a := \hat{E}^a$ forms, $\int_{W^{p+1}} \hat{E}^{\wedge(p+1)}$,

$$E^i = 0 \quad \Rightarrow \quad d^{p+1}\xi \sqrt{|g|} := \frac{\varepsilon_{a_0 \ldots a_p} \hat{E}^{a_0} \wedge \ldots \wedge \hat{E}^{a_p}}{(p+1)!} =: \hat{E}^{\wedge(p+1)} . \qquad (18.20)$$

Now, if one uses $\hat{E}^{\wedge(p+1)}$ instead of the Nambu–Goto term (18.20) in the standard super–p–brane action,[1]

$$S_p^{standard} = S_p^{N-G} + S_p^{WZ} := \int d^{p+1}\xi \sqrt{|g|} - p \int_{W^{p+1}} \hat{B}_{p+1} , \qquad (18.21)$$

one arrives at the so–called *moving frame* or *Lorentz harmonic* action

$$S_p = S_p^{LH} + S_p^{WZ} := \int_{W^{p+1}} \hat{E}^{\wedge(p+1)} - p \int_{W^{p+1}} \hat{B}_{p+1}$$

$$= \int_{W^{p+1}} \frac{1}{(p+1)!} \varepsilon_{a_0 \ldots a_p} \hat{E}^{a_0} \wedge \ldots \wedge \hat{E}^{a_p} - p \int_{W^{p+1}} \hat{B}_{p+1} , \qquad (18.22)$$

where $\hat{E}^a = \hat{E}^{\underline{b}} u_{\underline{b}}^a$, Eq. (18.15) and $u_{\underline{b}}^a$ are $(p+1)$ orthonormal D–vectors, $u^{\underline{b}a} u_{\underline{b}}^b = \eta^{ab}$ (see (18.10)). These are the auxiliary variable entering the action without derivatives. The last term of the standard action, $-p \int \hat{B}_{p+1}$, which remains in the same form in the spinor moving frame formulation, is the so–called Wess–Zumino (WZ) term. It is given by the integral of the pull–back the worldvolume W^{p+1} of the gauge $(p+1)$–superform B_{p+1} restricted by the superspace constraints imposed on its (super)field strength

$$H_{p+2} = dB_{p+1} = \propto \bar{\Gamma}_{\alpha\beta}^{(p)} \wedge E^\alpha \wedge E^\beta + \mathcal{O}(E^{\wedge(p+1)}) , \qquad (18.23)$$

$$\bar{\Gamma}_{\alpha\beta}^{(p)} := \frac{1}{p!} E^{\underline{a}_1} \wedge \ldots \wedge E^{\underline{a}_p} \Gamma_{\underline{a}_p \ldots \underline{a}_1 \, \alpha\beta} . \qquad (18.24)$$

The relation between coefficient for the first term in the r.h.s. of (18.23) (replaced by $\propto$ symbol in our schematic consideration) and the coefficient in front of the WZ term in the action is fixed by the requirement of κ–symmetry.

As it has been noticed above, on the surface of embedding equation (18.8) the moving frame action (18.22) coincides with the standard one, Eq.(18.21),

$$S_p|_{\hat{E}^i = 0} = S_p^{standard} . \qquad (18.25)$$

The proof of the classical equivalence will then be completed by showing that the embedding equation follows from the moving frame action (18.22).

This is indeed the case, the embedding equation appears as a result of varying the auxiliary moving frame variables in the action (18.22),

$$\delta_{\delta u} S_p = 0 \qquad \Rightarrow \qquad \hat{E}^i := \hat{E}^{\underline{a}} u_{\underline{a}}{}^i = 0 . \qquad (18.26)$$

As the harmonics are constrained variables, the variation in Eq. (18.26) requires some comments.

18.2.2.1. *Variations and derivatives of the harmonic variables*

Both the spaces of the variations δu of certain variables u and of the derivatives du of such variables can be identified with the elements of the fiber of the tangent bundle over the space of this variables, i.e. with elements of the linear space tangent to the space of the u variables. In the case of Lorentz harmonics the variables u are elements of the Lorentz group valued matrix U, Eq. (18.11) (see also (18.10)). The space tangent to the Lorentz group is isomorphic to the Lie algebra spanned by antisymmetric $D \times D$ matrices. This well known fact can be expressed by

$$dU_{\underline{a}}^{(\underline{b})} = U_{\underline{a}}^{(\underline{c})}\Omega_{(\underline{c})}^{\phantom{(\underline{c})}(\underline{d})} \equiv U_{\underline{a}\,(\underline{c})}\Omega^{(\underline{c})(\underline{b})} \quad \Leftrightarrow \quad \begin{cases} du_{\underline{a}}^{b} = u_{\underline{a}c}\Omega^{cb} + u_{\underline{a}}^{i}\Omega^{bi}\,, \\ du_{\underline{a}}^{i} = -u_{\underline{a}}^{j}\Omega^{ji} + u_{\underline{a}b}\Omega^{bi}\,, \end{cases}$$
(18.27)

which is just an equivalent representation of the definition of the Cartan forms $U^{-1}dU$ for the Lorentz group in which $U^{-1} = \eta U^T \eta$,

$$U^{\underline{c}(\underline{a})}dU_{\underline{c}}^{(\underline{b})} =: \Omega^{(\underline{a})(\underline{b})} \equiv -\Omega^{(\underline{b})(\underline{a})} = \begin{Bmatrix} \Omega^{ab} & \Omega^{aj} \\ -\Omega^{bi} & \Omega^{ij} \end{Bmatrix}\,.$$
(18.28)

As far as the harmonics are treated as homogeneous coordinates of the coset $\frac{SO(1,D-1)}{SO(1,p)\otimes SO(D-p-1)}$, the Cartan forms $\Omega^{ab} = -\Omega^{ba} = u^{\underline{c}a}du_{\underline{c}}^{b}$ and $\Omega^{ij} = -\Omega^{ji} := u^{\underline{c}i}du_{\underline{c}}^{j}$ have the properties of the connection under the $SO(1,p)$ and $SO(D{-}p{-}1)$ local gauge symmetries while the set of one–forms $\Omega^{ai} = u^{\underline{c}a}du_{\underline{c}}^{i}$ provide the vielbein for the coset $\frac{SO(1,D-1)}{SO(1,p)\otimes SO(D-p-1)}$.

One should notice that, in general, the transformation of $SO(1,p)$ and $SO(D{-}p{-}1)$ symmetries are local on the coset space itself; however, when harmonics are used to describe a p–brane, $U = U(\xi)$, this local $SO(1,p) \otimes SO(D{-}p{-}1)$ symmetry of Lorentz harmonic space (or superspace) gives rise to the worldvolume local gauge symmetries with ξ–dependent parameters.

The same line of reasoning can be applied to the variations of the Lorentz harmonic variables in some action functional. Formally, the corresponding equation can be derived by using the Lie derivative $\mathcal{L}_\delta := i_\delta d + d i_\delta$ where the second terms will give zero contributions for a zero–forms so that $\delta u = i_\delta du$. Applying this simple equation to (18.27) one finds

$$\delta U_{\underline{a}}^{(\underline{b})} = U_{\underline{a}}^{(\underline{c})}i_\delta\Omega_{(\underline{c})}^{\phantom{(\underline{c})}(\underline{b})} \quad \Leftrightarrow \quad \begin{cases} \delta u_{\underline{a}}^{b} = u_{\underline{a}c}i_\delta\Omega^{cb} + u_{\underline{a}}^{i}i_\delta\Omega^{bi}\,, \\ \delta u_{\underline{a}}^{i} = -u_{\underline{a}}^{j}i_\delta\Omega^{ji} + u_{\underline{a}b}i_\delta\Omega^{bi}\,, \end{cases}$$
(18.29)

where $i_\delta\Omega^{(\underline{c})(\underline{b})} = -i_\delta\Omega^{(\underline{b})(\underline{c})} = \{i_\delta\Omega^{cb}, i_\delta\Omega^{ij}, i_\delta\Omega^{bi}\}$ are parameters of independent variations which can be identified with the i_δ contractions of the Cartan forms (hence the notation). Clearly $i_\delta\Omega^{cb} = -i_\delta\Omega^{bc}$ parametrize the worldsheet Lorenz group $SO(1,p)$, $i_\delta\Omega^{cb} = -i_\delta\Omega^{bc}$ parametrize the transformations of the structure group $SO(D{-}p{-}1)$ of the normal bundle and $i_\delta\Omega^{bi}$ provides a basis for independent variations of the coset $\frac{SO(1,D-1)}{SO(1,p)\otimes SO(D-p-1)}$.

18.2.2.2. *Lorentz harmonics and generalized Cartan forms for superbrane in curved (super)space*

In the curved superspace one has to consider the local Lorentz $SO(1,D-1)$ symmetry and the Cartan forms as defined in (18.28) or (18.27) are not covariant. Their covariant counterparts are defined with the use of Lorentz covariant derivatives $(d + w)$, so that

$$\begin{cases} Du_{\underline{a}}^{\ b} := du_{\underline{a}}^{\ b} + u^{\underline{c}\,i}\hat{w}_{\underline{c}}^{\ \underline{a}} - u_{\underline{a}}^{\ \underline{c}}\Omega_{\underline{c}}^{\ b} = u_{\underline{a}}^{\ i}\Omega^{bi} \,, \\ Du_{\underline{a}}^{\ i} := du_{\underline{a}}^{\ b} + u^{\underline{b}\,i}\hat{w}_{\underline{b}}^{\ \underline{a}} + u_{\underline{a}}^{\ j}\Omega^{ji} = u_{\underline{ab}}\Omega^{bi} \,, \end{cases} \qquad (18.30)$$

or

$$\Omega^{(\underline{a})(\underline{b})} := U^{\underline{c}(\underline{a})}[(d + \hat{w})U]_{\underline{c}}^{(\underline{b})} \quad \Leftrightarrow$$

$$\begin{cases} \Omega^{ab} = u^{\underline{c}\,a}[(d + \hat{w})u]_{\underline{c}}^{\ b} = \Omega^{0\,ab} + (u\hat{w}u)^{ab} \,, \quad (a) \\ \Omega^{ij} := u^{\underline{c}\,i}[(d + \hat{w})u]_{\underline{c}}^{\ j} = \Omega^{0\,ij} + (u\hat{w}u)^{ij} \,, \quad (b) \quad (18.31) \\ \Omega^{ai} := u^{\underline{c}\,a}[(d + \hat{w})u]_{\underline{c}}^{\ i} = \Omega^{0\,ai} + (u\hat{w}u)^{ai} \,, \quad (c) \end{cases}$$

where $\hat{w} = d\hat{Z}^M w_{M\,\underline{b}}^{\ \underline{a}}(\hat{Z})$ is the pull–back of the tangent superspace spin connection $w_{\underline{b}}^{\ \underline{a}} = dZ^M w_{M\,\underline{b}}^{\ \underline{a}}(Z)$ and $\Omega^{0\,ai}$, $\Omega^{0\,ij}$, $\Omega^{0\,ai}$ are the Cartan forms as defined in Eqs. (18.28). Then D in (18.30) is the derivative covariant both with respect to the local Lorentz $SO(1, D - 1)$ transformations and worldsheet gauge symmetry $SO(1,p) \otimes SO(D - p - 1)$. This implies that

$$D\hat{E}^a := d\hat{E}^a - \hat{E}^b \wedge \Omega_b^{\ a} = \hat{T}^{\underline{b}}u_{\underline{b}}^{\ a} + \hat{E}^i \wedge \Omega^{ai} \,, \qquad (18.32)$$

where Ω^{ai} is the generalized Cartan form (18.31c) and $\hat{T}^{\underline{b}}$ is the pull–back of the superspace torsion $T^{\underline{a}} = DE^{\underline{a}} = dE^{\underline{a}} - E^{\underline{b}} \wedge w_{\underline{b}}^{\ \underline{a}}$. This is restricted by supergravity constraints

$$T^{\underline{a}} = DE^{\underline{a}} := dE^{\underline{a}} - E^{\underline{b}} \wedge w_{\underline{b}}^{\ \underline{a}} = -iE^\alpha \wedge E^\beta \Gamma^{\underline{a}}_{\alpha\beta} \qquad (18.33)$$

With this in mind one can further specify Eq. (18.32)

$$D\hat{E}^a = -i\hat{E}^\alpha \wedge \hat{E}^\beta \hat{\Gamma}^a_{\alpha\beta} + \hat{E}^i \wedge \Omega^{ai} \,, \qquad \hat{\Gamma}^a := \Gamma^{\underline{b}}_{\alpha\beta}\, u_{\underline{b}}^{\ a} \,. \qquad (18.34)$$

Using the formal i_δ symbol of the previous subsection [extending its definition by $i_\delta d\hat{Z}^M := \delta\hat{Z}^M$ and $i_\delta(\Omega_q \wedge \Omega_p) = \Omega_q \wedge i_\delta\Omega_p + (-)^p i_\delta\Omega_q \wedge \Omega_p$ for any q- and p-forms Ω_q and Ω_p] one can write the arbitrary variation of the 'tangential' supervielbein $\hat{E}^a$ (modulo the $SO(1,p)$ symmetry transformations) in the form

$$\delta\hat{E}^a = D(i_\delta\hat{E}^a) - 2i\hat{E}^\alpha\,\hat{\Gamma}^a_{\alpha\beta}\,i_\delta\hat{E}^\beta + \hat{E}^i\,i_\delta\Omega^{ai} - \Omega^{ai}\,i_\delta\hat{E}^i \,, \qquad (18.35)$$

where $i_\delta\Omega^{ai}$ are basic variation of the harmonic variables, Eq. (18.29), corresponding to the coset $\frac{SO(1,D-1)}{SO(1,p)\otimes SO(D-p-1)}$ and

$$i_\delta\hat{E}^a = \delta\hat{Z}^M\, E_M{}^{\underline{c}}(\hat{Z})\, u_{\underline{c}}{}^a(\xi)\,, \qquad i_\delta\hat{E}^i = \delta\hat{Z}^M\, E_{\underline{M}}^a(\hat{Z})\, u_{\underline{a}}{}^i(\xi)\,, \qquad (18.36)$$

$$i_\delta\hat{E}^\alpha = \delta\hat{Z}^M\, E_M^\alpha(\hat{Z})\,. \qquad (18.37)$$

These provide the covariant basis for the variations of the bosonic and fermionic coordinate functions $\delta\hat{Z}^M$.

18.2.2.3. *Equations of motion of the moving frame action*

In the above notation general variation of the action with respect to the coordinate functions and harmonic variables,

$$\delta S_p = \int_{W^{p+1}} [\hat{E}_a^{\wedge p} \wedge \delta\hat{E}^a - p\delta\hat{B}_{p+1}] \qquad (18.38)$$

(with $\hat{E}_a^{\wedge p} := \frac{1}{p!}\varepsilon_{aa_1\ldots a_p}\hat{E}^{a_1}\wedge\ldots\wedge\hat{E}^{a_p}$), can be written as

$$\delta S_p = \int_{W^{p+1}} \hat{E}_a^{\wedge p} \wedge [Di_\delta\hat{E}^a + i_\delta\hat{T}^a + \hat{E}^i i_\delta\Omega^{ai} - \Omega^{ai}i_\delta\hat{E}^i] - p\int_{W^{p+1}} i_\delta\hat{H}_{p+2}\,.$$
$$(18.39)$$

Using (18.33), (18.23) and the identity $\hat{E}^{a_1}\wedge\ldots\wedge\hat{E}^{a_p}\Gamma_{a_1\ldots a_p}\Gamma_a = \propto \hat{E}_a^{\wedge p}\,\bar{\hat{\Gamma}}$ with

$$\bar{\hat{\Gamma}} := i^{\frac{p(p-1)}{2}}\hat{\Gamma}_0\ldots\hat{\Gamma}_p \equiv \frac{i^{\frac{p(p-1)}{2}}}{(p+1)!}\,\varepsilon_{a_0\ldots a_p}u_{\underline{b}_0}^{a_0}\ldots u_{\underline{b}_p}^{a_p}\Gamma^{\underline{b}_0\ldots\underline{b}_p} \qquad (18.40)$$

obeying $\bar{\hat{\Gamma}}\hat{\bar{\Gamma}} = I$, $tr\bar{\hat{\Gamma}} = 0$, one finds[f]

$$\delta S_p = -2i\int_{W^{p+1}} \hat{E}_a^{\wedge p} \wedge \hat{E}^\beta[\hat{\Gamma}^a(I-\bar{\hat{\Gamma}})]_{\beta\alpha}\,i_\delta E^\alpha(\hat{Z}) + \int_{W^{p+1}} \hat{E}_a^{\wedge p} \wedge \hat{E}^i\, i_\delta\Omega^{ai}$$

$$-\int_{W^{p+1}} (\hat{E}_a^{\wedge p} \wedge \Omega^{ai} + p\, u^{ai}\, i_{\underline{a}}\hat{H}_{p+2})i_\delta\hat{E}^i + \mathcal{O}(i_\delta\hat{E}^a) \qquad (18.41)$$

The second term in (18.41) contains the basic variations of the harmonic variables and is used to obtain the embedding equation (18.8). The third

---

[f]We do not write explicitly the terms proportional to the 'tangential' bosonic varia-tions $i_\delta\hat{E}^a$, denoting them in (18.41) by $\mathcal{O}(i_\delta\hat{E}^a)$, as they do not produce any indepen-dent equation. This statement manifests a Noether identity which corresponds to the reparametrization gauge symmetry, *i.e.* the worldvolume diffeomorphism invariance.

term produces the bosonic equations of motion of the p–brane in the form

$$\Omega_a{}^{ai} = \propto \varepsilon^{a_1 \ldots a_{p+1}} H_{a_1 \ldots a_{p+1}}{}^i + fermion\ contributions\ , \qquad (18.42)$$

which generalizes the minimal surface equation $\Omega_a{}^{ai} = 0$ for the case of nonvanishing background flux (see sec. 18.3.3.4 for more details in p=1 case). Finally the first term in (18.41) contains the fermionic variation $i_\delta \hat{E}^\alpha := \delta \hat{Z}^M E_M{}^\alpha(\hat{Z})$ and produces the fermionic equation for super-p–brane

$$\Psi_{p+1\,\alpha}(\hat{Z}) := \hat{E}_a^{\wedge p} \wedge \hat{E}^\beta [\hat{\Gamma}^a (I - \bar{\hat{\Gamma}})]_{\beta\alpha} = 0\ . \qquad (18.43)$$

18.2.3. *Irreducible κ–symmetry. Spinor harmonics enter the game*

The presence of the projector $(I - \bar{\hat{\Gamma}})$ makes half of the fermionic equations (18.43) to be satisfied identically,

$$\Psi_{p+1}\,(\hat{Z})(I + \bar{\hat{\Gamma}}) \equiv 0\ . \qquad (18.44)$$

Eq. (18.44) is the Noether identity reflecting a fermionic gauge symmetry of the action (18.22), the κ–symmetry with the basic variations

$$i_\kappa \hat{E}^{\underline{a}} := \delta_\kappa \hat{Z}^M(\xi) E_M{}^{\underline{a}}(\hat{Z}(\xi)) = 0\ ,$$
$$i_\kappa \hat{E}^\alpha := \delta_\kappa \hat{Z}^M(\xi) E_M{}^\alpha(\hat{Z}(\xi)) = (I + \bar{\hat{\Gamma}})^\alpha{}_\beta \kappa^\beta(\xi)\ . \qquad (18.45)$$

These are *formally* the same as the ones for the infinitely reducible κ–symmetry of the standard action (18.21). However, the presence of additional variables makes the κ–symmetry of the action (18.22) *irreducible* in contradistinction to the κ–symmetry of the original action (18.21).

To see this one should notice that, allowing for additional variables, one can *factorize* the κ–symmetry projector. Within the Lorenz harmonic approach such a factorization reads

$$(I_{(+)}^{-} \bar{\hat{\Gamma}})^\alpha{}_\beta = 2 v_\beta^{\dot{\bar{\alpha}}\dot{q}}\, v_{\dot{\bar{\alpha}}\dot{q}}{}^\alpha\ , \qquad (I_{(-)}^{+} \bar{\hat{\Gamma}})^\alpha{}_\beta = 2 v_\beta^{\bar{\alpha}q}\, v_{\bar{\alpha}q}{}^\alpha\ , \qquad (18.46)$$

where $\bar{\alpha}$, $\dot{\alpha}$ are indices of the spinor representations of $SO(1,p)$ (the same or different depending on the values of D and p) and q, $\dot{q}$ are indices of the (same or different) representations of $SO(D - p - 1)$ and

$$V_\beta^{(\alpha)} := \left(v_\beta^{\bar{\alpha}q}, v_\beta^{\dot{\bar{\alpha}}\dot{q}} \right) \quad \in \quad Spin(1, D - 1) \qquad (18.47)$$

is the $Spin(1, D - 1)$–valued matrix of the *spinor moving frame variables* or *spinor harmonics*. These variables are, the 'square root' of the vector harmonics (18.11), (18.10) in the sense of that the following constraints hold

$$V\Gamma^{(\underline{a})}V^T = \Gamma^{\underline{b}}U_{\underline{b}}{}^{(\underline{a})} \quad \Rightarrow \quad \begin{cases} V\Gamma^a V^T = \Gamma^{\underline{b}}u_{\underline{b}}{}^a \,, \\ V\Gamma^i V^T = \Gamma^{\underline{b}}u_{\underline{b}}{}^i \,. \end{cases} \tag{18.48}$$

Eqs. (18.48) express the well known fact of that the gamma–matrices are Lorentz invariant. An equivalent form of these constraints is given by

$$(V^{-1})^T \tilde{\Gamma}^{\underline{a}} V^{-1} = \Gamma^{(\underline{b})}U_{(\underline{b})}{}^{\underline{a}} = \Gamma^b u_b{}^{\underline{a}} - \Gamma^i u^{i\underline{a}} \,, \tag{18.49}$$

where $V^{-1} := V_{(\underline{\alpha})}{}^{\underline{\beta}}$ is the matrix inverse to (18.47),

$$V_{(\underline{\alpha})}{}^{\underline{\beta}} := \left(v_{\bar{\alpha} q}{}^{\underline{\beta}} \,, v_{\dot{-}\dot{q}}{}^{\underline{\beta}} \right) \quad : \quad V_{(\underline{\alpha})}{}^{\underline{\gamma}} V_{\underline{\gamma}}^{(\underline{\beta})} = \delta_{(\underline{\alpha})}{}^{(\underline{\beta})} := \begin{pmatrix} \delta_{\bar{\alpha}}{}^{\bar{\beta}}\delta_q{}^p & 0 \\ 0 & \delta_{\dot{-}\dot{\alpha}}{}^{\beta}\delta_{\dot{q}}{}^{\dot{p}} \end{pmatrix} \,. \tag{18.50}$$

The spinor moving frame variables are also called *spinorial harmonics* because they provide the homogeneous coordinates for the coset of $Spin(1, D-1)$ group doubly covering the coset of Eq. (18.12),

$$\{V_{\underline{\beta}}^{(\underline{\alpha})}\} := \left\{ \left(v_{\underline{\beta}}^{\bar{\alpha} q} \,, v_{\underline{\beta}}^{\dot{-}\dot{\alpha}\dot{q}} \right) \right\} = \frac{Spin(1, D-1)}{Spin(1, p) \otimes Spin(D-p-1)} \,. \tag{18.51}$$

Now, using the factorization (18.46) one can write the κ–symmetry transformations (18.45) in the *irreducible* form

$$i_\kappa \hat{E}^\alpha := \delta_\kappa \hat{Z}^M E_M^\alpha(\hat{Z}) = 2v_{\bar{\alpha} q}{}^\alpha \kappa^{\bar{\alpha} q} \,, \tag{18.52}$$

where the *irreducible κ–symmetry parameter* is

$$\kappa^{\bar{\alpha} q} := v_{\underline{\beta}}^{\bar{\alpha} q} \kappa^\beta(\xi) \,. \tag{18.53}$$

18.2.4. *Spinor moving frame formulation of super-p-branes*

The spinor moving frame formulation of super-p-brane is described by the moving frame action (18.22),

$$S_p = \int_{W^{p+1}} \hat{E}^{\wedge(p+1)} - p \int_{W^{p+1}} \hat{B}_{p+1} \,, \tag{18.54}$$

in which the vector Lorentz harmonics $u_{\underline{b}}^a$ (moving frame variables), entering the definition of E^a in $\hat{E}^{\wedge(p+1)} := \frac{1}{(p+1)!}\varepsilon_{a_0 \ldots a_p}\hat{E}^{a_0} \wedge \ldots \wedge \hat{E}^{a_p}$, are composites of spinor harmonics as defined by the gamma–trace parts of the constraints (18.48),

$$\hat{E}^a := \hat{E}^{\underline{b}}u_{\underline{b}}^a \,, \qquad u_{\underline{b}}^a = \frac{1}{n}tr(\tilde{\Gamma}_{\underline{b}}V\Gamma^a V^T) \,, \tag{18.55}$$

where n the number of values of (minimal) D–dimensional spinor indices, $n = \delta_\alpha{}^\alpha$ (see footnote b). This composite nature does not change the variation of vector harmonics, which are expressed through the Cartan forms as in (18.29). This is the case because the variation of spinorial harmonics are expressed through the same Cartan forms,

$$V^{-1}\delta V = \frac{1}{4} i_\delta \Omega^{(\underline{a})(\underline{b})} \, \Gamma_{(\underline{a})(\underline{b})} \equiv \frac{1}{4}(U^{-1}\delta U)^{(\underline{a})(\underline{b})} \, \Gamma_{(\underline{a})(\underline{b})} \, . \tag{18.56}$$

This reflects the fact that locally the spinorial harmonics carry the same $D(D-1)/2$ degrees of freedom as the vector ones, which is tantamount to stating that the groups $Spin(1, D-1)$ and $SO(1, D-1)$ are locally isomorphic. In this sense *moving frame action can always be considered as spinor moving frame action.*

18.2.5. *Generalized action principle*

Generalized action principle for superbranes[18] gives an extended object counterpart of the rheonomic approach to supergravity.[29,51] It can be obtained from the spinor moving frame action by the following two steps.

First one replaces all the worldvolume *fields* dependent on $W^{(p+1)}$ local coordinates ξ^m by superfields, depending on the local bosonic and fermionic coordinates, ξ^m and η^q, of the worldvolume superspace $W^{(p+1|n/2)}$,

$$\hat{Z}^M(\xi) \mapsto \hat{Z}^M(\xi,\eta) \, , \quad V(\xi) \mapsto V(\xi,\eta) \quad \Rightarrow \quad u_{\underline{b}}{}^a(\xi) \mapsto u_{\underline{b}}{}^a(\xi,\eta) \, . \tag{18.57}$$

Secondly, one replaces the integral over worldvolume $W^{(p+1)}$ by integral over a surface $\widetilde{W}^{(p+1)}$ of maximal bosonic dimension in the worldvolume superspace. Its embedding into $W^{(p+1|n/2)}$ can be described by fermionic coordinate functions $\hat{\eta}^q(\xi)$, which provide the counterparts of the Volkov-Akulov Goldstone fermions (these would be $\theta^{\check{\alpha}}(x)$ in our notation),[61,62]

$$\widetilde{W}^{(p+1)} \in W^{(p+1|n/2)} \quad : \quad \eta^q = \hat{\eta}^q(\xi) \, . \tag{18.58}$$

This is tantamount to saying that the generalized action is given by Eq. (18.54) with an integral over the bosonic body of the worldvolume superspace $W^{(p+1|n/2)}$, which is defined by $\hat{\eta}^{\check{q}} = 0$ and is denoted by $W^{(p+1)}$, and with the Lagrangian form $\hat{E}^{\wedge(p+1)} - p\hat{B}_{p+1}$ constructed from superfields (18.57) pulled back to the surface $\widetilde{W}^{(p+1)}$ in the worldvolume superspace, *i.e.* from

$$\hat{Z}^M = \hat{Z}^M(\xi, \hat{\eta}(\xi)) \, , \quad V = V(\xi, \hat{\eta}(\xi)) \quad \Rightarrow \quad u_{\underline{b}}{}^a = u_{\underline{b}}{}^a(\xi, \hat{\eta}(\xi)) \, , \tag{18.59}$$

$$\hat{E}^a := E^{\underline{b}}(\hat{Z}(\xi, \hat{\eta}(\xi))) \, u_{\underline{b}}{}^a(\xi, \hat{\eta}(\xi)) \, . \tag{18.60}$$

To resume, the *generalized action functional* is given by

$$S_p = \int_{\widetilde{W}^{p+1}} (\hat{E}^{\wedge(p+1)} - p\hat{B}_{p+1}) := \int_{W^{p+1}} (\hat{E}^{\wedge(p+1)} - p\hat{B}_{p+1})|_{\eta=\hat{\eta}(\xi)} , \qquad (18.61)$$

where hat ($\hat{\ }$) implies pull–back to the worldvolume superspace, and also (spinor) moving frame variables are superfields, as in (18.57). Thus the original moving frame action (18.54) is a particular case of the generalized action for $\widetilde{W}^{p+1} = W^{p+1}$, *i.e.* for $\hat{\eta}^q(x) = 0$.

The set of equations of motion for this generalized action functional includes a counterpart of (18.8), but for the superfields pulled back to $\widetilde{W}^{p+1}$,

$$\hat{E}^i := E^{\underline{b}}(\hat{Z}(\xi, \hat{\eta}(\xi))) u_{\underline{b}}{}^i(\xi, \hat{\eta}(\xi)) = 0 , \qquad (18.62)$$

and also the dynamical equations of motion (18.42), (18.43) but for the superfields pulled back to $\widetilde{W}^{p+1}$.

Now the structure of the Lagrangian form guaranties that the action functional is independent on the choice of the surface $\widetilde{W}^{p+1}$. The arbitrary changes of this surface, which are described by arbitrary variations of the fermionic functions $\delta\hat{\eta}^q(\xi)$ are the gauge symmetry of the generalized action functional (18.61). More details on this symmetry can be found in Ref. 18 as well as in very recent Ref. 26 which uses a 'bottom-up' version of the generalized action principle proposed in Ref. 40. The consequence of this symmetry 'parametrized' by arbitrary $\delta\eta^q(\xi)$'s is that equations of motion, including (18.62) are valid *on an arbitrary surface* $\widetilde{W}^{p+1}$ in the worldvolume superspace $W^{(p+1|n/2)}$. As the set of such surfaces 'covers' the whole superspace $W^{(p+1|n/2)}$, it is *natural to assume* that the equations are valid in the whole superspace. This implies, in particular, lifting of Eq. (18.62) to the superembedding equation in its form of Eq. (18.6),

$$\hat{E}^i = E^{\underline{b}}(\hat{Z}(\xi, \eta)) u_{\underline{b}}{}^i(\xi, \eta) = 0 . \qquad (18.63)$$

This last stage, namely the *lifting* of the equations valid on an arbitrary surface in superspace to equations on the superspace, is the essence of the *rheonomic principle* of the group manifold approach to supergravity.[29,51]

Notice that such a rheonomic lifting does not follow from the action variation, but rather constitutes an additional stage in the procedure of the *generalized action principle*, which should be made separately after varying the generalized action *functional*. In particular, the lifted equations written in terms of complete superfields (not pulled back to M^{p+1}) should be checked on consistency, and the consistency is not guaranteed. It have to be checked case by case, see Ref. 15 for an example when the consistency does not hold.

The study of the selfconsistency condition for superembedding equation (18.63) can actually be used to derive equations of motion for D=10 type superstrings, D=11 supermembranes[16] as well as for M5-brane[42] and D=10 super-Dp-branes[41] for $p \leq 5$.[28] In the next section we will show explicitly how this happens in the case of superstring in general curved type IIB supergravity superspace.

18.3. Superembedding approach to $D = 10$ Green–Schwarz superstring in type IIB supergravity background

To discuss the superembedding approach to superstring in a general type IIB supergravity background, we need firstly to discuss the specific features of the stringy spinor moving frame formalism.

18.3.1. *Spinor moving frame action for superstring*

The special properties of the stringy (spinor) moving frame variables, *i.e.* of the $\frac{SO(1,D-1)}{SO(1,1)\otimes SO(D-2)}$ harmonics used to describe the D dimensional superstring, comes from the fact that the two dimensional $SO(1,1)$ pseudo rotations of the vectors $u_{\underline{a}}^a = (u_{\underline{a}}^0, u_{\underline{a}}^\#)$ (where the symbol $\#$ is used for $(D-1)$-the direction) is reducible and can be split onto the scaling of two light–like vectors $u_m^{++} := u_m^0 + u_m^\#$ and $u_m^{--} := u_m^0 - u_m^\#$ (the self-dual and anti-selfdual 2–vectors) by mutually inverse factors, $u_m^{\pm\pm} \mapsto e^{\pm 2\alpha} u_m^{\pm\pm}$. The constraints on the stringy moving frame variables (vector harmonics)

$$U_{\underline{a}}^{(b)} = (u_{\underline{a}}^{--}, u_{\underline{a}}^{++}, u_{\underline{a}}^j) \in SO(1,9) \tag{18.64}$$

read (*cf.* (18.10), (18.11))

$$U^T \eta U = \eta \quad \Leftrightarrow \quad \begin{cases} u_{\underline{a}}^{--} u^{a--} = 0, & u_{\underline{a}}^{++} u^{a++} = 0, \\ u_{\underline{a}}^{--} u^{a++} = 2, & u_{\underline{a}}^{\pm\pm} u^{a\,i} = 0, \\ u_{\underline{a}}^i u^{a\,j} = -\delta^{ij} \end{cases} \tag{18.65}$$

and also imply

$$U \eta U^T = \eta \quad \Leftrightarrow \quad \delta_{\underline{a}}^{\underline{b}} = \tfrac{1}{2} u_{\underline{a}}^{++} u^{\underline{b}--} + \tfrac{1}{2} u_{\underline{a}}^{--} u^{\underline{b}++} - u_{\underline{a}}^i u^{\underline{b}i}. \tag{18.66}$$

Then, induced worldvolume supervielbein (18.55) are

$$\hat{E}^{++} := \hat{E}^{\underline{a}} u_{\underline{a}}^{++}, \qquad \hat{E}^{--} := \hat{E}^{\underline{a}} u_{\underline{a}}^{--} \tag{18.67}$$

and the spinor moving frame action for superstring reads (see Ref. 6)

$$S_{IIB} = \frac{1}{2} \int_{W^2} \hat{E}^{++} \wedge \hat{E}^{--} - \int_{W^2} \hat{B}_2 \,, \qquad (18.68)$$

or, using the auxiliary worldvolume vielbein forms $e^{\pm\pm}$ (see Ref. 19), as

$$S_{IIB} = \frac{1}{2} \int_{W^2} \left(e^{++} \wedge \hat{E}^{--} - e^{--} \wedge \hat{E}^{++} - e^{++} \wedge e^{--} \right) - \int_{W^2} \hat{B}_2 \,. \quad (18.69)$$

Indeed, $\delta e^{\pm\pm}$ equations of motion express them through $\hat{E}^{\pm\pm}$ of (18.67),

$$e^{\pm\pm} = \hat{E}^{\pm\pm} := \hat{E}^{\underline{a}} u_{\underline{a}}^{\pm\pm} \,. \qquad (18.70)$$

Substituting the algebraic equations (18.70) back to the first order action (18.69) one arrives at the second order action (18.68).

As we discussed in Sec. 18.2.5, the above spinor moving frame action can be used to construct the generalized action.[18] This is given by formally the same functional (18.68) (or (18.69)) with the fields on W^2 replaced by the superfields and integration performed about an arbitrary surface $\widetilde{W}^2$ in the worldsheet superspace $W^{(2|8+8)}$. The generalized action principle for superstring produces in particular, the superembedding equation (18.6).[18]

18.3.2. *Stringy $\frac{Spin(1,9)}{SO(1,1)\otimes SO(8)}$ spinorial harmonics*

The D=10 stringy spinor harmonics are collected in $Spin(1,9)$ matrix

$$V_\alpha^{(\beta)} = (v_{\alpha q}^+ , v_{\alpha \dot{q}}^-) \in Spin(1,9) \,. \qquad (18.71)$$

The specific of string lays in that the spinor representation of $SO(1,1)$ is one dimensional and is described by sign indices $^+$ and $^-$ of $v_{\alpha q}^-$ and $v_{\alpha \dot{q}}^+$. For our D=10 case q and $\dot{q}$ are the s- and c-spinorial indices of $SO(8)$.

In the dynamical system with $SO(1,1) \otimes SO(8)$ symmetry, like our superstring described by the action (18.69) or (18.68), the harmonics are homogeneous coordinates of the coset $\frac{Spin(1,9)}{SO(1,1)\otimes SO(8)}$,

$$\{V_\alpha^{(\beta)}\} = \{(v_{\alpha q}^+ , v_{\alpha \dot{q}}^-)\} = \frac{Spin(1,9)}{SO(1,1) \otimes SO(8)} \,. \qquad (18.72)$$

The requirement for the matrix V to belong to $Spin(1,9)$ group (18.71) is imposed as the (reducible) constraint

$$\sigma^b U_{\underline{b}}^{(\underline{a})} = V \sigma^{(\underline{a})} V^T \,, \qquad (a) \qquad V^T \tilde{\sigma}_{\underline{b}} V = U_{\underline{b}}^{(\underline{a})} \tilde{\sigma}_{(\underline{a})} \,, \qquad (b) \qquad (18.73)$$

which express the Lorentz invariance of the D=10 sigma–matrices $\sigma_{\alpha\beta}^b, \tilde{\sigma}_{\underline{b}}^{\alpha\beta}$. These are symmetric, obey $\sigma^{\underline{a}}\tilde{\sigma}_{\underline{b}} + \sigma^{\underline{a}}\tilde{\sigma}_{\underline{b}} = \delta_\alpha{}^\beta$ and have $Spin(1,1) \otimes SO(8)$

invariant representation with which the constraints (18.73a) can be split into the following set of relations

$$u_{\underline{a}}^{++}\sigma_{\alpha\beta}^{a} = 2v_{\alpha q}^{+}v_{\beta q}^{+}\,, \qquad u_{\underline{a}}^{++}\tilde{\sigma}^{a\alpha\beta} = 2v_{\dot{q}}^{+\alpha}v_{\dot{q}}^{+\beta}\,, \tag{18.74}$$

$$u_{\underline{a}}^{--}\sigma_{\alpha\beta}^{a} = 2v_{\alpha\dot{q}}^{-}v_{\beta\dot{q}}^{-}\,, \qquad u_{\underline{a}}^{--}\tilde{\sigma}^{a\gamma\beta} = 2v_{q}^{-\gamma}v_{q}^{-\beta}\,, \tag{18.75}$$

$$u_{\underline{a}}^{i}\sigma_{\alpha\beta}^{a} = 2v_{(\alpha q}^{+}\gamma_{q\dot{q}}^{i}v_{\beta)\dot{q}}^{-}\,, \qquad u_{\underline{a}}^{i}\tilde{\sigma}^{a\gamma\beta} = -v_{q}^{-\gamma}\gamma_{q\dot{q}}^{i}v_{\dot{q}}^{+\beta} - v_{q}^{-\beta}\gamma_{q\dot{q}}^{i}v_{\dot{q}}^{+\gamma}\,. \tag{18.76}$$

These imply, in particular, that the spinor harmonics $v_{\alpha}^{+q}, v_{\alpha}^{-\dot{q}}$ can be treated as square roots from the light–like vectors $u_{\underline{a}}^{++}, u_{\underline{a}}^{--}$.

The second relations in (18.74)-(18.76) are written for inverse harmonics

$$V_{(\alpha)}{}^{\beta} = (v_{q}^{-\beta}\,, v_{\dot{q}}^{+\beta}) \in Spin(1,9)\,. \tag{18.77}$$

In the case of $D = 10$ Majorana–Weyl spinor representation, with $\alpha = 1,\ldots,16$, these cannot be constructed from the 'original' spinorial harmonics (18.72) and are defined by the constraints

$$V_{(\alpha)}{}^{\gamma}V_{\gamma}^{(\beta)} = \delta_{(\alpha)}{}^{(\beta)} := \begin{pmatrix} \delta_{q}^{p} & 0 \\ 0 & \delta_{\dot{q}}^{\dot{p}} \end{pmatrix} \tag{18.78}$$

(like *e.g.* inverse metric in general relativity). Eq. (18.78) implies

$$v_{p}^{-\alpha}v_{\alpha q}^{+} = \delta_{pq}, \qquad v_{p}^{-\alpha}v_{\alpha\dot{q}}^{-} = 0\,,$$
$$v_{\dot{p}}^{+\alpha}v_{\alpha q}^{+} = 0, \qquad v_{\dot{p}}^{+\alpha}v_{\alpha\dot{q}}^{-} = \delta_{\dot{p}\dot{q}}\,. \tag{18.79}$$

These relations can be used to factorize the projector and to get the irreducible form of the superstring κ–symmetry.[19] They are also necessary to develop the superembedding approach to superstrings.[6,16]

Finally, the split form of Eq. (18.73b) reads

$$v_{q}^{+}\tilde{\sigma}^{a}v_{p}^{+} = \delta_{qp}u_{\underline{a}}^{++}\,, \qquad v_{\dot{q}}^{+}\sigma^{a}v_{\dot{p}}^{+} = \delta_{\dot{q}\dot{p}}u_{\underline{a}}^{++}\,, \tag{18.80}$$

$$v_{\dot{q}}^{-}\tilde{\sigma}^{a}v_{\dot{p}}^{-} = \delta_{qp}u_{\underline{a}}^{--}\,, \qquad v_{q}^{-}\sigma^{a}v_{p}^{-} = \delta_{\dot{q}\dot{p}}u_{\underline{a}}^{--}\,, \tag{18.81}$$

$$v_{q}^{+}\tilde{\sigma}^{a}v_{\dot{q}}^{-} = \gamma_{q\dot{q}}^{i}u_{\underline{a}}^{i}\,, \qquad v_{q}^{-}\sigma^{a}v_{\dot{q}}^{+} = -\gamma_{q\dot{q}}^{i}u_{\underline{a}}^{i}\,. \tag{18.82}$$

18.3.3. *Superembedding approach to D=10 superstring in type IIB supergravity background*

The starting point is the superembedding equation in its form of Eq. (18.6),

$$\hat{E}^{i} := \hat{E}^{b}u_{\underline{b}}{}^{i}(\xi,\eta) = 0\,. \tag{18.83}$$

This has to be completed by the set of *conventional constraints* which includes (18.70) and the relations defining fermionic supervielbein forms

$$e^{\pm\pm} = \hat{E}^{\pm\pm} := \hat{E}^{\underline{a}} u_{\underline{a}}^{\pm\pm} \, , \tag{18.84}$$

$$e^{+q} = \hat{E}^{\alpha 1} v_{\alpha q}^{+} \, , \qquad e^{-\dot{q}} = \hat{E}^{\alpha 2} v_{\alpha \dot{q}}^{-} \, . \tag{18.85}$$

The superembedding equation (18.83) and the above set of conventional constraints can be collected in the following expressions for the pull–back of the supervielbein of target type IIB superspace,

$$\hat{E}^{\underline{a}} := \frac{1}{2} e^{++} u^{--\underline{a}} + \frac{1}{2} e^{--} u^{++\underline{a}} \, , \tag{18.86}$$

$$\hat{E}^{\alpha 1} = e^{+q} v_q^{-\alpha} + e^{\pm\pm} \chi_{\pm\pm}^{-\dot{q}} v_{\dot{q}}^{+\alpha} \, , \qquad \hat{E}^{\alpha 2} = e^{-\dot{q}} v_{\dot{q}}^{+\alpha} + e^{\pm\pm} \chi_{\pm\pm}^{+q} v_q^{-\alpha} \, . \tag{18.87}$$

Actually, Eqs. (18.87) contain a bit more than just Eq. (18.85): it also states that $\hat{E}_{+p}^{\alpha 1} v_{\alpha \dot{q}}^{-} = 0$ and $\hat{E}_{-\dot{p}}^{\alpha 2} v_{\alpha q}^{+} = 0$, and this excludes from consideration the case of D1-branes (see Refs. 6, 17).

18.3.3.1. *Other conventional constraints*

To complete the set of conventional constraints, let us notice that we use the $SO(1,1) \otimes SO(8)$ connection induced by embedding; this implies that the complete $SO(1,9) \otimes SO(1,1) \otimes SO(8)$ covariant derivatives of the vector harmonics read

$$\begin{cases} Du_{\underline{a}}^{++} = u_{\underline{a}}^i \Omega^{++i} \, , \\ Du_{\underline{a}}^{--} = u_{\underline{a}}^i \Omega^{--i} \, , \end{cases} \qquad Du_{\underline{a}}^i = \frac{1}{2} u_{\underline{a}}^{--} \Omega^{++i} + \frac{1}{2} u_{\underline{a}}^{++} \Omega^{--i} \, . \tag{18.88}$$

For the spinorial harmonics (18.72), (18.77) this connection gives

$$Dv_{\alpha q}^{+} = \frac{1}{2} \Omega^{++i} \gamma_{q\dot{p}}^i v_{\alpha\dot{p}}^{+} \, , \qquad Dv_{\dot{q}}^{+\alpha} = -\frac{1}{2} \Omega^{++i} \, v_p^{-\alpha} \gamma_{p\dot{q}}^i \, , \tag{18.89}$$

$$Dv_{\alpha\dot{q}}^{-} = \frac{1}{2} \Omega^{--i} \, v_{\alpha p}^{+} \gamma_{p\dot{q}}^i \, , \qquad Dv_q^{-\alpha} = -\frac{1}{2} \Omega^{--i} \, \gamma_{q\dot{p}}^i v_{\dot{p}}^{+\alpha} \, . \tag{18.90}$$

The integrability conditions for Eqs. (18.88) give the curved superspace generalization of the Peterson-Codazzi, Gauss and Ricci equations of the classical XIX-th century surface theory. These read

$$D\Omega^{\pm\pm i} = \hat{R}^{\pm\pm i} := \hat{R}^{\underline{ab}} u_{\underline{a}}^{\pm\pm} u_{\underline{b}}^i \, , \tag{18.91}$$

$$d\Omega^{(0)} = \frac{1}{4} \hat{R}^{\underline{ab}} u_{\underline{a}}^{++} u_{\underline{b}}^{--} + \frac{1}{4} \Omega^{--i} \wedge \Omega^{++i} \, , \tag{18.92}$$

$$\mathbb{R}^{ij} = \hat{R}^{\underline{ab}} u_{\underline{a}}^i u_{\underline{b}}^j - \Omega^{--[i} \wedge \Omega^{++j]} \, , \tag{18.93}$$

where $\Omega^{\pm\pm i}$ are the generalized Cartan forms (see Sec. 18.2.2.2),

$$\Omega^{\pm\pm\, i} := u^{\pm\pm\,\underline{a}}\left(du^i_{\underline{a}} + \omega_{\underline{a}}{}^{\underline{b}}u^i_{\underline{b}}\right)\,, \qquad (18.94)$$

$\Omega^{(0)} = \frac{1}{4}u^{\underline{a}--}((d+\hat{w})u^{++})_{\underline{a}}$ is the $SO(1,1)$ connection (the induced 2d spin connection) and $\mathbb{R}^{ij} = d\Omega^{ij} - \Omega^{ik} \wedge \Omega^{kj}$ is the curvature of normal bundle with $\Omega^{ij} = u^{\underline{a}i}((d+\hat{w})u^j)_{\underline{a}}$; finally, $\hat{w}_{\underline{a}}{}^{\underline{b}}$ is the pull back of the $D=10$ spin connection superform $\hat{w}_{\underline{a}}{}^{\underline{b}} = dZ^M w_{M,\underline{a}}{}^{\underline{b}}$ and $R^{\underline{ab}} = (dw - w \wedge w)^{\underline{ab}}$.

18.3.3.2. Torsion constraints

Below we will also need the type IIB torsion constraints,

$$T^{\underline{a}} = -i\mathcal{E}^{\alpha} \wedge \mathcal{E}^{\beta}\Gamma^{\underline{a}}_{\underline{\alpha\beta}} := -iE^{\alpha 1} \wedge E^{\beta 1}\sigma^{\underline{a}}_{\alpha\beta} - iE^{\alpha 2} \wedge E^{\beta 2}\sigma^{\underline{a}}_{\alpha\beta}\,, \qquad (18.95)$$

$$T^{\alpha 1} = -E^{\alpha 1} \wedge E^{\beta 1}\nabla_{\beta 1}e^{-\Phi} + \frac{1}{2}E^1\sigma^{\underline{a}} \wedge E^1 \tilde{\sigma}^{\alpha\beta}_{\underline{a}}\nabla_{\beta 1}e^{-\Phi}$$

$$+\, E^{\underline{a}} \wedge \mathcal{E}^{\underline{\beta}}T_{\underline{\beta a}}{}^{\alpha 1} + \frac{1}{2}E^{\underline{b}} \wedge E^{\underline{a}}T_{\underline{ab}}{}^{\alpha 1}\,, \qquad (18.96)$$

$$T^{\alpha 2} = -E^{\alpha 2} \wedge E^{\beta 2}\nabla_{\beta 2}e^{-\Phi} + \frac{1}{2}E^2\sigma^{\underline{a}} \wedge E^2 \tilde{\sigma}^{\alpha\beta}_{\underline{a}}\nabla_{\beta 2}e^{-\Phi}$$

$$+\, E^{\underline{a}} \wedge \mathcal{E}^{\underline{\beta}}T_{\underline{\beta a}}{}^{\alpha 2} + \frac{1}{2}E^{\underline{b}} \wedge E^{\underline{a}}T_{\underline{ab}}{}^{\alpha 2}\,, \qquad (18.97)$$

$$\mathcal{E}^{\underline{\alpha}} = (E^{\alpha\,1}\,,\ E^{\alpha\,2})\,, \qquad \begin{matrix}\alpha=1,...,16\,,\\ \underline{\alpha}=(\alpha i)=1,...,32\,,\end{matrix} \qquad \Gamma^{\underline{a}}_{\underline{\alpha\beta}} = diag\left(\sigma^{\underline{a}}_{\alpha\beta}\,,\ \sigma^{\underline{a}}_{\alpha\beta}\right)\,. \quad (18.98)$$

The fermionic torsions $T_{\underline{\beta a}}{}^{\underline{\alpha}} = (T_{\underline{\beta a}}{}^{\alpha 1}, T_{\underline{\beta a}}{}^{\alpha 2})$ can be read off from

$$\mathcal{E}^{\underline{\alpha}}T_{\underline{ab}}{}^{\underline{\beta}} = -\frac{1}{8}\left(H_{\underline{bcd}}\,\sigma^{\underline{cd}}\tau_3 + \sigma_{\underline{b}}\tilde{\not{R}}^{(1)}i\tau_2 - \sigma_{\underline{b}}\tilde{\not{R}}^{(3)}\tau_1 + \frac{1}{2}\sigma_{\underline{b}}\tilde{\not{R}}^{(5)}i\tau_2\right)_{\underline{\hat{a}}}{}^{\underline{\hat{\gamma}}}$$

$$= -\frac{1}{8}\mathcal{E}^{\underline{\alpha}}H_{\underline{bcd}}\,(i\tau_3 \otimes \sigma^{\underline{cd}})_{\underline{\alpha}}{}^{\underline{\beta}} + \frac{1}{16}\mathcal{E}^{\underline{\alpha}}\sum_{n=0}^{4}(\sigma_{\underline{b}}\tilde{\not{R}}^{(2n+1)} \otimes \tau_1(\tau_3)^n)_{\underline{\alpha}}{}^{\underline{\beta}}\,.$$
$$(18.99)$$

Here $\tau_3\sigma^{\underline{cd}} = \tau_3 \otimes \sigma^{\underline{cd}}$, etc. and $\tilde{\not{R}}^{(2n+1)} = \frac{1}{(2n+1)!}R_{\underline{c}_1...\underline{c}_{2n+1}}\tilde{\sigma}^{\underline{c}_1...\underline{c}_{2n+1}\,\alpha\beta}$ where $R_{\underline{a}_1...\underline{a}_{9-2n}}$ are the type IIB RR field strength. Notice that $R_{\underline{a}_1...\underline{a}_{9-2n}} = \frac{(-)^n}{(2n+1)!}\,\varepsilon_{\underline{a}_1...\underline{a}_{9-2n}\underline{b}_1...\underline{b}_{2n+1}}R^{\underline{b}_1...\underline{b}_{2n+1}}$, which describes, in particular, the self-duality of the 5-form field strength.

18.3.3.3. *Superstring equations of motion from superembedding*

The selfconsistency conditions for the superembedding equations is included in the integrability condition of Eqs. (18.86),

$$\hat{T}^{\underline{a}} = -i\hat{E}^{\alpha 1} \wedge \hat{E}^{\beta 1} \sigma^a_{\alpha\beta} - i\hat{E}^{\alpha 2} \wedge \hat{E}^{\beta 2} \sigma^a_{\alpha\beta} = \frac{1}{2} De^{++} u^{--}_{\underline{a}} + \frac{1}{2} De^{--} u^{++}_{\underline{a}}$$

$$+ \frac{1}{2} u^i_{\underline{a}}(e^{++} \wedge \Omega^{--i} + e^{--} \wedge \Omega^{++i}) . \tag{18.100}$$

Using (18.87) one finds, after some algebra, that the contraction of Eq. (18.100) with the light–like vectors $u^{\pm\pm}_{\underline{a}}$ determine the worldvolume bosonic torsion,

$$De^{++} = -2ie^{+q} \wedge e^{+q} - 4ie^{++} \wedge e^{--} \chi^{+q}_{++} \chi^{+q}_{--} , \tag{18.101}$$

$$De^{--} = -2ie^{-\dot{q}} \wedge e^{-\dot{q}} - 4ie^{++} \wedge e^{--} \chi^{-\dot{q}}_{++} \chi^{-\dot{q}}_{--} , \tag{18.102}$$

while the contraction with $u_{\underline{a}}^i$ gives the restriction for the covariant Cartan forms (18.94),

$$e^{++} \wedge \Omega^{--i} + e^{--} \wedge \Omega^{++i} = -4i\gamma^i_{q\dot{q}} e^{\pm\pm} \wedge e^{+q} \chi^{-\dot{q}}_{\pm\pm} - 4i\gamma^i_{q\dot{q}} e^{\pm\pm} \wedge e^{-\dot{q}} \chi^{+q}_{\pm\pm} . \tag{18.103}$$

To proceed further one needs to study the consistency (intergability) of the fermionic conventional constraints (18.87) which read

$$\hat{T}^{\alpha 1} = De^{+q} v^{-\alpha}_q + e^{+q} \wedge Dv^{-\alpha}_q + De^{\pm\pm} \chi^{-\dot{q}}_{\pm\pm} v^{+\alpha}_{\dot{q}} + e^{\pm\pm} \wedge D\chi^{-\dot{q}}_{\pm\pm} v^{+\alpha}_{\dot{q}}$$

$$+ e^{\pm\pm} \wedge \chi^{-\dot{q}}_{\pm\pm} Dv^{+\alpha}_{\dot{q}} , \tag{18.104}$$

$$\hat{T}^{\alpha 2} = De^{-\dot{q}} v^{+\alpha}_{\dot{q}} + e^{-\dot{q}} \wedge Dv^{+\alpha}_{\dot{q}} + De^{\pm\pm} \chi^{+q}_{\pm\pm} v^{-\alpha}_q + e^{\pm\pm} \wedge D\chi^{+q}_{\pm\pm} v^{-\alpha}_q$$

$$+ e^{\pm\pm} \wedge \chi^{+q}_{\pm\pm} Dv^{-\alpha}_q . \tag{18.105}$$

The right hand side of these equations can be specified by using Eqs. (18.89), (18.90). To specify the *l.h.s.*'s we need the explicit form of the fermionic torsion constraints for the type IIB superspace, Eqs. (18.96), (18.97).

Contracting (18.104) with $v^+_{\alpha q}$ and (18.105) with $v^-_{\alpha\dot{q}}$ one finds the fermionic torsion of the worldvolume superspace. These read

$$De^{+q} = -e^{+p} \wedge e^{+p'} (\delta^q_{(p} v^{-\beta}_{p')} - \delta_{pp'} v^{-\beta}_q) \widehat{D_{\beta 1} e^{-\Phi}} + \propto e^{\pm\pm} , \tag{18.106}$$

$$De^{-\dot{q}} = -e^{-\dot{p}} \wedge e^{-\dot{p}'} (\delta^{\dot{q}}_{(\dot{p}} v^{+\beta}_{\dot{p}')} - \delta_{\dot{p}\dot{p}'} v^{+\beta}_{\dot{q}}) \widehat{D_{\beta 2} e^{-\Phi}} + \propto e^{\pm\pm} , \tag{18.107}$$

where $\propto e^{\pm\pm}$ denotes the contributions from forms containing bosonic worldvolume supervielbein, which we will not need in this section.

Contracting (18.104) with $v_{\alpha\dot{q}}^{-}$ one arrives at

$$0 = -\frac{1}{2}e^{+p} \wedge e^{+q}\Omega_{+(q}^{-\dot{q}}\gamma_{p)\dot{q}}^{i} - 2ie^{+p} \wedge e^{+p}\chi_{++}^{-\dot{q}} - \frac{1}{2}e^{+q} \wedge e^{-\dot{p}}\Omega_{-\dot{p}}^{--i}\gamma_{q\dot{q}}^{i}$$
$$- 2ie^{-\dot{p}} \wedge e^{-\dot{p}}\chi_{--}^{-\dot{q}} + \propto e^{\pm\pm} \, . \tag{18.108}$$

The similar equation with $e^{+q} \leftrightarrow e^{-\dot{q}}$, $\pm \leftrightarrow \mp$ appears when contracting (18.105) with $v_{\alpha\dot{q}}^{+}$. An immediate consequence of these equations are

$$\chi_{--}^{-\dot{q}} := \hat{E}_{--}^{\alpha 1}v_{\alpha\dot{q}}^{-} = 0 \, , \qquad \chi_{++}^{+q} := \hat{E}_{++}^{\alpha 2}v_{\alpha q}^{+} = 0 \, , \tag{18.109}$$

$$\Omega_{-\dot{p}}^{--i} = 0 \, , \qquad \Omega_{+p}^{++i} = 0 \, . \tag{18.110}$$

Eqs. (18.109) are the equations of motion for the fermionic degrees of freedom of superstring. A simple way to be convinced in this is to observe that the linearized version of (18.109) can be written in the form similar to the light-cone gauge fermionic equations of the Green-Schwarz superstring, which define its 16 fermionic degrees of freedom as two chiral, namely one right-moving and one left-moving, 8 component fermions,

$$\partial_{--}\hat{\Theta}_{\dot{q}}^{-1} = 0 \, , \qquad \hat{\Theta}_{\dot{q}}^{-1}(\xi,\eta) := \hat{\theta}^{\alpha 1}v_{\alpha\dot{q}}^{-} \, , \tag{18.111}$$

$$\partial_{++}\hat{\Theta}_{q}^{+2} = 0 \, , \qquad \hat{\Theta}_{q}^{+2}(\xi,\eta) := \hat{\theta}^{\alpha 2}v_{\alpha q}^{+} \, . \tag{18.112}$$

The above $\hat{\Theta}_{\dot{q}}^{-1}$, $\hat{\Theta}_{q}^{+2}$ correspond to the light–cone gauge fermionic fields, but defined with the use of *moving* frame determined by the harmonics. The other half of the fermionic fields, $\hat{\Theta}_{q}^{+1} := \hat{\theta}^{\alpha 1}v_{\alpha q}^{+}$ and $\hat{\Theta}_{\dot{q}}^{-2} := \hat{\theta}^{\alpha 2}v_{\alpha\dot{q}}^{-}$ can be identified with the the fermionic coordinates of $W^{(2|8+8)}$,

$$\hat{\Theta}_{q}^{+1}(\xi,\eta) = \eta^{+q} \, , \qquad \hat{\Theta}_{\dot{q}}^{-2}(\xi,\eta) = \eta^{-\dot{q}} \, . \tag{18.113}$$

These superfield relations imply $\hat{\Theta}_{q}^{+1}(\xi,0) = 0$, $\hat{\Theta}_{\dot{q}}^{-2}(\xi,0) = 0$, and these equations give a covariant version of the conditions which might be fixed using the κ–symmetry of the standard Green–Schwarz action.

18.3.3.4. *Bosonic equations of motion and on-shell superembedding of the worldvolume superspace*

The fermionic equations of motion (18.109) simplify the expressions (18.87) for the pull-back of fermionic supervielbein forms, making them chiral,

$$\hat{E}^{\alpha 1} = e^{+q}v_{q}^{-\alpha} + e^{++}\chi_{++}^{-\dot{q}}v_{\dot{q}}^{+\alpha} \, , \qquad \hat{E}^{\alpha 2} = e^{-\dot{q}}v_{\dot{q}}^{+\alpha} + e^{--}\chi_{--}^{+q}v_{q}^{-\alpha} \, , \tag{18.114}$$

which means, in particular, left– and right–moving, but also containing the corresponding half of the fermionic coordinates. The bosonic worldsheet torsion (18.101), (18.102) and Eq. (18.103) also simplify,

$$De^{++} = -2ie^{+q} \wedge e^{+q} \, , \qquad De^{--} = -2ie^{-\dot{q}} \wedge e^{-\dot{q}} \, , \tag{18.115}$$

and Eq. (18.103) determines the generalized Cartan forms to be

$$\Omega^{--i} = -4ie^{+q}\gamma^i_{q\dot q}\chi^{-\dot q}_{++} + e^{++}\Omega^{--i}_{++} + e^{--}K^i , \qquad (18.116)$$

$$\Omega^{++i} = -4ie^{-\dot q}\chi^{+q}_{--}\gamma^i_{q\dot q} + e^{++}K^i + e^{--}\Omega^{++i}_{--} . \qquad (18.117)$$

Using the conventional constraints (18.70), one can write the mean curvature $\Omega^{--i}_{--} = \Omega^{++i}_{++} := K^i$ in the form

$$K^i := -2D_{--}E^b_{++}\,u^i_b = -2D_{++}E^b_{--}\,u^i_b . \qquad (18.118)$$

Its linearized version reads $K^i = \Box X^i$, where $X^i = x^\mu u^i_\mu$, so that one can expect the bosonic equations of motion to appear in the form of conditions for K^i.

This is indeed the case. The bosonic equations of motion appears as $\propto e^{--} \wedge e^{+q}$ component of Eq. (18.108). First one obtains $K^i\gamma^i_{q\dot q} = -\frac{1}{8}u^{a++}\hat H_{abc}v_q^{-\alpha}\sigma^{bc}{}_\alpha{}^\beta v_{\beta\dot q}^{-}$. Then, using Eqs. (18.81), (18.82), one finds that $v_q^{-\alpha}\sigma^{bc}{}_\alpha{}^\beta v_{\beta\dot q}^{-} = \gamma^i_{q\dot q}(u^{b--}u^{ci} - u^{c--}u^{bi})$ and arrives at

$$K^i = \tfrac{1}{4}\hat H^{--++i} , \qquad \hat H^{--++i} := \hat H_{abc}u^{a--}u^{b++}u^{ci} . \qquad (18.119)$$

Thus we have completed the derivation of superstring equations of motion from the superembedding equations.

18.4. Superembedding description of AdS superstring

18.4.1. *AdS superspace AdSS$^{(5,5|32)}$ as the solution of type IIB supergravity constraints*

The AdS superspace denoted by $AdSS^{(5,5|32)}$ (see Ref. 39) is the D=10 type IIB superspace the bosonic body of which is $AdS_5 \times S^5$. This is given by a solution of the type IIB supergravity constraints (18.95), (18.96), (18.97) with all but five form fluxes equal to zero, this is to say

$$H_{\underline{a_1 a_2 a_3}} = 0 , \qquad R_{\underline{a_1 a_2 a_3}} = 0 , \qquad C_0 = 0 = \Phi . \qquad (18.120)$$

The nonvanishing five form flux is characterized by a selfdual constant tensor

$$f^{\underline{a_1 a_2 a_3 a_4 a_5}} = \frac{1}{5!}\epsilon^{\underline{a_1 a_2 a_3 a_4 a_5 c_1 c_2 c_3 c_4 c_5}}f_{\underline{c_1 c_2 c_3 c_4 c_5}} , \qquad (18.121)$$

$$df^{\underline{a_1 a_2 a_3 a_4 a_5}} = 0 . \qquad (18.122)$$

The torsion and curvature two-forms of the $AdSS^{(5,5|32)}$ superspace are expressed through this constant tensor and σ-matrices by (see Refs. 10, 47)

$$T^{\underline{a}} = -i\mathcal{E} \wedge (I \otimes \sigma^{\underline{a}})\mathcal{E} \equiv -i \left(E^{\alpha 1} \wedge E^{\delta 1} + E^{\alpha 2} \wedge E^{\delta 2} \right) \sigma^{\underline{a}}_{\alpha\delta} , \qquad (18.123)$$

$$T^{\underline{\alpha}} = -\frac{1}{R} E^{\underline{a}} \wedge \mathcal{E}^{\underline{\beta}} \, f_{\underline{ab}_1 \ldots \underline{b}_4} (i\tau_2 \otimes \sigma^{\underline{b}_1 \cdots \underline{b}_4})_{\underline{\beta}}{}^{\underline{\alpha}} ,$$

$$\Leftrightarrow \quad \begin{cases} T^{\alpha 1} = \dfrac{1}{R} E^{\underline{a}} \wedge E^{\beta 2} \, f_{\underline{ab}_1 \ldots \underline{b}_4} (\sigma^{\underline{b}_1 \cdots \underline{b}_4})_{\beta}{}^{\alpha} , \\[2ex] T^{\alpha 2} = -\dfrac{1}{R} E^{\underline{a}} \wedge E^{\beta 1} \, f_{\underline{ab}_1 \ldots \underline{b}_4} (\sigma^{\underline{b}_1 \cdots \underline{b}_4})_{\beta}{}^{\alpha} , \end{cases} \qquad (18.124)$$

$$R^{\underline{ab}} = -\frac{1}{2R^2} E^{\underline{a}} \wedge E^{\underline{b}} - \frac{4i}{R} \mathcal{E}^{\alpha} \wedge \mathcal{E}^{\delta} (i\tau_2 \otimes \sigma_{\underline{c}_1 \underline{c}_2 \underline{c}_3})_{\alpha\delta} f^{\underline{abc}_1 \underline{c}_2 \underline{c}_3}$$

$$= -\frac{1}{2R^2} E^{\underline{a}} \wedge E^{\underline{b}} + \frac{8i}{R} E^{\alpha 2} \wedge E^{\delta 1} f^{\underline{abc}_1 \underline{c}_2 \underline{c}_3} \sigma_{\underline{c}_1 \underline{c}_2 \underline{c}_3 \alpha\delta} . \qquad (18.125)$$

To complete the definition of the $AdSS^{(5,5|32)}$ superspace we should add that, in a suitable frame one can split the set of bosonic supervielbein forms as $E^{\underline{a}} = (E^{\breve{a}}, E^{\breve{i}})$, with $\breve{a} = 0, 1, \ldots, 4$ and $\breve{i} = 1, \ldots, 5$ and find that the constant self-dual tensor (18.121) should have the form

$$f_{\breve{a}_1 \ldots \breve{a}_5} = \varepsilon_{\breve{a}_1 \ldots \breve{a}_5} , \qquad f_{\breve{i}_1 \ldots \breve{i}_5} = \varepsilon_{\breve{i}_1 \ldots \breve{i}_5} , \qquad (18.126)$$

with all other components vanishing. Then $\breve{a} = 0, 1, \ldots, 4$ is identified as the vector index of the 5d space tangent to AdS_5, and $i = 1, \ldots, 5$ – as the vector index of the space tangent to S^5. The constant R in (18.124), (18.125) defines the radius of AdS_5 or S^5 (these radii are equal).

Now the superembedding description of the AdS superstring can be obtained by specializing the equations describing superstring in general supergravity background to a particular form of this background given by Eqs. (18.123), (18.124), (18.125), (18.121), (18.122). But before turning to this, we describe some properties of constant selfdual tensor $f^{a_1 a_2 a_3 a_4 a_5} :=$ $f^{[5]} = \frac{1}{5!} \epsilon^{[5][5']} f_{[5']}$ as they are seen in stringy spinor moving frame.

18.4.2. Constant five form flux in stringy moving frame

Below we will mainly use a seemingly SO(1,9) covariant description by Eqs. (18.123), (18.124), (18.125) and (18.121), (18.122) so that a big part of our results are applicable for superstring in a generic constant 5-form flux background.

Although the distinction between self-duality and anti-self duality is conventional, the selfduality of the constant flux (18.121) is singled out by that we use the *sigma*–matrix representation with self-dual $\sigma^{[5]}_{\alpha\beta} := \sigma^{a_1a_2a_3a_4a_5}_{\alpha\beta}$ (which implies anti-selfdual $\tilde{\sigma}^{[5]\,\alpha\beta} := \tilde{\sigma}^{a_1a_2a_3a_4a_5\,\alpha\beta}$),

$$\sigma^{[5]}_{\alpha\beta} = \frac{1}{5!}\epsilon^{[5][5']}\sigma_{[5']\,\alpha\beta}\,, \qquad \tilde{\sigma}^{[5]\,\alpha\beta} = -\frac{1}{5!}\epsilon^{[5][5']}\tilde{\sigma}^{\alpha\beta}_{[5']}\,. \qquad (18.127)$$

It is convenient to introduce the spinor moving frame components of the constant selfdual tensor (which are not obliged to be constant but rather depend on the coordinate of the superstring worldsheet superspace),

$$f^{ijk} := f^{--++i_1i_2i_3} = f^{\underline{c}_1\underline{c}_2\underline{c}_3\underline{c}_4\underline{c}_5}u^{--}_{\underline{c}_1}u^{++}_{\underline{c}_2}u^{i_1}_{\underline{c}_3}u^{i_2}_{\underline{c}_4}u^{i_3}_{\underline{c}_5}\,,$$

$$f^{i_1i_2i_3i_4i_5} = f^{\underline{c}_1\underline{c}_2\underline{c}_3\underline{c}_4\underline{c}_5}u^{i_1}_{\underline{c}_1}u^{i_2}_{\underline{c}_2}u^{i_3}_{\underline{c}_3}u^{i_4}_{\underline{c}_4}u^{i_5}_{\underline{c}_5}\,,$$

$$f^{\pm\pm i_1i_2i_3i_4} = f^{\underline{c}_1\underline{c}_2\underline{c}_3\underline{c}_4\underline{c}_5}u^{\pm\pm}_{\underline{c}_1}u^{i_1}_{\underline{c}_2}u^{i_2}_{\underline{c}_3}u^{i_3}_{\underline{c}_4}u^{i_4}_{\underline{c}_5}\,. \qquad (18.128)$$

The self-dulaity equation (18.121) implies

$$f^{ijk} := f^{--++ijk} = \frac{1}{5!}\epsilon^{ijkl_1l_2l_3l_4l_5}f^{l_1l_2l_3l_4l_5}\,,$$

$$f^{++ijkl} = -\frac{1}{4!}\epsilon^{ijkli'j'k'l'}f^{++\,i'j'k'l'}\,,$$

$$f^{--ijkl} = \frac{1}{4!}\epsilon^{ijkli'j'k'l'}f^{--\,i'j'k'l'}\,. \qquad (18.129)$$

Now one can prove that, as a consequence of (18.121) and of the properties (18.80), (18.81), (18.11) of stringy harmonics, the following identities hold

$$f^{--\underline{abcd}}(v^-_q\,\sigma_{\underline{abcd}})^\alpha = 0\,, \qquad f^{++\underline{abcd}}(v^+_{\dot q}\,\sigma_{\underline{abcd}})^\alpha = 0\,, \qquad (18.130)$$

where

$$f^{\pm\pm\,\underline{b_1b_2b_3b_4}} = u^{\pm\pm}_{\underline{a}}f^{\underline{ab_1b_2b_3b_4}}\,. \qquad (18.131)$$

Notice by pass that tensors $f^{\pm\pm\,\underline{b_1b_2b_3b_4}}$ characterize the movement of superstring. For instance, when superstring moves in the AdS part of superspace and is frozen to a point on S^5, one can chose the frame in such a way that $u^{\pm\pm}_{\underline{a}} = \delta_{\underline{a}}{}^{\check{c}}u^{\pm\pm}_{\check{c}}$ and, then $f^{\pm\pm\,\underline{c}_1\underline{c}_2\underline{c}_3\underline{c}_4} = u^{\pm\pm}_{\check{a}}\epsilon^{\check{a}\check{b}_1\check{b}_2\check{b}_3\check{b}_4}\delta_{\check{b}_1}{}^{\underline{c}_1}\delta_{\check{b}_2}{}^{\underline{c}_2}\delta_{\check{b}_3}{}^{\underline{c}_3}\delta_{\check{b}_4}{}^{\underline{c}_4}$.

To prove (18.130), we observe that (18.121) and (18.127) imply $f_{[5]}\sigma^{[5]}_{\alpha\beta} = 0$ (but $f_{[5]}\tilde{\sigma}^{[5]\,\alpha\beta} \neq 0$) which implies $f^{\underline{abcde}}\sigma_{\underline{bcde}\,\beta}{}^\alpha = \frac{1}{10}\sigma^{\underline{a}}_{\beta\gamma}f_{[5]}\tilde{\sigma}^{[5]\,\gamma\alpha}$. Multiplying this by $u^{--}_{\underline{a}}v^{-\beta}_q$ and $u^{++}_{\underline{a}}v^{+\beta}_{\dot q}$ one finds that (18.130) holds due to the identities $u^{--}_{\underline{a}}(v^-_q\sigma^{\underline{a}})_\gamma = 0$ and $u^{++}_{\underline{a}}(v^+_{\dot q}\sigma^{\underline{a}})_\gamma = 0$, respectively.

Notice that in such a way we cannot prove

$$f^{--\underline{abcd}}(\sigma_{\underline{abcd}}v^-_{\dot q})_\alpha = 0\,, \qquad f^{++\underline{abcd}}(\sigma_{\underline{abcd}}v^+_q)^\alpha = 0\,. \qquad (18.132)$$

To show that these are true, let us observe first that, due to (18.130), we need only to prove the vanishing of their contractions with, respectively, $v_{\dot{q}}^{+\alpha}$ and $v_q^{-\alpha}$. For these one finds that

$$f^{--\underline{abcd}}(v_{\dot{p}}^+\sigma_{\underline{abcd}}v_{\dot{q}}^-) = f^{--i_1i_2i_3i_4}\tilde{\gamma}_{\dot{p}\dot{q}}^{i_1i_2i_3i_4} = 0 \,,$$

$$f^{++\underline{abcd}}(v_p^-\sigma_{\underline{abcd}}v_q^+) = f^{++i_1i_2i_3i_4}\gamma_{qp}^{i_1i_2i_3i_4} = 0 \,, \qquad (18.133)$$

where the last equalities follow from (18.128), (18.129) and the anti-self duality (self-duality) of $\tilde{\gamma}^{i_1i_2i_3i_4}$ ($\gamma^{i_1i_2i_3i_4}$),

$$\tilde{\gamma}_{\dot{q}\dot{p}}^{ijkl} = -\frac{1}{4!}\epsilon^{ijkli'j'k'l'}\tilde{\gamma}_{\dot{q}\dot{p}}^{i'j'k'l'} \,, \qquad \gamma_{qp}^{ijkl} = +\frac{1}{4!}\epsilon^{ijkli'j'k'l'}\gamma_{qp}^{i'j'k'l'} \qquad (18.134)$$

which correspond to the duality properties (18.127) of the 10D σ matrices.

To conclude, the only nonvanishing contribution to the expressions $f^{\pm\pm[4]}\sigma_{[4]} = f^{\pm\pm\underline{abcd}}(\sigma_{\underline{abcd}})_\beta{}^\alpha$ is

$$f^{--\underline{abcd}}(v_{\dot{q}}^+\sigma_{\underline{abcd}}v_q^+) = f^{++\underline{abcd}}(v_q^-\sigma_{\underline{abcd}}v_{\dot{q}}^-) = 4f^{ijk}\gamma_{q\dot{q}}^{ijk} \,. \qquad (18.135)$$

18.4.3. *Worldsheet superspace geometry of the superstring in AdS superspace*

The pull–backs of the AdS supervielbein on the worldsheet superspace have the form of (18.86) and (18.114),

$$\hat{E}^{\underline{a}} = \tfrac{1}{2}e^{++}u^{--\underline{a}} + \frac{1}{2}e^{--}u^{++\underline{a}} \,, \qquad (18.136)$$

$$\hat{E}^{\alpha 1} = e^{+q}v_q^{-\alpha} + e^{++}\chi_{++}^{-\dot{q}}v_{\dot{q}}^{+\alpha} \,, \qquad \hat{E}^{\alpha 2} = e^{-\dot{q}}v_{\dot{q}}^{+\alpha} + e^{--}\chi_{--}^{+q}v_q^{-\alpha} \,. \qquad (18.137)$$

As far as the H_3 flux in AdS superspace is equal to zero, Eq. (18.120), the string equations of motion (18.119) read

$$K^i = 0 \,, \qquad (18.138)$$

so that the generalized Cartan form, which provide the supersymmetric generalization of the second fundamental form for the AdS superstring worldvolume superspace, are given by Eqs. (18.116), (18.117)

$$\Omega^{--i} = -4ie^{+q}(\gamma^i\chi_{++}^-)_q + e^{++}\Omega_{++}^{--i} \,,$$

$$\Omega^{++i} = -4ie^{-\dot{q}}(\chi_{--}^+\gamma^i)_{\dot{q}} + e^{--}\Omega_{--}^{++i} \,. \qquad (18.139)$$

The worldvolume superspace torsion forms read (see Eqs. (18.115), (18.107))

$$De^{++} = -2ie^{+q} \wedge e^{+q} , \qquad De^{--} = -2ie^{-\dot{q}} \wedge e^{-\dot{q}} \qquad (18.140)$$

$$De^{+q} = -e^{++} \wedge e^{-\dot{p}} \left(2i(\chi_{--}{}^{+}\gamma^i)_{\dot{p}} (\gamma^i\chi_{++}{}^{-})_q + \tfrac{2}{R} f^{ijk}\gamma^{ijk}_{q\dot{q}} \right)$$
$$+ \frac{1}{2} e^{++} \wedge e^{--} \Omega_{--}^{++i}(\gamma^i\chi_{++}{}^{-})_q , \qquad (18.141)$$

$$De^{-\dot{q}} = -2ie^{--} \wedge e^{+p} \left(2i(\gamma^i\chi_{++}{}^{-})_p (\chi_{--}{}^{+}\gamma^i)_{\dot{q}} + \tfrac{2}{R} f^{ijk}\gamma^{ijk}_{q\dot{q}} \right)$$
$$- \frac{1}{2} e^{++} \wedge e^{--} \Omega_{++}^{--i}(\chi_{--}{}^{+}\gamma^i)_{\dot{q}} . \qquad (18.142)$$

To obtain these expressions the identities (18.130) have to be used; they are also needed to find that $D_{--}\chi_{++\dot{q}}^{-} = D_{-\dot{p}}\chi_{++\dot{q}}^{-} = 0$ and $D_{++}\chi_{--q}^{+} = D_{+p}\chi_{--q}^{+} = 0$ which can be summarized by

$$D\chi_{++\dot{q}}^{-} = -\frac{1}{2} e^{+q}\gamma^i_{q\dot{q}}\Omega_{++}^{--i} + e^{++}D_{++}\chi_{++\dot{q}}^{-} , \qquad (18.143)$$

$$D\chi_{--q}^{+} = -\frac{1}{2} e^{-\dot{p}}\gamma^i_{q\dot{q}}\Omega_{--}^{++i} + e^{--}D_{--}\chi_{--q}^{+} . \qquad (18.144)$$

The pull–back of the Riemann curvature two form of the AdSS super-space to the worldvolume superspace reads

$$\hat{R}^{\underline{ab}} = -\frac{1}{2}e^{++} \wedge e^{--} \left(\frac{1}{R^2} u^{--[\underline{a}} \wedge u^{\underline{b}]++} + \frac{16i}{R} f^{\underline{ab}ijk}(\chi_{--}{}^{+}\gamma^{ijk}\chi_{++}{}^{-}) \right)$$
$$- \frac{24i}{R} e^{--} \wedge e^{+q} f^{--\underline{ab}ij}(\gamma^{ij}\chi_{--}{}^{+})_q + \frac{24i}{R} e^{++}$$
$$\wedge e^{-\dot{q}} f^{++\underline{ab}ij}(\tilde{\gamma}^{ij}\chi_{++}{}^{-})_{\dot{q}} - \frac{8i}{R} e^{+q} \wedge e^{-\dot{q}} f^{\underline{ab}ijk}\gamma^{ijk}_{q\dot{q}} . \qquad (18.145)$$

In its derivation we have used the following consequences of the properties (18.74), (18.75), (18.76) of spinor moving frame variables (18.71)

$$f^{\underline{ab}[3]}v_{\dot{p}}^{-}\sigma_{[3]}v_q^{-} = -3f^{--\underline{ab}ij}\gamma^{ij}_{\dot{p}q} \qquad f^{\underline{ab}[3]}v_{\dot{p}}^{+}\sigma_{[3]}v_q^{+} = -3f^{++\underline{ab}ij}\tilde{\gamma}^{ij}_{\dot{p}q} ,$$
$$f^{\underline{ab}[3]}v_q^{-}\sigma_{[3]}v_{\dot{q}}^{+} = f^{\underline{ab}ijk}\gamma^{ijk}_{q\dot{q}} . \qquad (18.146)$$

Substituting Eqs. (18.145) and (18.139) into Gauss and Ricci equations (18.92) and (18.93), one finds the following expressions for the 2d Riemann curvature two form

$$d\Omega^{(0)} = e^{++} \wedge e^{--} \left(\frac{1}{4}\Omega_{++}^{--i}\Omega_{--}^{++i} + \frac{2i}{R} f^{ijk}(\chi_{--}{}^{+}\gamma^{ijk}\chi_{++}{}^{-}) \right)$$
$$+ i\, e^{--} \wedge e^{+q} \Omega_{--}^{++i}(\gamma^i\chi_{++}{}^{-})_q - i\, e^{++} \wedge e^{-\dot{q}} \Omega_{++}^{--i}(\chi_{++}{}^{-}\gamma^i)_{\dot{q}}$$
$$+ e^{+q} \wedge e^{-\dot{q}} \left(4(\gamma^i\chi_{++}{}^{-})_q(\chi_{++}{}^{-}\gamma^i)_{\dot{q}} + \frac{2i}{R} f^{ijk}\gamma^{ijk}_{q\dot{q}} \right) \qquad (18.147)$$

and for the curvature of the normal bundle

$$\mathbb{R}^{ij} = e^{++} \wedge e^{--} \left(\Omega_{--}^{++[i} \Omega_{++}^{--j]} - \frac{8i}{R} f^{ijk_1k_2k_3} (\chi_{--}^{+} \gamma^{k_1k_2k_3} \chi_{++}^{-}) \right)$$

$$+ 4i\, e^{--} \wedge e^{+q} \left(\Omega_{--}^{++[i} (\gamma^{j]} \chi_{++}^{-})_q - \frac{8}{R} f^{--ijkl} (\gamma^{kl} \chi_{--}^{+})_q \right)$$

$$- 4i\, e^{++} \wedge e^{-\dot{q}} \left(\Omega_{++}^{--[i} (\chi_{++}^{-} \gamma^{j]})_{\dot{q}} + \frac{8}{R} f^{++ijkl} (\tilde{\gamma}^{kl} \chi_{++}^{-})_{\dot{q}} \right)$$

$$- 8e^{+q} \wedge e^{-\dot{q}} \left(2(\gamma^{[i} \chi_{++}^{-})_q (\chi_{++}^{-} \gamma^{j]})_{\dot{q}} + \frac{i}{R} f^{ijk_1k_2k_3} \gamma_{q\dot{q}}^{k_1k_2k_3} \right).$$

(18.148)

The Peterson-Codazzi equations (18.91) for superstring in $AdS_5 \times S^5$ superspace read

$$D\Omega^{--i} = \hat{R}^{--i} = -\frac{8i}{R} e^{++} \wedge e^{--} f^{--ijkl} (\chi_{--}^{+} \gamma^{jkl} \chi_{++}^{-})$$

$$- \frac{24i}{R} e^{++} \wedge e^{-\dot{q}} f^{ijk} (\tilde{\gamma}^{jk} \chi_{++}^{-})_{\dot{q}} - \frac{8i}{R} e^{+q} \wedge e^{-\dot{q}} f^{--ijkl} \gamma_{q\dot{q}}^{jkl},$$

(18.149)

$$D\Omega^{++i} = \hat{R}^{++i} = -\frac{8i}{R} e^{++} \wedge e^{--} f^{++ijkl} (\chi_{--}^{+} \gamma^{jkl} \chi_{++}^{-})$$

$$- \frac{24i}{R} e^{--} \wedge e^{+q} f^{ijk} (\gamma^{jk} \chi_{--}^{+})_q - \frac{8i}{R} e^{+q} \wedge e^{-\dot{q}} f^{++ijkl} \gamma_{q\dot{q}}^{jkl}.$$

(18.150)

These do not give us new information after Eqs. (18.143) and (18.144) are taken into account.

Thus we have completed the construction of superembedding approach description of the superstring in $AdSS^{(5,5|32)}$ superspace.

18.5. Conclusion

In this contribution we have presented a brief review of spinor moving frame formulation, generalized action principle and superembedding approach to super-p-branes and (in Sec. 18.3) have elaborated in detail the superembedding approach to superstring in general type IIB supergravity background. On this basis we have given (in Sec. 18.4) the complete superembedding description of superstring in $AdS_5 \times S^5$ superspace. To our best knowledge, such a description of AdS superstring has not been developed before and we hope that it will be helpful in searching for new exact solutions of the

AdS superstring equations which, in its turn, might be useful for further study and applications of the AdS/CFT correspondence.

Acknowledgments

The author thanks José de Azcárraga and Dmitri Sorokin for discussions, useful comments and collaboration on early stages of this work which has been supported by the Basque Foundation for Science, *Ikerbasque*, and partially by research grants from the Spanish MICI (FIS2008-1980), the INTAS (2006-7928) as well as by the Ukrainian Academy of Sciences and Russian RFFI grant 38/50–2008.

References

1. Achúcarro, A., Evans, J., Townsend, P.K. and Wiltshire, D., *Super-p-branes*, *Phys. Lett.* **B198**, 441.
2. Akulov, V., Bandos, I., Kummer, W. and Zima, V.G., *Nucl. Phys.* **B527**, 61 (1998) [hep-th/9802032].
3. G. Arutyunov and S. Frolov, arXiv:0901.4937 [hep-th].
4. Bandos, I. A., *Sov. J. Nucl. Phys.* **51**, 906 (1990); *JETP. Lett.* **52**, 205 (1990).
5. Bandos, I., *Mod. Phys. Lett.* **A12**, 799 (1997).
6. Bandos, I.A., *Nucl. Phys.* **B599**, 197 (2001).
7. Bandos, I.A., *Phys. Lett.* **B558**, 197 (2003).
8. Bandos, I.A., de Azcárraga, J.A., Izquierdo, J.M. and Lukierski, J., *Phys. Rev. Lett.* **86**, 4451 (2001) [hep-th/0101113].
9. Bandos, I.A., de Azcárraga, J.A., Picón, M. and Varela, O., *Phys. Rev.* **D69**, 085007 (2004).
10. Bandos, I.A., Ivanov, E., Lukierski, J. and Sorokin, D., *JHEP* **0206**, 040 (2002) [hep-th/0205104].
11. Bandos, I.A. and Kummer, W., *Int. J. Mod. Phys.* **A14**, 4881-4914 (1999) [hep-th/9703099].
12. Bandos, I.A. and Kummer, W., *Phys. Lett.* **B413**, 311 (1997) [Err.-ibid. **B420**, 405 (1998)] [hep-th/9707110].
13. Bandos, I.A. and Kummer, W., *Phys. Lett.* **B462**, 254–264 (1999) [hep-th/9905144]; *Nucl. Phys.* **B565**, 291-332 (2000) [hep-th/9906041].
14. Bandos, I. and Lukierski, J., *Mod. Phys. Lett.* **14**, 1257 (1999) [hep-th/9811022]; *Lect. Not. Phys.* **539**, 195 (2000) [hep-th/9812074].
15. Bandos, I., Pasti, P., Sorokin, D. and Tonin, M., *Lect. Notes Phys.*, **509**, 79-91 (1998) [hep-th/9705064].
16. Bandos, I.A., Pasti, P., Sorokin, D.P., Tonin, M. and Volkov, D.V., *Nucl. Phys.* **B446**, 79 (1995) [hep-th/9501113].
17. Bandos, I.A., Sorokin, D.P. and Tonin, M., *Nucl. Phys.* **B497**, 275 (1997) [hep-th/9701127].

18. Bandos, I., Sorokin, D. and Volkov, D.V., *Phys. Lett.* **B352**, 269-275 (1995) [hep-th/9502141].

19. Bandos, I. A. and Zheltukhin, A.A., *JETP Lett.* **54**, 421 (1991); *Phys. Lett.* **B288**, 77 (1992); *D=10 superstring: Lagrangian and Hamiltonian mechanics in twistor-like Lorentz harmonic formulation*, Preprint IC/92/422, Trieste 1992. 81pp., *Phys. Part. Nucl.* **25**, 453 (1994).

20. Bandos, I.A. and Zheltukhin, A.A., *JETP Lett.* **55** (1992) 81; *Int. J. Mod. Phys.* **A8**, 1081 (1993); *Class. Quant. Grav.* **12**, 609 (1995) [hep-th/9405113].

21. Beisert, N., Ricci, R., Tseytlin, A.A. and Wolf, M., Phys. Rev. **D78**, 126004 (2008) [arXiv:0807.3228 [hep-th]].

22. Bergamin, L. and Kummer, W., *Phys. Rev.* **68**, 104005 (2003) [arXiv:hep-th/0306217].

23. Bergshoeff, E., Howe, P.S., Kerstan, S. and Wulff, L., *JHEP* **0710**, 050 (2007) [arXiv:0708.2722 [hep-th]].

24. Bergshoeff, E., Sezgin, E. and Townsend, P.K., *Phys. Lett.* **B189**, 75 (1987); *Ann. Phys.(NY)* **185**, 330 (1988).

25. Bergshoeff, E., Sezgin, E. and Townsend, P.K., *Phys. Lett.* **B209**, 451 (1988).

26. Berkovits, N. and Howe, P.S., *The cohomology of superspace, pure spinors and invariant integrals*, arXiv:0803.3024 [hep-th].

27. Berkovits, N. and Maldacena, J., *JHEP* **0809**, 062 (2008) [arXiv:0807.3196 [hep-th]].

28. Chu, C.S., Howe, P.S., Sezgin, E. and West, P.C., *Phys. Lett.* **B429**, 273 (1998) [hep-th/9803041].

29. Castellani, L. D'Auria, R. and Fré, P. *Supergravity and superstrings: a geometric perspective*, V.1,2,3, World Scientific, Singapore 1991.

30. Chryssomalakos, C., de Azcárraga, J. A., Izquierdo, J. M. and Pérez Bueno, J. C., *Nucl. Phys.* **B567**, 293 (2000) [hep-th/9904137].

31. de Azcárraga, J. A. and Lukierski, J., *Phys. Lett.* **B113**, 170–174 (1982).

32. Delduc, F., Galperin, A., Sokatchev, E., *Nucl. Phys.* **B368**, 143-171 (1992).

33. Galperin, A.S., Howe, P.S. and K. S. Stelle, K.S., *Nucl. Phys.* **B368**, 248-280 (1992) [hep-th/9201020].

34. Galperin, A., Ivanov, E., Kalitsyn, S., Ogievetsky, V. and Sokatchev, E., *Class. Quant. Grav.* **1**, 469-498 (1984); Galperin, A.S., Ivanov, E.A., Ogievetsky, V.I and Sokatchev, E. *Harmonic Superspace*, (CUP Cambridge, UK) 2001.

35. Gomis, J., Sorokin, D., and Wulff, L., *The complete AdS(4) x CP(3) superspace for the type IIA superstring and D-branes*, arXiv:0811.1566 [hep-th].

36. Green, M.B. and Schwarz, J.H., *Covariant description of superstrings*, *Phys. Lett.* **B136**, 367 (1984).

37. Grisaru, M.T., Howe, P.S., Mezincescu, L., Nilsson, B. and Townsend, P.K. *N=2 Superstrings In A Supergravity Background*, Phys. Lett. **B162**, 116 (1985).

38. Grumiller, D., Kummer, W., and Vassilevich, D.V., *Phys. Rept.* **369**, 327 (2002) [arXiv:hep-th/0204253].

39. Heslop, P. and Howe, P.S., *Chiral superfields in IIB supergravity*, *Phys. Lett.* **B502**, 259 (2001) [arXiv:hep-th/0008047].

40. Howe, P.S., Raetzel, O. and Sezgin, E., *On brane actions and superembeddings*, *JHEP* **9808**, 011 (1998).

41. Howe, P.S., and Sezgin, E., *Superbranes*, *Phys. Lett.* **B390**, 133 (1997) [hep-th/9607227].

42. Howe, P.S., and Sezgin, E., *D = 11, p = 5*, *Phys. Lett.* **B394**, 62–66 (1997) [hep-th/9611008].

43. Howe, P. S. and Tucker, R. W., *J. Phys.* **A10**, L155 (1977); *J. Math. Phys.* **19**, 981 (1978).

44. Kummer, W., *General treatment of all 2d covariant models*, *Rakhiv 1995, Methods in mathematical physics. Procs. 12th Hutsulian Workshop*, (Hadronic Pr., 1997) pp. 161-176 [gr-qc/9612016].

45. Kallosh, R. and Rahmanov, M., *Phys.Lett.* **B209** (1988) 233; **B214** (1988) 549.

46. Kallosh, R. and Rahmfeld, J., *The GS string action on AdS$_5$ × S^5*, *Phys. Lett.* **B443**, 143 (1998) [hep-th/9808038].

47. Kallosh, R. and Rajaraman, A., *Vacua of M-theory and string theory*, *Phys. Rev.* **D58**, 125003 (1998) [hep-th/9805041].

48. Metsaev, R.R. and Tseytlin, A.A., *Nucl. Phys.* **B533**, 109 (1998) [hep-th/9805028]; *J. Exp. Theor. Phys.* **91**, 1098 (2000).

49. Matone, M., Mazzucato, L., Oda, I., Sorokin, D. and Tonin, M., *Nucl. Phys.* **B639**, 182 (2002).

50. Myers, R.C., *JHEP* **9912**, 022 (1999) [hep-th/9910053].

51. Neeman, Y. and Regge, T, *Phys.Lett.* **B74**,54(1978); D'Auria, R., Fré, P. and Regge, T., *Riv. Nuovo Cim.* **3**, f12, 1 (1980);
Regge, T., *The group manifold approach to unified gravity*, in: *Relativity, groups and topology II: Les Houches, Session* **XL** 1983, B. S. DeWitt and R. Stora eds., Elsevier Sci. Pub. 1984, pp.993–1005, and refs therein.

52. Nissimov, E., Pacheva, S., Solomon, S., *Nucl. Phys.* **B296**, 462 (1988); **B299**, 183 (1988); **B297**, 349 (1988); **B317**, 344 (1989); *Phys. Lett.* **B228**, 181 (1989).

53. Schwarz, J.H. and West, P.C., *Phys. Lett.* **B126**, 301 (1983); Schwarz, J.H., *Nucl. Phys.* **B226**, 269 (1983).

54. Siegel, W., *Phys. Lett.* **B128**, 397 (1983).

55. Sokatchev, E., *Phys. Lett.* **B169**, 209 (1986); *Class. Quantum Grav.* **4**, 237 (1987).

56. Sorokin, D.P., *Phys. Rept.* **329**, 1 (2000) [hep-th/9906142].

57. Sorokin, D.P., Tkach, V.I., Volkov, D.V., *Superparticles, twistors and Siegel symmetry*, Preprint KFTI-88-31, Apr 1988, *Mod.Phys.Lett.* **A4**, 901 (1989).

58. Sorokin, D.P., Tkach, V.I., Volkov, D.V. and Zheltukhin, A.A., *Phys. Lett.* **B216**, 302 (1989).

59. Uvarov, D. V., *Phys. Lett.* **493**, 421 (2000), *Nucl. Phys. Proc. Suppl.* **102**, 120 (2001) [hep-th/0104235]; *Class. Quant. Grav.* **24**, 5383 (2007).

60. Van Holten, J.W., and Van Proeyen, A., *J. Phys.* **A15**, 3763 (1982).

61. Volkov, D.V. and Akulov, V.P., *JETP Lett.* **16**, 438 (1972); *Phys. Lett.* **B46**, 109 (1973).

62. Volkov, D.V. and Soroka, V.A., *JETP Lett.* **18**, 312-314 (1973).

63. Volkov, D.V. and Zheltukhin, A.A., *JETP Lett.* **48**, 63 (1988); *Lett. Math. Phys.* **17**, 141 (1989); *Nucl. Phys.* **B335**, 723 (1990).
64. Wiegmann, P.B., *Nucl. Phys.* **B323**, 311-329 (1989); **B323**, 330-336 (1989).
65. Zheltukhin, A.A., Uvarov, D.V., *JHEP* **0208**, 008 (2002) [hep-th/0206214].

Chapter 19

Heterotic (0,2) Gepner Models and Related Geometries

Maximilian Kreuzer

Institut für Theoretische Physik, Technische Universität Wien,
A-1040 Wien, Austria
E-mail: kreuzer@hep.itp.tuwien.ac.at

On the sad occasion of contributing to the memorial volume "Fundamental Interactions" for my teacher Wolfgang Kummer I decided to recollect and extend some unpublished notes from the mid 90s when I started to build up a string theory group in Vienna under Wolfgang as head of the particle physics group. His extremely supportive attitude was best expressed by his saying that one should let all flowers flourish. I hope that these notes will be useful in particular in view of the current renewed interest in heterotic model building.

The content of this contribution is based on the bridge between exact CFT and geometric techniques that is provided by the orbifold interpretation of simple current modular invariants. After reformulating the Gepner construction in this language I describe the generalization to heterotic (0,2) models and its application to the Geometry/CFT equivalence between Gepner-type and Distler-Kachru models that was proposed by Blumenhagen, Schimmrigk and Wisskirchen. We analyze a series of solutions to the anomaly equations, discuss the issue of mirror symmetry, and use the extended Poincaré polynomial to extend the construction to Landau-Ginzburg models beyond the realm of rational CFTs.

In the appendix we discuss Gepner points in torus orbifolds, which provide further relations to free bosons and free fermions, as well as simple currents in $N = 2$ SCFTs and minimal models.

19.1. Introduction

When a number of different constructions for heterotic string compactifications were developed in the late 1980s it soom became clear from the coincidence of spectra that Gepner models[1] and Calabi-Yau hypersurfaces in weighted projective spaces[2] should be closely related. The connection was found to be provided by Landau-Ginzburg models,[3] whose superpotential $W(\Phi_i)$ can be identified with the hypersurface equation $W(z_i) = 0$. A Fermat-type potential of the form $W = \sum \Phi_i^{K_i}$ then corresponds to a Gepner model with levels $k_i = K_i - 2$. The precise relation was later derived by Witten by virtue of his $N = 2$ supersymmetric gauged linear sigma model (GLSM),[4] which – in addition to the shape parameters (complex structure moduli) in the superpotential W – contains the size parameters (Kähler moduli) of the Calabi-Yau as D-terms.

The Gepner point thus turns out to be located at small values of the Kähler moduli, way outside the range of validity of sigma model perturbation theory, so that Gepner models provide an exactly solvable CFT stronghold inmidst the realm where strong quantum corrections invalidate any naive geometrical picture. This proved to be useful in many contexts like closed string mirror symmetry,[5] as well as homological mirror symmetry, where, for example, the transport of exact CFT boundary states to D-branes at large volume can be studies.[6]

In the context of perturbative heterotic strings the phenomenological condition of space-time supersymmetry in the RNS formalism implies that the (0,1) superconformal invariance that is left over from the gauge-fixed world-sheet supergravity[a] is extended to a (0,2) superconformal invariance plus quantization of the U(1) charges.[7,8] This is, in fact, an equivalence, because quantization of the N=2 superconformal U(1) charge implies locality of the spectral flow operator, which implements the space-time SUSY transformations on the internal CFT part of vertex operators.[8]

In the geometric context (0,2) models correspond to stable holomorphic vector bundles $V_1 \times V_2 \subset E_8 \times E_8$ on a Calabi-Yau manifold X with vanishing first Chern classes satisfying the anomaly cancellation condition $ch_2(V_1) + ch_2(V_2) = ch_2(TX)$. The notion of a (2,2) model then refers to the choice $V_1 = TX$ with trivial V_2, called standard embedding, so that the structure group $SU(3)$ of TX breaks $E_8 \times E_8$ to the gauge group $E_6 \times E_8$ in 4 dimensions. The name (2,2) originates from the CFT analog of this situation where we replace the compactification manifold by an abstract $N = (2,2)$ left-right symmetric superconformal field theory with central charge $c = 9$. This "internal sector" is combined with the right-moving space-time superfields and the left-moving bosonic space-time coordinates, augmented by a left-moving $SO(10) \times E_8$ current algebra, whose central

---

[a] The (0,1) superconformal invariance is hence required for a consistent coupling to the superghosts in BRST quantization.

charge 13 adds up with 4 non-compact dimensions and the internal $c = 9$ to the critical dimension 26 of the bosonic string.[b] The same spectral flow mechanism that generates space-time SUSY in the right-moving sector then extends the manifest $SO(10)$ times the $U(1)$ of the $N = 2$ superconformal algebra to the low energy E_6 gauge symmetry of the standard embedding.[c] In the geometric context this amounts to the GSO projection. For a general internal $N = 2$ SCFT with fractional charges it has to be augmented by charge quantization and is then refered to as "generalized GSO projection".

While the general (0,2) models have better phenomenological prospects, like featuring the more relalistic GUT gauge groups $SO(10)$ and $SU(5)$, the (2,2) case has been studied much more systematically. In the realm of σ models one reason for this was the discovery of world sheet instanton corrections,[9,10] which were believed to destabilize the vacua. A criterion for avoiding this problem was soon found by Distler and Greene;[11] see also Refs. 12–14. The technical difficulty of checking the 'splitting type' of the stable vector bundles, however, provided a powerful deterrent for further progress. The situation became much more secure with Witten's gauged linear sigma models,[4] the (0,2) version of which was used by Distler and Kachru[15,16] to generalize the construction introduced by Distler and Greene.[11] The resulting class of models is now believed to define honest (0,2) SCFTs at the infrared fixed point.[17] Somewhat ironically, with the recognition of the importance of moduli stabilization for model building world-sheet instantons can turn from an obstacle into a virtue, and one now has to work quite hard[18] to circumvent the cancellation mechanim that has been established for toric Calabi-Yau complete intersections by Beasley and Witten.[19] There is also much recent work on generalizations like heterotic M-theory[20] and heterotic compactification with H-field background flux,[21] but this is beyond the scope of the present note.

In the realm of exact methods a powerful generalization of Gepner's construction[1] was found by Schellekens and Yankielowicz,[22] who used simple currents[23] to produce a telephone book of (1,2) models[24] from tensor products of minimal models. For the (0,2) case their huge list of models apparently was so far from complete that it never was published. At the same time closely related methods were used by Font et al.[25,26] to construct pseudo-realistic models. On the CFT side the main problem is the arbitrariness in the selection of a reasonable subset from the huge set of

---

[b] The value $c = 9$ corresponds to 6 compact dimensions X^i plus the contribution from their right-moving fermionic superpartners ψ^i.

[c] More precisely, the mechanisms are mapped to one another by the bosonic string map and its inverse, the Gepner map, respectively (see below).

available models. A landmark in this effort was Schellekens' theorem on the conditions for the possibility of avoiding fraction electric charges.[27,28]

An interesting question is, of course, to what extent the geometric and the CFT approaches to (0,2) models overlap. The identification of models that are accessible to both constructions would provide further evidence for the stability of the σ model constructions, but most importantly allows to explore deformations of the rational models, which only live at certain points in moduli space. Originally based on a stochastic computer search for matching particle spectra Blumenhagen et al.[29,30] proposed a set of gauge bundle data on a complete intersection that is conjectured to describe the moduli space of a rational superconformal (0,2) cousin of the Gepner model on the quintic. Using the classification of simple current modular invariants[31] the product invariant that these authors employ can be translated into the canonical form[32] that exhibits its relation to orbifold twists and discrete torsion.[33] It turns out that the breaking of the gauge group from E_6 to $SO(10)$ is due to a certain twist of order 4 that acts on a minimal model factor of the internal conformal field theory (at odd level) and on an $SO(2)$ that is part of the linearly realized $SO(2) \times SO(8) \subset SO(10)$ gauge symmetry.[32] Assuming that the $\mathbb{Z}_4$ breaking mechanism does not care about the rest of the conformal field theory and only acts on a Fermat factor of a non-degenerate potential we analysed the anomaly matching conditions and proposed a whole series of identifications[32] that provides us with 3219 models, based on the list of 7555 weights for transverse hypersurfaces in weighted projective spaces,[34,35] and many more if we combine this with other constructions like orbifolding and discrete torsion.[36–38]

The purpose of this note is to collect the necessary ingredients for these constructions, where the concept of the extended Poincaré polynomial[39] is used to generalize the CFT approach to Landau-Ginzburg models beyond the exactly solvable case. In section 19.2 we discuss simple current modular invariants (SCMI)[23] and their geometric interpretation.[31] To set up the concepts we begin with recalling the geometric orbifolding idea and use it to motivate and interpret the formula for the most general invariant. In section 19.3 we use simple current techniques for the implementation of the generalized GSO projection and show how the Gepner construction generalizes to (0,2) models in general and, in particular, for gauge symmetry breaking in the proposed σ model connection. We discuss the counting of non-singlet spectra in terms of the information encoded in the extended Poincaré polynomial, thus extending the scope of the construction to arbitrary Landau-Ginzburg orbifolds. Since the charge conjugation of $N = 2$

minimal models is a simple current modular invariant, our discussion explains the observed (0,2) mirror symmetry[40,41] along the lines of Greene-Plesser orbifolds and their generalization due to Berglund and Hübsch,[42] which applies to the large class of transversal potentials that are minimal in a certain sense.[43] Section 19.4 briefly recollects the geometric side of the proposed identifications. Here we start with an ansatz for the base manifold and vector bundle data that are conjectured to describe the moduli spaces of our (0,2) models and find a unique solution to the anomaly equations. In section 19.5 we conclude with a number of topics for generalizations and further studies.

Some technical points are discussed in appendices. In appendix A we discuss Gepner points of torus orbifolds and exact CFT realizations for the extensions of $\mathbb{Z}_2 \times \mathbb{Z}_2$ orbifolds recently classified by Donagi and Wendland.[44] Appendix B discusses simple currents in $N = 2$ SCFTs and their use for explaining labels and field identifications of $N = 2$ minimal models.

19.2. Orbifolds and simple currents

The concept of an orbifold CFT originates from the geometric picture of closed strings on orbit spaces X/G where X is a smooth manifold with a discrete group action G, with or without fixed points. The modding out of G has two consequences: String states on the orbifold need to be invariant under the symmetry on the covering space X, which leads to a projection of the Hilbert space $\mathcal{H}_X$ on the covering space to G-invariant states. On the other hand, new closed string states emerge, whose 2∂-periodicity on X/G corresponds to periodicity up to a group transformation $g \in G$ on X. The Hilbert space has hence to be augmented by twisted sectors $\mathcal{H}_X^{(g)}$.

19.2.1. *Orbifold CFT and modular invariance*

For abstract conformal field theories $\mathcal{C}$ that are invariant under a group G of symmetry transformations the same result can be derived from modular invariance and factorization constraints on the partition function without relying on a geometric interpretation. Depicting the one-loop partition function by a torus that indicates the double-periodic boundary conditions imposed in the path integral, $Z_\mathcal{C} = \boxed{}$, the orbifold partition function can be obtained as a linear combination of partition functions with boundary conditions twisted by group transformations g and h,

$$Z_{\mathcal{C}}(g,h) = g\ \boxed{}_{h} \tag{19.1}$$

in the vertical and horizontal direction, respectively. If we interpret the horizontal direction as the spacial extension of a closed string and the vertical direction as Euclidean time then h amounts to twisted boundary conditions in the Hilbert space, while a normalized sum over twisted boundary conditions in periodic Euclidean time can be shown to be equivalent to a projector

$$\Pi_G = \frac{1}{|G|} \sum_{g \in G} g\ \boxed{}_{*} \tag{19.2}$$

onto G-invariant states. Under modular transformations

$$\tau \to \frac{a\tau+b}{c\tau+d}, \qquad \begin{pmatrix} a & b \\ c & d \end{pmatrix} \in \mathrm{PSL}(2,\mathbb{Z}) \tag{19.3}$$

boundary conditions are recombined: For the generators

$$S : \tau \to -1/\tau, \qquad T : \tau \to \tau + 1 \tag{19.4}$$

of $\mathrm{PSL}(2,\mathbb{Z})$ we observe

$$S : g\ \boxed{}_{h} \ \to \ h^{-1}\ \boxed{}_{g}, \qquad T : g\ \boxed{}_{h} \ \to \ gh\ \boxed{}_{h} \tag{19.5}$$

where T maps the double-periodicity $(1, \tau)$ to $(1, \tau + 1)$ and the action of S has been chosen as $(1, \tau) \to (\tau, -1)$. [d] The double-periodicities are consistently defined only if g and h commute so that we need to restrict to twists obeying $gh = hg$ in the case of non-abelian groups.

Since modular transformations mix up the twists of the periodicities along the homology cycles we expect an invariant to contain contributions from all combinations and it is easy to see that the simplest invariant solution is

$$Z_{\mathcal{C}/G} \equiv \frac{1}{|G|} \sum_{gh=hg} g\ \boxed{}_{h} \tag{19.6}$$

In the abelian case the sum over h corresponds to a sum over all twisted sectors. The sum over g then implements the projection onto invariant states; in accord with (19.2) the normalization ensures that the (invariant)

---

[d] While $S^2 = (ST)^3 = \mathbb{1} \in \mathrm{PSL}(2,\mathbb{Z})$ for modular group elements, the action of $S^2 = (ST)^3 : (1, \tau) \to (-1, -\tau)$ on the world sheet amounts to parity plus time reversal. Due to CPT invariance the action of S on the Hilbert space thus squares to a charge conjugation $S^2 = C$ of the conformal fields.

ground state contributes to the partition function with multiplicity one.[e] Our CFT result thus coincides with what we expect for closed strings on orbifolds X/G. But there might be further solutions.

19.2.2. *Discrete torsion and quantum symmetries*

Let us start with the more general ansatz

$$Z^\epsilon_{\mathcal{C}/G} \equiv \frac{1}{|G|} \sum_{gh=hg} \epsilon(g,h)\ g \begin{array}{|c|} \hline \\ \hline \end{array} h \tag{19.7}$$

with weight $\epsilon(g,h)$ for the (g,h)–twisted contribution. This modification can also be motivated from geometry and is called "discrete torsion"[33] because it is related to phase factors $\epsilon(g,h)$ due to B-field flux with only "discrete" values allowed by G-invariance (the field strength $H = dB$ of the 2-form B determines the "torsion" of the corresponding sigma model). With an analysis of the modular invariance and factorization constraints[f] Vafa[33] has shown that

$$\epsilon(g,g) = \epsilon(g,h)\epsilon(h,g) = 1, \quad \epsilon(g_1 g_2, h) = \epsilon(g_1,h)\epsilon(g_2,h). \tag{19.8}$$

Mathematically discrete torsion corresponds to an element of the group cohomology $H^2(G, U(1))$. For abelian groups $G = \mathbb{Z}_{n_1} \times \ldots \times \mathbb{Z}_{n_r}$ with generators g_i the most general solution is parametrized by an arbitrary choice of the phases $\epsilon(g_i, g_j)$ for $i < j$ obeying $\epsilon(g_i, g_j)^{\gcd(n_i, n_j)} = 1$.

The ambiguity of the orbifold CFT that is due to discrete torsion is quite easy to understand in the operator picture because the group action is originally defined only in the untwisted sector. For the twisted sectors we do know the group action on (untwisted) operators but the action on the twisted ground states (and on the corresponding twist fields) is a priory subject to a choice. We can thus think of $\epsilon(g,h)$ as an extra phase of the group action of g in the h-twisted sector.

While the symmetry of the original CFT is lost after orbifolding because of the projection to invariant states, a new symmetry emerges due to selection rules for operator products of twist fields $\Sigma_{h_1}(z)\Sigma_{h_2}(w)$, to which

---

[e] With the restriction to $gh = hg$ the formula also applies to the non-abelian case, where the sum can be interpreted to extend over conjugacy classes of twists followed by a projection onto states that are invariant under the respective normalizers.

[f] On a genus n surface the partition function depends on $2n$ twists along homology cycles, with a corresponding prefactor $\epsilon(g_1, g_2; \ldots; g_{2n-1}, g_{2n})$ that has to factorize into $\epsilon(g_1, g_2) \ldots \epsilon(g_{2n-1}, g_{2n})$. The only condition in the analysis that has to be used beyond the torus is a Dehn twist at genus 2.

we only expect contributions of fields twisted by $h_1 h_2$. The corresponding symmetry of the orbifold has been called quantum symmetry.[45] In the abelian case the quantum symmetry is dual of the original symmetry. Modding out the quantum symmetry of a $\mathbb{Z}_n$-orbifold just gives us back the original CFT.[46] If we mod out two commuting group actions $\langle g_1, g_2 \rangle$ in two steps then the freedom due to discrete torsion $\epsilon(g_2, g_1)$ can be recovered by combining the group action g_2 of the second orbifolding with an approprite power of the quantum symmetry q_1 that emerges from the g_1-twist in the first orbifold. These ideas can be used to extend the Green-Plesser mirror construction of Gepner models[g] to arbitrary orbifolds with discrete torsion.[37,46]

19.2.3. Simple currents

Simple currents are, in a sense, generalized free fields in rational conformal field theories. For free bosons there is a shift symmetry. When it is used for orbifolding the twisted sectors correspond to winding states. For free fermions a $\mathbb{Z}_2$ symmetry is provided by the fermion number. In this case the twisted sector is the Ramond sector, with a cut in the punctured complex plane, and the projection to invariant states is the GSO projection. Simple currents, as we will see, also come with discrete symmetries. Accordingly, they can be used to construct new conformal field theories, which turn out to be given in terms of the original characters but with a certain type of non-diagonal modular invariants.

We consider a rational conformal field theory, i.e. a CFT with left- and right-moving chiral algebras $\mathcal{A}_L$ and $\mathcal{A}_R$ such that the conformal fields are combined into a finite number of representations $\Phi_{i\bar{k}}$ of $\mathcal{A}_L \otimes \mathcal{A}_R$, where i labels the representation of $\mathcal{A}_L$. The chiral algebras contain the Virasoro algebra and possibly more. We may use the highest weight state, or primary field, in a conformal family as its representative. But it is important to keep in mind that the conformal weight h_i is well-defined for a primary field, but only defined modulo 1 for the conformal family.

The fusion algebra $\Phi_i \times \Phi_j = \mathcal{N}_{ij}{}^k \Phi_k$ of a rational CFT is the commutative associative algebra whose non-negative integral structure constants $\mathcal{N}_{ij}{}^k$ encode the fusion rules, i.e. the information of which representations

---

[g] More generally, we can consider arbitrary N=2 SCFTs for which mirror symmetry, i.e. right-moving charge conjugation, is equivalent to an orbifold.[46] This is the case for the large class of Landau-Ginzburg models for which a transversal potential exists whose number of monomials is equal to the number of fields,[43] as was discovered by Berglund and Hübsch.[42]

of the chiral algebra appear in operator product expansions $\Phi_i(z)\Phi_j(w)$.[h] A simple current J of a conformal field theory is a primary field that has a unique fusion product with all other primary fields,[23] i.e.

$$J \times \Phi_j = \Phi_{(Jj)}, \qquad j \to Jj \to J^2j \to J^3j \ldots, \qquad (19.9)$$

where we use the notation Jj for the label of the fusion product of J and Φ_j. A simple current thus decomposes the field content of the CFT into orbits, which have finite length for a rational theory.

Since the OPE $J(z)\Phi_j(w)$ contains only fields from a single conformal family, whose conformal weights can only differ by integers, all expansion coefficients $(z-w)^{h_{Jj}-h_J-h_j}$ have the same monodromy $e^{-2\partial i Q_J(\Phi_j)}$ with

$$Q_J(\Phi_j) \equiv h_J + h_j - h_{Jj} \quad \text{mod} \ 1 \qquad (19.10)$$

about the expansion point w. The monodromy of $J(z)$ for a big circle about the positions of $\Phi_j(w_j)$ and $\Phi_k(w_k)$ is the product of the two respective monodromies. Thus the phase transformation $e^{-2\partial iQ_J}$ is compatible with operator products and defines a symmetry of the CFT. Before we come to the resulting orbifold CFTs, which correspond to the simple current modular invariants, we need to collect some basic definitions and formulas for simple currents.

The order N_J of a simple current J is the length of the orbit of the identity $J^{N_J} = \mathbb{1}$. Because of associativity and commutativity of the fusion product the simple currents of a CFT form an abelian group $\mathcal{C}$, which is called the center. The definition of the monodromy charge implies $Q_{J \times K}(\Phi) \equiv Q_J(K\Phi) - Q_J(K) + Q_K(\Phi)$ modulo 1, so that

$$Q_{J \times K}(\Phi) \equiv Q_J(\Phi) + Q_K(\Phi), \qquad Q_{J^n}(\Phi) \equiv nQ_J(\Phi). \qquad (19.11)$$

$Q_J(\Phi)$ is hence a multiple of $1/N_J$. It can be shown that the charge quantum of Q_J is indeed $1/N_J$, so that a simple current J always comes with a discrete $\mathbb{Z}_{N_J}$ phase symmetry of the CFT (not every cyclic symmetry is generated by a simple current, though). The symbol $\equiv$ henceforth denotes equality modulo integers.

For the orbifolding of a CFT we may choose to mod out some subgroup of its full symmetry group. Similarly, we now choose some fixed subgroup $\mathcal{G}$ of the center $\mathcal{C}$ of a CFT that is generated by independent simple currents J_i of orders N_i. We use the notation $[\alpha] = \prod J_i^{\alpha_i}$ and $Q_i = Q_{J_i}$, where α_i are integers that are defined modulo N_i. Then we can parametrize the

---

[h] Multiplicities $\mathcal{N}_{ij}{}^k > 1$ indicate contributions from descendents in OPEs beyond the coefficients that are implied by the Ward identities of the chiral algebra.

conformal weights and monodromy charges of all simple currents in $\mathcal{G}$ in terms of a matrix R_{ij},[47]

$$R_{ij} = \frac{r_{ij}}{N_i} \equiv Q_i(J_j) = Q_j(J_i), \quad h_{[\alpha]} \equiv \frac{1}{2}\sum_i r_{ii}\alpha^i - \frac{1}{2}\sum_{ij}\alpha^i R_{ij}\alpha^j \quad (19.12)$$

with $r_{ij} \in \mathbb{Z}$. If N_i is odd we can always choose r_{ii} to be even. With this convention all diagonal elements R_{ii} are defined modulo 2 for both, even and odd N_i.[i] Using the definitions of Q and R we obtain

$$h_{[\alpha]\Phi} \equiv h_\Phi + h_{[\alpha]} - \alpha^i Q_i(\Phi), \quad Q_i([\alpha]\Phi) \equiv Q_i(\Phi) + R_{ij}\alpha^j. \quad (19.13)$$

It can be shown that S matrix elements for fields on the same orbits are related by phases,

$$S_{[\alpha]\Phi,[\beta]\Psi} = S_{\Phi,\Psi}\; e^{2\partial i(\alpha^k Q_k(\Psi)+\beta^k Q_k(\Phi)+\alpha^k R_{kl}\beta^l)}. \quad (19.14)$$

T-matrix elements only depend on conformal weights and, according to eq. (19.13), are related by phases $2\partial i(h_{[\alpha]} - \alpha^i Q_i(\Phi) - h_{[\beta]} + \beta^i Q_i(\Psi))$.

19.2.4. Simple current modular invariants and chiral algebras

The partition function of a rational CFT can be written as

$$Z(\tau) = \mathrm{Tr}\, e^{2\partial i\tau L_0} e^{-2\partial i\bar{\tau}\bar{L}_0} = \sum_{ij} M_{ij}\chi_i(\tau)\bar{\chi}_j(\bar{\tau}) \quad (19.15)$$

with a non-negative integral matrix M_{ij} that is called a modular invariant if

$$[M,S] = [M,T] = 0, \quad M_{11} = 1 \quad (19.16)$$

since under modular transformations $\chi_i(-1/\tau) = S_{ij}\chi_j(\tau)$ and $\chi_i(\tau+1) = T_{ij}\chi_j(\tau)$ so that $M \to S^t M S^*$ and $M \to T^t M T^*$ with symmetric unitary matrices S and T, respectively. Modular invariants of automorphism type are permutation matrices that uniquely map representation labels of the left movers to right movers, where the permutation is an automorphism of the fusion rules. Extension-type invariants, on the other hand, combine contributions of several characters to characters of extended chiral algebras while other representations of the original chiral algebra are projected out.

Simple current modular invariants (SCMIs) are modular invariants for which $M_{jk} \neq 0$ only if Φ_j and Φ_k are on the same orbit, i.e. if $k = Jj$ for some simple current $J \in \mathcal{C}$. T-invariance, which is also called "level matching", requires that $h_j - h_k \in \mathbb{Z}$. Using eq. (19.13) with the above notation $[\alpha] = \prod J_i^{\alpha^i} \in \mathcal{G} \subseteq \mathcal{C}$ we conclude that

$$h_j - h_{[\alpha]j} \equiv \alpha^i Q_i(\Phi_j) - h_{[\alpha]} \in \mathbb{Z}. \quad (19.17)$$

[i] It is easiest to first compute $R_{ij} \equiv Q_i(J_j)$ modulo 1 and then fix R_{ii} modulo 2 for the diagonal elements with even N_i by imposing that formula (19.12) for $h(J_i)$ has to hold.

We can think of $[\alpha]$ as the twist in the orbifolding procedure, which is in accord with the number $|\mathcal{G}|$ of twisted sectors as well as with the expected quantum symmetry due to twist selection rules. If the order N_i of J_i is even then eq. (19.17) implies that the twist J_i (like any odd power of J_i) can contribute to a modular invariant only if $r_{ii} = N_i R_{ii} \in 2\mathbb{Z}$. We henceforth assume that all generators of $\mathcal{G}$ satisfy this condition. [j]

With the orbifolding procedure in mind it is now not difficult to guess that the SCMI should impose a projection $\delta_{\mathbb{Z}}(Q_i + X_{ij}\alpha^j)$ where $\delta_{\mathbb{Z}}$ is one for integers and zero otherwise and the linear ansatz $X_{ij}\alpha^j$ for the phase shift in the projection is suggested by comparing eq. (19.17) with $h_{[\alpha]} \equiv -\frac{1}{2}\alpha^i R_{ij}\alpha^j$ as well as by the expected quantum symmetry. Using regularity assumptions[k] it can be shown[31] that the most general SCMI reads

$$M_{\Phi,[\alpha]\Phi} = \mu(\Phi) \prod_i \delta_{\mathbb{Z}}\left(Q_i(\Phi) + X_{ij}\alpha^j\right), \qquad (19.18)$$

where T-invariance implies $X + X^T \equiv R$ modulo 1 for off-diagoal and modulo 2 for diagonal matrix elements, X is quantized by $\gcd(N_i, N_j)X_{ij} \in \mathbb{Z}$, and $\mu(\Phi)$ denotes the multiplicity of the primary field Φ on its orbit, i.e. $\mu(\Phi) = |\mathcal{G}|/|\mathcal{G}_\Phi|$ where $|\mathcal{G}_\Phi|$ is the size of the orbit of the action of $\mathcal{G}$ on Φ. While the symmetric part $X_{(ij)} \equiv \frac{1}{2}R_{ij}$ of X is fixed by level matching, the ambiguity due to the choice of a properly quantized antisymmetric part $E_{ij} \equiv X_{ij} - \frac{1}{2}R_{ij}$ corresponds to the discrete torsion of the orbifolding procedure.

We can now briefly discuss different types of invariants. If $X = 0$ we have a pure extension invariant because all fields with non-integral charges are projected out while all fields on a simple current orbit are combined to new conformal families. $X = 0$ is only possible if the conformal weights of all simple currents $J \in \mathcal{G}$ are integral and since these currents are in the orbit of the identity they extend the chiral algebras $\mathcal{A}_L$ and $\mathcal{A}_R$ so that we obtain a new rational symmetric and diagonal CFT.

Let us define the kernel $\mathrm{Ker}_{\mathbb{Z}} X$ as the set of integral solutions $[\alpha]$ of $X_{ij}\alpha^j \in \mathbb{Z}$ with α_j definded modulo N_j. If this kernel is trivial then $\left(Q_i(\Phi) + X_{ij}\alpha^j\right) \in \mathbb{Z}$ has a unique solution $[\alpha]$ for each charge, which defines a unique position $[\alpha]\Phi$ on the orbit that only depends on the charge $Q_i(\Phi)$ of Φ. We then obtain an automorphism invariant. In general, the extension of the right-moving chiral algebra $\mathcal{A}_R$ is give by the kernel $\mathrm{Ker}_{\mathbb{Z}} X$

---

[j] The maximal subgroup of $\mathcal{C}$ that can contribute to a SCMI is called "effective center".
[k] 'Regularity' requires that $M_{\Phi,[\alpha]\Phi}$ only depends on $Q_i(\Phi)$.[47] Discrete Fourier sum and 2-loop modular invariance imply that the 'phases' are bilinear and antisymmetric.

and, since

$$M_{[\alpha]\Phi,\Phi} = \mu(\Phi) \prod_i \delta_{\mathbb{Z}} \left(Q_i(\Phi) + \alpha^j X_{ji} \right), \qquad (19.19)$$

the extension of the left-moving chiral algebra $\mathcal{A}_L$ is give by the kernel $\mathrm{Ker}_{\mathbb{Z}} \, X^T$ of the transposed matrix. While the extensions are of the same size, they need not be isomorphic. For example, an extension of $\mathcal{A}_R$ by $\mathbb{Z}_9$ can occur together with an extension of $\mathcal{A}_L$ by $\mathbb{Z}_3 \times \mathbb{Z}_3$.

19.3. Gepner-type (0,2) models

The right-moving sector of a heterotic string consists of four space-time co-ordates and their superpartners (X^μ, ψ^μ), a ghost plus superghost system (b, c, β, γ), and a supersymmetric sigma model on a Calabi-Yau, whose abstract version is an $N = 2, c = 9$ SCFT $\mathcal{C}_{int}$. Equivalently, we can use light-cone gauge, which amounts to ignoring the ghosts and restricting space-time indices to transverse directions. The left-moving sector is a bosonic string with space-time plus ghost part (X^μ, b, c) and the same internal sector $\mathcal{C}_{int}$ with $c = 9$, whose central charges add up to $4+9-26 = -13$ so that we need to add a left-moving CFT with central charge 13 for criticality. Modular invariance requires this CFT to be either an $E_8 \times SO(10)$ or $SO(26)$ level 1 affine Lie algebra (we will henceforth ignore the $SO(26)$ case because it is phenomenologically less attractive). In the geometric context of a sigma model on a Calabi-Yau the superstring vacuum is then obtained by aligning space-time spinors and tensors with internal Ramond and Neveu-Schwarz sectors, respectively, and performing the GSO projection. For abstract $N = 2$ SCFTs U(1) charges may be quantized in fractional units so that, in addition, a projection to integral charges (generalized GSO) is required for space-time supersymmetry.

All of these operations can be understood as SCMIs of extension type.[22,39] To see this let us first discuss the simple currents in the relevant CFTs. For the $D_n \cong SO(2n)$ current algebra the center $\mathcal{C}_n$ has order 4 and consists of the spinor representation s, its conjugate $\bar{s}$, and the vector v with

$$sv = \bar{s}, \quad s^2 = \bar{s}^2 = v^n, \quad v^2 = \mathbb{1} \quad \Rightarrow \quad \mathcal{C}_n \cong \begin{cases} \mathbb{Z}_4 & \text{for } n \notin 2\mathbb{Z} \\ \mathbb{Z}_2 \times \mathbb{Z}_2 & \text{for } n \in 2\mathbb{Z} \end{cases}. \quad (19.20)$$

The conformal weights and monodromies are

$$h_s = \tfrac{n}{8}, \quad h_v = \tfrac{1}{2}, \quad R_{vv} = 1, \quad R_{vs} = 1/2, \quad R_{ss} = \begin{cases} 3n/4 & \text{for } n \notin 2\mathbb{Z} \\ n/4 & \text{for } n \in 2\mathbb{Z} \end{cases} \quad (19.21)$$

since $s^2 = v^n$ so that $N_s = 4$ for n odd and $N_s = 2$ for n even.

For the internal $N = 2$ SCFT $\mathcal{C}_{int}$ the center always contains the supercurrent J_v with $h = 3/2$ and $J_v^2 = \mathbb{1}$ and the spectral flow current J_s with $h = c/24$ and $J_s^{2M} = J_v^k$ where $c = 3k/M$ and $1/M$ is the charge quantum in the NS sector (see appendix B).[39] The monodromy charge Q_v is 0 in the NS sector and $1/2$ in the Ramond sector. $J_s = e^{i\sqrt{c/12}X}$ is the Ramond ground state of maximal $U(1)$ charge $c/6$ and can be written as a vertex operator in terms of the bosonized $U(1)$ current $J(z) = \sqrt{c/3}\partial X(z)$ so that $Q_{J_s} \equiv -\frac{1}{2}Q$ and $Q_{J_s}(J_s) \equiv -c/12$ modulo 1.

19.3.1. *The (2,2) case and the generalized GSO projection*

In order to apply simple current techniques it is convenient to start from a left-right symmetric theory. This can be achived by applying the bosonic string map to the right-movers,[22]

$$SO(2)_{LC} \to D_5 \times E_8, \qquad (0, v) \to (v, 0), \qquad (s, \bar{s}) \to -(\bar{s}, s), \qquad (19.22)$$

which maps modular invariant partition functions of heterotic strings to modular invariant partition functions of bosonic strings. The inverse map will be called Gepner map. For simplicity we discuss the spectrum in terms of light-cone space-time $SO(2)_{LC}$ representations rather than using the equivalent $SO(4) \otimes (b, c, \beta, \gamma)$, which would necessitate superghosts contributions with the benefit of manifest Lorentz invariance.

Consistent quantization of the gauge fixed N=1 supergravity theory requires that the Ramond and NS sectors of the space-time and internal sectors are aligned. After the bosonic string map this implies that $SO(10)$ spinor representations are aligned with the Ramond sector of the internal SCFT. This can be implemented by a SCMI that extends the chiral algebra with the current $J_{RNS} = J_v \otimes v$, which has conformal weight 2 because $Q_{J_v} \equiv 1/2$ for Ramond fields and $Q_v \equiv 1/2$ for $SO(10)$ spinors. Similarly, in the case of a Gepner model, where $\mathcal{C}_{int} = \mathcal{C}_{k_1} \otimes \ldots \otimes \mathcal{C}_{k_l}$ is a tensor product of $N = 2$ SCFTs, the alignment can be implemented as a SCMI extending the chiral algebra by all bilinears of the respective supercurrents $J_{ij} = J_{v_i}J_{v_j}$ with $h_{ij} = 3$. Rather then defining a "superconformal tensor product" with an implicit alignment we keep the alignment procedure explicit because we will later be interested in (0,2) models for which the chiral algebra extension that implements the alignment only takes place in the right-moving sector, where it is needed for consistency.

Space-time supersymmetry now requires that the spectral flow in the internal sector is combined with an $SO(10)$ spin field s after the bosonic string map so that space-time bosons/fermions in the heterotic string have

NS/R contributions from the internal N=2 SCFT.[8] This is implemented by the simple current $J_{GSO} = J_s \otimes s$, which has integral conformal weight $h_{GSO} = c/24 + n/8 = 3/8 + 5/8 = 1$ and hence can be used for a SCMI of extension type. Inspection of the massless spectrum (see below) shows that the 2×16 states in $(J_{GSO})^{\pm 1}$ together with the $U(1)$ current of the $N = 2$ SCFT lead to the 33 massless vector bosons that extend the 45_{adj} of D_5 to the 78_{adj} of the gauge group E_6 that is familiar from the standard embedding $SU(3) \subseteq E_8$. The mechanism that implements space-time SUSY in the fermionic string is hence related by the bosonic string map to the mechanism that extends $E_8 \times SO(10)$ to the gauge group $E_8 \times E_6$ of a $(2,2)$ compactification. Since $Q_{GSO} = -\frac{1}{2}Q$ this "generalized GSO projection" implies a projection to even $U(1)$ charges in the bosonic string and, according to eq. (19.22), to odd $U(1)$ charges in the Gepner construction of the superstring,[1] where the space-time contribution is taken into account.

For sigma models on CY manifolds the charges are already quantized in (half)integral units in the (R)NS sector. The standard GSO projection can hence be regarded as a generalized GSO projection with $M = 1$. In order to simplify the comparison between abstract and geometrical constructions of $N = 2$ SCFTs it has been suggested[48] to define an intermediate projection which extends the chiral algebra only by simple currents that have no contributions from the spacetime/gauge sector. The corresponding subgroup $\mathcal{G}_{CY}$ of the center contains all alignment currents of the building blocks of the internal SCFT plus the current $J_{CY} = J_{GSO}^2 J_{RNS}^{c/3} = J_s^2 J_v^{c/3}$.[1]

In order to set up the enumeration of massless states of the heterotic string we recall the relevant vertex operators. On the bosonic side, where the NS vacuum has $h = -1$, there are the universal operators

$$\left(\partial X^\mu \times \mathbb{1}_{D_5 \times E_8} + \mathbb{1}_{st} \times J_{-1}^{(D_5 \times E_8)} \right) \times \mathbb{1}_{int} \qquad (19.23)$$

and the model-dependent contributions

$$\mathbb{1}_{st} \times \mathbb{1}_{D_5 \times E_8} \times \sum_{\substack{0 \le r < 4 \\ h_{int} = 1 - h_{D_5}(s^r)}} (s)^r \times \Phi_{int} . \qquad (19.24)$$

For the right-movers the NS vacuum has $h = -1/2$ and the relevant vertex operators are

$$\sum_{\substack{0 \le r < 4 \\ \bar{h}_{int} = 1/2 - h_{D_1}(s^r)}} (\bar{s})_{st}^r \times \bar{\Phi}_{int} . \qquad (19.25)$$

---

[1] The discussion in Ref. 48 attemts independence of the space-time dimension $2n = 10 - 2c/3$. Note, however, that standard compactifications on K3's have internal $N = 4$ SCFTs so that the bosonic analog of $\mathcal{N}=2$ space-time SUSY in 6-dimensional (4,4) models is the extension of the gauge group $E_8 \times D_6$ to $E_8 \times E_7$, where the $3 = 133 - 66 - 2 \cdot 32$ D_6-singlet gauge bosons come from the $SU(2)$ R-symmetry currents of the $N = 4$ SCFT.

The enumeration of the non-universal states can therefore be organized according to the following data,

$D_5^{(B)}$	h_{int}	Q_{int}
$0 = \mathbb{1}$	1	$\pm 2, 0$
$s = \underline{16}$	$\frac{3}{8}$	$\frac{3}{2}, -\frac{1}{2}$
$v = \underline{10}$	$\frac{1}{2}$	± 1
$\bar{s} = \underline{\overline{16}}$	$\frac{3}{8}$	$\frac{1}{2}, -\frac{3}{2}$

$$
\begin{array}{ccccc}
 & & 1 & & \\
 & y & & x & \\
y & & a & & x \\
1 & g & & g & 1 \\
 & x & & a & \\
x & & y & & \\
Q & & 1 & & \bar{Q}
\end{array}
$$

$D_5 \to D_1^{(F)}$	$\bar{h}_{int}$	$\bar{Q}_{int}$
$0 \to \Psi^\mu = v$	0	0
$s \to \underline{\bar{\Sigma}} = \bar{s}$	$\frac{3}{8}$	$\frac{3}{2}, -\frac{1}{2}$
$v \to \mathbb{1}_{st} = 0$	$\frac{1}{2}$	± 1
$\bar{s} \to \underline{\Sigma} = s$	$\frac{3}{8}$	$\frac{1}{2}, -\frac{3}{2}$

where the entries of the "Hodge diamond" are multiplicities of internal fields with (left,right) charges $(Q, \bar{Q})$.

Since spectral flow relates (anti)chiral primary states with Ramond ground states the counting can be performed in any of these sectors, with an appropriate shift of charges. For CY compactifictions Hodge duality further implies $x = y$ where $y = 1$ corresponds to extended $\mathcal{N} = 2$ space-time SUSY and $y = 3$ yields $\mathcal{N} = 4$. The bosonic (left-moving) analogs of these extensions are gauge groups E_7 and E_8, respectively, where $x \neq y$ is possible for orbifolds with discrete torsion.[37,38,46] The $h_{12} = a$ complex structure deformations (we call them *anti-generations* of charged particles) correspond to chiral primary fields with symmetric charges $Q = \bar{Q} = 1$, while the $h_{11} = g$ *generations* count Kähler moduli, so that effectively the CY Hodge diamond is rotated by $\partial/2$ as compared to the diamond of left/right charge multiplicities of the $N = 2$ SCFT.

19.3.2. *The extended Poincaré polynomial*

The aim of the extended Poincaré polynomial (EPP) is to encode all information about an $N = 2$ superconformal theory that is necessary to compute the massless spectrum of any tensor product with $c = 9$ containing this model as one factor. It takes advantage of the fact that the generalized GSO-projection corresponds to an extension invariant so that we may, in a first step, disregard the projection to integral charge in the expression (19.18) and consider the 'unprojected orbifold'. Eventually, to obtain the projected orbifold, we just need to omit the contributions with non-integral monodromy charges.

For $N = 2$ SCFTs the Poincaré polynomial encodes charge degeneracies,

$$
P(t, \bar{t}) = \text{tr}_{(c,c)}\, t^Q\, \bar{t}^{\bar{Q}} = (t\bar{t})^{c/6}\, \text{tr}_{R_{gs}}\, t^Q\, \bar{t}^{\bar{Q}}, \tag{19.26}
$$

where we assume locality of spectral flow. In order to be able to combine the information of the factors of a tensor product we need to encode information on the twists. We thus define the 'full extended' Poincaré polynomial as

$$\mathcal{P}(t,\bar{t},x,\sigma) = \sum_{l \geq 0} \sum_{k=0}^{1} x^l \sigma^k P_{l,k}(t,\bar{t}), \qquad (19.27)$$

where $P_{l,k}(t,\bar{t})$ is the Poincaré polynomial of the unprojected sector twisted by $J_s^{2l} J_v^k$. Hence, $P_{l,k}$ is obtained by looking for all Ramond ground states Φ_{ij} with $j = J_s^{2l} J_v^k i$ and the $U(1)$ charges of i and j are encoded by the exponents of t and $\bar{t}$, respectively.

For a tensor product with alignment of Ramond/NS sectors we obtain

$$\mathcal{P}(t,\bar{t},x,\sigma) = \sum_{l \geq 0} x^l \left(\sum_{k=0}^{1} P_{l,k}^{(1)}(t,\bar{t}) P_{l,k}^{(2)}(t,\bar{t}) + \sigma \sum_{k=0}^{1} P_{l,k}^{(1)}(t,\bar{t}) P_{l,1-k}^{(2)}(t,\bar{t}) \right).$$

By iteration (19.27) thus indeed encodes all information from the factor theories of a Gepner model that enters the computation of the charged massless spectrum. In fact, this information is still redundant: Consider a R ground states Φ_{ij} whose contribution to $P_{l,k}$ is $t^{Q+\frac{c}{6}} \bar{t}^{\bar{Q}+\frac{c}{6}}$. Then eqs. (19.13) and (B.6) imply for the $U(1)$ charges

$$\bar{Q} = Q + l\,c/3 - k \quad \Rightarrow \quad k(19.13) + l\,c/3 - \bar{Q} \quad \text{mod } 2. \qquad (19.28)$$

As the exponent of σ is fixed in terms of the other exponents we set

$$\sigma \to -1 \quad \Rightarrow \quad \mathcal{P}(t,\bar{t},x) := \mathcal{P}(t,\bar{t},x,-1). \qquad (19.29)$$

The negative sign is convenient for index computations since it implies opposite signs for contributions to generations and anti-generations.[m] For minimal models at level $k = K - 2$ one finds[39]

$$\mathcal{P}^{(MM)}(x;t^K,\bar{t}^K) = \sum_{l=1}^{K-1} (t\bar{t})^{l-1} \frac{1-(-x)^l \ \bar{t}^{K-2l}}{1-(-x)^K} = \frac{P(t\bar{t}) - \sum_{l=1}^{K-1}(-x)^l t^{l-1}\bar{t}^{K-1-l}}{1-(-x)^K}$$

$$(19.30)$$

where the ordinary Poincaré polynomial is $P(t^K) = \frac{1-t^{K-1}}{1-t}$.

Since the numbers of (anti)chiral primaries and of Ramond ground states are finite also in non-rational SCFTs extended Poincaré polynomials can be defined in a more general context and explicit formulas have been given for Landau-Ginzburg orbifolds.[39]

---

[m] In the original definition of the extended Poincaré polynomial[49] Schellekens, in addition, puts $\bar{t} = 1$. For diagonal theories we have shown[39] that, for a given Q, all states contribute with the same sign, so that it is indeed sufficient to drop the $\bar{Q}$-dependence in applications to heterotic $(2,2)$ string vacua built from diagonal theories.

19.3.3. *Gauge/SUSY breaking and (0,2) models*

While the chiral algebra extension of a SCMI based on J_{GSO} and alignment currents can be reduced by switching on discrete torsion $X \neq X^T$ this would not only break the left-moving E_6 but also the right-moving space-time SUSY of the heterotic string. We hence need to increase the twist group $\mathcal{G}$ at least by one additional generator of even order. While there are many possibilities for this type of models we would always end up with at least $SO(10)$. For smaller gauge groups, like the "exceptional" series $E_5 = D_5 = SO(10)$, $E_4 = A_4 = SU(5)$ and $E_3 = SU(3) \times SU(2)$ that is familiar from geometric/sigma model constructions, we have to start with smaller building blocks and use asymmetric extensions that rebuild the $D_5 \times E_8$ needed for the Gepner map only in the right-moving sector.

A natural implementation of this idea can be motivated by the free fermion construction of $D_n = SO(2n)$ in terms of $2n$ Majorana fermions with aligned spin structures. The extension of $SO(2m) \otimes SO(2n)$ to $SO(2m + 2n)$ is achived by aligment of all spin structures and can be implemented by a SCMI of extension type with the current $J = v_{D_m} \otimes v_{D_n}$, in complete analogy to the alignment of spin structures for a tensor product of SCFTs. The exceptional series is thus obtained by starting with a gauge sector $SO(2l) \otimes SO(2)^{5-l} \otimes E_8$ and a generalized GSO projection[29]

$$J_{GSO} = J_s \otimes s_{SO(2l)} \otimes (s_{SO(2)})^{5-l} \tag{19.31}$$

as is illustrated in the table:

l	E_{l+1}	$D_l \times D_1^{5-l}$	$|E_{l+1}| - |D_l| - |U(1)|$	currens $(J_{GSO})^{\pm 1}$	
5	E_6	SO_{10}	$32 = 78 - 45 - 1$	$|s| = 16$	$h = \frac{5}{8} + \frac{3}{8}$
4	$E_5 = SO_{10}$	$SO_8 \times SO_2$	$16 = 45 - 28 - 1$	$|s| = 8$	$h = \frac{4}{8} + \frac{1+3}{8}$
3	$E_4 = SU_5$	$SO_6 \times (SO_2)^2$	$8 = 24 - 15 - 1$	$|s| = 4$	$h = \frac{3}{8} + \frac{2 \times 1 + 3}{8}$
2	$SU_3 \times SU_2$	$SO_4 \times (SO_2)^3$	$4 = 11 - 6 - 1$	$|s| = 2$	$h = \frac{2}{8} + \frac{3 \times 1 + 3}{8}$

For the rest of this paper we restrict to the case $l = 4$, i.e. to $SO(10)$ models based on a CFT of the form $\mathcal{C}_{int} \times SO(8) \times SO(2) \times E_8$ with $c = 26 - 4$.

Blumenhagen and A. Wißkirchen[29] performed a computer search for spectra of heterotic models of this type that agree with Distler-Kachru models and came up with a small list, the most promising candidate of which is an $SO(10)$ model with 80 generations. They used the original approach of Schellekens and Yankielowicz constructing SCMIs as products of invariants for cyclic subgroups of the center.[22] Translating their data into our language we find, in addition to J_{GSO} and the alignment currents, a $\mathbb{Z}_4$ twist whose simple current generator $J_B = (J_s^{k=3})^5 \times s_{SO(2)}$ is the product of the spinor of $SO(2)$ times the 5th power of the spectral flow of one of the minimal model factors of the quintic.

We call J_B, which squares to the alignment current $J_B^2 = J_v^{k=3} \otimes v_{SO(2)}$, *Bonn twist*. Since only one minimal model enters this construction it appears natural to generalize the discussion to an internal SCFT of the form[32] $\mathcal{C}_{int} = \mathcal{C}' \otimes \mathcal{F}_K$, where $\mathcal{F}_K$ is a minimal model whose level $k = K - 2$ needs to be odd in order that $J_s^{2K} = J_v$. In the Landau-Ginzburg discription $\mathcal{F}_K$ has a Fermat-type potential $W = \Phi^K$ and is hence referred to as *Fermat factor*. The Bonn twist thus generalizes to

$$J_B = (J_s^{\mathcal{F}})^K \times s_{SO(2)}, \qquad N_B = 4, \qquad J_B^2 = J_v^{\mathcal{F}} \otimes v_{SO(2)} \qquad (19.32)$$

so that the resulting $(0,2)$ model can be defined by a SCMI based on the generators J_B, J_{GSO} and two more alignment currents

$$J_A = v_{SO(8)} \otimes v_{SO(2)}, \qquad J_C = J_v^{\mathcal{C}'} \otimes v_{SO(8)}. \qquad (19.33)$$

The nonvanishing monodromies are $R_{BB} \equiv \frac{K-1}{2} \bmod 2$, $R_{AB} \equiv \frac{1}{2} \bmod 1$ and $R_{B,GSO} \equiv \frac{K-1}{4} \bmod 1$. We need J_{GSO} and the alignment currents J_A, J_B^2 and J_C in the chiral algebra on the right-moving side, i.e. in the kernel of X, so that the corresponding rows of the matrix X must be 0 (in the case of J_B modulo $1/2$). This fixes all discrete torsions $X - X^T$ and implies

R	J_{GSO}	J_A	J_B	J_C
J_{GSO}	0	0	$\frac{K-1}{4}$	0
J_A	0	0	$\frac{1}{2}$	0
J_B	$\frac{K-1}{4}$	$\frac{1}{2}$	$\frac{K-1}{2}$	0
J_C	0	0	0	0

X	J_{GSO}	J_A	J_B	J_C
J_{GSO}	0	0	$\frac{K-1}{4}$	0
J_A	0	0	$\frac{1}{2}$	0
J_B	0	0	$\frac{K-1}{4}$	0
J_C	0	0	0	0

In the present extension of $SO(8) \times U(1)$ to $SO(10)$ the massless matter representations are assembled by the orbits of J_{GSO} in the following way,[29]

$$16 = 8_{-1}^{\bar{s}} + 8_1^v, \qquad \overline{16} = 8_{-1}^v + 8_1^{\bar{s}}, \qquad 10 = 1_{-2} + 8_0^s + 1_2, \qquad (19.34)$$

where the subscripts denote the $U(1)$ charges $Q \equiv -2Q_{GSO} \bmod 2$. For a field $\Phi_{a,Ja}$ that is twisted by

$$J = J_{GSO}^{2n} J_A^\alpha J_B^{2\beta - \rho} J_C^\gamma, \qquad \alpha, \beta, \gamma, \rho = 0, 1 \qquad (19.35)$$

this leads to the following charge projections for the monodromy charges

$$Q_{GSO} \equiv -\tfrac{1}{2} Q_{U(1)} \equiv 0, \qquad Q_A \equiv \tfrac{1}{2}\rho, \qquad Q_B \equiv \tfrac{K-1}{4}\rho, \qquad Q_C \equiv 0 \qquad (19.36)$$

modulo 1. Equivalently, $\bar{Q}_{GSO} \equiv \bar{Q}_A \equiv \bar{Q}_C \equiv 0$ and $\bar{Q}_B \equiv \frac{1}{2}\alpha + \frac{K-1}{4}\rho$.

The massless matter representations (chiral superfields) as well as possible gauge group extensions (vector superfields) can now be enumerated straightforwardly. Space-time quantum numbers come from representations of the right-moving chiral algebra while the gauge group representations follow from left-moving CFT quantum numbers. The correspondences have been worked out for $E_5 = SO(10)$, $E_4 = SU(5)$ and $E_3 = SU(3) \times SU(2)$ by Blumenhagen and Wisskirchen[29] (tables in section 6). Only gauge-singlet representations can depend on non-topological information, i.e. uncharged fields with $r = 0$ and $h_{int} = 1$ in eq. (19.24). All charged matter fields and non-abelian gauge group extensions can hence be determined in terms of the data encoded in the extended Poincaré polynomial of $\mathcal{C}'$. Our construction can thus be used for all Landau-Ginzburg orbifolds based on $N = 2$ SCFTs of the form $\mathcal{C}' \otimes \mathcal{F}$ with a Fermat factor $\mathcal{F} \sim \Phi^K$ with $K \in 2\mathbb{Z} + 1$.

19.4. Geometry and vector bundle data

Witten's gauged linear sigma model[4] made it possible to construct a large class of $(0,2)$ string vacua.[15] The starting point is a supersymmetric abelian gauge theory that leads in the Calabi-Yau phase to a σ model described by an exact sequence (monad)

$$0 \rightarrow V \rightarrow \bigoplus_{i=1}^{r+1} \mathcal{O}(n_i) \overset{F_i}{\rightarrow} \mathcal{O}(m) \rightarrow 0 \qquad (19.37)$$

defining a bundle V of rank r over a complete intersection Calabi-Yau X. F_i are homogeneous polynomials of degrees $m - n_i$ not vanishing simultaneously on X. For weighted projective ambient spaces we can write this data as

$$V_{n_1 \ldots, n_r+1}[m] \longrightarrow \mathbb{P}_{w_1, \ldots, w_{N+4}}[d_1, \ldots, d_N], \qquad (19.38)$$

where $r = 4, 5$ corresponds to unbroken gauge groups $SO(10)$ and $SU(5)$, respectively. The Calabi-Yau condition $c_1(X) = 0$ and the condition $c_1(V) = 0$, which guarantees the existence of spinors, read

$$\sum d_l - \sum w_j = m - \sum n_i = 0 \qquad (19.39)$$

and the cancellation of gauge anomalies $ch_2(V) = ch_2(TX)$ with $ch_2 = \frac{1}{2}c_1^2 - c_2$ implies the quadratic diophantine constraint

$$\sum d_l^2 - \sum w_j^2 = m^2 - \sum n_i^2. \qquad (19.40)$$

For a Calabi-Yau hypersurface $W = 0$ the choice of $m = d = \sum w_j$ with $n_i = w_i$ solves these equations and $F_i = \partial_i W$ corresponds to the $(2,2)$ case.

The suggested CFT/geometry correspondence[29] assosiates the vector bundle $V_{1,1,1,1,1}[5]$ over $\mathbb{P}_{1,1,1,1,2,2}[4,4]$ to the (0,2) cousin of the Gepner model 3^5. Since the twist J_B that defines the (0,2) model only acts on one of the Fermat factors we expect that this is part of a larger picture, where the Gepner model data directly translate into vector bundle data $V_{n_1,\ldots,n_5}[d]$ with $k_i = d/n_i - 2$. For the base manifold the doubling of the respective weight seems to correspond to the increase of the order of the twist group. We hence make the ansatz

$$V_{n_1,\ldots,n_5}[m] \to \mathbb{P}_{n_1,\ldots,n_4,2n_5,w_6}[d_1,d_2], \qquad (19.41)$$

i.e. $w_i = n_i$ for $i < 5$ and $w_5 = 2n_5$, which has to obey (19.39) and (19.40) or

$$d_1 + d_2 = m + n_5 + w_6, \qquad d_1^2 + d_2^2 = m^2 + 3n_5^2 + w_6^2. \qquad (19.42)$$

It is quite non-trivial and encouraging that this non-linear system has a linear solution $w_6 = (m - n_5)/2 = d_1/2$ and $d_2 = (m + 3n_5)/2$. We hence conjecture a correspondence between the (0,2) models defined in the previous section with the Distler-Kachru models defined by the data[39]

$$V_{n_1,\ldots,n_5}[m] \to \mathbb{P}_{n_1,\ldots,n_4,2n_5,\frac{m-n_5}{2}}[m - n_5, (m + 3n_5)/2]. \qquad (19.43)$$

The increase of the codimension of the Calabi-Yau may be interpreted as providing an additional field of degree $w_6 = d_1/2$ describing the twisted sectors for the $\mathbb{Z}_2$ orbifolding due to J_B.

In the Calabi–Yau phase a toric approach to the resolution of singularities appears to be most natural.[50] For the (2,2) model the Newton polytope Δ of a generic transversal degree m polynomial is reflexive and its polar polytope Δ^* provides a desingularization of the hypersurface in the weighted projective space $\mathbb{P}_{n_1,\ldots,n_5}$.[51] For the complete intersection (19.43) the Batyrev-Borisov construction[52] suggests to consider the Minkowski sum $\Delta = \Delta_1 + \Delta_2$ of the Newton polytopes Δ_l of degree d_l polynomials w.r.t. the weights w_j. If Δ is reflexive then a natural resolution of singularities can again be based on a triangulation of the fan over Δ^*. A useful collection of tools and formulas for further studies of this class of models can be found in a paper by Blumenhagen.[53]

19.5. Conclusion

We discussed the construction of a large class of heterotic (0,2) Gepner-type models in terms of simple current techniques and their generalization to Landau-Ginzburg models based on the topological information encoded in the extended Poincaré polynomial. Already without further orbifolding the 7555 transversal potentials lead to 3219 models, 220 of which are of Fermat type. For a large subclass of the potentials the mirrors of the $(2,2)$ models can be constructed as orbifolds. In this case our analysis provides the ingredients for an orbifold mirror construction also for the $(0,2)$ version, thus explaining the mirror symmetry that has been observed in orbifold spectra.[40,41]

In addition to the phenomenological interest of heterotic models it would be interesting to generalize the proposed identifications with Distler-Kachru models and to test them by comparing spectra in geometrical phases[53] and Yukawa couplings at the Landau-Ginzburg points.[54]

Appendix A. Gepner models, torus orbifolds & mirror symmetry

In accord with the three weighted projective spaces $\mathbb{P}_{111}[3]$, $\mathbb{WP}_{112}[4]$ and $\mathbb{WP}_{123}[6]$ that admit a transversal CY equation of degree $d = 3, 4, 6$, there are three Gepner models with levels $k = (1, 1, 1)$, $k = (2, 2, 0)$ and $k = (4, 1, 0)$, and superpotentials $W = X^3 + Y^3 + Z^3$, $W = X^4 + Y^4 + Z^2$ and $W = X^6 + Y^3 + Z^2$, respectively, that describe 2d tori. While the Kähler modulus is fixed at the Landau-Ginzburg point at a value that is consistent with the $\mathbb{Z}_d$ quantum symmetry originating in the GSO projection, the complex structure deformation corresponds to a deformation of W by λXYZ. At the Gepner point $\lambda = 0$ the complex structure moduli are $\tau = e^{2\partial i/d}$, where $e^{2\partial i/3}$ and $e^{2\partial i/6}$ are related by $\tau \to \tau + 1$.

We focus on $\mathbb{Z}_2 \times \mathbb{Z}_2$ orbifolds, whose abelian extensions were recently classified and compared to free fermion models by Donagi and Wendland.[44] Since we want to realize the $\mathbb{Z}_2$'s as symmetries of Gepner models we consider $\mathbb{WP}_{112}[4]$ and $\mathbb{WP}_{123}[6]$, for which a phase rotation of the first homogeneous coordinate corresponds to a phase rotation by $2\partial/d$ of the flat double-periodic torus coordinate $z \in T^2$ (this can be checked by counting fixed points and orders of stabilizers). The $\mathbb{Z}_2$ orbifold $z \to -z$ hence corresponds to the phase symmetry $\rho = \mathbb{Z}_2 : 1\,0\,0$ in both cases.

With the notation of[44] as subscript and the Hodge numbers as supersript, the four inequivalent orbifolds by a $\mathbb{Z}_2 \times \mathbb{Z}_2$ twist group G_T are $X_{0-1}^{51,3}$, $X_{0-2}^{19,19}$, $X_{0-3}^{11,11}$, and $X_{0-4}^{3,3}$. They differ by the number of shifts $z \to z + \frac{1}{2}$ that are included and we can choose the following generators,[44]

$$X_{0-1}^{51,3} : \begin{array}{l} \theta^{(1)}(z_1, z_2, z_3) = (-z_1, z_2, -z_3) \\ \theta^{(2)}(z_1, z_2, z_3) = (z_1, -z_2, -z_3) \end{array} \tag{A.1}$$

$$X_{0-2}^{19,19} : \begin{array}{l} \theta^{(1)}(z_1, z_2, z_3) = (-z_1, z_2, -z_3) \\ \theta^{(2)}(z_1, z_2, z_3) = (z_1, -z_2, \frac{1}{2}-z_3) \end{array} \tag{A.2}$$

$$X_{0-3}^{11,11} : \begin{array}{l} \theta^{(1)}(z_1, z_2, z_3) = (-z_1, z_2+\frac{1}{2}, -z_3) \\ \theta^{(2)}(z_1, z_2, z_3) = (z_1, -z_2, \frac{1}{2}-z_3) \end{array} \tag{A.3}$$

$$X_{0-4}^{3,3} : \begin{array}{l} \theta^{(1)}(z_1, z_2, z_3) = (z_1+\frac{1}{2}, -z_2, -z_3) \\ \theta^{(2)}(z_1, z_2, z_3) = (-z_1, z_2+\frac{1}{2}, \frac{1}{2}-z_3) \end{array} . \tag{A.4}$$

Only $\mathbb{P}_{112}[4]$ admits a second independent $\mathbb{Z}_2$ action, namely $\sigma = \mathbb{Z}_2 : 1\,0\,1$, which has no fixed points and hence corresponds to a shift $z \to z+\frac{1}{2}$ of order 2. The product $\rho \circ \sigma = \mathbb{Z}_2 : 0\,0\,1$ also has 4 fixed points and corresponds to the rotation $z \to \frac{1}{2} - z$ about $z = \frac{1}{4}$, which is equivalent to ρ. For the realization of X_{0-n} in terms of Gepner models we hence need at least $n-1$ factors of $\mathbb{P}_{112}[4]$. This can be confirmed by computing the Hodge numbers with the program package PALP.[55] In a UNIX shell environment the required input data can be organized as follows,

```
Weight1="6 1 2 3 1 2 3 1 2 3 "
TorusQ1="/Z6: 1 2 3 0 0 0 0 0 /Z6: 0 0 0 1 2 3 0 0 0"
Weight2="12 2 4 6 2 4 6 3 3 6 "
TorusQ2="/Z6: 1 2 3 0 0 0 0 0 0 /Z6: 0 0 0 1 2 3 0 0 0"
Weight3="12 2 4 6 3 3 6 3 3 6 "
TorusQ3="/Z6: 1 2 3 0 0 0 0 0 /Z4: 0 0 0 1 1 2 0 0 0"
Weight4="4 1 1 2 1 1 2 1 1 2 "
TorusQ4="/Z4: 1 1 2 0 0 0 0 0 /Z4: 0 0 0 1 1 2 0 0 0"
X01="$Weight1 $TorusQ1 /Z2: 1 0 0 0 0 0 1 0 0 /Z2: 0 0 0 1 0 0 1 0 0"
X02="$Weight2 $TorusQ2 /Z2: 1 0 0 0 0 0 1 0 0 /Z2: 0 0 0 1 0 0 0 0 1"
X03="$Weight3 $TorusQ3 /Z2: 1 0 0 1 0 1 1 0 0 /Z2: 0 0 0 1 0 0 0 0 1"
X04="$Weight4 $TorusQ4 /Z2: 1 0 1 1 0 0 1 0 0 /Z2: 1 0 0 1 0 1 0 0 1"
echo -e "$X01 \n$X02 \n$X03 \n$X04" | poly.x -lf
```

where "Weight*" includes a sufficient number of $\mathbb{P}_{112}[4]$ factors for the shift symmetries, "TorusQ*" provides two GSO projections for torus factors (the overall GSO is automatic) and "X0*" completes the input line for the respective $\mathbb{Z}_2 \times \mathbb{Z}_2$ orbifold $X_{0-1}, \ldots, X_{0-4}$. The last line pipes the input into the executable poly.x contained in PALP,[55] with flags "-l" and "-f" for "Landau-Ginzburg" and "filter" (i.e. read input from pipe), respectively.

The mirror models can now be constructed using the Green-Plesser orbifold construction. In[44] it was observed that discrete torsions often provide the mirrors. This is special to $\mathbb{Z}_2$-torsions, however, for which a discrete torsion between two phase symmetries of even order of the LG superpotential can be switched on/off by redefinition of the action on massive fields Z^2, as has been discussed in detail in.[46] For general orders of the generators, the mirror models of orbifolds with discrete torsion again have discrete torsion[46] and we do not know of any indications that mirror symmetry and discrete torsion are related for $\mathbb{Z}_n$ twists with $n \neq 2$.[37,38]

In the classification of extensions $G_S \to G \to G_T$ of the twist group,[44] G_S is the subgroup of shifts. Only $\mathbb{P}_{112}[4]$ admits a symmetry that corresponds to a second independent shift σ' of order 2, which however cannot be diagonalized simultaneously with σ. It exchanges X and Y and reverses the sign of Z. The mirror construction in this case proceeds by first taking the Green-Plesser mirror for the diagonal subgroup and then performing the mirror moddings of the remaining twists on the mirror CFT, which may involve quantum symmetries. It would be interesting to use examples from[44] with non-trivial fundamental groups to further test the conjecture that mirror symmetry exchanges torsion in $H^2(X, \mathbb{Z})$ with torsion in $H^3(X, \mathbb{Z})$.[56]

Appendix B. N=2 SCFT, simple currents & minimal models

The $N = 2$ superconformal algebra[57] is generated by the Fourier modes of $T(z)$, of its fermionic superpartners $G^\pm$, and of a $U(1)$ current $J(z)$

$$\{G_r^-, G_s^+\} = 2L_{r+s} - (r - s)J_{r+s} + \tfrac{c}{3}(r^2 - \tfrac{1}{4})\delta_{r+s}, \tag{B.1}$$

$$[L_n, G_r^\pm] = (\tfrac{n}{2} - r)G_{n+r}^\pm, \qquad [J_n, G_r^\pm] = \pm G_{n+r}^\pm, \tag{B.2}$$

$$[L_n, J_m] = -mJ_{m+n}, \qquad [J_m, J_n] = \tfrac{c}{3}m\delta_{m+n}, \tag{B.3}$$

where $r, s \in \mathbb{Z} + \tfrac{1}{2}$ in the NS sector. According to (B.1) the Ramond gound states $G_0|\alpha\rangle_R = 0$ have $h_\alpha = c/24$. The analogous unitarity bound in the NS sector is saturated by the chiral primary fields[57] $G_{-\frac{1}{2}}^+|\Phi\rangle = 0$, which obey $\{G_{\frac{1}{2}}^-, G_{-\frac{1}{2}}^+\}|\Phi\rangle = (2L_0 - J_0)|\Phi\rangle = 0$ and hence $h = Q/2$. Their conjugate anti-chiral states saturate the BPS bound $h = -Q/2$.

The N=2 algebra admits the continous spectral flow

$$L_n \xrightarrow{\mathcal{U}_\theta} L_n + \theta J_n + \tfrac{c}{6}\theta^2\delta_n, \qquad J_n \xrightarrow{\mathcal{U}_\theta} J_n + \tfrac{c}{3}\theta\delta_n, \qquad G_r \xrightarrow{\mathcal{U}_\theta} G_{r\pm\theta}^\pm \tag{B.4}$$

which for $\theta = \pm\tfrac{1}{2}$ maps Ramond ground states into chiral and antichiral primary fields, respectively. Spectral flow is best understood by bosonization of the $U(1)$ current $J(z) = i\sqrt{c/3}\,\partial X(z)$ in terms of a free field X. A charged operator $\mathcal{O}_q$ can thus be written as a normal ordered product of a vertex operator with a neutral operator $\mathcal{O}_0$,

$$\mathcal{O}_q = e^{i\sqrt{3/c}\,qX}\,\mathcal{O}_0(\partial X, \ldots, \psi, \ldots). \tag{B.5}$$

The contribution of the vertex operator to h is $\tfrac{3q^2}{2c}$ so that in unitary theories the maximal charges of Ramond ground states and chiral primary states are $c/6$ and $c/3$, respectively. In particular, the Ramond ground state

$J_s = e^{i\sqrt{c/12}\,X}$ with maximal charge $c/6$ is a simple current. A short calculation shows that its monodromy charge is $Q_s = -\frac{1}{2}Q$. If the $U(1)$ charges Q are quantized in units of $1/M$ in the NS sector then $c = 3k/M$ for some integer k. Since the $U(1)$ charges are shifted by $-c/6 = -k/2M$ in the Ramond sector the order N_s of J_s is $2M$ if $k \in 2\mathbb{Z}$ and $4M$ if $k \notin 2\mathbb{Z}$.

Already for $N = 1$ SCFTs the supercurrent G is a universal simple current, which we denote by $J_v = G$. Its monodromy charge is $Q_v = 0$ for NS fields and $Q_v = 1/2$ for Ramond fields since $h_v = 3/2$ and the conformal weights of superpartners differ by integers in the Ramond sector and by half-integers for NS states. Putting the pieces together we find the matrix of monodromies

$$R_{v,v} = 0, \qquad R_{v,s} = 1/2, \qquad R_{s,s} = n - c/12 \quad \text{with } n = \begin{cases} 0 & k \in 4\mathbb{Z} \\ 1 & k \notin 4\mathbb{Z} \end{cases} \quad \text{(B.6)}$$

where we used $h_s = c/24$ and $Q_s(J_s) = -c/12$. Note that $J_s^{2M} = J_v^k$ (since the monodromy charges agree) so that the center is $\mathbb{Z}_{2M} \times \mathbb{Z}_2$ for $k \in 2\mathbb{Z}$ and $\mathbb{Z}_{4M}$ for $k \notin 2\mathbb{Z}$.

B.1. $N = 2$ minimal models

Minimal models have a number of different realizations. Here we use the coset construction for the $N = 2$ superconformal series $\mathcal{C}_k$

$$(SU(2)_k \times U(1)_4)/U(1)_{2K}, \qquad c = 3k/K \quad \text{with } K = k + 2 \quad \text{(B.7)}$$

as a quotient of $SU(2)$ level k for $k \in \mathbb{N}$ times $U(1)_4 \cong SO(2)_1$ by $U(1)_{2K}$. Primary fields Φ_m^{ls} are labelled accordingly by $0 \le l \le k$, $s \bmod 4$ and $m \bmod 2K$ with the branching rule $l + m + s \in 2\mathbb{Z}$. The fusion rules are

$$\Phi_{m_1}^{l_1 s_1} \times \Phi_{m_2}^{l_2 s_2} = \sum_{l=|l_1-l_2|}^{\min(l_1+l_2,k)-|k-l_1-l_2|} \Phi_{m_1+m_2}^{l,s_1+s_2} \quad \text{(B.8)}$$

so that Φ_m^{0s} and $\Phi_{m+K}^{k,s+2}$ are simple currents. The conformal weights and the $U(1)$ charges obey

$$h \equiv \frac{l(l+2)-m^2}{4K} + \frac{s^2}{8} \bmod 1, \quad Q \equiv \frac{s}{2} - \frac{m}{K} \bmod 2 \quad \text{exact for } \begin{Bmatrix} |m-s| \le l \\ -1 \le s \le 1 \end{Bmatrix} \quad \text{(B.9)}$$

and the NS and Ramond sectors correspond to even and odd s, respectively. The formulas (B.9) are exact in the standard range $|m - s| \le l$, $-1 \le s \le 1$ and otherwise sufficient to determine the monodromy charges of simple currents. In particular, the selection rule $l + m + s \in 2\mathbb{Z}$ is implemented by integrality of the monodromy charge Q_K^{k2} of the simple current Φ_K^{k2}, which has integral conformal weight. According to the rules for modular invariance the branching rule thus necessitates the field identification

$$\Phi_m^{ls} \sim \Phi_{m+K}^{k-l,s+2} \quad \text{with } J_{id} = \Phi_K^{k2}, \quad Q_{id} \equiv (l + m + s)/2 \quad \text{(B.10)}$$

due to an extension of the chiral algebra by the "identification current" J_{id}. The center of the minimal model at level k is hence of order $4K$ and generated by the spectral flow current $J_s := \Phi_1^{01} \sim \Phi_{1-K}^{k3}$ and the supercurrent $J_v := \Phi_0^{02} \sim \Phi_K^{k0}$ with $J_s^{2K} = J_v^k$; more generally all above formulas for $N = 2$ SCFTs apply with $M = K$. Ramond ground states and (anti)chiral primary fields are now easily identified as follows,

anti-chiral primary	Ramond ground states	chiral primary
$\Phi_l^{l0} \sim \Phi_{K+l}^{k-l,2} \rightarrow \|l\rangle_a$	$\Phi_{\pm(l+1)}^{l,\pm1} \sim \Phi_{\mp(k-l+1)}^{k-l,\mp1} \rightarrow \|l_\pm\rangle_R$	$\Phi_{-l}^{l0} \sim \Phi_{K-l}^{k-l,2} \rightarrow \|l\rangle_c$
$Q = -\frac{l}{K}, \quad h = -\frac{Q}{2}$	$Q = \pm(\frac{c}{6} - \frac{l}{K}), \quad h = \frac{c}{24}$	$Q = \frac{l}{K}, \quad h = \frac{Q}{2}$

The Landau-Ginzburg description of the minimal model with the diagonal modular invariant has superpotential $W = X^K$ with $X \sim \Phi_l^{l0}$.

In order to determine the conformal weights and multiplicities of all fields relevant for massless string spectra we follow the discussion in ref.[22] and first note that the supercurrent J_v acts as $J_v\Phi_m^{ls} = \Phi_m^{l,s+2} \sim \Phi_{m\pm K}^{k-l,s}$. Choosing m such that $-K < m \leq K$ we find that $m \rightarrow m - K$ for $m > 0$ and $m \rightarrow m + K$ for $m \leq 0$. It is then straightforward to check that $l + 1 - |m| \rightarrow -(l + 1 - |m|)$, i.e. the fields inside the cone $|m| \leq l + 1$ are mapped to the outside and vice versa.

In the NS sector we choose $s = 0$. Then (B.9) gives the correct value of h inside the cone, i.e. for $|m| \leq l$. The conformal weight of the respective superpartners is $h + \frac{1}{2}$; their multiplicity is 2 unless $G_{-1/2}^+$ or $G_{-1/2}^-$ vanishes. This happens for $|m| = l$ for which the multiplicity of the superpartner is 1 for $l > 0$. For $l = m = 0$, i.e. the superpartner J_v of the identity, the lowest states occur at $h = 3/2$ with multiplicity 2.

In the R sector highest weight states are annihilated by G_0^+ or G_0^-. They thus come in pairs $\Phi_m^{l,\pm1}$ that are related by the action of $G_0^\pm$. Usually we can fulfill $|m| < l$ by field identification, in which case h is degenerate and given correctly by (B.9). The only exception is $|m| = l + 1$ where $G_0^+ = G_0^- = 0$. In that case one has to make a choice of chirality: The Ramond ground states have $h = c/24$ in accordance with (B.9), and their superpartners have $h = 1 + c/24$. The choice $m = l + 1$ and $s = 1$ leads to the standard range given in (B.9). The only descendent that plays a role for the massless spectrum of strings is the descendent $J_{-1}|0\rangle$ of the vacuum.

Acknowledgments

I would like to thank Ron Donagi and Emanuel Scheidegger for helpful discussions. This work is supported in part by the *Austrian Research Funds FWF* under grant Nr. P18679.

References

1. D. Gepner, *Space-time supersymmetry in compactified string theory and superconformal models*, Nucl. Phys. **B296** (1988) 757; *N = 2 string theory*, in: Proceedings of the Trieste Spring School Strings 1989, eds. M.Green et al. (World Scientific, Singapore, 1990).
2. P. Candelas, M. Lynker and R. Schimmrigk, *Calabi-Yau manifolds in weighted* $\mathbb{P}_4$, Nucl. Phys. **B341** (1990) 383.
3. C. Vafa, *String vacua and orbifoldized LG models*, Mod. Phys. Lett. **A4** (1989) 1169; *Superstring vacua*, HUTP-89/A057 preprint.
4. E. Witten, *Phases of N=2 theories in two dimensions*, Nucl. Phys. **B403** (1993) 159 [arXiv:hep-th/9301042].
5. P. Candelas, X.C. De la Ossa, P.S. Green, L. Parkes, *An exactly soluble superconformal theory from a mirror pair of Calabi-Yau manifolds*, Phys. Lett. **B258** (1991) 118.
6. M. Herbst, K. Hori, D. Page, *Phases of N=2 theories in 1+1 dimensions with boundary*, arXiv:0803.2045 [hep-th].
7. C.M. Hull, E. Witten, *supersymmetric sigma models and the heterotic string*, Phys. Lett. **B160** (1985) 398.
8. T. Banks, L.J. Dixon, D. Friedan, E. Martinec, *Phenomenology and conformal field theory or can string theory predict the weak mixing angle?*, Nucl. Phys. **B299** (1988) 613.
9. X.G. Wen, E. Witten, *World sheet instantons and the Peccei-Quinn symmetry*, Phys. Lett. **B166** (1986) 397.
10. M. Dine, N. Seiberg, X.G. Wen, E. Witten, *Nonperturbative effects on the string worldsheet I and II*, Nucl. Phys. **B278** (1986) 769, **B289** (1987) 319.
11. J. Distler, B. Greene, *Aspects of (2,0) string compactifications*, Nucl. Phys. **B304** (1988) 1.
12. J. Distler, *Resurrecting (2,0) compactifications*, Phys. Lett. **B188B** (1987) 431.
13. M. Cvetič, *Exact construction of (0,2) Calabi-Yau manifolds*, Phys. Rev. Lett. **59** (1987) 2829.
14. B.R. Greene, *Superconformal compactifications in weighted projective space*, Commun. Math. Phys. **130** (1990) 335.
15. J. Distler, S. Kachru, *(0,2) Landau-Ginzburg theory*, Nucl. Phys. **B413** (1994) 213 [arXiv:hep-th/9309110].
16. S. Kachru, *Some three generation (0,2) Calabi-Yau models*, Phys. Lett. B **349** (1995) 76 [arXiv:hep-th/9501131].
17. E. Silverstein, E. Witten, *Criteria for conformal invariance of (0,2) models*, Nucl. Phys. **B444** (1995) 161 [arXiv:hep-th/9503212].
18. V. Braun, M. Kreuzer, B. A. Ovrut and E. Scheidegger, *Worldsheet Instantons, Torsion Curves, and Non-Perturbative Superpotentials*, Phys. Lett. B **649** (2007) 334 [arXiv:hep-th/0703134]; *Worldsheet instantons and torsion curves. Part A: Direct computation*, JHEP **0710** (2007) 022 [arXiv:hep-th/0703182]; *Part B: Mirror Symmetry*, JHEP **0710** (2007) 023 [arXiv:0704.0449].

19. C. Beasley, E. Witten, *Residues and world-sheet instantons,* JHEP 10 (2003) 065, [arXiv:hep-th/0304115].

20. R. Donagi, A. Lukas, B.A. Ovrut, D. Waldram, *Holomorphic vector bundles and non-perturbative vacua in M-theory* JHEP 06 (1999) 034 [arXiv:hep-th/9901009].

21. K. Becker, M. Becker, J.X. Fu, L.S. Tseng, S.T. Yau, *Anomaly cancellation and smooth non-Kaehler solutions in heterotic string theory,* Nucl. Phys. B **751** (2006) 108 [arXiv:hep-th/0604137].

22. A.N. Schellekens, S. Yankielowicz, *New Modular Invariants for N = 2 Tensor Products and Four-Dimensional Strings,* Nucl. Phys. **B330** (1990) 103.

23. A.N.Schellekens and S.Yankielowicz, *Simple currents, modular invariants and fixed points,* Int. J. Mod. Phys. **A5** (1990) 2903.

24. A.N. Schellekens, S. Yankielowicz, *Tables Supplements* to ref.[22] CERN-TH.5440S/89 and CERN-TH.5440T/89 preprints (unpublished).

25. A. Font, L.E. Ibáñez, M. Mondragon, F. Quevedo, G.G. Ross, *(0,2) heterotic string compactifications from N = 2 superconformal theories,* Phys. Lett. **B227** (1989) 34.

26. A. Font, L.E. Ibáñez, F. Quevedo, A. Sierra, *Twisted N = 2 coset models, discrete torsion and asymmetric heterotic string compactification,* Nucl. Phys. **B337** (1990) 119.

27. A.N. Schellekens, *Electric charge quantization in string theory,* Phys. Lett. **B237** (1990) 363.

28. J.D. Lykken, *String model building in the age of D-branes,* [arXiv:hep-th/9607144].

29. R. Blumenhagen, A. Wißkirchen, *Exactly solvable (0,2) supersymmetric string vacua with GUT gauge groups,* Nucl. Phys. **B454** (1995) 561 [arXiv:hep-th/9506104]; *Exploring the moduli space of (0,2) strings,* Nucl. Phys. **B475** (1996) 225 [arXiv:hep-th/9604140].

30. R. Blumenhagen, R. Schimmrigk, A. Wißkirchen, *The (0,2) exactly solvable structure of chiral rings, Landau–Ginzburg theories and Calabi–Yau manifolds,* Nucl. Phys. **B461** (1996) 460 [arXiv:hep-th/9510055].

31. M. Kreuzer, A.N. Schellekens, *Simple currents versus orbifolds with discrete torsion – a complete classification,* Nucl. Phys. **B411** (1994) 97 [arXiv:hep-th/9306145].

32. M. Kreuzer, M. Nikbakht-Tehrani, *(0,2) string compactification,* Nucl. Phys. B (Proc. Suppl.) **56B** (1997) 136 [arXiv:hep-th/9611130].

33. C.Vafa, *Modular invariance and discrete torsion on orbifolds,* Nucl. Phys. **B273** (1986) 592.

34. M. Kreuzer, H. Skarke, *No mirror symmetry in Landau-Ginzburg spectra!,* Nucl. Phys. **B388** (1992) 113 [arXiv:hep-th/9205004].

35. A. Klemm, R. Schimmrigk, *Landau–Ginzburg string vacua,* Nucl. Phys. **B411** (1994) 559 [arXiv:hep-th/9204060].

36. M. Kreuzer, H. Skarke, *All abelian symmetries of Landau–Ginzburg potentials,* Nucl. Phys. **B405** (1993) 305 [arXiv:hep-th/9211047].

37. M. Kreuzer, H. Skarke, *ADE models with discrete torsion,* Phys. Lett. **B318** (1993) 305 [arXiv:hep-th/9307145.

38. M. Kreuzer, H. Skarke, *Landau–Ginzburg orbifolds with discrete torsion,* Mod. Phys. Lett. **A10** (1995) 1073 [arXiv:hep-th/9412033].
39. M. Kreuzer, C. Schweigert, *On the extended Poincaré polynomial,* Phys. Lett. **B352** (1995) 276 [arXiv:hep-th/9503174].
40. R. Blumenhagen, R. Schimmrigk, A. Wißkirchen, *(0,2) mirror symmetry,* Nucl. Phys. **B486** (1997) 598 [arXiv:hep-th/9609167].
41. R. Blumenhagen, S. Sethi, *On orbifolds of (0,2) models,* Nucl. Phys. **B491** (1997) 263 [arXiv:hep-th/9611172]; R. Blumenhagen, M. Flohr, *Aspects of (0,2) orbifolds and mirror symmetry,* Phys. Lett. **B404** (1997) 41 [arXiv:hep-th/9702199].
42. P. Berglund and T. Hubsch, *A generalized construction of mirror manifolds,* Nucl. Phys. B **393** (1993) 377 [arXiv:hep-th/9201014].
43. M. Kreuzer, *The Mirror map for invertible LG models,* Phys. Lett. B **328** (1994) 312 [arXiv:hep-th/9402114].
44. R. Donagi and K. Wendland, *On orbifolds and free fermion constructions,* [arXiv:0809.0330].
45. C. Vafa, *Quantum symmetries of string vacua,* Mod. Phys. Lett. **A4** (1989) 1615.
46. M. Kreuzer, H. Skarke, *Orbifolds with discrete torsion and mirror symmetry,* Phys. Lett. **B357** (1995) 81 [arXiv:hep-th/9505120].
47. B. Gato-Rivera, A.N. Schellekens, *Complete Classification of Simple Current Automorphisms,* Nucl. Phys. **B353** (1991) 519; *Complete Classification of Simple Current Modular Invariants for $(\mathbb{Z}_p)^k$,* Commun. Math. Phys. **145** (1992) 85.
48. J. Fuchs, C. Schweigert, J. Walcher, *Projections in string theory and boundary states for Gepner models,* Nucl. Phys. B **588** (2000) 110 [arXiv:hep-th/0003298].
49. A.N. Schellekens, *Field identification fixed points in N = 2 coset theories,* Nucl. Phys. **B366** (1991) 27.
50. J. Distler, B.R. Green, D.R. Morrison, *Resolving singularities in (0,2) models,* Nucl. Phys. **B481** (1996) 312 [arXiv:hep-th/9605222].
51. V. V.Batyrev, *Dual polyhedra and mirror symmetry for Calabi–Yau hypersurfaces in toric varieties,* J. Alg. Geom. **3** (1994) 493 [alg-geom/9310003].
52. V.V.Batyrev, L.A.Borisov, *On Calabi-Yau complete intersections in toric varieties* [arXiv:alg-geom/9412017]; *Mirror duality and string-theoretic Hodge numbers* [arXiv:alg-geom/9509009].
53. R. Blumenhagen, *(0,2) target-space duality, CICYs and reflexive sheaves,* Nucl. Phys. **B514** (1998) 688 [arXiv:hep-th/9707198].
54. I.V. Melnikov, *(0,2) Landau-Ginzburg models and residues,* arXiv:0902.3908.
55. M. Kreuzer, H. Skarke, *PALP: A package for analyzing lattice polytopes* with applications to toric geometry, Computer Physics Commun. **157** (2004) 87 [arXiv:math.SC/0204356].
56. V. Batyrev, M. Kreuzer, *Integral cohomology and mirror symmetry for Calabi-Yau 3-folds*[math.AG/0505432].
57. W.Lerche, C.Vafa, N.Warner, *Chiral rings in N=2 superconformal theories,* Nucl. Phys. **B324** (1989) 427.

Chapter 20

Canonical Analysis of Cosmological Topologically Massive Gravity at the Chiral Point

Daniel Grumiller[1], Roman Jackiw[2] and Niklas Johansson[3]

[1,2] *Massachusetts Institute of Technology,*
77 Massachusetts Ave., Cambridge, MA 02139, USA

[3] *Institutionen för Fysik och Astronomi, Uppsala Universitet,*
Box 803, S-751 08 Uppsala, Sweden

Wolfgang Kummer was a pioneer of two-dimensional gravity and a strong advocate of the first order formulation in terms of Cartan variables. In the present work we apply Wolfgang Kummer's philosophy, the 'Vienna School approach', to a specific three-dimensional model of gravity, cosmological topologically massive gravity at the chiral point. Exploiting a new Chern–Simons representation we perform a canonical analysis. The dimension of the physical phase space is two per point, and thus the theory exhibits a local physical degree of freedom, the topologically massive graviton.

20.1. Introduction

Gravity in lower dimensions provides an excellent expedient for testing ideas about classical and quantum gravity in higher dimensions. The lowest spacetime dimension where gravity can be described is two, and Wolfgang Kummer contributed significantly to research on two-dimensional gravity, see Ref. 1 for a review. Those who knew Wolfgang will recall that one of his main points was to advocate a gauge theoretic approach towards gravity, see Ref. 2 for his last proceedings contributions. Instead of using the metric, $g_{\mu\nu}$, as fundamental field he insisted on employing the Cartan variables, Vielbein e^a_μ and connection $\omega^a{}_{b\,\mu}$. His approach greatly facilitated the canonical analysis and the quantization of the theory.

---

[1] E-mail: grumil@hep.itp.tuwien.ac.at

[2] E-mail: jackiw@mit.edu

[3] E-mail: Niklas.Johansson@fysast.uu.se

In the present work we shall study gravity in three dimensions along similar lines. We start by collecting a few well-known features of gravity in three dimensions. Pure Einstein–Hilbert gravity exhibits no physical bulk degrees of freedom.[3–5] If the theory is deformed by a negative cosmological constant it has black hole solutions.[6] Another possible deformation is to add a gravitational Chern–Simons term. The resulting theory is called topologically massive gravity (TMG) and, remarkably, contains a massive graviton.[7] Including both terms yields cosmological topologically massive gravity[8] (CTMG), a theory that exhibits both gravitons and black holes. Parameterizing the negative cosmological constant by $\Lambda = -1/\ell^2$ the (second order) action is given by

$$I_{\mathrm{CTMG}}[g] = \int d^3x\sqrt{-g}\left[R + \frac{2}{\ell^2} + \frac{1}{2\mu}\,\varepsilon^{\lambda\mu\nu}\Gamma^\rho{}_{\lambda\sigma}\left(\partial_\mu\Gamma^\sigma{}_{\nu\rho} + \frac{2}{3}\Gamma^\sigma{}_{\mu\tau}\Gamma^\tau{}_{\nu\rho}\right)\right].$$
$$(20.1)$$

In Ref. 9 it was advocated to study the theory (20.1) at the chiral point

$$\mu\ell = 1\,,\qquad\qquad\qquad (20.2)$$

where the theory exhibits very special properties. We abbreviate this theory by the acronym CCTMG ('chiral cosmological topologically massive gravity'). By imposing the Brown–Henneaux boundary conditions Ref. 9 argued that CCTMG exhibits no bulk degrees of freedom. On the other hand Ref. 10 found that CCTMG exhibits one bulk degree of freedom. By slightly relaxing the Brown–Henneaux boundary conditions — still requiring spacetime to be asymptotically AdS — Ref. 11 demonstrated that indeed a physical degree of freedom exists in CCTMG: the topologically massive graviton. The analyses in Refs. 9–11 were focused on the linearized level, i.e., perturbing around an AdS$_3$ background.

In the present work we go beyond the linearized approximation and perform a non-perturbative (classical) canonical analysis of CCTMG (see also Refs. 12–14).[a] Our main goal is to derive the dimension of the physical phase space, which allows us to deduce the number of physical bulk degrees of freedom.

This paper is organized as follows. In Section 20.2 we present a new Chern–Simons formulation of cosmological topologically massive gravity. In Section 20.3 we focus on the chiral point and establish the Hamiltonian formulation, identifying all primary, secondary and ternary constraints. In Section 20.4 we perform a constraint analysis and check the first/second class properties of all constraints, which allows us to establish the dimension of the physical phase space. In Section 20.5 we conclude.

---

[a]For further recent literature related to CCTMG see Refs. 15–21.

Our conventions are as follows. We use Greek spacetime indices and Latin frame indices. The former are raised and lowered with the spacetime metric $g_{\mu\nu}$ and the latter with the flat metric η_{ab}. Both have signature $-, +, +$. For the Dreibein e^a_μ we choose $\mathrm{sign}\,(\det e) = 1$. When writing p-forms we usually suppress the spacetime indices, e.g. e^a denotes the 1-form $e^a = e^a_\mu dx^\mu$. We disregard boundary terms in the present work, so equivalences between actions have to be true only up to total derivatives.

20.2. Chern–Simons formulation

Instead of the action (20.1) which functionally depends on the metric one can equivalently use the action

$$I_{\mathrm{CTMG}}[e] = \int \left[2e^a \wedge R_a(\omega) + \frac{1}{3\ell^2} \varepsilon_{abc}\, e^a \wedge e^b \wedge e^c - \frac{1}{\mu} \mathrm{CS}(\omega) \right] \quad (20.3)$$

which functionally depends on the Dreibein. The gravitational Chern–Simons term

$$\mathrm{CS}(\omega) := \omega^a \wedge d\omega_a + \frac{1}{3} \varepsilon_{abc}\, \omega^a \wedge \omega^b \wedge \omega^c \quad (20.4)$$

and the (dualized) curvature 2-form

$$R_a(\omega) := d\omega_a + \frac{1}{2} \varepsilon_{abc}\, \omega^b \wedge \omega^c \quad (20.5)$$

depend both exclusively on the (dualized) connection defined by $\omega^a := \frac{1}{2}\varepsilon^{abc}\omega_{bc}$. Note that the connection is not varied independently in the formulation (20.3), but rather it is the Levi-Civita connection, i.e., metric compatible $\omega^{ab} = -\omega^{ba}$ and torsion-free, $T^a = 0$, where

$$T_a := de_a + \varepsilon_{abc}\, \omega^b \wedge e^c \quad (20.6)$$

is the torsion 2-form. This means that ω^a in (20.3) has to be expressed in terms of e^a (and derivatives thereof) before variation.

For our purposes it is very convenient to employ a formulation where we can vary independently the Dreibein and the connection.[22] This is achieved by supplementing the action (20.3) with a Lagrange multiplier term enforcing the torsion constraint,

$$I_{\mathrm{CTMG}}[e, \omega, \lambda] = \int \left[2e^a \wedge R_a + \frac{1}{3\ell^2} \varepsilon_{abc}\, e^a \wedge e^b \wedge e^c - \frac{1}{\mu} \mathrm{CS}(\omega) + \lambda^a \wedge T_a \right].$$

$$(20.7)$$

The first order action (20.7) is classically equivalent[22] to the second order action (20.1). This can be shown as follows. Varying (20.7) with respect to λ_a and ω_a establishes the condition of vanishing torsion (20.6) and an algebraic relation for λ_a,

$$\frac{1}{2}\,\varepsilon_{abc}\,\lambda^a \wedge e^b = \frac{1}{\mu}\,R_c - T_c = \frac{1}{\mu}\,R_c\,, \tag{20.8}$$

in terms of Dreibein, connection and derivatives thereof. Thus, both ω_a and λ_a can be expressed in terms of the Dreibein, and first and second derivatives thereof. Varying (20.7) with respect to the Dreibein and plugging into that equation the relations for λ_a and ω_a in terms of e_a yields a set of third order partial differential equations in e_a. Using the defining relation between Dreibein and metric, $g_{\mu\nu} = e^a_\mu e^b_\nu \eta_{ab}$, finally establishes

$$G_{\mu\nu} + \frac{1}{\mu}\,C_{\mu\nu} = 0\,, \tag{20.9}$$

where

$$G_{\mu\nu} = R_{\mu\nu} - \frac{1}{2}\,g_{\mu\nu}R - \frac{1}{\ell^2}\,g_{\mu\nu} \tag{20.10}$$

is the Einstein tensor (including cosmological constant) and

$$C_{\mu\nu} = \frac{1}{2}\,\varepsilon_\mu{}^{\alpha\beta}\,\nabla_\alpha R_{\beta\nu} + (\mu \leftrightarrow \nu) \tag{20.11}$$

is essentially the Cotton tensor. The equations of motion (20.9) also follow directly from varying the second order action (20.1) with respect to the metric.

We make now some field redefinitions to further simplify the action (20.7). We shift the Lagrange multiplier $\lambda^a \to \lambda^a - e^a/(\mu\ell^2)$ and obtain

$$I_{\text{CTMG}}[e,\omega,\lambda] = \int \Big[2e^a \wedge R_a + \frac{1}{3\ell^2}\,\varepsilon_{abc}\,e^a \wedge e^b \wedge e^c - \frac{1}{\mu}\,\text{CS}(\omega) + \big(\lambda^a - \frac{e^a}{\mu\ell^2}\big)\wedge T_a\Big]. \tag{20.12}$$

In the absence of the $\lambda^a \wedge T_a$-term in (20.12), the well-known field redefinitions

$$A^a := \omega^a + e^a/\ell\,, \qquad \tilde{A}^a := \omega^a - e^a/\ell \tag{20.13}$$

turn the action into a difference of two Chern–Simons terms.[23–26] Curiously, under the same redefinitions (20.13) the Lagrange multiplier term can be recast into a difference of two Einstein–Hilbert terms, where λ plays the role of the Dreibein:

$$\frac{2}{\ell}\,I_{\text{CTMG}}[A,\tilde{A},\lambda] = \big(1 - \frac{1}{\mu\ell}\big)I_{\text{CS}}[A] + I_{EH}[\lambda,A] - \big(1 + \frac{1}{\mu\ell}\big)I_{\text{CS}}[\tilde{A}] - I_{EH}[\lambda,\tilde{A}]. \tag{20.14}$$

We have introduced here the abbreviations

$$I_{CS}[A] := \int CS(A) \qquad (20.15)$$

and

$$I_{EH}[\lambda, A] := \int \lambda^a \wedge R_a(A) \qquad (20.16)$$

and similarly for $\tilde{A}$.

The reformulation (20.14) of the action (20.7) as difference of Chern–Simons and Einstein-Hilbert terms seems to be new. It is worthwhile repeating that in both Einstein–Hilbert terms the Lagrange multiplier λ^a formally plays the role of a 'Dreibein'. This suggests that λ^a should be invertible. We have checked that for pure AdS$_3$ [which obviously solves the field equations (20.9)] the symmetric tensor $\lambda_{\mu\nu} = e^a_{(\mu}\lambda_{\nu)a}$ is proportional to the metric. Thus, requiring invertibility of λ^a is necessary in general to guarantee invertibility of the metric.

The advantage of the formulation (20.14) is twofold. Because the action contains only first derivatives (linearly) a canonical analysis is facilitated. Moreover, at the chiral point $\mu^2 \ell^2 = 1$ one of the Chern–Simons terms vanishes.

20.3. Hamiltonian action at the chiral point

We focus now on the theory at the chiral point and assume for sake of specificity $\mu\ell = 1$. The action (20.14) simplifies to

$$I_{CCTMG}[A, \tilde{A}, \lambda] = \frac{\ell}{2} I_{EH}(\lambda, A) - \ell I_{CS}(\tilde{A}) - \frac{\ell}{2} I_{EH}(\lambda, \tilde{A}) = \int d^3 x \, \mathcal{L} \, . \qquad (20.17)$$

To set up the canonical analysis one could now declare the 27 fields λ^a, A^a, $\tilde{A}^a$ to be canonical coordinates and calculate their 27 canonical momenta.[13] In this way one produces many second class constraints which have to be eliminated by the Dirac procedure.[27] However, this is not the most efficient way to start the canonical analysis. As realized in Ref. 28 if an action is already in first order form a convenient short-cut exists. In the present case this short-cut consists basically of picking the appropriate sets of fields as canonical coordinates and momenta, respectively.

We use the 18 fields $\lambda^a_\mu, \tilde{A}^a_0, \tilde{A}^a_1, A^a_0$ as canonical coordinates and introduce the notation

$$q^a_1 = \lambda^a_1 \, , \ q^a_2 = \lambda^a_2 \, , \ q^a_3 = \tilde{A}^a_1 \, , \ \bar{q}^a_1 = \lambda^a_0 \, , \ \bar{q}^a_2 = \tilde{A}^a_0 \, , \ \bar{q}^a_3 = A^a_0 \, . \qquad (20.18)$$

Like in electrodynamics or non-abelian gauge theory the momenta $\bar{p}_i^a$ of the zero components $\bar{q}_i^a$ are primary constraints. The simplest way to deal with them is to exclude the pairs $\bar{q}_i^a, \bar{p}_i^a$ from the phase space and to treat the $\bar{q}_i^a$ as Lagrange multipliers for the secondary constraints ("Gauss constraints"). This reduces the dimension of our phase space to 18. The 9 momenta p_i^a,

$$\frac{\partial \mathcal{L}}{\partial \partial_0 \lambda_{1\,a}} = p_1^a = \frac{\ell}{2}(A_2^a - \tilde{A}_2^a) = e_2^a \qquad (20.19)$$

$$\frac{\partial \mathcal{L}}{\partial \partial_0 \lambda_{2\,a}} = p_2^a = -\frac{\ell}{2}(A_1^a - \tilde{A}_1^a) = -e_1^a \qquad (20.20)$$

$$\frac{\partial \mathcal{L}}{\partial \partial_0 \tilde{A}_{1\,a}} = p_3^a = -2\ell\,\tilde{A}_2^a \qquad (20.21)$$

depend linearly on the fields $A_1^a, A_2^a, \tilde{A}_2^a$. These fields are not contained in our set of canonical coordinates.

The Hamiltonian action is now determined as

$$I_{\text{CCTMG}}[q, p; \bar{q}] = \int d^3x \left(p_{i\,a}\dot{q}_i^a - \mathcal{H}\right), \qquad (20.22)$$

where the Hamiltonian density

$$\mathcal{H} = \bar{q}_{i\,a}\,G_i^a \qquad (20.23)$$

is a sum over secondary constraints $G_i^a \approx 0$, as expected on general grounds.[b] They are given by

$$G_1^a = -\frac{\ell}{2}\,R^a + \frac{\ell}{2}\,\tilde{R}^a, \qquad (20.24)$$

$$G_2^a = \frac{\ell}{2}\,\tilde{D}\lambda^a + 2\ell\,\tilde{R}^a, \qquad (20.25)$$

$$G_3^a = -\frac{\ell}{2}\,D\lambda^a. \qquad (20.26)$$

We have introduced the following abbreviations

$$R^a := \left(\partial_1 A_2^a - \partial_2 A_1^a\right) + \frac{1}{2}\varepsilon^a{}_{bc}\left(A_1^b A_2^c - A_2^b A_1^c\right) \qquad (20.27)$$

and

$$D\lambda^a := \left(\partial_1 \lambda_2^a - \partial_2 \lambda_1^a\right) + \varepsilon^a{}_{bc}\left(A_1^b \lambda_2^c - A_2^b \lambda_1^c\right) \qquad (20.28)$$

and similarly for $\tilde{R}$ and $\tilde{D}\lambda$, with A replaced by $\tilde{A}$ in the definitions (20.27) and (20.28), respectively.

---

[b]The notation $\approx$ means 'vanishing weakly',[27] i.e., vanishing on the surface of constraints.

We focus now on the first/second class properties of the constraints and on their Poisson bracket algebra. We have found 9 secondary constraints G_i^a. If all of them were first class then the physical phase space would be zero-dimensional, because each first class constraint eliminates two dimensions from the phase space, and the dimension of the phase space spanned by q_i^a, p_i^a is 18.

20.4. Constraint analysis

With the canonical Poisson bracket

$$\{q_i^a(x), p_j^b(x')\} = \{q_i^a, p_j'^b\} = \delta_{ij}\, \eta^{ab}\, \delta^{(2)}(x - x') \tag{20.29}$$

we can now calculate the Poisson brackets of the constraints G_i^a with each other and with the Hamiltonian density. The latter,

$$\{G_i^a, \mathcal{H}'\} = \bar{q}_{j\,b}'\, \{G_i^a, G_j'^{\,b}\} \tag{20.30}$$

reduce to a sum over brackets between the secondary constraints. We calculate now these brackets explicitly.

To this end we express the secondary constraints (20.24)-(20.26) in terms of canonical coordinates and momenta:

$$G_1^a = -\partial_1 p_1^a - \partial_2 p_2^a - \varepsilon^a{}_{bc}\left(\frac{2}{\ell} p_1^b p_2^c + \frac{1}{2\ell} p_2^b p_3^c + q_3^b p_1^c\right) \tag{20.31}$$

$$\widehat{G}_2^a = G_2^a + G_3^a = -\partial_1 p_3^a - 2\ell\, \partial_2 q_3^a + \varepsilon^a{}_{bc}\, p_i^b q_i^c \tag{20.32}$$

$$G_3^a = -\frac{\ell}{2}(\partial_1 q_2^a - \partial_2 q_1^a) + \varepsilon^a{}_{bc}\left(p_1^b q_1^c + p_2^b q_2^c - \frac{1}{4} p_3^b q_1^c - \frac{\ell}{2} q_3^b q_2^c\right). \tag{20.33}$$

Note that instead of G_2^a we use for convenience the linear combination $\widehat{G}_2^a = G_2^a + G_3^a$. Straightforward calculation obtains:

$$\{G_1^a, G_1'^{\,b}\} = Z_{11}^{ab}\, \delta^{(2)}(x - x') \tag{20.34}$$

$$\{\widehat{G}_2^a, \widehat{G}_2'^{\,b}\} = -\varepsilon^{ab}{}_c\, \widehat{G}_2^c\, \delta^{(2)}(x - x') \approx 0 \tag{20.35}$$

$$\{G_3^a, G_3'^{\,b}\} = -\varepsilon^{ab}{}_c\, G_3^c\, \delta^{(2)}(x - x') + Z_{33}^{ab}\, \delta^{(2)}(x - x') \tag{20.36}$$

$$\{G_1^a, \widehat{G}_2'^{\,b}\} = \varepsilon^{ab}{}_c\, G_1^c\, \delta^{(2)}(x - x') \approx 0 \tag{20.37}$$

$$\{\widehat{G}_2^a, G_3'^{\,b}\} = -\varepsilon^{ab}{}_c\, G_3^c\, \delta^{(2)}(x - x') \approx 0 \tag{20.38}$$

$$\{G_1^a, G_3'^{\,b}\} = -\varepsilon^{ab}{}_c\left(G_1^c - \frac{1}{4}\widehat{G}_2^c\right)\delta^{(2)}(x - x') + Z_{13}^{ab}\, \delta^{(2)}(x - x'). \tag{20.39}$$

We have used here the abbreviations

$$Z_{11}^{ab} = \frac{1}{2\ell} \left(p_2^a p_1^b - p_2^b p_1^a \right) \tag{20.40}$$

$$Z_{33}^{ab} = \frac{\ell}{8} \left(q_2^a q_1^b - q_2^b q_1^a \right) \tag{20.41}$$

$$Z_{13}^{ab} = -\frac{1}{4} \left(p_1^a q_1^b + p_2^a q_2^b \right) + \frac{1}{4} \eta^{ab} \left(p_1^c q_{1\,c} + p_2^c q_{2\,c} \right) \tag{20.42}$$

or, equivalently,

$$Z_{11}^{ab} = -\frac{1}{2\ell} \left(e^a \wedge e^b \right)_{12} \tag{20.43}$$

$$Z_{33}^{ab} = -\frac{\ell}{8} \left(\lambda^a \wedge \lambda^b \right)_{12} \tag{20.44}$$

$$Z_{13}^{ab} = \frac{1}{4} \left(e^a \wedge \lambda^b \right)_{12} - \frac{1}{4} \eta^{ab} \eta_{cd} \left(e^c \wedge \lambda^d \right)_{12} . \tag{20.45}$$

If the quantities Z_{ij}^{ab} were all vanishing then all secondary constraints would be first class. Since some of them are non-vanishing we have a certain number of second class constraints. Namely, not all entries of Z_{11}^{ab} can vanish because this would lead to a singular Dreibein e^a. Similarly, not all entries Z_{33}^{ab} can vanish because this would lead to a singular Lagrange multiplier 1-form λ^a. Since the algebra of constraints does not close we shall encounter ternary constraints from consistency requirements, namely the vanishing of the Poisson brackets (20.30).

In the analysis below, the 9×9-matrix

$$M_{ij}^{ab} := \int_{x'} d^2 x' \left\{ G_i^a, G_j'^b \right\} \tag{20.46}$$

evaluated on the surface of constraints will play a crucial role. First, note that before imposing the ternary constraints we can establish an upper bound on the dimension $2n$ of the physical phase space in terms of the rank r_M of M_{ij}^{ab}. We started with a phase space of dimension 18 and accounted for 9 constraints. The rank r_M counts how many of these that are second class. Thus, before additional constraints are introduced we have

$$2n \leq 18 - r_M - 2 * (9 - r_M) = r_M . \tag{20.47}$$

Now we turn to the ternary constraints. We note that after imposing these we are done, since the consistency conditions analog to (20.30) arising from the T_i^a do not generate quaternary constraints. Since the algebra (20.34)–(20.39) closes on δ-functions, requiring vanishing of the brackets (20.30) is equivalent to requiring

$$T_i^a := M_{ij}^{ab} \bar{q}_{j\,b} \approx 0 . \tag{20.48}$$

Because the ternary constraints T_i^a contain the canonical partners of the primary constraints $\bar{p}_i^a$ complications arise, since some of the latter may lose their status as first class constraints. Thus we have to include the $\bar{q}_i^a$ as canonical variables, giving a phase space of dimension 36 before imposing the constraints. We determine now the rank of the 27×27 matrix

$$\widehat{M}_{ij}^{ab} := \int_{x'} d^2x' \left\{ C_i^a, C_j'^{\,b} \right\} \tag{20.49}$$

evaluated on the surface of constraints using the order $C_i^a = (\bar{p}_i^a, G_i^a, T_i^a)$. Because of (20.48) we have

$$\{T_i^a, \bar{p}_j^b\} = M_{ij}^{ab}, \tag{20.50}$$

and thus $\widehat{M}$ has the block form

$$\widehat{M} \approx \begin{pmatrix} \mathbb{O} & \mathbb{O} & -M^T \\ \mathbb{O} & M & B \\ M & -B^T & C \end{pmatrix}, \tag{20.51}$$

where all the blocks are 9×9 matrices. The form of the non-vanishing matrices B and C is not needed for determining a lower bound for the rank of $\widehat{M}$. We can put all copies of M and M^T on lower triangular form by row-operations that do not spoil the block structure of (20.51). This makes $\widehat{M}$ lower triangular with $3r_M$ non zero anti-diagonal elements. Thus, a lower bound for the rank $r_{\widehat{M}}$ of $\widehat{M}$ is $3r_M$.

We are now in a position to count the number of linearly independent first- and second-class constraints. We have $r_{\widehat{M}}$ second class constraints. The total number of constraints is 9(primary) + 9(secondary) + 9(ternary) = 27, but out of the nine ternary constraints T_i^a, only r_M are linearly independent. This is so because of (20.48).

Thus, the total number of linearly independent constraints is $9 + 9 + r_M = 18 + r_M$, and $r_{\widehat{M}}$ of these are second class. The dimension $2n$ of the physical phase space is therefore bounded by

$$2n = 36 - r_{\widehat{M}} - 2 * (18 + r_M - r_{\widehat{M}}) = r_{\widehat{M}} - 2r_M \geq r_M. \tag{20.52}$$

The two inequalities (20.47) and (20.52) establish $2n = r_M$.

Thus, all that remains is to determine the rank of M. Using the order $G_i^a = (G_1^a, G_3^a, \hat{G}_2^a)$, M has the block form

$$M \approx \begin{pmatrix} A_{6\times6} & \mathbb{O}_{6\times3} \\ \mathbb{O}_{3\times6} & \mathbb{O}_{3\times3} \end{pmatrix}, \qquad A_{6\times6} := \begin{pmatrix} Z_{11} & Z_{13} \\ -Z_{13}^T & Z_{33} \end{pmatrix}. \tag{20.53}$$

The block entries $\mathbb{O}_{x \times y}$ contain x rows and y columns of zeros. From (20.53) we deduce that the rank of the antisymmetric matrix M_{ij}^{ab} must be either six, four or two. Its nine Eigenvalues $n_1 \ldots n_9$ are given by

$$n_1 \ldots n_5 = 0, \qquad n_{6,7} = \pm \frac{i}{4} \left(e^a \wedge \lambda_a \right)_{12}, \qquad n_{8,9} = \pm \frac{i}{4} \sqrt{P}. \qquad (20.54)$$

Therefore its rank equals (at most) four, and not six as suggested by naive counting. The polynomial under the square root in the last expression in (20.54) is given by

$$P = \frac{2}{\ell^2} \left(e^a \wedge e^b \right)_{12} \left(e_a \wedge e_b \right)_{12} + \frac{\ell^2}{8} \left(\lambda^a \wedge \lambda^b \right)_{12} \left(\lambda_a \wedge \lambda_b \right)_{12} + \left(e^a \wedge \lambda^b \right)_{12} \left(e_a \wedge \lambda_b \right)_{12}.$$
$$(20.55)$$

The rank of (20.53) is four in general and two if in addition the condition

$$\left(e^a \wedge \lambda_a \right)_{12} = -p_1^a q_{1\,a} - p_2^a q_{2\,a} = 0 \qquad (20.56)$$

holds. Because of (20.8) on-shell we obtain

$$e^a \wedge \lambda_a \propto e^a \wedge \mathrm{Ric}_a \propto R_{\mu\nu} \, dx^\mu \wedge dx^\nu = 0 \qquad (20.57)$$

where Ric_a is the Ricci 1-form with respect to the Levi-Civita connection (we recall that on-shell torsion vanishes). Thus, the constraint (20.56) must hold on all classical solutions. Therefore, in the physically relevant sector, $2n = r_M = 2.^{\mathrm{c}}$ This completes our constraint analysis.[d]

To summarize, the dimension of the physical phase space is two and therefore CCTMG exhibits one physical bulk degree of freedom, which at the linearized level coincides with the topologically massive graviton.

20.5. Conclusions

In this paper we have reformulated cosmological topologically massive gravity at the chiral point as a Chern–Simons action plus the difference between

---

[c]It is possible, although not necessary, to impose (20.56) as a further constraint. This does not change anything essential about the counting procedure.

[d]As a consistency check we investigate now what happens when the torsion constraint is dropped in (20.12). In the current formulation this can be achieved by imposing the constraints

$$G_4^a = q_1^a \approx 0, \qquad G_5^a = q_2^a \approx 0, \qquad G_6^a = \bar{q}_1^a \approx 0. \qquad (20.58)$$

These constraints render the constraints G_3^a and T_i^a superfluous. Thus, we have now 24 linearly independent constraints, $\bar{p}_i^a$, G_1^a, $\widehat{G}_2^a$, G_4^a, G_5^a, G_6^a. The rank of the 24 × 24 matrix analog to (20.51) turns out to be equal to twelve. Therefore, we have now twelve first class and twelve second class constraints, which eliminates all dimensions from the phase space. Thus no physical bulk degrees of freedom remain. This is the anticipated result.

two Einstein–Hilbert actions, see (20.17). We have performed a canonical analysis and recovered the anticipated[e] result of one physical bulk degree of freedom, which at the linearized level corresponds to the topologically massive graviton.

We have also encountered sectors of our first order theory that are not related to the second order formulation with regular field configurations, but that may be worthwhile studying in their own right. For instance, if one imposes by hand the constraints $\bar{q}_1^a = 0 = \bar{q}_3^a$ then no ternary constraints arise, but the Dreibein and Lagrange multiplier fail to be invertible.

Finally, we mention that the Poisson bracket algebra of the secondary constraints (20.34)-(20.39) closes with δ-functions rather than with first derivatives thereof because of our gauge theoretic reformulation of CCTMG. The same happens in $1+1$ dimensions (see the contribution by L. Bergamin and R. Meyer in this volume, eq. (11.20)), where this feature was exhibited and exploited by Wolfgang Kummer and his 'Vienna School'.[1,2]

Acknowledgments

We thank Steve Carlip, Stanley Deser, Mu-In Park and Andy Strominger for correspondence. NJ thanks the CTP at MIT for its kind hospitality during parts of this work. This work is supported in part by funds provided by the U.S. Department of Energy (DoE) under the cooperative research agreement DEFG02-05ER41360. DG is supported by the project MC-OIF 021421 of the European Commission under the Sixth EU Framework Programme for Research and Technological Development (FP6). The research of NJ was supported in part by the STINT CTP-Uppsala exchange program.

References

1. D. Grumiller, W. Kummer, and D. V. Vassilevich, *Phys. Rept.* **369**, 327–429, (2002).
2. W. Kummer, Progress and problems in quantum gravity. (2005).

---

[e]A recent canonical analysis in the first order formulation[13] obtains a 2-dimensional physical phase space 'for each internal index a', i.e., a 6-dimensional physical phase space. This result disagrees with ours and with previous literature, but it is then interpreted as a single graviton degree of freedom, concurrent with our result. Correspondence with the author revealed that he found additional constraints after posting his e-print and that currently he is reconsidering the constraint algebra. Another recent analysis[14] agrees with our results.

3. S. Weinberg, *Gravitation and cosmology: principles and applications of the general theory of relativity.* (Wiley, New York, 1972).

4. S. Deser, R. Jackiw, and G. 't Hooft, *Ann. Phys.* **152**, 220, (1984).

5. S. Deser and R. Jackiw, *Annals Phys.* **153**, 405–416, (1984).

6. M. Banados, C. Teitelboim, and J. Zanelli, *Phys. Rev. Lett.* **69**, 1849–1851, (1992).

7. S. Deser, R. Jackiw, and S. Templeton, *Phys. Rev. Lett.* **48**, 975–978, (1982). *Ann. Phys.* **140**, 372–411, (1982). *Erratum-ibid.* **185**, 406, (1988).

8. S. Deser, Cosmological Topological Supergravity. Print-82-0692 (Brandeis).

9. W. Li, W. Song, and A. Strominger, *JHEP* **0804** 082, (2008).

10. S. Carlip, S. Deser, A. Waldron, and D. K. Wise, Cosmological Topologically Massive Gravitons and Photons. (2008).

11. D. Grumiller and N. Johansson, *JHEP* **0807** (2008) 134.

12. S. Deser and X. Xiang, *Phys. Lett.* **B263**, 39–43, (1991).

13. M.-I. Park, Constraint Dynamics and Gravitons in Three Dimensions. (2008).

14. S. Carlip, *JHEP* **0810** (2008) 078.

15. R. Banerjee, S. Gangopadhyay, and S. Kulkarni, Black Hole Entropy from Covariant Anomalies. (2008).

16. G. Compere and D. Marolf, Setting the boundary free in AdS/CFT. (2008).

17. M. Alishahiha and F. Ardalan, Central Charge for 2D Gravity on AdS(2) and AdS(2)/CFT(1) Correspondence. (2008).

18. K. Hotta, Y. Hyakutake, T. Kubota, and H. Tanida, Brown-Henneaux's Canonical Approach to Topologically Massive Gravity. (2008).

19. W. Li, W. Song, and A. Strominger, Comment on 'Cosmological Topological Massive Gravitons and Photons'. (2008).

20. I. Sachs and S. N. Solodukhin, *JHEP* **0808** (2008) 003.

21. D. A. Lowe and S. Roy, Chiral geometries of (2+1)-d AdS gravity. (2008).

22. P. Baekler, E. W. Mielke, and F. W. Hehl, *Nuovo Cim.* **107B**, 91–110, (1991).

23. A. Achucarro and P. K. Townsend, *Phys. Lett.* **B180**, 89, (1986).

24. E. Witten, *Nucl. Phys.* **B311**, 46, (1988).

25. M. Blagojevic and M. Vasilic, *Phys. Rev.* **D68**, 104023, (2003).

26. S. L. Cacciatori, M. M. Caldarelli, A. Giacomini, D. Klemm, and D. S. Mansi, *J. Geom. Phys.* **56**, 2523–2543, (2006).

27. P. A. M. Dirac, *Lectures on Quantum Mechanics.* (Belfer Graduate School of Science, Yeshiva University, New York, 1996).

28. L. D. Faddeev and R. Jackiw, *Phys. Rev. Lett.* **60**, 1692, (1988).

PART III

Wolfgang Kummer and the Physics Community

Chapter 21

Wolfgang Kummer at CERN

Herwig Schopper

CERN, 1211 Geneva, Switzerland
E-mail: Herwig.Schopper@cern.ch

Wolfgang Kummer was not only a great theorist but also a man with a noble spirit and extensive education, based on a fascinating long-term Austrian cultural tradition. As an experimentalist I am not sufficiently knowledgeable to evaluate his contributions to theoretical physics – this will certainly be done by more competent scientists. Nevertheless I admired him for not only being attached to fundamental and abstract problems like quantum field theory, quantum gravity or black holes, but for his interest in down to earth questions like electron-proton scattering or the toponium mass. I got to know Wolfgang Kummer very well and appreciate his human qualities during his long attachment to CERN, in particular when he served as president of the CERN Council, the highest decision taking authority of this international research centre, from 1985 to 1987 falling into my term as Director-General.

Kummer's career was intimately connected with CERN from very early on. Walter Thirring had obtained for him a Ford scholarship to CERN for the academic year 1961/62. There he got to know Victor Weisskopf, then Director General of CERN, who invited him to come back as a CERN fellow and his scientific assistant for the two years 1963–1964.

From 1966 to 1971 Kummer became the first director of the Institute for High Energy Physics of the Austrian Academy of Sciences, while being at the same time professor at the Technical University of Vienna. Simultaneously he became the Austrian delegate to the CERN Council, where he was soon elected to chair the Finance Committee. This was the glorious period when the proton-proton collider ISR was built, at a time when the CERN budget was still increasing rapidly. In 1980, Kummer returned to the CERN Council as its Vice-President. In this function he could con-

tribute very positively to the development of CERN not only because of
his diplomatic skill but also since at this critical point in time this function
was more suitably filled by a physicist than a diplomat or administrator.
Several issues were at stake. The conversion of the proton accelerator SPS
into a proton-antiproton collider was getting into shape, a project which
had not been formally approved by Council but which brought the Nobel
Prize to C. Rubbia and S. van der Meer. In 1980 the first proposal for the
construction of LEP was presented to Council. It had also been decided
that the regime of two Director-Generals with Leon van Hove and J. Adams
should be followed by a new structure with only one Director-General which
resulted in my appointment. Although the deliberations in the Committee
of Council are confidential, I am sure that Kummer's influence in convincing
the other delegates to take positive decisions was crucial.

In the meeting of the CERN Council of 12/13 December 1984, Kummer
was elected as its president for the period 1985 to 1987 following Sir Alec
Merrison. One of his first duties was to welcome the accession of Portugal
as a new Member State which after a certain period of negotiations became
effective in 1985. After this pleasant occurrence time became much more
dramatic because of expected but also unforeseen events. It was a crucial
period for the construction of LEP which had to be realised within a constant budget – a completely new course of action at CERN. In the past
all new projects had received additional funds. Hence it was unavoidable
that financial problems came up leading to long and difficult discussions
in the Finance Committee and Council. These were aggravated by serious
difficulties in the excavation of the LEP tunnel requiring additional funds.
Another issue was my extension as Director-General which demanded diplomatic ingenuity of the President of Council. It was a difficult decision to
take since according to the tradition of CERN it never happened before or
after that the term of 5 years of a Director General was extended. For-
tunately Kummer could attend also some happy events. On 4 June 1987
the first magnet of LEP was installed in the presence of J. Chirac, Prime
Minister of France at that time, and P. Aubert, President of the Swiss
Confederation. To make life more complicated, at the insistence of the UK,
the CERN Council decided to set up an evaluation committee chaired by
the renowned French physicist Anatole Abragam. There were no doubts
about the scientific excellence of the CERN programme but this committee
including some high level industrialists, was charged to review the management of CERN. The main criticism concerned the personnel policy of
CERN claiming that the flexibility and the turn over were not satisfac-

tory. They recommended an early retirement programme asking Council to provide additional funds to compensate the pension fund. Indeed Council decided to establish such a programme, however, without any financial compensation! During these difficult discussions and negotiations Kummer always defended the interests of CERN and its staff. Publicly he said in an interview[1] 'Schopper achieved something unbelievable in shuffling 1000 of the 3500 staff members to build LEP', and he believed that 'compared to other international organisations CERN management is doing very, very well', especially considering that other international organisations operate within 'much more comfortable budgets'. Kummer took his task as President of Council very seriously. He did not limit himself, as many other had done, just to chair the Council meetings, but he got intimately involved in the preparation of the major decisions. To obtain all the necessary detailed information he attended regularly the meetings of the Finance Committee and the Scientific Policy Committee which was chaired consecutively by Italo Manneli and Don Perkins with outstanding international members including Nobel Laureates.

Unfortunately Kummer's presidency was cruelly interrupted by an attack of three terrorists on 26 December 1985 at the counter of the Israeli airline El Al at the Vienna airport. Three people were killed and 30 injured and Kummer was one of the victims, suffering from severe injuries from hand grenade splinters and shrapnel. He was about to leave for Christmas vacations with his wife Lore. She remained unhurt since she was just buying a journal. After only 11 days in intensive care (and in rather critical condition), Kummer recovered quickly and certainly his regular sporting activities, like skiing, were crucial for his amazingly quick recovery. I was really surprised to see him back at CERN after a few weeks after the incident resuming his job as Council President, a proof of his enormous feeling of responsibility.

But Kummer was not only a sportsman in his spare time. He loved music and he was in particular an excellent pianist and trained tenor. While in Geneva he participated in chamber music evenings with his colleagues such as Volker Soergel and Jack Steinberger.

Of course, he kept close relations with CERN after his term as President of Council. For example, I invited him to speak in honour of Viki Weisskopf at his 80th birthday on 1 September 1988 at a CERN Fest-colloquium and indeed he gave a a talk which was appreciated by Viki and all the listeners. While Kummer had numerous academic and administrative positions, such as being first Secretary and then President of the High Energy Board of the

European Physical Society from 1995–1999, he was especially responsible for building up a group of theoretical high energy physics at the Technical University of Vienna. Thus Kummer became an essential element in continuing a long tradition of outstanding Austrian physicists.

Kummer was also an excellent teacher and therefore he had been invited by many institutions as a guest professor, for example at CERN, but also at Princeton University, Brookhaven National Laboratory BNL and the University of Cambridge. We all felt very sorry when we learned about his serious illness and we were shocked when he passed away much too early in July 2007. Knowing him personally as an eminent man full of warm kindness I became aware of the great loss to all his friends, to physics and to CERN in particular.

References

1. W. Sweet, "Abragam and Rubbia Reports Chart Future for CERN", *Physics Today* **40**, issue 9, p. 73 (September 1987).

Chapter 22

Wolfgang Kummer and the Little Lost Lane Boy

Kenneth Lane

Department of Physics, Boston University
590 Commonwealth Avenue, Boston, Massachusetts 02215
E-mail: lane@physics.bu.edu

I relate how Wolfgang and I met, collaborated on our very first paper on spontaneously broken gauge theories, and how he found my lost son in San Francisco.

This is the story of how Wolfgang Kummer found my 6-year old son, Greg, when he became totally lost at San Francisco's Fisherman's Wharf.

In 1972 Wolfgang Kummer was a Max Kade Foundation Fellow on leave at the University of Pennsylvania where I was post-doc, my first job after getting my Ph. D. in 1970. In those days, two-year postdocs were the norm and I was preparing to leave Penn to take up a second post-doc at Berkeley. While I loved working at Penn, I found Philadelphia an oppressive place to live and was eagerly looking forward to moving to the Bay Area (in 1972 the air out there was still crystal clear and the turmoil of the 60s had all but died out). The theory groups at Penn — high energy, nuclear and condensed matter — were not only very distinguished, they were extraordinarily collegial, with offices on the same corridor, lunches together in the theory common room, etc. Wolfgang was a welcome visitor and fit well into our very democratic society. He and I were friendly, but I don't recall interacting much with him until our collaboration on "divergence cancellations in spontaneously broken gauge theories".[1]

High energy physics was at the threshold of a revolution in early 1972. Late in '71 Gerard 't Hooft had published his world-shaking papers on renormalizing gauge theories[2-5] and he and Tini Veltman gave us dimensional regularization[6] in 1972. Before dimensional regulation became available, Steve Weinberg re-published his "Model of leptons"[7,8] arguing through several examples that it was a finite theory of the weak interactions when formulated in the "unitary gauge". In a paper that Wolfgang and I found very educational, Tom Appelquist and Helen Quinn constructed an abelian version[9] of Steve's model and, again in the unitary gauge, showed it to be renormalizable at one loop.[a] In the unitary gauge, in common use in many early papers (and strongly advocated by Weinberg in Ref. 8), the unphysical Higgs-Goldstone bosons are gauged away, leaving behind, like the smile on the Cheshire Cat, a non-polynomial term in the Lagrangian, $\propto -i\delta^4(0)\ln(1 + \phi/v)$. Here, v is the vacuum expectation value (VEV) of the physical Higgs field and ϕ is that field after it is shifted to have no VEV. The gauge's name derived from the fact that it was manifestly unitary (unlike other early so-called R gauges which kept the unphysical Goldstone bosons and employed Fadeev-Popov ghosts), though not manifestly renormalizable because the gauge boson propagators had the "classical" form,

$$\Delta_{\mu\nu}(p) = \frac{p_\mu p_\nu/M^2 - g_{\mu\nu}}{p^2 - M^2}, \qquad (22.1)$$

with its apparently nonrenormalizable bad high-energy behavior.[b]

I have a strong memory that in one of Weinberg's papers he referred to the divergence cancellations in the new spontaneously broken gauge theories (SBGTs) as "miraculous". I haven't been able to find this statement in his early papers, and it is possible that I just heard him say it in a talk, that someone else wrote it or said it, or that I am a victim of false memory. Whichever is true, I also remember thinking that there must be a good reason for the "miracle". And, today, I doubt that 't Hooft and Veltman, Ben Lee, Jean Zinn-Justin, and Steve himself really thought there was a miracle; they surely understood what was going on. In any case, in 1972, the divergence cancellations certainly appeared almost miraculous; as Wolfgang and I later wrote,[1] "A common feature of these theories is the fact that, to a given order of perturbation theory, individual Feynman diagrams contributing to a specific process contain nonrenormalizable divergences. We shall call these infinities 'spurious' because they happen to cancel when all

---

[a]There were a number of other very important early papers, but they are not as central to my story as these two were.

[b]Strictly speaking, proofs of renormalizabilty in the unitary gauge of spontaneously broken gauge theories apply to S-matrix (on-shell) amplitudes. Correspondingly, proofs of unitarity and the absence of unphysical particle singularities in R gauges applied to the S-matrix elements.

individual diagrams are added. Cancellations of this type have been shown to occur in Refs. 7–9 by means of a simple cutoff prescription to regulate the integrals involved." But those papers provided no simple underlying mechanism for the cancellations.

Having come from the culture of current algebra and PCAC of the late '60s, in which one obtained Ward identities by applying momenta to matrix elements of vector and axial-vector currents, as in

$$p_1^\lambda p_2^\mu \int e^{i(p_1 \cdot x_1 + p_2 \cdot x_2)} \langle B|T(j_\mu^a(x_1) j_\nu^b(x_2))|A\rangle, \qquad (22.2)$$

I started playing around with similar forms that arise in the amplitudes of SBGT from the $p_\mu p_\nu$ parts of the U gauge boson propagators, Eq. (22.1). Here my memory does fail me on the order of events, but once I started doing this, two things happened pretty quickly. One was that it became clear that it was very easy to see the divergence cancellations and the reason for them in the Appelquist-Quinn and Weinberg models by working in coordinate space rather than momentum space. The other was that I started talking to Wolfgang about this, he quickly became enthusiastic about it, and we began collaborating.

To get to our main point (while avoiding a lot of equations and diagrams that appeared in our paper), we realized that the "nonrenormaliz-able" $\partial_\mu \partial_\nu / M^2$ parts of the $\Delta_{\mu\nu}(x - y)$ could, by integrating by parts, be made to act on (pieces of) gauge currents appearing in amplitudes containing $\langle B|T(j_\mu^a(x)j_\nu^b(y)j_\chi^c(z)\cdots)|A\rangle$, and that these derivatives gave rise to "renormalizable" terms involving, for example,

$$\langle B|T(\partial_x^\mu j_\mu^a(x)\, \partial_y^\nu j_\nu^b(y)j_\lambda^c(z)\cdots)|A\rangle \qquad (22.3)$$

and "nonrenormalizable" terms such as

$$\langle B|T([j_0^a(x), \partial_y^\nu j_\nu^b(y)]\delta(x_0 - y_0)j_\lambda^c(z)\cdots)|A\rangle \qquad (22.4)$$

and

$$\langle B|T([j_0^a(x), j_\lambda^c(z)]\delta(x_0 - z_0)\partial_y^\nu j_\nu^b(y)\cdots)|A\rangle. \qquad (22.5)$$

The form of the equal-time commutators in these nonrenormalizable terms was always dictated by the *gauge symmetry* and either they vanished (as in the abelian Appelquist-Quinn model) or they were a gauge current or a "σ-term". And, in the latter case they were *completely* canceled by other matrix elements containing the same gauge currents or σ-terms and the same propagators. I emphasize: the cancellation was complete, finite as well as divergent pieces. Thus, we understood that the "miraculous" cancellation of the unitary gauge nonrenormalizable terms was a simple, direct consequence of the gauge algebra and the Ward identities for divergences of the matrix elements of the gauge currents, $j^a_\mu(x)$.

Finally we stated "[W]e have considered one-loop diagrams only. While our method works very well in these cases, yielding great insight into the question of renormalization in the U gauge, it goes without saying that a generalization of the procedure to an arbitrary number of loops is needed before a program like that carried out in the R gauge [by 't Hooft and Lee and Zinn-Justin] is obtained." Obviously, this never happened and anyway was made completely unnecessary by the introduction of the R_ξ gauge[10] and dimensional regularization[6] and the magnificent work of Becchi, Rouet, Stora[11] and Tyutin.

In the end, I am very proud of Wolfgang's and my work together and our paper. Still, it received only 17 citations. Perhaps some that were intended for our paper ended up in the pot of 2525 citations received by the one which immediately preceded ours in Phys. Rev. D.[12]

Now, what has all this got to do with Wolfgang finding my lost son, Greg? In the summer of 1972 we — my then wife Mary-Ann, 2-year old daughter Buffy, Greg and I — moved to Berkeley. (Ah, Berkeley, where we actually thought George McGovern would win the presidential election of 1972. And where, thanks to Ronald Reagan, a condition of employment at the University of California was signing a loyalty oath swearing allegiance to the Constitution of the United States *and* of the State of California — two inconsistent documents! And where we still called the lab on the hill "The Rad Lab".) Wolfgang and I had not quite finished our work, so he made a trip out to Berkeley. (Remember, this was before cheap long distance telephone and e-mail, LaTex, Skype, etc.)

On a brilliant California Sunday, the family went across the bay to San Francisco and met up with Wolfgang to sight-see at Fisherman's Wharf. In the course of our meandering around the Wharf in a sea of tourists, we did not notice that Greg turned left while we went straight. Suddenly, we realized that here we were chatting happily, pushing Buffy along in her stroller, surrounded by throngs of people, and Greg was nowhere in sight. Panicked, we retraced our steps, looking everywhere we had been for the past 20–30 minutes. No Greg. We were frightened out of our wits. Mary-Ann and I became convinced that Greg had been kidnapped, because we couldn't otherwise understand why we couldn't find him on the street. Thus, we decided to find some police and report a missing-or-kidnapped little boy. At that moment, Wolfgang either went into a restaurant near where we were standing, or he saw Greg coming out of it. Whatever it was, he took Greg by the hand and presented him to his two frantic parents. It seems that Greg had just walked into the restaurant following two big people he thought were his parents. He was, as I recall, a lot less scared than we were. Our feeling was one of overwhelming relief and gratitude to dear Wolfgang for keeping cool and finding Greg.

In all the years since then, I saw Wolfgang only one more time, at some conference. I was never invited to the Schladming Winter Conference, but that was fine with me. I am a miserable skier and probably would have ripped my knee out even earlier than I ultimately did at Aspen in 1985. Worse, Wolfgang, a marvelous skier I was told, would have been very disappointed in my clumsiness. When I met him that time, I was older and a lot more seasoned than I had been as a young postdoc at Penn. But Wolfgang was still the same charming and handsome fellow I had known years before. It is pretty much the way I have always remembered him, except that I always see Wolfgang coming out of a restaurant on Fisherman's Wharf with my son's hand in his.

Acknowledgments

I thank the organizers of Wolfgang Kummer's memorial volume for inviting me to contribute. And I thank my friend Tom Appelquist for reading my contribution and his kind remarks. My research is supported by the U. S. Department of Energy under Grant No. DE-FG02-91ER40676.

References

1. W. Kummer and K. D. Lane, Divergence cancellations in spontaneously broken gauge theories, *Phys. Rev.* **D7**, 1910–1923, (1973). doi: 10.1103/PhysRevD.7.1910.
2. G. 't Hooft, Renormalization of Massless Yang-Mills Fields, *Nucl. Phys.* **B33**, 173–199, (1971). doi: 10.1016/0550-3213(71)90395-6.
3. G. 't Hooft, RENORMALIZABLE LAGRANGIANS FOR MASSIVE YANG-MILLS FIELDS, *Nucl. Phys.* **B35**, 167–188, (1971). doi: 10.1016/0550-3213(71)90139-8.
4. G. 't Hooft, Predictions for neutrino - electron cross-sections in Weinberg's model of weak interactions, *Phys. Lett.* **B37**, 195, (1971). doi: 10.1016/0370-2693(71)90050-5.
5. G. 't Hooft, The Renormalization procedure for Yang-Mills Fields.
6. G. 't Hooft and M. J. G. Veltman, Regularization and Renormalization of

Gauge Fields, *Nucl. Phys.* **B44**, 189–213, (1972). doi: 10.1016/0550-3213(72)90279-9.

7. S. Weinberg, A Model of Leptons, *Phys. Rev. Lett.* **19**, 1264–1266, (1967). doi: 10.1103/PhysRevLett.19.1264.

8. S. Weinberg, Physical Processes in a Convergent Theory of the Weak and Electromagnetic Interactions, *Phys. Rev. Lett.* **27**, 1688–1691, (1971). doi: 10.1103/PhysRevLett.27.1688.

9. T. Appelquist and H. R. Quinn, Divergence cancellations in a simplified weak interaction model, *Phys. Lett.* **B39**, 229–232, (1972). doi: 10.1016/0370-2693(72)90783-6.

10. K. Fujikawa, B. W. Lee, and A. I. Sanda, Generalized Renormalizable Gauge Formulation of Spontaneously Broken Gauge Theories, *Phys. Rev.* **D6**, 2923–2943, (1972). doi: 10.1103/PhysRevD.6.2923.

11. C. Becchi, A. Rouet, and R. Stora, Renormalization of Gauge Theories, *Annals Phys.* **98**, 287–321, (1976). doi: 10.1016/0003-4916(76)90156-1.

12. S. R. Coleman and E. J. Weinberg, Radiative Corrections as the Origin of Spontaneous Symmetry Breaking, *Phys. Rev.* **D7**, 1888–1910, (1973). doi: 10.1103/PhysRevD.7.1888.

Chapter 23

Mitigation of Fossil Fuel Consumption and Global Warming by Thermal Solar Electric Power Production in the World's Deserts

Jack Steinberger

CERN, 1211 Geneva, Switzerland
E-mail: Jack.Steinberger@cern.ch

23.1. Introduction

Wolfgang Kummer was a valued colleague. He is remembered for his contributions to the present understanding of particle physics, his contributions to the creation of a school of theoretical particle physics in Austria, and his long and important support of CERN, in particular as member and leader in the CERN council, its president from 1985 to 1987. I have the privilege of remembering Wolfgang also as friend and musician, whose singing and piano playing at chamber music evenings in my house allowed us to enjoy some of the most precious human cultural heritage together.

In contributing to this memoire in Wolfgang's honour, unfortunately I no longer can offer something on particle physics. I hope that it will be excused that these lines concern a societal question, that of the future of energy and climate of our society and planet, as well as the outstanding role which thermal solar electricity production in the world's deserts could play in mitigating this threat. It might be easier to excuse me for this, if it is remembered that also Wolfgang was very concerned about environmental issues, and Lore Kummer still is much involved.

23.2. Short summary of present status of fossil fuel use and climate change

The last 150 years have witnessed a remarkable increase in population, in the production of goods and the exploitation of our planet's resources, based on consumption of ever bigger amounts of energy, largely. About

80% of this energy is presently produced in the burning of fossil fuels.

The last century has seen a population increase of a factor three, from ~2 billion to ~6 billion, (Fig. 23.1). The present growth rate of the global population is about 1%/year.

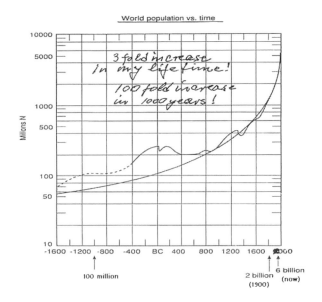

Figure 23.1   Evolution of the world population. 100 fold increase during the last four millennia, a 3 fold increase in the last century (IPCC).

The consequent increase in the effective greenhouse gases, measured in atmospheric CO_2 concentration was 35%, from 280 to 380 ppm (Fig. 23.2).

The consequent temperature increase has been about 0.8 degrees Celsius (Fig. 23.3), and can be expected to increase to about 8 degrees by the end of the century if we continue as at present; the sea level has risen 20 cm.

Present yearly increase in energy consumption is ~ 1% in the developed countries, 3.6% in Asia, 1.9% globally (Fig. 23.4).

How long will fossil fuels last? On the assumption of "business as usual", with present population growth of 1%/year and present per capita energy consumption growth of 1.9%/year, and that natural gas will replace the oil when it is gone, and coal will replace gas when it is gone, the lifetimes of the known low cost reserves will be: about 25 years for oil (7,500 EJ[a]), 35 years for natural gas (7,500 EJ), 60 years for coal (27,000 EJ).

---

[a]1 EJ = 10^{18} Joules.

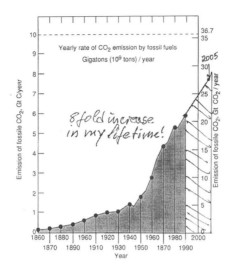

Figure 23.2   Evolution of the atmospheric greenhouse gas concentrations (IPCC).

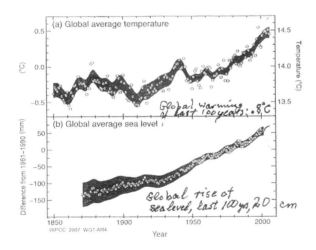

Figure 23.3   Evolution of the atmospheric temperature and of the sea level (IPCC).

This is an incredibly short time on the scales of human civilization, or the time it has taken our planet to accumulate the fossil fuel resources. The atmospheric CO_2 level would then be $\sim$ 700 ppm; the temperature rise would be perhaps 8°C; the sea level rise would be several meters. The main disaster regions would be in Africa, Asia, and Central America. What

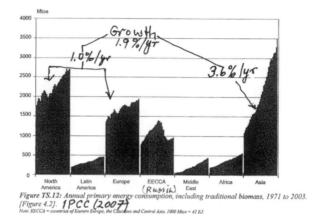

Figure TS.12: Annual primary energy consumption, including traditional biomass, 1971 to 2003.
[Figure 4.2]. **IPCC (2007)**
Note: EECCA = countries of Eastern Europe, the Caucasus and Central Asia. 1000 Mtoe = 42 EJ.

Figure 23.4   Annual primary energy consumption, by regions, 1971 to 2003 (IPCC 2007).

would be the consequent migrations and bloody conflicts?

23.3. How can widespread catastrophe be mitigated?

It is obvious and fundamental that effective mitigation requires a global response. This, given our organization into independent nations, is a formidable challenge, since mitigation efforts by a particular state penalize its economy with respect to its competitors. Perhaps the clearest demonstration of this problem is the "carbon tax". Such a tax, at an adequate level, perhaps \$200 per ton of CO_2, would be an effective means of discouraging the use of fossil fuels and would encourage the development and use of alternatives, but no country has installed this, despite the fact that most governments are very conscious of the threat to our society posed by diminishing fossil fuel supplies and consequent climate change.

It must also be kept in mind that the time scale for substantial change cannot be less than of the order of 30 years, the lifetime of power generators or buildings. Given that it is already late, it is essential to take the appropriate steps urgently.

Mitigative possibilities:

Reduce the birth-rates and populations.

Reduce consumption, especially in the developed world. We can be quite as happy consuming a lot less. A dominant problem with respect to reducing consumption is the global, competitive economy, in which, presently, a country's economic survival depends on constant economic growth.

Increase energy efficiency. Substantial savings in the heating of buildings, perhaps less in transport, are possible,

Carbon Capture and Storage (CCS). This has been widely proposed in the last few years as a mitigation to global warming. It requires the collection of emitted CO_2, its compression, liquefaction, and transportation to possible deep underground or undersea reservoirs, and its secure storage in these for many hundreds of years. It can be imagined for concentrated centres of fossil fuel burning, such as power plants and certain industries, but is not conceivable for the transportation or building sectors. For electric power plants, to be economically conceivable, it requires that before combustion, the oxygen is separated from the air. The increase in electricity cost is estimated at about a factor two. Before CCS could be seriously planned, the long term storage security of the various proposed storage technologies would have to be seriously researched, which has not been done until now. A further, very serious deficiency of this proposed solution is that it does nothing to conserve the global fossil fuel resources.

Nuclear Power. This has the merit of conserving the fossil fuel resources as well as avoiding greenhouse gas production, and is reasonably economical. It now provides $\sim 20\%$ of global electricity. Disadvantages of the nuclear are the problem of the storage of the nuclear waste, the vulnerability to terror attacks, as well as the possible encouragement of nuclear weapon proliferation. Also, the known uranium resources are quite limited, and the possible extension to viable, safe thorium breeders has not yet been demonstrated.

Sustainable energy sources. As discussed below, this is quite possible in the electric power sector, but much more difficult in the transportation sector. The main, well known possible renewable electric power sources, with my own estimate of their viability, are:

Hydroelectric. At present it is about 7% of global electric power. It is clean, available when needed, perfect. It is however limited, and very large extensions of present hydro-power cannot be expected.

Geothermal. This can be used effectively, but is possible only in certain regions, and very limited. At present it contributes a very small part, $\sim 0.2\%$, of the global electricity supply.

Wind. Wind-power is economically acceptable in certain regions. At present it accounts for $\sim 1.5\%$ of electric power production, and is expanding rapidly. One negative feature of large scale wind-power is that the wind does not blow all the time, and storing the electricity produced, for instance in hydro reservoirs, is not in general economically feasible.

Photo-voltaic. As best I can understand, photo-voltaic electric power production, although widely supported, cannot furnish a significant part of the world's energy demand, given the cost of the installations and the fact that, as with wind-power, the electricity cannot be economically stored for the times when the sun does not shine.

Bio-fuels. These, for instance alcohol, have recently been seriously promoted as a large possible replacement for fossil fuels. They solve the climate change problem, since the CO_2 produced in their use is reabsorbed in their production (but not the greenhouse gases resulting from the production of the fertilizer). However, I am on the side of the many sceptics, who fear the tensions that large scale bio-fuel production would cause in the world's food supply.

Solar-thermal electric power from the world's deserts. The sun's radiation is concentrated on a thermal vector, a liquid. This in turn is used to produce the steam to drive the turbine which drives the generator. For production when the sun is not shining, in particular at night, the heat is stored in thermal reservoirs. The electric power must be transmitted for thousands of kilometres, which requires using high voltage, DC networks. There is plenty of sun for the global energy need; a few percent of the deserts suffice. Although the R+D pilot plants for the three essential technologies: solar concentration, thermal storage and power transmission, are in a preliminary stage, there is good reason to believe that costs comparable to present fossil fuel costs can be achieved. Given these properties, this technology seems the most promising source of energy for the future.

23.4. Thermal solar power from deserts with overnight storage

23.4.1. Requirements

(1) Large fields of concentrators. Present technologies are:

    (a) Parabolic troughs focusing the sunlight on a tube containing the thermal vector liquid. This requires rotation of the reflectors in one dimension to follow the sun. Typical dimensions of parabolic reflectors in present fields are ~ 5 m across and ~ 1km in length.

    (b) Fresnel lenses. In this technology the parabolic trough is replaced by several, narrow cylindrical reflectors mounted on a flat surface. This technique is probably more economical in construction, and has smaller vulnerability to wind damage.

(c) Solar tower fields. The sunlight is focused onto a central collector, now typically ~ 10m diam., 10 m high, on a tower, now typically ~ 100m high, from a field, now typically a few hundred meters in radius, of focusing reflectors, now typically ~ 20 m^2. These must be rotated in two dimensions to follow the sun.

(2) Thermal fluid. To achieve high efficiency, this should work at high temperature. In present pilot plants the thermal fluid is largely oil, at a maximum temperature of ~ 400°C, but many present pilot plant proposals use "Direct Steam", that is steam under pressure exceeding 100 atmospheres, to go to perhaps close to 600°C.

(3) Thermal storage. The thermal storage medium in present pilot plants is a molten salt mixture, 60% NaNO$_3$, 40% KNO$_3$, with melting point of 300°C. Proposed "Direct Steam" plants require more complicated systems, to deal effectively with the three thermal stages: water, evaporation and steam.

(4) Electric power generation. Conventional, commercially available steam turbine driven generators can be used.

(5) Electric power transmission. The losses of alternating current transmission lines are too high, but high voltage direct current lines, have acceptable losses of ~5% per thousand km. This technology is tested and already in use, in particular in Brazil, in a 2000 km line.

23.4.2. *Pilot Projects*

(1) In the USA. In the Mojava desert in California, 8 plants, with total power of 450 MW (for comparison, a typical modern gas or coal plant has 500–1000 MW) were constructed in the 80's and are still in operation. They use parabolic troughs, with oil at 390°C as thermal vector, and have a total solar collection surface of 2.3 km^2. They feed power into the local network. They do not incorporate overnight storage, are constructed by private enterprises, are subventioned by a price for the power delivered which is about twice the normal (fossil fuel) power cost. One of the challenges is the maintenance of the optical quality despite desert sand storms. This has now been demonstrated. Figure 23.5 is a view of one of the collector fields, and Fig. 23.6 shows their cumulative power delivery over two decades.

(2) Andasol in Spain. Three 50 MW Parabolic trough power plants, with 7 hour molten salt storage, are under construction near Granada. The

Figure 23.5   View of one of the Mojava desert parabolic trough receiver fields.

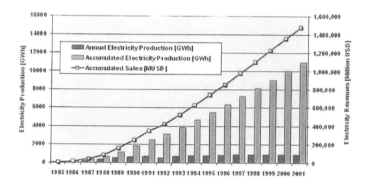

Figure 23.6   Cumulative power delivered by the 8 Mojava desert parabolic plants.

first is expected to come into operation in the fall of 2008. For each
plant there are two storage tanks, the hot one at 386°C, the cold one
at 292°C, each is 14 m high and 38.5 m in diameter. Between them
is the heat exchanger with the thermal vector (see Fig. 23.9). They
are constructed by private companies, with cost subvention by Spain
at about 3 times the market prize for the electricity. The electric power
is injected into the local network.

(3) Solar Tres in Spain. Solar Tres is a 15 MW solar tower pilot project
with 15 hour thermal storage, so that it should be able to operate at
full power day and night, 6500 hours per year. The tower height is 120
m. The heat vector is molten salt, $300°C < T < 560°C$. The collector
field consists of 2500 focusing mirrors, each 115 m^2. Total estimated
cost is € 200 M.

Figure 23.7   Site of Andasol 1, southern Spain.

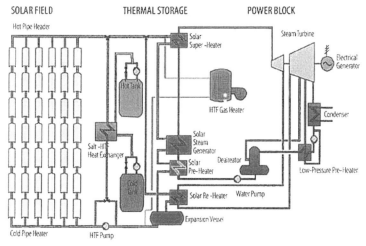

Figure 23.8   Schematic of the Andasol plants.

(4) Abengoa Solar, Spain. Since 2007 two 50 MW pilot plants are under construction in Sanlucar, using parabolic troughs, 300,000 m^2 per plant, and oil as thermal vector, without storage. It is intended to extend this to a total of 300 MW.

(5) Archimede, Italy, south of Syracuse. The 5 MW project, a collaboration of the Italian energy agency ENEA and the electric power company ENEL, explores the use of molten salt as thermal vector, $300^\circ\text{C} < T < 540^\circ\text{C}$, with parabolic trough collectors. Collector loops have been successfully tested.

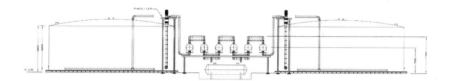

Figure 23.9   Andasol storage system, hot storage tank, cold storage tank, heat exchanger in between.

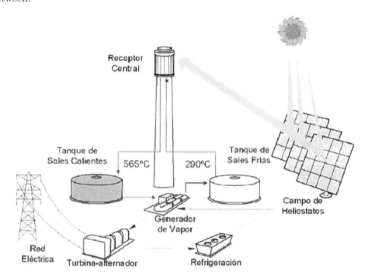

Figure 23.10   Schematics of Solar Tres.

23.5. Future Projects

New legal requirements in California, stipulating that 20 % of electric power production by 2010 be renewable, has produced some interesting proposals by new companies, which have been accepted by the power companies, in agreements in which the power will be accepted into the local networks at something less than twice the present cost of electricity. I am aware of three such projects, each $\sim$ 500 MW total, and particularly interesting, because all three propose direct steam as thermal vector, as well as other technological changes. The advantage of direct steam over oil as thermal vector is that higher temperature, and consequently higher efficiencies, can be achieved. Unfortunately, the California law does not encourage storage, so that these projects do not include overnight storage.

(1) The new company, Ausra, led by the Australian engineer D. Mills, with a substantial career of innovative thermal solar projects in Australia, has an agreement with Pacific Gas and Electric (PG&E) for plants using Fresnel lenses (Fig. 23.11) instead of parabolic troughs, and direct steam as thermal vector. The first such plant, 177 MW, is to be constructed in Carrizo, CA.

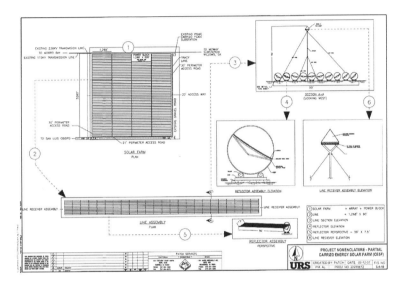

Figure 23.11   Layout of solar field and diagrams of Fresnel collectors for proposed Ausra solar farm in Carrizo, CA.

(2) Bright Source Energy (LuzII) has a contract with PG&E for 500 MW with option of additional 400 MW, in Ivanpah, CA, starting with a 100 MW plant, which is to be ready in 2011, followed by 2 (or 4) 200 MW plants, one per year.. The technology is Solar Tower, with direct steam thermal vector. The 200 MW plants are to have three towers each, and 150,000 mirrors (heliostats), of 7 m^2. This relatively small area of the individual mirrors makes it possible to fabricate this from flat glass, and achieve the curvature by tensioning.

(3) Founded in 2007 by one man, eSolar is under contract with Southern California Edison, for 245 MW of solar power. It also uses Solar towers technology, small heliostats and direct steam. eSolar is focused on "scalability" for cost efficiency.

Figure 23.12   Layout of proposed eSolar plant.

For me, the most interesting feature of these new projects is that the proposed technologies differ substantially from those which have been tested most extensively. It is clear evidence that the technologies need further studies. It has been demonstrated that thermal solar energy is practicable at a reasonable cost. The remaining question is, what technologies, for concentration, for the thermal vector and for storage, are the most economical.

23.6. Conclusion

Our present economy is based on energy from fossil fuels. For future generations this has the two very disturbing consequences that it produces climate change and that the planet's readily accessible fossil fuel resources, already half used, will be gone altogether in 50 or 100 more years, a tiny period on the time scales of either the planet or humanity. If our society has a sense of responsibility for its children, fossil fuel use must be severely curtailed, and as quickly as possible. The unique, adequately available source of energy which can replace fossil fuels at a reasonable cost is solar electric power production in the world's deserts, using a thermal fluid, with overnight storage. The practical possibility has been demonstrated by pilot projects in California and Spain, but much more should be done in the immediate future to study the solar concentration, the storage and the thermal vector technologies to arrive at the most economical methods.

Chapter 24

(My) Life with Wolfgang Kummer

M. Schweda

Institute for Theoretical Physics, Vienna University of Technology,
Wiedner Hauptstrasse 8-10, A-1040 Vienna, Austria
E-mail: mschweda@tph.tuwien.ac.at

My laudation speech held in the office of the rector of the TU Wien at the occasion of the official retirement of W. Kummer on Oct. 1, 2004

Rector Magnificus, dear Peter [Skalicky], dear Mrs. Kummer, dear Wolfgang and dear Joachim [Burgdörfer], "(My) Life with Wolfgang Kummer" is the title of my laudation within this small circle and I warmly welcome all of you who are here today. Actually, after my last laudatory speech, and having gone through three extremely difficult years healthwise (with three operations), I had decided never to deliver a laudation again. But I had forgotten that you, my dear Wolfgang, would be retiring this year — a both sad and joyous occasion, which induced me to give one more speech here. I, who have been your colleague here longer than anyone else — we have worked at this university together since 1968 — have, of course, a number of things to relate. It is above all thanks to you that I found my way back to theoretical physics. Therefore I may perhaps take the liberty of making this talk a little too personal and chronological.

When I began my studies in the winter semester of 1960, here in this very building, you had already received your doctorate and were an assistant at the Institute for Theoretical Physics. At that time, the Sputnik had already been shot into space and Austria had joined CERN — but the study of physics here was pretty much a disaster. There were only four professors — Hittmair, Lihl, Regler and Ortner — and the selection of courses offered was correspondingly lean and motivation was thin on the ground. I remember Regler's first lecture in experimental physics very clearly: I was sitting in the front row at the time, and after giving his official welcome, Regler approached the first row and started numbering the excited students sitting there: 1, 2, 3 and so on. When he got to the number 10, he informed us that only every tenth student was going to be needed in the field.

My first dealings with theoretical physics were not very inspiring. Otto Hittmair had a habit of tossing the chalk up into the air and catching it again whenever he stumbled during his lecture. The mathematicians did not orient their lectures to the needs of physics; in fact, at some point I was assigned to give a talk in a theoretical physics seminar about the operators of quantum mechanics — and still had no idea what an operator was. We had to learn a great deal in private study. The four sets of lecture notes for the theoretical physics cycle were very helpful, but after completing my diploma exams in theory, I still did not know of the existence of the Dirac equation. Nevertheless, in December 1965 I completed my degree. Months before that, I had already realised that there was no position open to me here in theoretical physics at that time, and so I looked for alternative options, because I wanted, at all costs, to continue working at our university — which was then still called the Technische Hochschule — and receive my doctorate. So I became a numerical mathematician and took up my position at the TH on 1December 1965. In the midst of my joy at having found regular employment at the TH, I received the next shock when Professor Stetter, Chairman of the newly founded Institute for Numerical Mathematics, informed me that a doctorate under his supervision would take about five to six years. This was totally unacceptable to me, and so I began searching for new possibilities.

At this time it was generally known here that a brilliant new star had risen in the firmament of theoretical physics. It was the incredibly talented Wolfgang Kummer, who, always clad in a white lab coat, having earned his habilitation and established a reputation through his CERN sojourns as a Ford scholarship holder and CERN Fellow, had become a cult figure even at this early stage in his career. Your lectures on the introduction to the theory of elementary particles, dear Wolfgang, had fascinated me to such an extent that one day, on the spur of the moment, I went to your office at the Atomic Institute and asked you to give me a dissertation subject. Without hesitating, you handed me a precisely defined problem about semileptonic decays within the SU(6) symmetry group and a helpful textbook by Okun about weak interactions.

Two years of hard work now lay before me. During the day I was busy with numerical mathematics, and in the evenings, usually until midnight, I worked on my dissertation — at first secretly and later, after a discussion with Stetter, officially. At the end of the two years, my dissertation was finished.

In the meantime you had been climbing the career ladder with amazing speed. On 1 January 1966 you became Director of the Institute for High Energy Physics — known by the acronym HEPHY. There they called you the "wolf" — heaven knows why! Many colleagues were able to work in physics at this newly established institute: Majerotto, Flamm, Otter, Regler, and many others.

Your extraordinary talent as a theoretician and the numerous publications associated with this galvanised our sleepy old, time-honoured TH — on 1 October 1968 you became a full professor of theoretical physics and were awarded a newly established second chair. At that time you and Herbert Matis of the Hochschule für Welthandel (today's University of Economics and Business Administration) were the youngest university professors in Austria. Since I was at the right place at the right time, you appointed me, your first doctoral student, as your university assistant. Now I was perfectly happy, for I could do research and teach physics all day long under your direction — an inconceivable, new situation for me. Here I have to insert a nice little anecdote: When I said goodbye to Professor Stetter on my last working day at the Institute for Numerical Mathematics, he said to me: "Now at least you can devote yourself completely to your dissertation." I replied, "Professor, it is already finished!"

But to return to our story: At first our institute had no premises of its own, and so we moved to the HEPHY on Nikolsdorfergasse. By this time, W. Konetschny and W. Kainz had joined us. Suddenly I was working with three W. K.s at the same institute — a quite remarkable coincidence! We published together, ate together and played tennis together. Apropos tennis: all four of us were tennis fanatics and the result was that the Union of Students once jestingly reported that you looked more like a tennis trainer than a full professor of theoretical physics. You used to sweep past me on the ski slopes in Schladming, too. Whether in physics, in sports or in cultural domains, you were always a length ahead — but to be honest, I must say this never bothered me — on the contrary, it was wonderful to be able to follow in your footsteps. As sportsmen, we know that in every race someone always has to be second, and your lead was always so great that no one could have caught up to you.

My own career in this house was strongly influenced by your thoughtful guidance. For this I thank you today from the bottom of my heart. As the father of three children, I was, of course, not as mobile as you were. You had, in the meantime, gone to Pennsylvania to teach as a guest professor for a year, but still you managed to arrange for me to finally go to Marseille. For six consecutive years, I spent almost all of every three-month summer break at CNRS in Marseille. There I became acquainted with Carlo Becchi, Alain Rouet, Raymond Stora and Olivier Piguet. Now I had a second excellent teacher in addition to yourself, namely Carlo. Thanks to you and Carlo, I learned quantum field theory and understood it. — But now it is time to put my own memories aside and talk only about you.

At the beginning of the 1970s we found a new home on Argentinier-straße; later we returned to the main building — and once again the good-will of the upper echelon was yours. You were given the largest and most beautiful room in the entire university — Professor Inzingers former office — I think it had an area of about 100 square metres and seemed as big as a riding hall. From this presidential sanctum you made pivotal contributions to the development of our university from that time on. With your all-round talents, you were naturally in great demand inside and outside our own walls. You became the director of the inter-university computation centre. The number of your memberships in various associations has been immense. Some people have jokingly called you the "king of presidents and vice-presidents".

But I must mention a couple of things in particular. You were the Austrian delegate to the CERN Council, Chairman of its Finance Committee, later its Vice-President and then President. You have been honoured with an abundance of prizes and your scientific competence has been widely recognised: In 1971 you were awarded the Culture Prize of the Province of Lower Austria, in 1981 you received the Cardinal Innitzer Prize, the Schrödinger Prize was conferred upon you in 1988, and in 2000 you were presented with the Walter Thirring Prize of the Austro-Ukrainian Institute for Science and Technology.

Since we are going to honour your outstanding scientific achievements in the field of theoretical physics at a separate ceremony, I have almost finished. I only want to say two more things: Behind every competent man there is a competent woman. You have had your wife Lore at your side for 42 years, giving you strength for the lifes work of which you can be so proud. And a final sentence: Both of us are devout Catholics. — In this spirit, I wish you all the very best and much good health.

PS: For our profession one needs a great deal of imagination and inventiveness — which we both possess in adequate quantity.

Chapter 25

Schubert in Stony Brook and Kinks in Vienna

P. van Nieuwenhuizen

C.N. Yang Institute for Theoretical Physics
Stony Brook University, Stony Brook, NY 11794-3840, USA
E-mail: vannieu@insti.physics.sunysb.edu

Wolfgang Kummer became my friend during the many years I have been visiting Vienna in January. He created a marvellous Institute for Theoretical Physics with an enthusiastic group of students, postdocs and colleagues. Great Institutes need a central person to grow around. Two such Institutes I have been attached to are Utrecht where M. Veltman created a school of field theorists, and Stony Brook where C.N. Yang was the natural leader. In Vienna Wolfgang played a similar role. His mere presence during seminars gave a sense of enthusiasm and direction, and his active participation was an example for younger physicists not to be shy and also ask further clarifications when needed. In particular I was impressed by his collaboration with students. Almost every afternoon I would find him sitting with one or two students at his desk or in the conference room, trying to understand new problems in dilaton gravity, and discussing and calculating with them on equal footing. I invited Wolfgang to visit Stony Brook, but he was noncommittal, and I soon found out that he was seriously ill. Yet, he would come in everyday, and be the life of the Institute. Every year with his wife Lore we would go out one evening in Vienna, either to a play or a concert. These were great evenings as I will describe below. Near the end, I would phone him at his home, and we would talk physics. He remained optimistic about his chances, but then suddenly all went wrong and he died. I miss a dear friend. Below is the story of how we became friends.

In 1991 I was involved in the organization at Stony Brook of the big annual conference on string theory. It was called "Strings and Symmetries 1991" (Warren Siegel had flipped the second 9 to make the title more symmetrical), and one of the participants was Wolfgang Kummer from Vienna who would talk on "Non-Einsteinian gravity with torsion at $d = 2$".[1] I did not personally know Wolfgang then, but his work with Kainz, Schweda, and Konetschny[2] on the axial gauge in Yang-Mills theory was well-known. The conference had decided to have parties at the homes of the organizers on a night in the middle of the conference. One such party was at my home, and since the weather was nice that May, I had put torches in the garden near the swimming pool. One novelty of that conference was that we had hired string quartets of students from our music department (quite a well known music department), who played classical music during the breaks between talks. That night I also had a quartet playing in the garden. By 8 o'clock the about 80 participants who were assigned to my house, arrived by bus and the party started. As usual, a lot of shop talk was going on. By 10 o'clock the student-musicians left, and some students and conference participants started playing, or trying to play, some tunes on the piano in my house. I was standing outside, which was a good thing as much of the music was only entertaining to who was playing it. And then it happened. I heard somebody singing and playing one of the beautiful songs of Schubert. People stopped talking and listened. Curiously I went inside, and there I found Wolfgang sitting behind the piano. After much urging by everybody, he played some more, and afterwards we drifted to the room with drinks. I asked him where he had learned to play and sing so well, and he told me that for a while, before becoming a physicist, he had played with the idea of becoming a professional musician. (It occurs to me that another great Austrian physicist, Walter Thirring, who was also a friend of Wolfgang, went through the same steps. In fact, Walter's brother was supposed to become the physicist, and Walter the musician, but when his brother died in WW2, Walter's father, the Thirring of the Lense-Thirring effect, persuaded Walter to become a physicist. Finally, after his retirement as physics professor, Walter could devote more of his time to music.) Since I love music by Schubert, I got into an animated discussion with Wolfgang.

After our first contact at Stony Brook, I met Wolfgang at other String conferences. At some point (I no longer recall who mentioned it first), we concluded that it would be nice for me to visit Vienna. I had only been once in my life in Vienna, coming from the States on my way to the first Marcel Grossmann meeting in Trieste. That first stay had stayed in my mind for a remarkable event. From the airport, I had gone to Vienna for sightseeing, but being tired, I had gone to one of the famous coffee houses and had ordered a Wiener Kaffee. After I had drunk it, sleep overwhelmed me, and I must have slept for hours, when, awakening and opening my eyes, I saw in front of me a waiter who asked politely "Wünschen Sie noch etwas?" (Would you like something else?). Quite civilized!

Anyhow, I gladly accepted, and in January 1997 I visited the TU (Technische Hochschule) for the first time. Wolfgang and Lore drove me that first day around Vienna, to find a nice place to stay, and we found a most wonderful place (reader take note: the Benediktus Haus, a convent in the middle of Vienna, with Spartan beds, no TV, and thick stone walls which keep out any noise. Very nice. Tel. 43-1-53498-900). In the Institute, I told Wolfgang about problems I had discovered in the standard articles on quantum kinks and he made some useful remarks. But he was at that time in the middle of exciting work with his students on dilatons in string theory, so he invited me to join him and his students, because he had no time to start yet another project. I wanted to solve my problems, so I told others about them, and to my ever-lasting luck, Toni Rebhan promised to look at them. A few days later Toni returned, and said that he believed there were indeed serious errors in the literature on loop corrections to super-symmetric (and nonsupersymmetric) kinks, and we started to collaborate. That collaboration is now in its 12th year. Every year I visit the TU and study with Toni and a student new problems, but always having to do with quantum solitons, and we have gone from one theoretical roller coaster to another. Wonderful.

During these years, Wolfgang asked me to give each time a series of lectures on a topic of my choosing, and these seminars I will always remember for the electric atmosphere and the friendly sense of humor. The room used to be completely filled, and I would intersperse my technical discussions with historical flashbacks. Wolfgang directed these gathering in a unique way, commenting on history, and encouraging discussions of scientific points. He was simply irreplaceable. Of course, due to our daily discussions, we got to know each other well, and so a tradition was born. One evening each visit, I would take Wolfgang and Lore out, and in addition at other times they would drive me to the countryside to see Austria. I recall a trip to the famous convent in Melk, passing the place where Austrians had kidnapped Richard Lionheart on his way back from the Crusades. Another time we ate asparagus in a small restaurant overlooking the Donau. (Wolfgang loved asparagus.) One time I took them out to a play called "Love letters" by A.R. Gurney. Wolfgang and Lore had stopped going to the Burg Theater because the then-present artistic director had embarked on a modernistic direction which they did not like at all, but after that evening they resolved they would resume their visits. Another time I took them to a play in the Schwarzenberg palace, about diplomats, and wives of diplomats who hate diplomatic life and leave their diplomatic husbands. I looked on with great interest, but I had absolutely no clue what was really going on. But Wolfgang soon got the point: the diplomat-actors were real diplomats, and the diplomat-wive-actors were women who had been married to real diplomats, but left their husbands to go back to Vienna! An enlightening evening.

We also went to concerts, and dinners together. We even made plans for a winter vacation together with our wives in a hot country. Unfortunately, these plans were canceled by his disease.

Wolfgang was a warm human being, with whom I loved talking physics and art. There is no replacement for him.

References

1. W. Kummer and D. J. Schwarz, in Proceedings of the conference "Strings and Symmetries 1991", Stony Brook, May 20-25, 1992, edited by N. Berkovits, H. Itoyama, K. Schoutens, A. Sevrin, W. Siegel, P. van Nieuwenhuizen and J. Yamron (World Scientific, Singapore, 1991) p. 168.
2. W. Kainz, W. Kummer and M. Schweda, Nucl. Phys. B **79**, 484 (1974); W. Konetschny and W. Kummer, Nucl. Phys. B **100**, 106 (1975); Nucl. Phys. B **108**, 397 (1976); Nucl. Phys. B **124**, 145 (1977).

Author Index

Bandos, I. A., 303
Bergamin, L., 183
Blaschke, D. N., 145

Delduc, F., 145
Deser, S., 25

Faustmann, C., 277
Freund, P. G. O., 33

Gegenberg, J., 231
Gieres, F., 145
Grosse, H., 75
Grumiller, D., vii, 363

Hansen, M., 115

Jackiw, R., 363
Johansson, N., 363

Katanaev, M. O., 249
Kreuzer, M., 335
Kunstatter, G., 231

Landshoff, P. V., 3, 175
Lane, K., 381

Mann, R. B., 215
Meyer, R., 183

Neufeld, H., 277

Rauch, H., 91
Rebhan, A., vii, 41, 175

Schopper, H., 377
Schwarz, D. J., 267
Schweda, M., 145, 399
Sorella, S. P., 145
Steinberger, J., 387
Strobl, T., 115

Thirring, W., 277

Vairo, A., 9
van Nieuwenhuizen, P., 41, 403
Vassilevich, D. V., vii, 293

Wimmer, R., 41
Wohlgenannt, M., 75